PARTICLE GLOSSARY

(All *charged* particles have electromagnetic interactions as well as the interactions indicated.)

HADRONS

Strongly Interacting (also have weak interactions)

Baryons (spin = 1/2, 3/2, . . .) Fermions (obey Pauli exclusion principle) — **bound state of three quarks**
Examples:
nucleon (proton, neutron); hyperons (usually Greek letter names)

Mesons (spin = 0, 1, . . .) Bosons — **bound state of quark-antiquark**
Examples:
pions; heavier mesons named with letters

Quarks (spin = 1/2) Fundamental fermions whose bound states form all the baryons and mesons

LEPTONS

Weakly Interacting (but no strong interaction)

(Spin = 1/2) Fermions (obey Pauli exclusion principle)
Only 6 are known to exist:
e (electron), ν_e (electron neutrino); μ (muon), ν_μ (muon neutrino); τ (tau), ν_τ (tau neutrino)

GAUGE BOSONS†
(quanta of the force field, which "transmit" the force)

Force	Quantum of the Force Field	Mass	Spin
Strong	gluon	0	1
Electromagnetic	photon	0	1
Weak	$W^\pm, Z^0$	~80, 91 GeV	1
Gravitational	graviton††	0	2

†The spin 1 gauge bosons are also referred to as *intermediate vector bosons*.
††No experiments have yet detected gravitational waves, let alone their quanta.

The Fundamental Particles and Their Interactions

William B. Rolnick
Wayne State University

With an appendix on Particle Detectors
by Joey Huston, Michigan State University

▲▼ ADDISON-WESLEY PUBLISHING COMPANY

Reading, Massachusetts • Menlo Park, California • New York
Don Mills, Ontario • Wokingham, England • Amsterdam • Bonn
Sydney • Singapore • Tokyo • Madrid • San Juan • Milan • Paris

To Ellen, Stefan, Aaron & Corky (Quark(s)(y)) Beast
and all those who have given me encouragement and support

Sponsoring Editor: Stuart Johnson
Assistant Editor: Jennifer Duggan
Senior Production Supervisor: Jim Rigney
Cover Designer: Leslie A. Haimes
Text designed and produced by Lachina Publishing Services

Cover Photo: When particles collide at very high energies secondary particles are produced in a highly collimated form known as "jets." The photograph on the cover shows four jets produced from a proton-antiproton collision at a center of mass energy of 630 GeV. (*Courtesy CERN Media Service*)

Library of Congress Cataloging-in-Publication Data

Rolnick, William B.
    The fundamental particles and their interactions / William B.
Rolnick.
        p.   cm.
    Includes index.
    ISBN 0-201-57838-7
    1. Particles (Nuclear physics)   2. Standard model (Nuclear
physics)   I. Title.
QC793.2.R65   1993
539.7–dc20   93-31895
                                                              CIP

1 2 3 4 5 6 7 8 9 10   MA   989796959493

Preface

The purpose of this book is to present the concepts of particle physics, including some of the experimental evidence of their discovery and verification, at an introductory level. It provides the insights and background necessary to understand and appreciate the ideas as well as the interpretation of experimental results. An introductory overview of the apparati of high-energy experiments is included as well. Brief sketches of several of the most popular speculative ideas are presented in the latter part of the book, to give the reader a qualitative knowledge of those ideas even though their detailed formalisms are beyond the level of this book. This book is intended for graduate students or advanced undergraduate seniors who have had senior-level courses in electrodynamics and modern physics and have some familiarity with matrices. The text of the book provides an overview appropriate for those in areas of physics other than particle physics as well as those in related sciences (e.g., engineering, chemistry, etc.) who would like to become acquainted with the underlying concepts and be equipped to follow future breakthroughs in this field. When the problems are incorporated, the presentation of the ideas is extended to a depth appropriate for an introductory course.

This book has been designed from my lecture notes for a survey of elementary particle physics, a required course for our Ph.D. candidates. The course has no listed prerequisites and assumes no knowledge beyond undergraduate physics. Elementary particle theory is explained *ab initio* in terms of gauge theories, since our understanding of nature at this time is clearly in terms of these beautiful structures. Furthermore, the usual historical approach is as confusing to beginning students as it was to those of us who lived through that period. We do not teach mechanics, electromagnetic theory, etc. by showing all the false directions of thought that were pursued during their development and, although we have lived through the excitement of many fundamental discoveries in particle physics, I believe that a coherent treatment of the accepted theories should be presented to beginning students. Some of the historical flow, where it has particular significance, *is* included in the theoretical discussions to emphasize the wonder of discovery, both experimental and theoretical. New, more physical explanations of some concepts are presented to make them accessible to beginning students and others not in the field. Introductory qualitative descriptions of more advanced material are included in the later chapters to give the reader a glimpse of those provocative ideas.

The importance of and the creativity involved in the planning and interpretation of experiments are also acknowledged in this book. However, the approach usually used in texts on particle physics, which incorporates many experimental results and dilemmas with the explanation of the theory, is not pedagogically successful for beginning students. They often can't see the forest when presented with so many trees, some of them false trees. Some significant experimental results *are* given with the exposition of the theoretical structure, with further evidence presented later in the book. Therefore, the instructor can cover the full theoretical structure of the standard model (through Chapter 12) followed by the explanations of the experimental methods (Chapters 13 and 14), and has the option to include as much or as little of the experimental results, presented in Chapter 15, as desired. That chapter, which describes experimental evidence for the standard model with three generations, can also serve as a reminder of the theoretical material already studied.

Several important experiments are described in this book, and more are cited, to give the student an appreciation for the experimental foundation of our understanding of elementary particles. However, this book is not intended to be a compendium or document of record for experiments or experimental results; consequently, many important, equally noteworthy experiments are not mentioned. The Particle Properties Data Book of the Particle Data Group (available from Berkeley and CERN) and *The Experimental Foundations of Particle Physics* by R. N. Cahn and G. Goldhaber are recommended as good starting points for those who want more complete information, along with some of the other references listed in this book's bibliography. Likewise, there have been many people involved in the theoretical developments discussed here, but most are not cited by name. Several more advanced texts, listed in the bibliography, contain additional information. The book *Inward Bound* by A. Pais describes the early developments in a historical way, and *The Second Creation* by R. P. Crease and C. C. Mann and *Constructing Quarks* by A. Pickering include the more recent years.

With a minimum of formalism or detail, this book provides the reader with:

(a) an *introduction to the generalized concepts of quantum mechanics*, which includes a discussion of relativistic quantum mechanics in the form of the Dirac equation, its interpretation, and how it is connected to the ideas incorporated in quantum field theories. The relevant *qualitative aspects of quantum field theories* are introduced in the form of creation and annihilation operators, with a clear motivation for their application to particle interactions.

(b) an *introduction to the concepts of group theory*, which are relevant to particle physics, so that the students can appreciate the power and beauty of the gauge theories. Group theory is a field of mathematics which mathematicians teach from the general to the specific. Even physics courses in group theory are somewhat abstract, although they cover only those parts of the field (mainly representation theory) relevant to physics. In this book, group theoretic concepts are explained from examples already familiar to the students. This approach provides them with sufficient understanding to grasp the essence and beauty of the gauge theories.

(c) the *underlying concepts*, which motivate the design and interpretation *of experiments* with an intuitive, "from the ground up" explanation of what a cross section is, etc.

(d) a survey of *experimental equipment and facilities*, which utilizes pictures to impress the reader with the large scale and summarizes the different detection techniques employed. The impact of future plans for "big science" projects on science as a whole and on our society in general is indicated, and how the superconducting supercollider (SSC) fits into that picture is discussed.

(e) an *appendix on particle detectors*, written by Joey Huston, who is an experimental physicist at Michigan State University doing experimental work at Fermilab, which is provided for those who would like more detail.

(f) *sketches of theoretical conjectures, whose full exposition is beyond the level of this book*, which are provided as introductions to those concepts.

Although the Feynman rules for calculating amplitudes of processes are of considerable importance for obtaining quantitative results, I have omitted them since an *understanding* of them is beyond the level of most of the readers, and they are not essential to an understanding of the underlying concepts at this level. I have also omitted other calculational details, which often confuse beginning students and make the subject matter appear overwhelming. Furthermore, they are of little or no interest to most of those in other fields.

Problems are provided in order to expand the students' perceptions of the material and to provide depth beyond the material in the chapters. Those reading the book for cultural reasons may skip the problems; however, it is recommended that students do not. The problems vary in difficulty, and some may not be appropriate for students with minimal physics or math backgrounds. Some of the footnotes can also be omitted by beginning students.

The book is divided into five parts, following an introductory chapter:

Part A — THEORETICAL BACKGROUND: Chapter 2, A General Introduction to Quantum Mechanics; Chapter 3, Angular Momentum and Rotations; Chapter 4, Relativistic Quantum Mechanics and Quantum Field Theory; Chapter 5, Groups and Their Representations; Chapter 6, Internal Symmetries and Conserved Numbers

Part B — THE STANDARD MODEL: Chapter 7, Quantum Electrodynamics as a Gauge Theory; Chapter 8, Yang-Mills Theories; Chapter 9, Quantum Chromodynamics (QCD); Chapter 10, The Electroweak Theory — I. The Gauge Theory; Chapter 11, The

Electroweak Theory—II. Breaking the Symmetry; Chapter 12, The Discrete Symmetries P, C, and T
Part C — EXPERIMENTAL CONCEPTS AND EVIDENCE: Chapter 13, Experimental Concepts; Chapter 14, Experimental Devices and Facilities; Chapter 15, Evidence for the Standard Model with Three Generations
Part D — BEYOND THE STANDARD MODEL: Chapter 16, QCD Revisited; Chapter 17, Larger Symmetries; Chapter 18, Neutrino Mass, Mixing and Oscillations; Chapter 19, Further Developments: Brief Introductions to Technicolor, Horizontal Symmetries, String Theories and Composite Models

Many different courses can be designed using this book by choosing and/or omitting chapters in Part A, based upon the students' backgrounds. Some of the material in Part C is descriptive, especially Chapters 14 and 15, and can be assigned as readings at appropriate points in the course. Part D need not be covered in a one-semester course; however, each chapter is self-contained so that any of them can be included. In a course I taught in 1992, using a draft of this book, paired students were required to select topics from Chapter 12 and Part D to learn and present to the rest of the class. (They were encouraged to ask me for clarifications before their presentations and were not graded.) In this way the students learned some of the material and heard about more of it. The students felt that they benefited from this experience that required them to use the jargon and convey the ideas of particle physics. I owe a special thank you to those students of that class who expended time and effort to point out unclear portions of the book and the usual minus-sign errors, etc. Thank you Yang Chen, Eric Koistinen, Ale Lukaszew, Yongxing Wang, and Grace Yong.

I am pleased to thank my colleague, W. W. Repko of Michigan State University, who carefully read one draft of this book and has provided me with useful advice and criticism of subsequent changes as well. I also wish to thank James M. Johnson for reading the manuscript and making a number of interesting suggestions. Special thanks to Joey Huston of Michigan State University, who brought me up to date about modern-day detectors and wrote the Appendix on them for this book. I am indebted to T. Cormier for providing me with reading material and several informative discussions on quark-gluon plasmas and relativistic heavy ion experiments.

In order to ensure that the book would be understandable to those not in this field, I asked some colleagues, whose expertise is in other areas, to read the first draft of this manuscript. I am indebted to J. J. Chang, H. H. Denman, P. H. Keyes, and M. G. Stewart, who read various parts of the manuscript, for their time and useful suggestions. Thanks also to the reviewers whose constructive criticisms have led to the final form of this book.

Finally, I must thank the secretaries who patiently drew and redrew (on the computer) the figures I gave them and typed some of the tables; they are Beverly Jacobs, Michelle Caprara, and Vance Briceland. Thank you also to Beverly for running off all those copies.

Bill Rolnick
December, 1992

The final manuscript was corrected and modified thanks to: D. Schramm for useful comments on the discussion of cosmological evidence for three generations; J. S. Nico for the latest information on solar neutrino experiments; T. Cormier for an update on pions in heavy ion collisions; H. Bohm for a discussion of the Argonne bubble chamber; S. Payson for pointing out some discrepancies, etc.; G. Dunifer for comments on cosmology; J. Johnson for many useful comments; and those students of the first class to take the course from the final manuscript who pointed out several sign errors and other discrepancies: Dah-Chin Ling, Punya Talagala, and Dankun Yang.

April, 1993

Contents

CHAPTER 3
Angular Momentum and Rotations   **32**

CHAPTER 4
Relativistic Quantum Mechanics and
Quantum Field Theory   **48**

CHAPTER 5
Groups and Their Representations   **80**

CHAPTER 6
"Internal" Symmetries and Conserved Numbers   **99**

PART B   The Standard Model

PART C   Experimental Concepts and Evidence

CHAPTER 15
Evidence for the Standard Model with
Three Generations   296

PART D   Beyond the Standard Model

CHAPTER 16
QCD Revisited   337

CHAPTER 17
Larger Symmetries   349

CHAPTER 18
Neutrino Mass, Mixing, and Oscillations   371

CHAPTER 19
Further Developments   **380**

Chapter 1

Introduction

1.1 IN THE BEGINNING

In front of the Detroit Institute of Arts is one of the several original versions of the sculpture *The Thinker* by Rodin. Whenever I see that sculpture, I envision an animal (one of our distant ancestors) suddenly and unexpectedly sitting down in the forest in that position to contemplate. What did that "first" thinker think about? Well, perhaps that thinker was the first elementary particle theorist. Elementary particle physics began when humans first started wondering what we and everything around us are made of, and what is happening to hold matter together. Certainly throughout recorded history man has searched for the basic building blocks of matter and their interactions. From Aristotle's "earth, air, fire and water" to Mendeleev's periodic table of elements (about 90 of them) we have taken many small steps in our pursuit of such understanding.

Could there be as many as 90 different basic building blocks from which matter is composed? Why not? Many do not find it aesthetically pleasing to have so many; they are disturbed by the notion. "We seek an underlying simplicity," they say, for which 90 is too big a number. Well, nature may not comply! BUT before 1900, J. J. Thomson already had shown the existence of a very low mass particle that is a common *constituent* of many (and today we know that it is of all) atoms, namely the **electron**. Rutherford, in his classic scattering experiments, showed that the atom has a tiny positively charged nucleus ($\sim 10^{-5}$ times the size of the atom) and that the lightest element has a singly (positively) charged nucleus, which is now known as the **proton** . . . and then there were two.

However, there was some difficulty understanding the many different nuclei as combinations of only those two building blocks.[1] In the early 1930s, an electrically neutral particle was discovered which was heavy and tiny like the pro-

---

1. For example, the nucleus of N^{14} would be made of 14 protons and 7 electrons which is 21 spin $\frac{1}{2}$ particles. The observed spin of the ground state of this nucleus was found to be 1, and an odd number of spin $\frac{1}{2}$ particles cannot form a spin 1 state. N^{14} also obeys the symmetry properties (under interchange) appropriate to an *even* number of spin $\frac{1}{2}$ constituents.

1

ton. At first it was conjectured that this particle might be a composite made of a proton and electron, like the hydrogen atom, but as small as the nucleus. This idea failed close theoretical scrutiny, and so this new particle would have to be another elementary particle — its name? — the **neutron** . . . and then there were three.

It has been noted that we physicists call something "elementary" when we know nothing about its structure, i.e., this label expresses our ignorance! Towards the end of the nineteenth century, atoms were thought of as elementary and the "elementary particle physics" of that era is now called chemistry. The discoveries of the electron and the nucleus and the study of their interactions led to the development of quantum mechanics, and that era's elementary particle physics is now known as atomic physics. The study of the structure of nuclei, composites made of protons and neutrons, was the elementary particle physics of the 1930s and 1940s and is called nuclear physics. This book will explore the elementary particle physics which developed beyond nuclear physics.

It is apparent that we have explored the structure of ever smaller particles in this intriguing quest for the basic building blocks of matter. Since the ground-breaking scattering experiments of Rutherford in the early part of this century, it has become usual to employ scattering experiments as the means of exploring for structure in particles. By firing one particle at another and analyzing the resulting debris, we should be able to "see" what is inside. (Using light to look inside something is analogous to firing photons at a particle under investigation.) HOWEVER, when the proton was investigated in this way, it did not break into its "constituents"; instead new particles were *created*! This creation process is reminiscent of the creation of an electron-positron pair from a photon, observed earlier in this century and understood in terms of quantum electrodynamics. We certainly do not think of the photon as being composed of an electron and positron. Now, however, we are discussing a reaction in which the particles created are new ones, previously unknown. A typical scattering reaction of this sort is:

$$p + p \rightarrow p + \Sigma^+ + K^0$$

REST ENERGY (mc^2) in MeV: 938   938   938   1189   493

where the latter two particles (whose electric charges are indicated as superscripts) are not found in ordinary matter. Many other final states occur as well, as seen in reactions such as:

$$p + p \rightarrow n + \Sigma^+ + K^+$$

$$p + p \rightarrow \Delta^{++} + \Sigma^{*0} + K^0,$$

with the appearance of many new particles (Σ, K, Σ^*, Δ, etc.). Considering the first reaction alone, $m_{\Sigma^+} > m_p$ (1189 > 938) and the Σ^+ (sigma plus) and K^0 (K nought or zero) together have almost twice the mass of the proton. It is extremely unlikely that they are constituents of the proton, since an extremely strong binding force (to produce the very large binding energy needed) would be required. Furthermore, here we have seen a few of the many other final states which occur, with new particles appearing (as well as our old friend the neutron). The existence of all these different "channels" is convincing evidence that we are not dealing with a simple "shattering" of protons. A very large number of new particles have been created in the scatterings of the types of particles we find around us (viz., protons, neutrons, electrons, and photons), and beams of some of the longer-lived, newly discovered particles have been used to produce more new particles. In a typical experiment of this sort, a beam of one such particle (π or pion) has been used to produce reactions such as:

$$\pi^- + p \rightarrow \Lambda^0 + K^0,$$

REST ENERGY (mc^2) in MeV: 140   938   1116   493

and

$$\pi^- + p \rightarrow \Sigma^- + K^+.$$

Here also, it can be concluded that neither the Λ^0 (lambda nought or zero), the Σ^-, nor the K particles were in the π^- or p.

Hundreds of such particles have been created in this way (and we shall learn how to categorize them) . . . and then there were many? Very many! Might some of them be composed of the others?

One might ask whether these reactions violate long-venerated conservation laws since more mass comes out than goes in. The answer is no; there is no conservation of mass law! The special theory of relativity has taught us that mass is a form of energy, and so, as long as there is enough energy initially, the reactions can occur. In these reactions we always find that energy is conserved when the kinetic energy is taken into account.[2] Linear and angular momenta have also been found to be conserved, and many other conservation laws are obeyed as well. Many millions of such reactions have been seen and analyzed in cosmic rays and at the large particle accelerators built in the latter part of the twentieth century.

If we were going to study the tree, we might begin by studying the properties of roots and then proceed to leaves, etc. But we all have seen trees, so we appreciate the significance of their various parts. Before we begin our discussions of the "roots," "leaves," etc., of elementary particle physics, I must ask the reader to become familiar with *our* "tree," namely the chart of Basic Interactions on the inside front cover of this book that shows the nature of the interactions and the theories which attempt to explain them.

There are four basic interactions in nature, in terms of which all observations of natural processes can be explained. They are (1) the **strong interaction,** which was first detected as nuclear forces, (2) the **electromagnetic interaction,** which is responsible for chemical reactions as well as electrical and magnetic phenomena, (3) the **weak interaction,** which manifests itself in radioactivity and

---

2. The incoming and emerging particles are so far apart that potential energies are negligible.

β-decays, and (4) the **gravitational interaction**, which is responsible for gravity. The relative coupling strengths, shown in the Basic Interactions chart for the effective interactions of a proton or neutron, span many orders of magnitude. The basic gravitational force is very weak ($\sim 10^{-39}$ times the strong interaction strength), despite the large gravitational forces we see around us and the major role that gravity plays on the cosmological scale. These large gravitational forces are a result of the solely attractive nature of the gravitational force and the large sizes of the objects involved. In the vast majority of the interactions considered in particle physics, the gravitational interaction is so small as to be unobservable (in the presence of the other interactions).

There appear to be two qualitatively different size ranges of these basic forces. The electromagnetic and gravitational interactions produce forces of long range (e.g., the Coulomb force and the gravitational force, respectively), whereas the strong interaction range is of the order of the sizes of nuclei, and the weak interaction range is one thousandth of that size. We shall see that the range of a force is usually directly related to the Compton wavelength of the gauge boson "transmitting" that force, so that the masslessness of the photon and graviton is directly related to the long (∞) range of the corresponding interactions. Likewise, the very short range of the weak interaction is directly related to the extremely large masses of the $W^{\pm}$ and the Z^0 gauge bosons which transmit this interaction. This relationship between the mass of the gauge boson and the range of the interaction does not apply to the strong interaction. The gluons, the gauge bosons of the strong interaction, have no mass, yet the *observed* range of the strong interaction (e.g., nuclear forces) is very short. There is a much more subtle mechanism operating here, which we shall discuss in later chapters. The theory that has been developed to account for the strong interaction is quantum chromodynamics (QCD), and that which accounts for the electromagnetic interaction is quantum electrodynamics (QED). In order to understand the weak interaction, it was necessary to incorporate QED into the structure, which is called the electroweak theory. QCD plus the electroweak theory comprise the standard model, *a theory* which explains just about every observation and experimental result to date. As you move right under Theories in the chart of Basic Interactions, you will encounter more comprehensive theories, which often include those to the left within them. Those to the right of the standard model are still very conjectural at this point. Grand unified theories (GUTs) are attempts to build a more unified theory, which contains both QCD and the electroweak theory in a nontrivial mathematical structure. (Note that grand unified theories is written as a plural, indicating that *there is no one accepted theory*.) Supergravity theories are ingenious attempts to unify grand unified theories with the gravitational interaction, and string theories take us to the outer limits of these attempts.

It is also important that the reader be acquainted with the vocabulary of particle classifications alluded to earlier, which are listed and explained in the Particle Glossary (on the inside front cover). There we see that there are three categories of particles: particles that have strong interactions (and possibly weak and electromagnetic interactions as well) are called **hadrons**, particles that have weak interactions (but no strong interactions) are called **leptons**, and quanta of the force fields are called **gauge bosons**. The **hadrons are** not fundamental particles, but instead are either **bound states of three quarks**, called **baryons** (having odd half-integral spin) or **bound states of a quark and an antiquark**, called **mesons** (with integral spin).

1.2   THE FUNDAMENTAL PARTICLES

Let us now take our first look at the fundamental particles as understood in this last decade of the twentieth century.

1.2.1   Basic Fermions (Spin $\frac{1}{2}$)

Quarks   There are **six** kinds or **flavors** of quarks, each of which carries a color charge[3] of the strong interactions (QCD). Each flavor comes in **three colors**. The different flavors fall naturally into **three** sets of two members (doublets) which are called **families** or **generations**. The doublets and the corresponding flavor names are shown below for the first, second, and third generations, respectively.

Generation:	*1*		*2*		*3*		
	$\begin{pmatrix} u \\ d \end{pmatrix}$	up down	$\begin{pmatrix} c \\ s \end{pmatrix}$	charm strange	$\begin{pmatrix} t \\ b \end{pmatrix}$	top* bottom	$q = \frac{2}{3}e$ $q = -\frac{1}{3}e$

*Not yet found experimentally.

The electric charge common to the upper row and that common to the lower row are indicated at the far right. The first generation (up and down) form the (stable) particles in ordinary matter, e.g., neutrons and protons. The higher generations of quarks decay into the lower ones.

It is useful to keep track of the number of baryons in a state, and this is accomplished by assigning a baryon number (B) to each. The baryons observed in nature (e.g., the proton and the neutron) are all assigned $B = 1$ (with their antiparticles having $B = -1$). Since each baryon (neutron, proton, hyperons, etc.) is a bound state of three quarks, the quarks each have baryon number $B = \frac{1}{3}$. It has been found experimentally that baryon number is conserved. This was first realized by chemists in the nineteenth century, for the concept of baryon number appeared in the study of chemistry and nuclear physics as *atomic weight*. No deep theoretical reason for the conservation of baryon number is known. This is unusual for, as we shall discuss, most conservation laws are the results of symmetries.

---

3. The designation of the charge-like property in QCD as color is indeed colorful, but has nothing to do with actual colors.

The masses of the quarks increase from generation 1 to generation 3 and, in the second and third generations, from lower to upper. The definition of mass for quarks is, however, more subtle than you can imagine at this point as will become apparent in the next paragraph.

All observed particles have a net color charge of zero, and it is believed to be a property of nature that no colorful state can ever be produced. This is called **(color) confinement.** Quantum chromodynamics is mathematically very difficult to handle, and a rigorous proof of confinement has not been constructed; however, several proofs exist in simplified versions of QCD, and strong semiquantitative arguments have been constructed in the full theory. We shall see that the short range of the strong interaction, despite the masslessness of the gluons, is a consequence of color confinement. Furthermore, we shall never be able to produce quarks singly (as free particles) since they are colored. Consequently, the concept of rest mass has no precise meaning for quarks. The reader might conclude that, if we cannot produce them, it would be more accurate to consider them to be internal degrees of freedom rather than particles. Within the quantum field theories of the standard model and beyond, however, the quarks are treated as particles, and so it is considered appropriate to generalize the definition of "particle" to include them. You may wonder how the existence of the quarks and their properties were determined, since they have never appeared as free particles. It is indeed an amazing and fascinating story as we shall see.

Leptons   The first leptons discovered were the electron and its neutrino, the latter in the days when nuclear physics was our elementary particle physics. They form the first generation of leptons, and two more generations of leptons have been discovered. So we have three generations of leptons (just as for quarks), and they fall into doublets just as the quarks do:

Generation:	*1*	*2*	*3*	
	$\begin{pmatrix} \nu_e \\ e^- \end{pmatrix}$	$\begin{pmatrix} \nu_\mu \\ \mu^- \end{pmatrix}$	$\begin{pmatrix} \nu_\tau \\ \tau^- \end{pmatrix}$	$q = 0$ $q = -1e$

$m_e = 0.51$ MeV, $m_\mu = 106$ MeV, $m_\tau = 1784$ MeV, and $m_\nu = 0$ (as far as is known). The generations of leptons do not make transitions from one to another, and so a separate conservation law exists for each generation. Thus, besides (overall) **lepton number conservation** there is a **conservation of the number of each family or generation type.** These conservation laws are not understood in any deep theoretical way, just as we have noted for baryon number conservation.

At the present time, there is no theoretical understanding of the existence of the two heavier copies of the lowest generation of particles. Experimental evidence does exist, as we shall discuss, which makes it very unlikely that there are higher generations that are copies of the ones already found.

TABLE 1.1
Selected Particles and Their Antiparticles

Particle	Antiparticle
e^-	e^+ or $\bar{e}$
p	$\bar{p}$
ν	$\bar{\nu}$
γ	γ
π^0	π^0 $u\bar{u}+d\bar{d}$
W^+	W^-

1.2.2   Gauge Bosons

These are the photon (γ), Z^0, $W^\pm$, gluons, and gravitons as listed in the Particle Glossary.

1.2.3   Higgs Scalar (spin 0) Boson

This is a particle (or perhaps a composite) required to account for the masses of the Z^0 and the $W^\pm$. It is their large masses which are responsible for the short range of the weak interaction. Such a particle has yet to be found. The search goes on.

1.2.4   Antiparticles

For each particle there exists an antiparticle of the same mass, lifetime, and spin but with opposite charge and (baryon, lepton, etc.) numbers. The earliest such particle found was the positron, which is the antiparticle of the electron and is usually designated as e^+. A list of some particles and their antiparticles appears in Table 1.1 that displays the usual notation for antiparticles. The antiparticles of those in the antiparticle column are the corresponding particles, so that the antiparticle of the W^- is the W^+, etc. Notice that the π^0 and the γ (photon) are their own antiparticles.

1.3   NATURAL UNITS

In our theoretical and phenomenological notation, both $\hbar$ and c occur over and over again. Using an arbitrarily chosen length (the meter) and time interval (the second) as units is neither facilitative (for handling the theory) nor natural. It has become standard to choose more natural units that make $\hbar = c = 1$; thus, we need never write $\hbar$ or c again in our expressions. You are already familiar with part of this from your study of special relativity where β ($\equiv v/c$) is used to describe speeds in terms of c. There the speed of light corresponds to $\beta = 1$. We have already begun to use natural units in this book. Recall that we expressed the masses of the leptons in MeV, which is correct in natural units, since mc^2 is the same as m where c is 1. Likewise, *all* energies will appear as pure numbers times a mass. If we do not write $\hbar$ or c any more, then the units for a length will appear as follows: The Compton wavelength $\hbar/mc$ (which occurs in photon scattering experiments) would appear as m^{-1}, and *all* lengths will show up in our calculations as pure numbers times an inverse mass (i.e., if the electron is present, the result might be $5.3m_e^{-1}$).

When m appears in a result, how do we distinguish between an inverse length and an energy? To change to SI units, for example, we must know what quantity we are dealing with. If it is an inverse length we multiply the m that appears explicitly by c and divide by $\hbar$, whereas for an energy we multiply it by c^2. Then we evaluate everything in the new set of units (where $\hbar$ and c are no longer 1); the inverse length will be in inverse meters, and the energy will be in Joules. For any quantity, the numbers of factors of $\hbar$ and c to be inserted are uniquely determined by the type of physical quantity it is.

It is not difficult to transform a calculation of one such quantity into another. Suppose a mass M appears in a calculation done in natural units. It could represent an energy, which means that it is actually Mc^2 in SI units, or it could be an inverse length, which means that it is $Mc/\hbar$. Suppose we initially treat it as an energy and obtain $Mc^2 = x$ Joules, and subsequently would like it to be an inverse length $Mc/\hbar = y$ meters^{-1}. Comparing these two equations we see that

$$x \text{ Joules} \leftrightarrow y \text{ (meters)}^{-1} \text{ so that } 1 \text{ Joule} \leftrightarrow \frac{y}{x} \text{ (meters)}^{-1}$$

$$\text{and } \frac{y}{x} = \frac{1}{\hbar c} \approx 3.16 \times 10^{25} \text{ in SI units.}$$

This yields

$$1 \text{ GeV} = 1.6 \times 10^{-10} \text{ Joule} \leftrightarrow 5.06 \times 10^{15} \text{ (meters)}^{-1}.$$

In this book we shall employ these conversions from natural units in several places.

PROBLEMS

1. A calculation in natural units yields $1.0m_p^{-1}$, where m_p is the proton mass: $m_p = 1.67 \times 10^{-27}$ kg.
   a. Suppose we were calculating a distance. What distance in meters would this be? How does it compare to the sizes of nuclei?
   b. Suppose we were calculating a time. What time in seconds would this be? How does it compare to the time it would take a high-speed ($v \sim c$) particle to cross the distance (in the lab) you calculated in part a?
   c. Suppose it were a number per unit energy. Using the value of m_p in kg, find the result in (MeV)$^{-1}$. How does it compare to 1/938 MeV?

Part A

Theoretical Background

Chapter 2

A General Introduction to Quantum Mechanics

The paradigm of modern physics is quantum mechanics. In this chapter we shall provide a generalization of the introduction to quantum mechanics that students receive in a senior-level undergraduate course.

2.1 SYSTEMS AND STATES

A system is described by a wavefunction, which is a function of position and time, whose behavior with time is governed by the Schrödinger equation (for non-relativistic situations[1]):

$$H\Psi(\vec{r},t) = i\,\frac{\partial\Psi(\vec{r},t)}{\partial t},$$

(2.1)

where the operator $H = -(1/2m)\nabla^2 + V(\vec{r})$,[2] which has solutions of definite energy E of the form

$$\Psi_E(\vec{r},t) = \psi_E(\vec{r})e^{-iEt}.$$

A system can be in any one of the different energy states or its wavefunction may be a linear combination of the Ψ_E. (The Ψ_E form a complete set.) For the case of a particle in a rigid box of length L in one dimension, the states are designated by a discrete set of numbers:

$$\psi_n(x) = A\sin k_n x \text{ (where } k_n \text{ is } n\pi/L,\ n = 1, 2, \ldots).$$

(2.2)

Similarly, for the states of the hydrogen atom:

$$\psi_{n,\ell,m}(\vec{r}) \text{ are functions of } r,\ \theta,\ \text{and } \phi.$$

(2.3)

---

1. For relativistic equations, see Chapter 4.
2. Recall that $\nabla^2\psi \equiv \partial^2\psi/\partial x^2 + \partial^2\psi/\partial y^2 + \partial^2\psi/\partial z^2$.

We can take any of these functions of position and Fourier analyze it to find its momentum content. Alternatively, the wavefunction of a state can be reconstructed if we know the "amount" of each Fourier component it contains. This amount is a function of $\vec{p}$ and we can designate it as $\chi_n(\vec{p})$ or $\chi_{n,\ell,m}(\vec{p})$, respectively, for the examples above. It has as much information about the state of the system as the original spatial wavefunction. A general notation, originated by Dirac, allows for either of these descriptions (and others as well), and we now proceed to describe and use it.

Suppose a system is designated by a (where a is a set of numbers); in the above examples a is n or n, ℓ, m, respectively. We represent the state by the symbol $|a\rangle$, which is called a *ket*. This is analogous to, but not the same as, the wavefunction for the state. Recall that we always multiply the wavefunction by its complex conjugate (or sometimes by that of another wavefunction) to compare with experimental results. In this generalized notation, the entity analogous to the complex conjugate of a wavefunction is represented by $\langle a|$, which is called a *bra*. In quantum mechanics, all observables are represented by operators, such as $(-i(\partial/\partial x))$ for the momentum p_x, $\vec{r}$ for position, etc.; let us represent any of these operators by $\mathbf{O}$. The expectation values of these operators are $\int \psi_i^* \, \mathbf{O} \, \psi_i \, d^3r$, where i is the state being considered. In fact, all observable results are obtained from integrals of the form $\int \psi_a^* \, \mathbf{O} \, \psi_i \, d^3r$, where a may be different from i for transitions (as we shall see in Sections 2.2 and 2.10). The quantity $\mathbf{O}\psi_i$ is another function, which we may call ψ_b, so that the form may be written $\int \psi_a^* \psi_b \, d^3r$. In our generalized notation, this is represented by $\langle a|b\rangle$, which is analogous to the dot product of $|b\rangle$ with $|a\rangle$ and may be thought of as the overlap of the states. It is obvious, from the form in terms of wavefunctions, that

$$(\langle a|b\rangle)^* = \langle b|a\rangle. \tag{2.4}$$

For a state of one particle in a potential, let $|\vec{r}_1\rangle$ represent the state in which the particle is located only at $\vec{r}_1$.[3] The amount of state $|a\rangle$ which has the particle located at $\vec{r}_1$ would then be $\langle \vec{r}_1|a\rangle$, so that

$$\psi_a(\vec{r}_1) = \langle \vec{r}_1|a\rangle,$$

where ψ_a is the usual wavefunction. Since this is true for arbitrary $\vec{r}_1$ we can drop the subscript 1 and find that the wavefunction is obtainable from $|a\rangle$:

$$\psi_a(\vec{r}) = \langle \vec{r}|a\rangle. \tag{2.5}$$

Likewise, the state $|\vec{p}\rangle$ may be defined as a state of definite momentum $\vec{p}$, and we can obtain the Fourier function, which is usually called the wavefunction in momentum space, for $|a\rangle$:

$$\chi_a(\vec{p}) = \langle \vec{p}|a\rangle. \tag{2.6}$$

---

3. To be rigorous, one should consider an infinitesimal region around $\vec{r}_1$.

The wavefunction for a state of momentum $\vec{p}$ is a plane wave $(\sim e^{i\vec{p}\cdot\vec{r}})$, so that Eq. (2.5) tells us

$$\langle\vec{r}|\vec{p}\rangle \propto e^{i\vec{p}\cdot\vec{r}}. \tag{2.7}$$

This connection between kets and wavefunctions is easily generalized to more complicated systems. We see that the generalized state contains all the information required to describe the system.

We have learned that it is very useful to consider a complete set of orthogonal vectors as a basis for ordinary vectors. (In three-dimensional space they are usually the unit vectors i, j, and k.) All vectors can then be expressed in terms of them. In the "space" of possible states, one can also choose a complete set of orthogonal states $|\alpha\rangle$,[4] with the property that

$$\langle\alpha|\beta\rangle = 0 \text{ unless } |\alpha\rangle = |\beta\rangle. \tag{2.8}$$

For isolated systems we usually use the states of definite energy as the complete orthogonal set. For scattering situations we generally choose two complete sets, designated $|a\rangle_{in}$ and $|a\rangle_{out}$, where a runs through all possible states of a simple basis set. An example of such a state would be one in which one particle is sitting still ($\vec{p} = 0$) and the other has a definite non-zero momentum. The state $|a\rangle_{in}$ is a state in which the simple situation described by a occurs initially (i.e., at $t \to -\infty$). In ordinary quantum mechanics (as opposed to quantum field theory), the particles present do not change, i.e., there is no creation or annihilation. However, there can be an interaction between the particles, and so, the state can become more complicated as time evolves. The state $|a\rangle_{out}$ is a state which becomes the simple configuration described by a as $t \to \infty$.

2.2   THE SCATTERING AMPLITUDE AND TRANSITION PROBABILITIES

A description of the scattering is most easily obtained by calculating how much $|j\rangle_{out}$ there is in the state $|i\rangle_{in}$:

$$S_{ji} = {}_{out}\langle j|i\rangle_{in}, \tag{2.9}$$

which is called the scattering amplitude or **S-matrix element**. For about two centuries, we have known that the intensities in optical phenomena are proportional to the square of the resultant amplitudes of the waves. Likewise, in quantum mechanics we find that the probability of scattering from an incident state i to a final state j is equal to the absolute square of the scattering amplitude S:

$$P(i \to j) = |S_{ji}|^2,$$

---

4. A standard method for accomplishing this is the Gram-Schmidt orthogonalization procedure.

usually referred to as the transition probability. This is analogous to the expression: $|\int \psi_j^{(out)*} \psi_i^{(in)} d^3r|^2$ of the conventional formulation. The analogy between quantum mechanics and optics is not an accident, but is a consequence of the fact that optics is the classical limit of the quantum theory of photons. In fact, considering the waveparticle duality of the entities of nature, a photon is as much a particle as is an electron, albeit with zero rest mass. In fact, in quantum electrodynamics, the vector potential $\vec{A}$ plays the same role for photons as does the wavefunction for electrons.

2.3   CONSERVATION OF CHARGE

Consider the wavefunction $\psi(\vec{r}, t)$ for a simple system of a particle in a potential. It is referred to as the probability *amplitude* and the probability of finding the particle in a volume of size *dxdydz* around the point $\vec{r}$ is

$$P(\vec{r}, t)dxdydz = |\psi(\vec{r}, t)|^2 dxdydz, \tag{2.10}$$

where $P(\vec{r}, t)$ is the probability density. Of course, if we place detectors *everywhere*, the particle must be found, and we express this **conservation of probability** by writing, for the integral over all space:

$$\iiint P(\vec{r}, t) \, d^3r = 1, \tag{2.11}$$

which is true for all t. Let us consider a limited region of space $\Delta\tau$ and calculate the time rate of change of the probability of finding the particle in this region. Taking the time derivative we obtain

$$\iiint_{\Delta\tau} \frac{\partial}{\partial t} P(\vec{r}, t) \, d^3r = \iiint_{\Delta\tau} \left\{ \left(\frac{\partial}{\partial t} \psi^* \right) \psi + \psi^* \frac{\partial}{\partial t} \psi \right\} d^3r.$$

Let us insert $(-i$ times) the time-dependent Schrödinger equation (Eq. 2.1):

$$\frac{\partial}{\partial t} \psi = (-i) \left\{ \frac{-1}{2m} \nabla^2 \psi + V\psi \right\},$$

and its complex conjugate, into the right-hand side to obtain

$$\iiint_{\Delta\tau} \frac{\partial}{\partial t} P(\vec{r}, t) \, d^3r$$

$$= i \iiint_{\Delta\tau} \left\{ \left(\frac{-1}{2m} \nabla^2 \psi^* + V\psi^* \right) \psi - \psi^* \left(\frac{-1}{2m} \nabla^2 \psi + V\psi \right) \right\} d^3r.$$

Since V only multiplies $\psi^*\psi$ (i.e., it is only a multiplicative operator), the terms with V cancel out yielding

$$\iiint_{\Delta\tau} \frac{\partial}{\partial t} P(\vec{r}, t) \, d^3r = \frac{i}{2m} \iiint_{\Delta\tau} \{ -(\nabla^2 \psi^*)\psi + \psi^* \nabla^2 \psi \} \, d^3r.$$

Notice that the integrand on the right-hand side is an exact divergence:

$$-(\nabla^2 \psi^*)\psi + \psi^* \nabla^2 \psi = \vec{\nabla} \cdot (-(\vec{\nabla}\psi^*)\psi + \psi^* \vec{\nabla}\psi).$$

Substituting yields

$$\iiint_{\Delta\tau} \frac{\partial}{\partial t} P(\vec{r},t)\, d^3r = -\iiint_{\Delta\tau} \vec{\nabla} \cdot \vec{J}_P\, d^3r, \tag{2.12}$$

where $\vec{J}_P = (1/2im)(\psi^* \vec{\nabla}\psi - (\vec{\nabla}\psi^*)\psi)$. Using the divergence theorem (studied in electrodynamics), the volume integral on the right-hand side can be recast into a surface integral to yield

$$\frac{\partial}{\partial t} \iiint_{\Delta\tau} P(\vec{r},t)\, d^3r = -\iint_{S_{\Delta\tau}} \vec{J}_P \cdot \hat{n}\, dS, \tag{2.13}$$

where $S_{\Delta\tau}$ is the surface bounding the region $\Delta\tau$ and $\hat{n}$ is the outward normal. This is *also* an expression of the conservation of probability as it says that the change of the probability in the region is due to a flow of probability out of that region, with $\vec{J}_P$ playing the role of **probability-current density**. If we let $\Delta\tau$ become infinitesimal, we may remove the integration signs in Eq. (2.12) to obtain **the continuity equation**:

$$\frac{\partial}{\partial t} P(\vec{r},t) + \vec{\nabla} \cdot \vec{J}_P = 0. \tag{2.14}$$

For N charged particles, each with charge q and definite momentum $\vec{p}$, the charge density (ρ) and current density ($\vec{J}$) would be directly related to the probability density and probability-current density, respectively: $\rho = qNP$ and $\vec{J} = qN\vec{J}_P$. It is left to the reader (in Problem 4) to verify that $qN\vec{J}_P = \rho\vec{v}$, which is the correct expression for electric current. Thus the continuity equation is identical to that for ordinary charge:

$$\frac{\partial}{\partial t} \rho(\vec{r},t) + \vec{\nabla} \cdot \vec{J} = 0. \tag{2.15}$$

In fact, this continuity equation for electric charge and current may be obtained in the general case, i.e., in the presence of interactions, and with many kinds of particles.

2.4   MEASUREMENT AND OPERATORS

Recall that measurable quantities, like position and momentum, are associated with operators which act on the wavefunctions of states: for position and momentum they are multiplication by x and differentiation with respect to x($p_x \to -i(\partial/\partial x)$), respectively. **All observable quantities are represented by operators.** The effect of these operators on a wavefunction is to change the function into another function. In the general formulation, we represent observables by operators (A) such that for any ket $|\psi\rangle$, $A|\psi\rangle$ is also a ket. For a state of a basis set $|i\rangle$, $A|i\rangle$ is a ket, although in general not one of the simple basis kets. One can take the scalar product of this ket with any ket $|j\rangle$ in the basis to get $\langle j|A|i\rangle$. This product is then referred to as the matrix A with matrix elements:

$$A_{ji} = \langle j|A|i\rangle. \tag{2.16}$$

The bra associated with $A|i\rangle$ is written $\langle i|A^\dagger$.

Consider $(\langle j|A|i\rangle)^*$; the "braket" inside the parentheses may be thought of as the scalar product of the ket $A|i\rangle$ with the bra $\langle j|$. From Eq. (2.4), we have

$$(\langle j|A|i\rangle)^* = (\langle j|[A|i\rangle])^* = [\langle i|A^\dagger]|j\rangle = \langle i|A^\dagger|j\rangle, \tag{2.17}$$

where we have inserted square brackets to aid in following the reasoning. (In this equation we may now consider the $A^\dagger$ to be operating on the state $|j\rangle$.) Reading this as a matrix equation we see that

$$A_{ji}^* = A_{ij}^\dagger, \tag{2.18}$$

which, when read from right to left, is the correct definition of the Hermitian conjugate $(A^\dagger)$ of a matrix; hence the notation.

The average value of an observable is obtained by taking its expectation value in the state under consideration, e.g.,

$$\bar{p}_x = \int \psi^* \left(-i\, \frac{\partial}{\partial x} \right) \psi\, dx. \tag{2.19}$$

Although quantum mechanics involves complex quantities, a real answer will certainly be obtained when something is measured.[5] Let us express this reality property for the expectation values of operators which represent observables:

$$\langle \psi|A|\psi\rangle = a_\psi, \text{ the expectation value in the state } \psi. \tag{2.20}$$

Let us write a^* explicitly,

$$a_\psi^* = (\langle \psi|A|\psi\rangle)^* = \langle \psi|A^\dagger|\psi\rangle,$$

using Eq. (2.17). The reality requirement ($a_\psi = a_\psi^*$ for all ψ) becomes

$$\langle \psi|A|\psi\rangle = \langle \psi|A^\dagger|\psi\rangle \text{ for all possible } \psi.$$

Since this is true for all possible ψ we obtain

For any observable A: $A = A^\dagger$, i.e., A is Hermitian. (2.21)

EXAMPLE: Angular Momentum

Angular momentum provides a simple and important example of the connection between operators and matrices. Understanding the angular momentum

---

5. I will leave it as an exercise for the reader to check that the above expression for $\bar{p}_x$ is real.

operators is so important that we shall devote an entire chapter to them (Chapter 3).

Recall that the angular momentum operator ($\vec{L}$) is defined as

$$\vec{L} = \vec{r} \times \vec{p} = \vec{r} \times (-i\vec{\nabla}). \tag{2.22}$$

Using spherical coordinates (r, θ, ϕ), it can be shown that

$$L_z = -i \frac{\partial}{\partial \phi}. \tag{2.23}$$

There exists a set of orthogonal and normalized (orthonormal)[6] functions $Y_\ell^m(\theta, \phi)$ (where ℓ is a positive integer and m can be any integer between $+\ell$ and $-\ell$) such that

$$L_z Y_\ell^m(\theta, \phi) = m Y_\ell^m(\theta, \phi) \quad \text{and} \quad L^2 Y_\ell^m(\theta, \phi) = \ell(\ell + 1) Y_\ell^m(\theta, \phi). \tag{2.24}$$

We say that $Y_\ell^m(\theta, \phi)$ is an eigenfunction of both L_z and L^2. For a fixed ℓ there are $(2\ell + 1)$ possible values for m. Let us use these $(2\ell + 1)$ Y_ℓ^m's as our basis states. We can then calculate the matrix elements as outlined in Eq. (2.16):

$$\vec{L}_{m_2 m_1} = \int (Y_\ell^{m_2})^* \vec{L} Y_\ell^{m_1} \, d\Omega.^7 \tag{2.25}$$

Since L_z yields only the same Y_ℓ^m, we have

$$(L_z)_{m_2 m_1} = m_1 \delta_{m_2 m_1} \text{ (where } \delta_{m_2 m_1} \text{ is 1 for } m_1 = m_2 \text{ and 0 otherwise.)} \tag{2.26}$$

This is what it looks like in matrix form:

$$L_z = \begin{pmatrix} \ell & 0 & & \cdots & & ..0 \\ 0 & \ell-1 & & 0... & & ..0 \\ \vdots & \vdots & \ddots & & 0 & ..0 \\ & & & & & \ddots \\ 0 & 0 & & \cdots & & -\ell \end{pmatrix}. \tag{2.27}$$

In Appendix A we display the L_x, L_y, and L_z matrices for $\ell = 1$. ∎

2.5   COORDINATE TRANSFORMATIONS AND UNITARY OPERATORS

Consider the transition from an initial state $|a\rangle$ to a final state $|b\rangle$. We wish to change the coordinate system used to observe this event (by a rotation, translation, reflection, or whatever). We know that, in the new coordinate system,

---

6. $\int (Y_{\ell'}^{m'})^* Y_\ell^m \, d\Omega = \delta_{\ell'\ell} \delta_{m'm}$. These functions may be found in the Mathematical Tables at the back of the book.

7. In fact, it can be shown that the $\vec{L}$ operators acting on the Y_ℓ^m produce linear combinations of Y_ℓ^m's with the same ℓ.

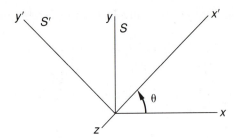

FIGURE 2.1
The coordinate system S' is obtained from S by a rotation about the z axis by angle θ.

the states will be described differently. For example, in a coordinate system S' rotated by θ about the z axis with respect to S, a wavefunction $\psi(x)$, which was independent of y and z in the original system S, is no longer independent of the new y', since $x = x' \cos \theta - y' \sin \theta$. (See Fig. 2.1.) The states in the new system $|a\rangle'$ are linearly related to those in the original $|a\rangle$, which we express by writing

$$|a\rangle' = U|a\rangle, \text{ and for the corresponding bras } '\langle a| = \langle a|U^\dagger. \quad (2.28)$$

Here U is an operator (in the space of states[8]) which changes the kets to those for the new coordinate system, and $U^\dagger$ is its corresponding Hermitian conjugate operator. The event being observed has the same probability of occurring whether we use the original or the new coordinate system! If we look at one set of rulers or the other, at both sets at once, or at none, the event will still take place at the same rate. Let us express this in mathematical terms:

$$P(a' \to b') = P(a \to b),$$

which then yields

$$|'\langle b|a\rangle'|^2 = |\langle b|a\rangle|^2. \quad (2.29)$$

Substituting Eq. (2.28) into Eq. (2.29) yields

$$|\langle b|U^\dagger U|a\rangle|^2 = |\langle b|a\rangle|^2. \quad (2.30)$$

This must be true for all states, and it can be shown that, except for a time-reversed system, this implies that[9]

$$U^\dagger U = 1 = UU^\dagger. \quad (2.31)$$

An operator which has this property is called *unitary*. Thus, **coordinate transformations are represented by unitary operations.**

---

8. Mathematicians call this a Hilbert space.

9. Some subtleties in the choices of phases of states must be considered. See, e.g., *Group Theory* by Wigner.

When we consider the operators for observables, such as currents, momenta, etc., we expect that, in general, a change of basis results in a change of the (observed) expectation values of the corresponding operators. For example, the currents and momenta are vectors under rotations, and so the expectation values of their components in the new rotated frame, with the new basis set of states, would be a linear combination of those in the old frame with the original basis. It is not hard to find the general rule for any matrix element of any such operator $\mathcal{O}$. Before we begin, let us clarify our notation. A measurement of the position of a particle is usually represented by the operator $\vec{r}$. However, this operator yields $\vec{r}$ for the position in frame S and $\vec{r}'$ in frame S'. To avoid any confusion, we shall refer to the position operator as $\vec{r}_{op}$, the momentum operator as $\vec{p}_{op}$, etc., and the new frame as S^{new} in this discussion.

First let us consider an apparatus set up to measure the x component of the momentum of a particle (p_x) in the old basis (S frame). Someone in a rotated frame (S^{new}) would see this measurement and the numerical result, but would say that the apparatus does not measure p_x (the analogous quantity) in the rotated frame, but a *linear combination* of the components of $\vec{p}$. Therefore, the observer in S^{new} would not use p_{xop}, but p_{xop}^{new} to describe the measurement made by that apparatus. Let us now generalize this argument, by using $\mathcal{O}$ for the operator corresponding to the apparatus in the old basis and $\mathcal{O}^{new}$ in the new basis, to find the precise relationship between $\mathcal{O}^{new}$ and $\mathcal{O}$. We must demand that

$$^{new}\langle b| \ \mathcal{O}^{new} \ |a\rangle^{new} = \langle b| \ \mathcal{O} \ |a\rangle,$$

i.e., the measurement made by the apparatus must be the same, but **the apparatus corresponds to the measurement of a different quantity in the transformed frame**. Substituting from Eq. (2.28), this becomes

$$\langle b|U^\dagger \ \mathcal{O}^{new} \ U|a\rangle = \langle b| \ \mathcal{O} \ |a\rangle.$$

For this to be true for all states $|a\rangle$ and $|b\rangle$, the operators on both sides of the equation must be equal, viz.,

$$U^\dagger \ \mathcal{O}^{new} \ U = \mathcal{O}.$$

Multiplying both sides of this equation by U on the left and $U^\dagger$ on the right (utilizing Eq. 2.31) we obtain

$$\mathcal{O}^{new} = U \ \mathcal{O} \ U^\dagger, \text{ looking at the same apparatus,} \qquad (2.32)$$

which is called a **unitary similarity transformation**. It corresponds to looking at **one apparatus** giving **one result**.

On the other hand, to measure the *analogous* **quantity in the new (e.g., rotated) frame**, we need a **new apparatus, reoriented (e.g., rotated)** with respect to the old one; the measurement of the analogous quantity must be represented by the *same* operator in the new frame. The *new* measurement is

$$^{new}\langle b| \ \mathcal{O} \ |a\rangle^{new} = \langle b|U^\dagger \ \mathcal{O} \ U|a\rangle, \qquad (2.33)$$

where we have used Eq. (2.28), and, in what follows, we shall be more interested in the transformation $U^\dagger \, \mathcal{O} \, U$, than that of Eq. (2.32). Very often, a prime is used to represent the new frame, but care must then be exercised to distinguish the operators and the measured values, as we have pointed out in our discussion of the $\vec{r}_{op}$ above.

In the problems of Chapter 3, we shall see that, for the rotation of a vector operator $\vec{V}$,

$$R^\dagger V_i R = \sum_j R_{ij} V_j,$$

where R is the rotation operator and R_{ij} is the corresponding rotation matrix.[10] Using Eq. (2.33) and reverting to the more usual notation of S' for the new frame, we find that measurements in the new frame are related to those in the old as expected for a vector quantity:

$$'\langle b|V_i|a\rangle' = \langle b|R^\dagger V_i R|a\rangle = \sum_j R_{ij} \langle b|V_j|a\rangle.$$

2.6 STATIONARY STATES

States of definite energy are called stationary states and have a simple time dependence:

$$|\psi_E(t)\rangle = e^{-iEt}|\psi_E\rangle, \tag{2.34}$$

just as in the usual wavefunction formulation, where $|\psi_E\rangle$ obeys

$$H|\psi_E\rangle = E|\psi_E\rangle, \tag{2.35}$$

which is the form of the time-independent Schrödinger equation. H (which is an observable) is called the Hamiltonian, and E is the energy. We say that **$|\Psi_E\rangle$ is the eigenket and E is the eigenvalue for the operator H.** Our discussion of observable operators has taught us (see Eq. 2.21) that they must be Hermitian, and so **H is Hermitian.** This follows since **E is real.**

In general, the other observable operators have eigenkets, with accompanying **eigenvalues** as well; the latter are often referred to as **quantum numbers.** In an obvious notation we then write

$$A|\psi_a\rangle = a|\psi_a\rangle, \tag{2.36}$$

where the eigenvalue a is a real number. A state is usually described by many quantum numbers related to various observables, e.g., atomic states are described by n, ℓ, m_ℓ, m_s for each electron.

---

10. For example, for a rotation in two dimensions $R_{ij} = \begin{pmatrix} \cos\theta & \sin\theta \\ -\sin\theta & \cos\theta \end{pmatrix}$ as we shall discuss in Chapters 3 and 5.

2.7   CONNECTION BETWEEN SYMMETRY OPERATIONS AND OBSERVABLES

The stationary states of a system form a complete set, meaning that a general state of the system at $t = 0$ can be expressed as a linear combination of them:

$$|\psi(0)\rangle = \sum_E a_E |\psi_E(0)\rangle. \tag{2.37}$$

Each of these components evolves with time with its particular value of E, as shown in Eq. (2.34) above, so that

$$|\psi(t)\rangle = \sum_E a_E e^{-iEt} |\psi_E(0)\rangle. \tag{2.38}$$

Thus, the general state does not have a simple behavior with time, but we may formally write

$$|\psi(t)\rangle = e^{-iHt} |\psi(0)\rangle, \tag{2.39}$$

since for each term in the sum in Eq. (2.37), $e^{-iHt}|\psi_E(0)\rangle = e^{-iEt}|\psi_E(0)\rangle$. Equation (2.39) is of the form:

$$|\psi(t)\rangle = U(t)|\psi(0)\rangle, \tag{2.40}$$

where $U(t) = e^{-iHt}$. Equations (2.39) and (2.40) show that $U(t)$ **is the time-translation operator** since it takes states from time 0 to t. Since H is Hermitian, we find that $U^\dagger(t) = e^{+iHt}$.[11] In fact, the Hermiticity of H implies that U is unitary:

$$U^\dagger U = UU^\dagger = 1. \tag{2.41}$$

This may be seen as follows:

$$U^\dagger U|\psi_E\rangle = e^{iHt}e^{-iHt}|\psi_E\rangle = e^{iHt}e^{-iEt}|\psi_E\rangle = e^{-iEt}e^{iHt}|\psi_E\rangle$$
$$= e^{-iEt}e^{iEt}|\psi_E\rangle = 1|\psi_E\rangle.^{12} \text{ (Likewise, for } UU^\dagger.)$$

Equation (2.41) then follows because the $|\psi_E\rangle$ form a complete set. Here we see a connection between the unitary operator $U(t) = e^{-iHt}$, which produces a time translation, and the Hermitian observable called the Hamiltonian. We express this connection by saying that **the Hamiltonian (H) is the generator of time translations.**

The laws of nature are the same today as they were yesterday, and we expect them to be unchanged tomorrow. (The identical experiment will give identical results whether it is done today, tomorrow or next week.) We have tacitly assumed this by considering a time-independent Hamiltonian. This time inde-

---

11. We have used $(f(A))^\dagger = f^*(A^\dagger)$, where f^* means change any i appearing explicitly in f to $-i$.

12. We can interchange e^{iHt} and e^{-iEt} because e^{-iEt} is not an operator but a number.

pendence led to states of definite energy, the stationary states, as solutions of the Schrödinger equation. It is left to the reader to verify that these solutions, $|\psi_E(t)\rangle$, have the same energy for all t, so that energy is conserved. Thus, time-translation invariance leads directly to the conservation of energy. When we discuss the equations of motion in Section 2.9, we shall see, more generally, how a continuous symmetry implies conservation of the physical quantities corresponding to its generators. Symmetries such as translational invariance (moving the origin of our coordinate system) and rotational invariance lead to the conservation of momentum and angular momentum, respectively, and many conservation laws have this deep connection with the symmetries of nature. It should be pointed out that any given state will look different in a different coordinate system; it is the *invariance of the laws of nature* to these transformations that leads to the conservation laws. This is also true for time translations; the states change (they evolve in time), but the laws of nature are unchanged as time goes on (as far as we know).

EXAMPLE: *Translations and (Linear) Momentum*

Consider observer S, with origin O, viewing a system described by the wavefunction $\psi(\vec{r})$. Suppose another observer S', with axes parallel to those of S but origin O' shifted from O by the vector $\vec{a}$ as shown in Fig. 2.2, is also viewing this system. The physical point P has its location described as $\vec{r}$ by S and $\vec{r}'$ by S', and $\vec{a} + \vec{r}' = \vec{r}$, as is evident from Fig. 2.2.

The value of the wavefunction at P will certainly be the same[13] to both observers since it is the probability amplitude at that physical point. Thus, the wavefunction (ψ') in the S' frame must be related to ψ by

$$\psi'(\vec{r}') = \psi(\vec{r}) = \psi(\vec{a} + \vec{r}') = \psi(x' + a_x, y' + a_y, z' + a_z). \qquad (2.42)$$

Since the transformation under consideration does not involve moving the physical system, it is referred to as a **passive transformation**.

Let us proceed assuming that the function is smooth enough to do a Taylor series expansion of the right-hand side of Eq. (2.42):

$$\psi'(\vec{r}') = \psi(\vec{r}') + \left(a_x \frac{\partial}{\partial x'} + a_y \frac{\partial}{\partial y'} + a_z \frac{\partial}{\partial z'} \right) \psi(\vec{r}') + \text{higher order derivatives};$$

$$(2.43)$$

the second term on the right may be written:

$$\vec{a} \cdot \vec{\nabla}' \psi(\vec{r}').$$

Recall that for ordinary numbers $e^x = 1 + x + x^2/2! + \text{higher powers}$. By solving Problem 13, you can show that the coefficients of the powers of x are iden-

---

13. There is always the freedom to change the overall phase of the wavefunction by the same amount everywhere, so that we can choose the phases in S and S' to match.

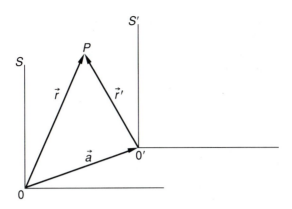

FIGURE 2.2
The position vectors to point P in two coordinate systems (S and S') whose origins are separated by a translation ($\vec{a}$).

tical to the coefficients of the derivative terms appearing in the expansion of $\psi'(\vec{r}')$. Consequently, we may rewrite Eq. (2.43) in the form:

$$\psi'(\vec{r}') = e^{\vec{a}\cdot\vec{\nabla}'}\psi(\vec{r}'), \tag{2.44}$$

where the exponential of the operator stands for the power series. Recall that $-i\vec{\nabla}'$, operating on a function of $\vec{r}'$, is the momentum operator $\vec{p}$ in quantum mechanics, so we may write

$$\psi'(\vec{r}') = e^{i\vec{p}\cdot\vec{a}}\psi(\vec{r}') \qquad \text{(passive transformation).} \tag{2.45}$$

Thus **the momentum ($\vec{p}$) is the generator of translations**.

A similar expansion of $\psi(\vec{r}')$ shows that

$$\psi(\vec{r}') = \psi(\vec{r} - \vec{a}) = e^{-i\vec{p}\cdot\vec{a}}\psi(\vec{r}) \qquad \text{(passive transformation),} \tag{2.46}$$

where the left-hand side is the old function of the new position variable. ∎

In a similar way, it has been shown (see Problem 16) that if the S' frame is rotated with respect to the S frame (by θ counterclockwise about an axis, whose direction we will call $\hat{\theta}$), then

$$\psi'(\vec{r}') = e^{i\vec{J}\cdot\vec{\theta}}\psi(\vec{r}') \text{ and } \psi(\vec{r}') = e^{-i\vec{J}\cdot\vec{\theta}}\psi(\vec{r}) \qquad \text{(passive transformation),}$$

$$\tag{2.47}$$

where $\vec{J}$ is the (Hermitian) angular momentum operator,[14] and the transformation operator is

$$U = e^{i\vec{J}\cdot\vec{\theta}}, \text{ which is unitary.} \tag{2.48}$$

---

14. $\vec{J} = \vec{L} + \vec{S}$, where $\vec{S}$ is the spin, which will be discussed later.

Thus **the angular momentum ($\vec{J}$) is the generator of rotations**. Note that here we have considered a rotated observer, but we might just as well look at **a physical system rotated in the opposite direction still viewed in S**. This is referred to as an **active transformation**, for which

$$\psi'(\vec{r}') = e^{-i\vec{J}\cdot\vec{\theta}}\psi(\vec{r}') \text{ and } \psi(\vec{r}') = e^{i\vec{J}\cdot\vec{\theta}}\psi(\vec{r}) \qquad \text{(active transformation)},$$

$$(2.49)$$

where θ is the angle of rotation of the physical system and, in the second expression, $\vec{r}'$ is related to $\vec{r}$ by the rotation.

The symmetries we have been discussing are continuous symmetries in the obvious sense that we can "move" continuously between two transformations, e.g., a rotation by $45°$ can be reached (and is usually accomplished) by continuously changing the angle of rotation from $0°$. All the rotations, by angles between the two, exist as physical operations. In fact, the validity of the Taylor series arguments above is due to the continuity of the operators with respect to the transformation parameter ($\vec{a}$ for translations and $\vec{\theta}$ for rotations). In general, such transformations have the form:

$$V = e^{iG\alpha},$$

$$(2.50)$$

where G is an operator called the generator (e.g., $\vec{J}$ for rotations and $\vec{p}$ for translations) and α is a real parameter (e.g., $\vec{\theta}$ for rotations and $\vec{a}$ for translations). We have learned that V must be unitary and that observables must be Hermitian. Furthermore, the generators considered so far are observables, so they *must* be Hermitian. These two requirements *are* compatible, and it can be shown from Eq. (2.50) (see Problem 15) that all **the generators must be Hermitian**.

2.8 COMMUTATORS OF OPERATORS

Suppose some observable, represented by **the operator A**, is **unchanged by rotations**, i.e., it is a scalar. Let us consider a (passive) rotation of coordinate system. Then for any state $|\psi\rangle$ we must have

$$'\langle\psi|A|\psi\rangle' = \langle\psi|A|\psi\rangle,$$

where $|\psi\rangle'$ is the state in the rotated system:

$$|\psi\rangle' = U|\psi\rangle \text{ with } U = e^{i\vec{J}\cdot\vec{\theta}} \text{ and } '\langle\psi| = \langle\psi|U^\dagger.$$

$$(2.51)$$

Substitution then yields

$$\langle\psi|U^\dagger A U|\psi\rangle = \langle\psi|A|\psi\rangle \text{ for all } |\psi\rangle.$$

This implies that

$$U^\dagger A U = A,$$

$$(2.52)$$

and if we multiply each side of this equation by U (from the left), we obtain $AU = UA$ (since $UU^\dagger = 1$); this can be written:

$$[A, U] = 0, \tag{2.53}$$

where $[A, U]$ stands for $(AU - UA)$ and is called the **commutator of A and U**. We have discussed functions $f(\mathbf{O})$ of operators, at first by saying that when operating on an eigenfunction of $\mathbf{O}$, they become the number $f(o)$, where o is the corresponding eigenvalue. From our construction of such a function in Eqs. (2.43) and (2.44), we may infer that when operating on a general state they are defined as the power series expansions of the functions $f(\mathbf{O})$, in which $\mathbf{O}^n$ signifies that the operator $\mathbf{O}$ should be applied n times. In fact, $U = e^{i\vec{J}\cdot\vec{\theta}}$ may be obtained from a power series (analogous to that in Eq. 2.43) in the vector operators J_i. Consequently, Eq. (2.53), with $U = e^{i\vec{J}\cdot\vec{\theta}}$, leads to the requirement that

$$[A, \vec{J}] = 0 \text{ for rotationally invariant operators } A. \tag{2.54}$$

(See Problem 19.) If the interactions are unchanged by rotations, then the Hamiltonian would be rotationally invariant and, from the preceding, we have

$$[H, \vec{J}] = 0 \text{ for rotational symmetry.} \tag{2.55}$$

For other continuous symmetries, these same arguments follow with U replaced by $V = e^{iG\alpha}$ (see Eq. 2.50), where G is the generator of the continuous symmetry (e.g., $\vec{p}$ for translation invariance, etc.). If the **interactions** are **invariant to the continuous symmetry operations generated by G**, we can conclude that

$$[H, G] = 0. \tag{2.56}$$

2.9   EQUATIONS OF MOTION AND CONSTANTS OF THE MOTION

The time-dependent Schrödinger equation is

$$i \frac{\partial}{\partial t} |\psi\rangle = H|\psi\rangle. \tag{2.57}$$

Treating H as if it were a number, a formal solution is

$$|\psi(t)\rangle = e^{-iHt}|\psi(0)\rangle, \tag{2.58}$$

which is exactly what we had in Eq. (2.39).

Let us discuss *operators which have no explicit time dependence*. In general, the matrix elements of such an observable A (with no explicit t in it)[15]

---

15. An example is the momentum of a particle moving in a potential.

would *change* with time, because of the time evolution of the states. Using Eq. (2.58), we may write

$$a_{ij}(t) = \langle \psi_i(t)|A|\psi_j(t)\rangle = \langle \psi_i(0)|e^{iHt}A\,e^{-iHt}|\psi_j(0)\rangle. \qquad (2.59)$$

If $[A,H] = 0$, then we can move e^{iHt} across A in Eq. (2.59), to obtain

$$a_{ij}(t) = \langle \psi_i(0)|A|\psi_j(0)\rangle = a_{ij}(0), \text{ independent of time.}$$

Conversely, given $a_{ij}(t) = a_{ij}(0)$ (all i and j) for arbitrary t, then Eq. (2.59) shows that we must have $e^{iHt}A\,e^{-iHt} = A$ for any t; this implies that $[A,H] = 0$. Thus,

$$[A,H] = 0 \qquad (2.60)$$

if and only if A represents a constant of the motion (with no explicit time behavior).

Notice that **the state contains the time evolution within it (for no explicit time behavior in A)**. But since the observable quantities are the only physical entities, we might create a completely different "picture" than this familiar **"Schrödinger picture."** We can let the space of states consist of time-independent states $|\psi(0)\rangle$ and, in order to obtain the same physical result ($a_{ij}(t)$), put **the time behavior in the operator**. This is called the **"Heisenberg picture,"** whose operators A_H are related to their simple, time-independent forms in the Schrödinger picture (we now call the A above A_S) as follows:

$$A_H = e^{iHt}A_S e^{-iHt}. \qquad (2.61)$$

Since the states in the Heisenberg picture are independent of time, the time behavior of physical quantities may be seen directly by looking at the corresponding operators. Let us calculate the time derivative of A_H for A_S independent of time

$$\frac{\partial}{\partial t} A_H(t) = iHe^{iHt}A_S e^{-iHt} + e^{iHt}A_S e^{-iHt}(-iH),$$

where we have used the chain rule for differentiation, but have been careful not to interchange the operators H and A_S. We can rewrite the right-hand side to obtain

$$\frac{\partial}{\partial t} A_H(t) = i[H, A_H]. \qquad (2.62)$$

Thus, in the Heisenberg picture, we see a direct connection between the time behavior of operators and their commutators with the Hamiltonian. It follows from Eq. (2.61) (see Problem 20) that if $[H, A_S]$ is zero, then $A_H = A_S$, so that $[H, A_H]$ is zero and vice versa. But Eq. (2.60) shows that $[H, A_S] = 0$ is equivalent to A being a constant of the motion; we conclude (using Eq. 2.62) that A_H is independent of time if and only if A represents a constant of the motion (with no explicit time dependence). Alternatively, if A_H commutes with H, Eq. (2.62) shows that it is independent of time; consequently, its matrix elements are independent of time as well since the states do not carry any time dependence in the Heisenberg picture.

Thus, in either the Schrödinger or Heisenberg picture, **operators which commute with the Hamiltonian represent constants of the motion (conserved quantities) and vice versa.**

From the previous discussion of rotationally invariant interactions, we now see that the **conservation of angular momentum follows from the fact that $\vec{J}$ commutes with the Hamiltonian** (Eq. 2.55). An identical argument can be made for any symmetry of the interactions (Eq. 2.56). We conclude then that **the generator of any symmetry of the Hamiltonian is a constant of the motion** (since it commutes with H).[16]

2.10   THE *S*-MATRIX AND INTERACTIONS

We have learned in Section 2.1 that there are two complete (orthonormal) sets of states employed to discuss scattering experiments, viz., $|i\rangle_{out}$ and $|i\rangle_{in}$. The scattering amplitudes were described as overlaps of *in* and *out* states. Here we shall extend that discussion.

There exists a unitary operator that connects these two complete sets, which is called the *S*-matrix (usually written S); its operation is often expressed in terms of its transformations of the corresponding bras:

$$_{in}\langle j|S = {}_{out}\langle j|. \tag{2.63}$$

Substituting this into Eq. (2.9), we obtain

$$S_{ji} = {}_{in}\langle j|S|i\rangle_{in}. \tag{2.64}$$

Thus, the *S*-matrix element defined in Eq. (2.9) is the matrix element of the *S*-matrix operator in the *in* state basis. Since S changes *in* states to *out* states, it must contain an operator representing the interaction. **When there is no interaction**, the *in* states and the *out* states are identical, so that **we expect S to be the unit operator.** It can be shown, from the treatment of potential scattering using the Schrödinger equation,[17] that the matrix elements of the *S*-matrix may be written:

$$S_{ji} = \delta_{ji} - i\int d^3r\,dt\,\phi_j^*(\vec{r},t)V(\vec{r},t)\psi_i(\vec{r},t), \tag{2.65}$$

where ψ_i is the wavefunction of the incoming *interacting* state and ϕ_j is the wavefunction of a set of *noninteracting* particles which matches ψ_{out} at $t \to +\infty$. The appearance of V in the second term shows that it is an interaction term,

---

16. In Appendix C, we show the connection between continuous symmetries and conserved quantities in classical field theories.

17. This discussion is developed in several advanced quantum mechanics books.

which describes the actual scattering; in the absence of an interaction ($V = 0$), that term disappears. Furthermore, the wavefunction ψ_i of the incoming *interacting* state depends on the interaction V as well. In fact, the wavefunction ψ_i can be expressed as an infinite series of terms, each successive term depending on a higher power of V:

$$\psi_i(\vec{r},t) = \phi_i(\vec{r},t) + \int d^3r' \int dt' \, G(\vec{r} - \vec{r}', \, t - t') V(\vec{r}',t') \, \psi_i(\vec{r}',t').$$

Expressed symbolically, this is

$$\psi_i = \phi_i + GV\psi_i, \tag{2.66}$$

where G is called a Green's function, and successive iterations of this relationship yield

$$\psi_i = \phi_i + GV\phi_i + GVGV\phi_i + GVGVGV\phi_i + \cdots. \tag{2.67}$$

Using Eq. (2.67) in Eq. (2.65), the interaction term of the S-matrix can be expressed as a power series in the interaction. This is called a **perturbation expansion**.

The form for the S-matrix appearing in Eq. (2.65) may be written:

$$S = 1 + kM, \tag{2.68}$$

where k contains kinematic factors and delta-functions for conservation of energy and momentum, and **M is the interaction matrix**. The same form for the S-matrix also appears in relativistic quantum mechanics and quantum field theory (both of which we shall discuss in Chapter 4). From Section 2.2, we may infer that when the final state is different than the initial state, transition probabilities will be proportional to $|M|^2$:

$$P(i \rightarrow j) = |M_{ji}|^2 \text{ for } i \neq j. \tag{2.69}$$

In the realm of quantum field theory, we shall see in Chapter 4 that interactions among particles need not be merely scatterings, but creation and annihilation of particles also occur. Equation (2.69) will also apply to those transitions. In particular, **the rate of decay of a particle is proportional to the absolute square of an appropriate interaction matrix element $|M_{ji}|^2$**.

2.11 SEMICLASSICAL ARGUMENTS AND THE CORRESPONDENCE PRINCIPLE

In an attempt to obtain a qualitative understanding of a physical situation or reaction, we often resort to "semiclassical" arguments, which give us a more intuitive feeling (and sometimes insight) than does a detailed quantum mechanical calculation. What is a semiclassical argument and of what value is it? A semiclassical argument or picture is a classical argument or picture in which our *intuitive* understanding is incorporated, but which is modified in a seemingly ad hoc manner so as not to qualitatively conflict with the general requirements of quantum mechanics. It is neither rigorous nor exact, but serves as a guide and as a trigger for our intuitive grasp of the world. It aids us in thinking about new approaches to outstanding problems, both theoretical and experimental. The validity of a semiclassical argument always awaits a rigorous derivation from quantum mechanics. However, in macroscopic situations our intuition is based upon observation so that there are many circumstances in which the exact (and usually laborious) quantum mechanical calculation *must* coincide, quantitatively, with the semiclassical result. This is the basis of what has become known as *the correspondence principle*.

PROBLEMS

1. Show that for a discrete (countable) orthonormal **complete** basis $|i\rangle$, the operator $\mathbf{1} \equiv \sum_i |i\rangle\langle i|$ is equal to the unit operator, i.e., $\mathbf{1}|\psi\rangle = |\psi\rangle$ and $\langle\psi|\mathbf{1} = \langle\psi|$. Hint: Completeness implies that all $|\psi\rangle$ can be expanded as a linear combination of $|i\rangle$: $|\psi\rangle = \sum_j y_j|j\rangle$, where y_j are complex numbers, and $(|i\rangle\langle i|)|\psi\rangle = |i\rangle(\langle i|\psi\rangle)$.
2. Show that $\langle a|b\rangle = \int d^3r\, \psi_a^*(\vec{r})\psi_b(\vec{r})$. Hint: The generalization of $\mathbf{1}$ for the continuous complete set $|\vec{r}\rangle$ is $\mathbf{1} = \int d^3r\, |\vec{r}\rangle\langle\vec{r}|$. Use it on the left to obtain the expression on the right.
3. a. Using the completeness criterion expressed in Problem 1, show that the definition of the S matrix elements in Eq. (2.9) implies that S is unitary, i.e., show that $S^\dagger S = 1 = SS^\dagger$. Don't forget that matrix multiplication involves row by column **sums** (of products).
   b. What operator changes an *in* ket to an *out* ket?
   c. For the physical situation of no scattering at all, show that the S matrix would be the unit matrix.
4. Verify that $qN\vec{J}_P = \rho\vec{v}$, which is the correct expression for electric current (for N particles, each of charge q moving with velocity $\vec{v}$), by noting that $\psi \sim e^{i\vec{p}\cdot\vec{r}}$ and substituting this into the expression for $\vec{J}_P$.
5. a. From the definition of Hermitian conjugate in Eq. (2.18), show that $(A_1A_2)^\dagger = A_2^\dagger A_1^\dagger$.
   b. Show that $(A_1A_2\ldots A_n)^\dagger = A_n^\dagger\ldots A_2^\dagger A_1^\dagger$. You may use your result in part a.
6. Check that the expression for $\bar{p}_x$ in Eq. (2.19) is real by taking its complex conjugate and using integration by parts. (Note that ψ vanishes at $\pm\infty$.) What does this tell us about the operator $-i(\partial/\partial x)$?
7. Show that the property of $Y_\ell^m: L_z Y_\ell^m(\theta,\phi) = mY_\ell^m$ implies that Y_ℓ^m has the form $Y_\ell^m(\theta,\phi) = P_\ell^m(\theta)e^{im\phi}$, where P_ℓ^m is some function of θ alone.
8. Check that the matrices representing the angular momenta for an $\ell = 1$ system, displayed in Appendix A, are Hermitian.
9. Show that the orthonormality of basis states is preserved by unitary transformations.
10. A useful exact mathematical formula, known as the Baker-Hausdorff lemma, may be expressed as

$$e^{i\alpha G} A\, e^{-i\alpha G} = A + i\alpha[G,A] + \frac{(i\alpha)^2}{2!}[G,[G,A]] + \cdots$$

$$+ \frac{(i\alpha)^n}{n!}[G,[G,\ldots[G,A]]\ldots] + \cdots,$$

where there are n successive commutators with G in the nth term.

   a. Verify this lemma through $n = 3$, by expanding the exponentials on the left-hand side.

   b. Prove this lemma. A hint is provided at the end of this problem set.

11. a. What is the operator for the translation of the coordinate system by D in the x direction?

   b. A particle's position ($\vec{r}$ is the operator) is measured and shown on a dial to be $\vec{R}_0$ in the original coordinate system. Use the translation operator you found in part a to construct, to $\mathcal{O}(D^2)$, the corresponding operator which an observer using the translated system should employ to explain the result on the dial.

   c. Redo part b exactly, using the Baker-Hausdorff lemma shown in Problem 10.

   d. If momentum is conserved, what symmetry exists? What is the connection between the momentum operators and the symmetry operators (U)?

12. Consider the matrix

$$
V = \begin{pmatrix} \frac{1}{2} & -\frac{1}{\sqrt{2}} & \frac{1}{2} \\ -\frac{1}{\sqrt{2}} & 0 & \frac{1}{\sqrt{2}} \\ \frac{1}{2} & \frac{1}{\sqrt{2}} & \frac{1}{2} \end{pmatrix}.
$$

   a. Show that it is unitary.

   b. Use it as a transformation on the basis (Y_1^m) states represented by the columns

$$
Y_1^{+1} = \begin{pmatrix} 1 \\ 0 \\ 0 \end{pmatrix}, \quad Y_1^0 = \begin{pmatrix} 0 \\ 1 \\ 0 \end{pmatrix}, \text{ and } Y_1^{-1} = \begin{pmatrix} 0 \\ 0 \\ 1 \end{pmatrix}
$$

to express these states in the transformed system. How do your results compare to the columns of V? Check the orthonormality of your new expressions directly.

   c. Express L_x in the transformed system. Note that in the new basis, L_x is diagonal. How do its diagonal elements compare with those of L_z in the original basis?

   d. Using your result in part c, express Y_1^{+1} in terms of $\mathcal{Y}_1^m$, where the latter are orthonormal functions corresponding to $L_x = m$.

13. Show that the coefficients of the powers of x in the expansion of e^x are identical to the coefficients of the derivative terms appearing in the expansion of $\psi'(\vec{r}')$.

14. Derive the expression for $\psi(\vec{r}')$ in terms of $\psi(\vec{r})$ shown in Eq. (2.46).

15. a. If $|g\rangle$ is an eigenket of G with eigenvalue g, show that $\langle g|G^\dagger = g^*\langle g|$.

   b. Consider an operator of the form $V = e^{iG\alpha}$, with α real. Show that V is unitary if and only if G is Hermitian. You may assume that the eigenstates of G form a complete orthonormal set.

16. Consider a passive rotation about the z axis by an angle θ. Show that the operator

$$
L_z = (\vec{r} \times \vec{p})_z = -i\left(x\frac{\partial}{\partial y} - y\frac{\partial}{\partial x}\right)
$$

is the generator of this rotation. Hint: For an infinitesimal angle ϵ, create steps analogous to those used for translations by expressing (x, y, z) in terms of (x', y', z'), as is done in Eq. (2.42). From that result show that $\psi'(\vec{r}') = e^{iL_z\theta}\psi(\vec{r}')$ for finite θ.

17. Show that multiplication by $f(r)$, where $r = \sqrt{x^2 + y^2 + z^2}$, commutes with the operator $\vec{L} = \vec{r} \times \vec{p} = \vec{r} \times (-i\vec{\nabla})$, as expected for a rotationally invariant operator, i.e., compare $\vec{L}f\psi$ to $f\vec{L}\psi$.

18. Show that commutation relations are preserved by unitary similarity transformations, i.e., show that $[A',B'] = C'$ if and only if $[A,B] = C$.

19. a. Show that if $[A,e^{iG\alpha}] = 0$ for all real values of α, where A and G are operators, then $[A,G] = 0$.

    b. Show that if $[A,G] = 0$, then $[A,f(G)] = 0$ where f is any function which can be expanded in a power series.

20. Show that if $[H,A_S]$ is zero, then $A_H = A_S$, so that $[H,A_H]$ is zero and vice versa.

21. In Appendix A, do the problems intertwined with the text, which start with the italicized words: (a) *Check*, (b) *verify*, (c) *check this*, (d) *Calculate*, (e) *Find out*.

22. Suppose the Hamiltonian for a particle has (besides a rotationally invariant part) an interaction of the form: (constant) $\vec{M}\cdot\vec{p}$, where $\vec{M}$ is a *fixed* external uniform field. (We are considering only the particle as a dynamical system.)

    a. Find the commutators of this interaction with each of L_x, L_y, and L_z, calling the direction of $\vec{M}$ the z direction (i.e., $\vec{M} = M_0\hat{z}$ with M_0 a constant). Remember that $\vec{p}$ is an operator.

    b. Is this a rotationally invariant interaction? Explain how your answer is related to your result for part a.

    c. From your result in part a would you say that any component of $\vec{L}$ is a constant of the motion? Explain briefly. How is your answer related to the symmetry of the system?

23. Show that the matrix elements of operators which contain time explicitly obey:

$$\frac{d}{dt}(\langle\psi_1(t)|A|\psi_2(t)\rangle) = i\langle\psi_1(t)|[H,A]|\psi_2(t)\rangle + \left\langle\psi_1(t)\left|\frac{\partial A}{\partial t}\right|\psi_2(t)\right\rangle.$$

Hint for Problem 10: In the nth term on the right-hand side, the commutators are done from the inside out. It amounts to placing a G before or $(-G)$ after A at each step and keeping all the terms. It is equivalent to expanding $(b - a)^n$ in monomials and interpreting b to mean place a G before A and a to mean place a G after A. The number of times we obtain the term $G^k AG^{n-k}$, including the number of factors of (-1), is the same as the coefficient of $b^k a^{n-k}$ in the expansion, which is $(-1)^{n-k}n!/(k!(n-k)!)$. Assume that the commutator in the nth term on the right-hand side equals $\sum_{k=0}^{n} G^k AG^{n-k}(-1)^{n-k}(n!/(k!(n-k)!))$ and compare it to the $\mathcal{O}(\alpha^n)$ term on the left-hand side.

Chapter 3

Angular Momentum
and Rotations

As we have learned in Chapter 2, the angular momentum operators are the generators of rotations. We shall see in later chapters that operators analogous to them arise as generators of other symmetries as well; furthermore, the eigenvalues associated with these operators are used to label states. Therefore it is important to study certain features of the angular momentum operators in some detail, which we shall do in this chapter.

3.1 ANGULAR MOMENTUM OPERATORS

Take two books and orient them the same way, as shown in the left column of Fig. 3.1. Let us call the horizontal plane the x-y plane, with your line of sight as the $+x$ axis and the $+y$ axis towards the left. Following Fig. 3.1: Rotate one book 90° clockwise about the $+x$ axis and then 90° clockwise about the $+y$ axis. Rotate the second book 90° clockwise about the $+y$ axis first and then 90° clockwise about the $+x$ axis. The final orientations of these books are different! Let us write this in terms of operators:

$$e^{iJ_x(\pi/2)} e^{iJ_y(\pi/2)} \neq e^{iJ_y(\pi/2)} e^{iJ_x(\pi/2)}. \tag{3.1}$$

Recall that a function of an operator is to be understood as a power series expansion of the function, in which A^n signifies that the operator A should be applied n times. If J_x and J_y commuted with one another, the two sides of Eq. (3.1) would be equal, since we could pull all powers of one through those of the other. Therefore the inequality implies that

$$[J_x, J_y] \neq 0. \tag{3.2}$$

A rotation about the z axis, in place of the x or y rotation, will yield a similar conclusion concerning any two of the three components; thus, **the components of the $\vec{J}$ operator do not commute.**

In fact, we can calculate the commutators from the definition of $\vec{J}$ as the *total* angular momentum. Since the detailed treatment of spin has not been

32

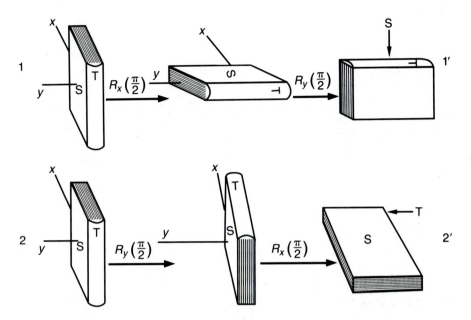

FIGURE 3.1
Two books (1 and 2) begin in the same orientation and are then each rotated successively about the x and y axes. The order of these rotations is reversed for 2 compared to 1, which leads to a different final orientation, showing the noncommutativity of these rotations.

explored here, we will show this explicitly for the orbital angular momentum. We have

$$L_x = (\vec{r} \times \vec{p})_x = yp_z - zp_y, \tag{3.3}$$

and similar expressions for L_y and L_z. Recall that $p_x = -i(\partial/\partial x)$. Using the fact that $[\partial/\partial x, x] = 1$ (which you can verify by remembering that this is an operator equation, viz., $[\partial/\partial x, x] f(x, y, z) = f(x, y, z)$), we leave it to the reader to derive the following results:

$$[L_x, L_y] = iL_z, \qquad [L_y, L_z] = iL_x, \qquad \text{and } [L_z, L_x] = iL_y. \tag{3.4}$$

Notice that the latter two equations may be obtained from the first by replacing x by y, y by z, and z by x (a cyclic permutation of x, y, and z) once and then again. For states with spin, it can also be shown that

$$[J_x, J_y] = iJ_z \text{ (and cyclic).}[1] \tag{3.5}$$

Operators whose commutators give back linear combinations of those same operators are said to form a **Lie algebra** (where multiplication is replaced by tak-

---

1. "And cyclic" means use two successive $x \to y \to z \to x$ replacements to get the other two equations.

ing commutators). The group of transformations they generate are called **Lie groups**.

We have previously pointed out that physical states are usually designated by their quantum numbers, which are the eigenvalues of physical quantities; these states are eigenstates of those physical operators.

Suppose a state is an eigenstate of two operators A and B, with eigenvalues a and b respectively. Let us call that state $|\psi_{ab}\rangle$. We have

$$A|\psi_{ab}\rangle = a|\psi_{ab}\rangle \text{ and } B|\psi_{ab}\rangle = b|\psi_{ab}\rangle, \qquad (3.6)$$

where a and b are just numbers. Let us now apply these operators to the state in two different orders:

$$AB|\psi_{ab}\rangle = Ab|\psi_{ab}\rangle = bA|\psi_{ab}\rangle = ba|\psi_{ab}\rangle,$$

likewise

$$BA|\psi_{ab}\rangle = ab|\psi_{ab}\rangle.$$

The right-hand sides of these two equations are identical, since $ab = ba$ (a and b are ordinary numbers). Thus, we have

$$[A,B]|\psi_{ab}\rangle = 0.$$

We can take this argument one step further. The eigenstates of these operators form a complete set, so that the commutator itself must be zero: **we must have**

$$[A,B] = 0, \qquad (3.7)$$

if A and B are to be used together to label a complete set of states. This tells us that we *cannot* use J_x, J_y, and J_z together to label our states, even though they are conserved physical quantities, since *they do not commute among themselves*.

However, let us consider the operator $J^2 = \vec{J}\cdot\vec{J}$. It is a scalar (rotationally invariant) operator, as is evident from its structure. As we have seen in Chapter 2, rotationally invariant operators commute with all the components of $\vec{J}$, so that

$$[J_i, J^2] = 0 \qquad \text{for } i = 1,2,3. \qquad (3.8)$$

(This result also follows from the commutation relations in Eq. 3.5; see Problem 3.) So if we take the set J_z and J^2, we have a set of commuting operators, and we can find states for our basis which are eigenstates of these operators. The quantum numbers ℓ, from the quantized values of L^2 which are $\ell(\ell + 1)$, and m_ℓ, the z component of the orbital angular momentum, are often used to label the states of atomic systems. Now we see the rationale behind that choice.

When spin is present, we may use the operator J^2 with eigenvalues $j(j + 1)$ and the operator J_z with eigenvalues m:

$$J^2|j,m\rangle = j(j + 1)|j,m\rangle \text{ and } J_z|j,m\rangle = m|j,m\rangle. \qquad (3.9)$$

It is left to the reader (in Problem 3), using Eq. (3.8), to show that J_x and J_y, when operating on one of these states, do not change the j value of the states. In fact, they produce instead linear combinations of such states, which differ in their m values. It is convenient to consider the following linear combinations of these two operators:

$$J_+ \equiv J_x + iJ_y \text{ and } J_- \equiv J_x - iJ_y, \qquad (3.10)$$

because (as may be shown by solving Problem 4) J_+, when operating on one of these states, produces only one such state whose m is raised by 1 (and gives zero if the state has $m = j$) and J_- lowers m by 1 (giving zero when acting on the lowest state):

$$J_\pm |j,m\rangle = \sqrt{j(j+1) - m(m \pm 1)} |j, m \pm 1\rangle.[2] \qquad (3.11)$$

(For J_+ read the equation with the upper signs and for J_- the lower signs.) These operators are often referred to as **ladder operators**, since they move us up and down the states (arranged in order of their m values), or as **raising and lowering operators**, respectively.

Let us consider the case of rotational invariance. We have seen that rotational invariance implies that

$$[H, J_x] = 0 = [H, J_y] = [H, J_z], \qquad (3.12)$$

which immediately yields

$$\mathbf{[H, J_\pm] = 0,} \qquad (3.13)$$

i.e., the ladder operators commute with the Hamiltonian. The states $|j,m\rangle$ may be taken as the stationary states, i.e., they are eigenstates of the Hamiltonian (states of definite energy) as well as J^2 and J_z.[3] Consider first the "highest" state, which has $m = j$. Since it is an eigenstate of H we write

$$H|j,j\rangle = E_j|j,j\rangle, \qquad (3.14)$$

where E_j is the energy of this state. Now let us apply the lowering operator to this equation:

$$J_- H|j,j\rangle = E_j J_-|j,j\rangle. \qquad (3.15)$$

Using the fact that J_- commutes with H ($HJ_- = J_-H$), Eq. (3.15) becomes

$$HJ_-|j,j\rangle = E_j J_-|j,j\rangle. \qquad (3.16)$$

We have learned that J_- lowers the m value by 1, as shown in Eq. (3.11), e.g.,

$$J_-|j,j\rangle = (\text{constant}) |j, j - 1\rangle; \qquad (3.17)$$

substituting into Eq. (3.16) and canceling the constant from both sides, we find

$$H|j, j - 1\rangle = E_j|j, j - 1\rangle. \qquad (3.18)$$

This shows us that the state with m lowered by 1 has the same energy (E_j) as the highest state. Proceeding to lower the m value in the same way again and again, we find that all the states in this set have the same energy. Thus, we

---

2. See Appendix A for a discussion of raising and lowering operators for $j = 1$ (or $\ell = 1$).
3. They can be eigenstates of J^2, J_z, and H, since all three operators commute.

have **(2j + 1) degenerate states, i.e., states of the same energy** E_j (with different m values) for the same j, when the interactions are **rotationally invariant**.

The set of rotation operators possesses certain properties, to be described
in Chapter 5, which make it a mathematical group. The states $|j,m\rangle$ with fixed
j and m running from $-j$ to $+j$ are said to **span a (2j + 1)-dimensional representation of the rotation group.** This is a succinct way of expressing the fact that
square matrices of dimension (2j + 1) can be constructed for the operators,
using these states in the manner shown in Chapter 2. In fact the representation
is **irreducible**, since no subset of these states would be inclusive enough to allow
us to represent J_x, J_y, and J_z, i.e., when applied to the states of a subset, these
operators take us out of that subset. This follows for J_x and J_y from Eq. (3.11),
which shows that $J_\pm$ connect each m state to the next one, since J_x and J_y are
linear combinations of them. (Notice that in the $\ell = 1$ example in Appendix A,
both L_x and L_y operating on one of the states produces another in such a manner that all are needed to form a matrix. This is true for general j.)

The generalization of what we have learned here is that **all the states, span-
ning an irreducible representation of a symmetry (of the interaction) group, have
the same energy**, i.e., they are **degenerate**. This result, in fact, is true of all continuous symmetry groups.

A familiar situation, in which we can see a manifestation of these rules,
is the Zeeman effect. The states of an isolated atom are described by a rotation-
ally invariant Hamiltonian, so that they may be designated as $|j,m\rangle$. For a given
j, all the states (of different m) are degenerate. However, when we put the atom
in an external magnetic field ($\vec{B}$), the degeneracy of the states with different
values of m is broken. The interaction ($-\vec{\mu}\cdot\vec{B}$) involves the magnetic field,
whose direction we can take as our z direction, i.e., $\vec{B} = B\hat{k}$. The Hamiltonian
for the *atom* (in the *fixed* $\vec{B}$ field) is not invariant to rotations of the atom
alone about the x or y axes, but is still invariant to rotations about the z axis.
Hence, H does not commute with J_x or J_y. This implies that

$$[H, J_\pm] \neq 0, \text{ but we still have } [H, J_z] = 0. \tag{3.19}$$

Since H commutes with J_z the stationary states can still be designated by their
m values, but the argument above, which showed their degeneracy, is no longer valid. Thus, we expect that their energies will be shifted away from one
another. This is what is observed in the Zeeman effect.

EXAMPLE: An Active Rotation of a Vector ($\vec{V}$)

Let us consider a counterclockwise rotation *of a vector* $\vec{V}$ about the z axis by
angle ϕ, as shown for a two-dimensional vector in Fig. 3.2. For $\vec{V}$ in the x-y
plane, from Fig. 3.2, we see that the components of the new vector $\vec{V}'$ are

$$V'_x = V\cos(\theta + \phi) = V\cos\theta\cos\phi - V\sin\theta\sin\phi = V_x\cos\phi - V_y\sin\phi$$
$$V'_y = V\sin(\theta + \phi) = V\cos\theta\sin\phi + V\sin\theta\cos\phi = V_x\sin\phi + V_y\cos\phi.$$

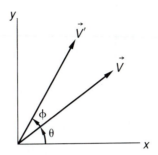

FIGURE 3.2
Rotation about the z axis of the two-dimensional vector $\vec{V}$ by angle ϕ to the vector $\vec{V}'$. θ is the angle that $\vec{V}$ makes with the x axis.

If $\vec{V}$ had a z component, it would be unchanged by a rotation about the z axis, viz., $V'_z = V_z$. Thus, we may express the new components of a three-dimensional vector in terms of the old:

$$V'_x = V_x \cos \phi - V_y \sin \phi,$$
$$V'_y = V_x \sin \phi + V_y \cos \phi,$$
$$V'_z = V_z. \tag{3.20}$$

We may write this as a matrix equation, to show explicitly the matrix which represents the rotation:

$$\begin{pmatrix} V'_x \\ V'_y \\ V'_z \end{pmatrix} = \begin{pmatrix} \cos \phi & -\sin \phi & 0 \\ \sin \phi & \cos \phi & 0 \\ 0 & 0 & 1 \end{pmatrix} \begin{pmatrix} V_x \\ V_y \\ V_z \end{pmatrix}, \qquad \text{active rotation.} \tag{3.21}$$

(A rotation of the coordinate system by ϕ (a passive rotation) has opposite signs for the $\sin \phi$ terms.) ■

Suppose we operate on the basis set used in this example, viz. V_x, V_y, and V_z, with the unitary matrix U:

$$U = \begin{pmatrix} \dfrac{1}{\sqrt{2}} & \dfrac{i}{\sqrt{2}} & 0 \\ 0 & 0 & 1 \\ \dfrac{1}{\sqrt{2}} & \dfrac{-i}{\sqrt{2}} & 0 \end{pmatrix}. \tag{3.22}$$

We obtain a new set:

$$\begin{pmatrix} V_+ \\ V_z \\ V_- \end{pmatrix} = U \begin{pmatrix} V_x \\ V_y \\ V_z \end{pmatrix}, \tag{3.23}$$

where

$$V_\pm = \tfrac{1}{\sqrt{2}} (V_x \pm iV_y),^4$$

and it is straightforward (using the equations above which express V'_x and V'_y in terms of V_x and V_y) to verify that an active rotation yields

$$V'_+ = e^{+i\phi} V_+, \; V'_- = e^{-i\phi} V_-, \text{ and of course } V'_z = V_z. \tag{3.24}$$

Here we are comparing the *same* function of the old and new variables for active rotations for which $\psi(\vec{r}') = e^{i\vec{J}\cdot\vec{\theta}}\psi(\vec{r})$, as shown in Eq. (2.49). Therefore, the form should be $V'_\alpha = e^{+iJ_z\phi} V_\alpha$ (for $\alpha = +, 0$, or $-$), so that J_z acting on V_+ yields $+1$, on V_z yields 0, and on V_- yields -1. Thus, you might expect that

$$V_+ \sim |1,+1\rangle, \; V_z \sim |1,0\rangle, \text{ and } V_- \sim |1,-1\rangle. \tag{3.25}$$

(We may also check this for the vector $\vec{V} = \vec{r}$ by applying the L_z operator directly to $V_\pm$ and V_z. See Problem 10.) Considerations of general rotations (about any axis) verify these assignments. Thus, we see that a vector (in three-space) behaves like a $j = 1$ object under rotations.

Using Eq. (3.11) for these states, we find

$$J_+|1,m\rangle = \sqrt{2 - m(m + 1)}|1, m + 1\rangle;$$

more explicitly:

$$J_+|1,-1\rangle = \sqrt{2}|1,0\rangle,$$
$$J_+|1,0\rangle = \sqrt{2}|1,1\rangle$$
$$J_+|1,+1\rangle = 0. \tag{3.26}$$

Equation (3.11) will yield analogous results for J_-.

The $|1,m\rangle$ states are orthonormal, which means that they are mutually orthogonal and normalized so that a state multiplied by itself yields 1. Employing this property, we obtain, from Eq. (3.26),

$$(J_+)_{--} \equiv \langle 1,-1|J_+|1,-1\rangle = 0,$$
$$(J_+)_{0-} \equiv \langle 1,0|J_+|1,-1\rangle = \sqrt{2},$$
$$(J_+)_{+0} \equiv \langle 1,+1|J_+|1,0\rangle = \sqrt{2},$$
$$(J_+)_{+-} \equiv \langle 1,+1|J_+|1,-1\rangle = 0, \text{ etc.} \tag{3.27}$$

---

4. Some authors use the negative of our V_+, which facilitates calculations, but for our purposes it is simpler to omit that minus sign, which is merely a change of phase.

Using these results to construct the matrix for J_+ we obtain

$$m = \quad + \quad 0 \quad -$$

$$J_+ = \begin{pmatrix} 0 & \sqrt{2} & 0 \\ 0 & 0 & \sqrt{2} \\ 0 & 0 & 0 \end{pmatrix}, \tag{3.28}$$

where the rows and columns are in the order $m = +1, 0, -1$ as indicated. Since $J_- = J_+^\dagger$ (as J_x and J_y are Hermitian), we may immediately write

$$J_- = \begin{pmatrix} 0 & 0 & 0 \\ \sqrt{2} & 0 & 0 \\ 0 & \sqrt{2} & 0 \end{pmatrix}. \tag{3.29}$$

We may express J_x and J_y in terms of J_+ and J_- using Eq. (3.10), and it is left as an exercise to write the J_x and J_y matrices utilizing those for $J_\pm$ displayed in Eqs. (3.28) and (3.29). (The reader should then compare those results with the matrices L_x and L_y for $\ell = 1$, shown in Appendix A.) We also have

$$J_z = \begin{pmatrix} +1 & 0 & 0 \\ 0 & 0 & 0 \\ 0 & 0 & -1 \end{pmatrix}, \tag{3.30}$$

since $|1, m\rangle$ are eigenstates of J_z with eigenvalue m.

Thus, we have constructed the irreducible three-dimensional (matrix) representation of the generators of rotations. Furthermore, it is possible to use these matrices directly upon the states' kets represented by column matrices (and the associated bras by row matrices which are the adjoints of the corresponding columns):

$$|1, +1\rangle \rightarrow \begin{pmatrix} 1 \\ 0 \\ 0 \end{pmatrix}, \ |1, 0\rangle \rightarrow \begin{pmatrix} 0 \\ 1 \\ 0 \end{pmatrix}, \text{ and } |1, -1\rangle \rightarrow \begin{pmatrix} 0 \\ 0 \\ 1 \end{pmatrix}. \tag{3.31}$$

(See Problem 12.)

3.2 SPIN ANGULAR MOMENTUM

We have seen, in Chapter 1, that the fundamental particles of matter have spin $\frac{1}{2}$. This is expressed in the present context by saying that they possess an intrinsic angular momentum corresponding to $j = \frac{1}{2}$. There are only two possible values for m, viz., $+\frac{1}{2}$ and $-\frac{1}{2}$ (corresponding to "spin up" and "spin down," respectively). Here $(2j + 1) = 2$, so the angular momentum operators J_i are to be represented by 2×2 Hermitian matrices which obey

$$[J_x, J_y] = iJ_z \text{ (and cyclic)}. \tag{3.5}$$

It can be shown (see Problem 13), using the same method as followed above (for $j = 1$), that

$$\vec{J} = \tfrac{1}{2}\vec{\sigma} \text{ for } j = \tfrac{1}{2}, \tag{3.32}$$

where the $\vec{\sigma}$, called the "Pauli matrices," are

$$\sigma_x \equiv \begin{pmatrix} 0 & 1 \\ 1 & 0 \end{pmatrix}, \ \sigma_y \equiv \begin{pmatrix} 0 & -i \\ i & 0 \end{pmatrix}, \text{ and } \sigma_z \equiv \begin{pmatrix} 1 & 0 \\ 0 & -1 \end{pmatrix}. \tag{3.33}$$

The (spin) up and down states which are called **spinors** are

$$|\text{up}\rangle \equiv |\uparrow\rangle \sim \begin{pmatrix} 1 \\ 0 \end{pmatrix} \text{ and } |\text{down}\rangle \equiv |\downarrow\rangle \sim \begin{pmatrix} 0 \\ 1 \end{pmatrix}. \tag{3.34}$$

They span the space of states, i.e., a state with any orientation of the spin of the fundamental particles of matter can be represented as a linear combination of the $|\uparrow\rangle$ and $|\downarrow\rangle$ states.

Consider a passive counter-clockwise rotation about the z axis by ϕ. We have seen in Chapter 2 that the description of the state in the new system ($|\Phi\rangle'$) is related to the original description ($|\Phi\rangle$) by

$$|\Phi\rangle' = e^{i(\sigma_z/2)\phi}|\Phi\rangle. \tag{3.35}$$

Let us expand the operator $e^{i(\sigma_z/2)\phi}$:

$$e^{i(\sigma_z/2)\phi} = \sum_n \frac{\left(i\sigma_z \dfrac{\phi}{2}\right)^n}{n!}. \tag{3.36}$$

The matrix σ_z, shown in Eq. (3.33), obeys

$$\sigma_z^2 = \sigma_z\sigma_z = 1 \text{ (the } 2 \times 2 \text{ unit matrix)}.$$

From this it follows that

$$(\sigma_z)^n = 1 \text{ for } n \text{ an even integer},$$

$$= \sigma_z \text{ for } n \text{ an odd integer}. \tag{3.37}$$

Thus, we may write

$$e^{i(\sigma_z/2)\phi} = \sum_{n \text{ even}} \frac{\left(i\dfrac{\phi}{2}\right)^n}{n!} + \sigma_z \sum_{n \text{ odd}} \frac{\left(i\dfrac{\phi}{2}\right)^n}{n!}, \tag{3.38}$$

where we have pulled the common factor σ_z out of the second sum. Let us recall that

$$e^{i\alpha} = \cos \alpha + i \sin \alpha = \sum_{n \text{ even}} \frac{(i\alpha)^n}{n!} + \sum_{n \text{ odd}} \frac{(i\alpha)^n}{n!}, \tag{3.39}$$

where α is a real number. We know that $\cos \alpha$ is even (i.e., it remains the same number if we replace α by $-\alpha$) and $\sin \alpha$ is odd (it changes sign for the same replacement). Thus the first sum must be $\cos \alpha$ and the second is $i \sin \alpha$. Our equation for the rotation, Eq. (3.38), then may be written:

$$e^{i\sigma_z(\phi/2)} = \cos \frac{\phi}{2} + i\sigma_z \sin \frac{\phi}{2}, \tag{3.40}$$

(and for an axis in the direction $\hat{n}$ instead, we need merely replace σ_z by $\vec{\sigma} \cdot \hat{n}$; see Problem 15). Notice that half the angle appears in Eq. (3.40).

Let us consider a rotation by 2π. This should leave the coordinate system unchanged; however, we see that the rotation operator for a spinor (Eq. 3.40) becomes

$$e^{i\sigma_z(2\pi/2)} = \cos \pi + i\sigma_z \sin \pi = -1. \tag{3.41}$$

Inserting this into Eq. (3.35) shows that **spinors change sign under a 2π rotation**.

However, all measurements are proportional to a quantity of the form $|\langle \psi_f | A | \psi_i \rangle|^2$ for some operator A (which represents the quantity or transition being measured), where $\langle \psi_f |$ depends on the measuring apparatus. A change in the phase of $| \psi_i \rangle$ would not show up in the measurement, since this is an absolute square. In other words, a change in the overall phase (sign, in this case) of a wavefunction has no physical significance. So is this property of spinors undetectable and therefore of no consequence? It would be a nice test of quantum mechanics and the existence of spinor states if we could devise an observable consequence of this calculation! You should be pleased(!) to know that such an experiment *has* been devised, in which the interference of two neutron beams is studied. The experiment is then repeated with the spin of *one* of the beams rotated by 360° by precessing it in a magnetic field.[5] This effect shows up because, although we measure only the absolute squares of matrix elements, in this experiment the *relative* phase between the two (coherent) parts of the wavefunction has been changed:

$$|\psi\rangle = |\psi_1\rangle + |\psi_2\rangle \rightarrow |\psi\rangle = |\psi_1\rangle - |\psi_2\rangle. \tag{3.42}$$

The measured quantities are proportional to

$$|\langle \psi_f | A(|\psi_1\rangle \pm |\psi_2\rangle)|^2$$
$$= |\langle \psi_f | A | \psi_1 \rangle|^2 + |\langle \psi_f | A | \psi_2 \rangle|^2 \pm 2\text{Re}\{(\langle \psi_f | A | \psi_1 \rangle)^*(\langle \psi_f | A | \psi_2 \rangle)\}. \tag{3.43}$$

Notice that the sign change shows up in the cross-term.

---

5. This experiment is discussed in *Modern Quantum Mechanics* by J. J. Sakurai.

We see here a correspondence between rotations and 2×2 unitary matrices ($e^{i(\frac{1}{2}\vec{\sigma}\cdot\hat{n}\phi)}$). However, although a *rotation* by $360°$ is the same as a rotation by $0°$, the 2×2 matrix corresponding to $360°$ is

$$\begin{pmatrix} -1 & 0 \\ 0 & -1 \end{pmatrix} = -\mathbf{1},$$

whereas that corresponding to $0°$ (no rotation) is

$$\begin{pmatrix} +1 & 0 \\ 0 & +1 \end{pmatrix} = +\mathbf{1}$$

(it leaves the spinor unchanged). Furthermore, we can follow any rotation ($\vec{\phi} = \phi\hat{n}$) by a rotation of $360°$ and still have the same rotation ($\vec{\phi}$). However, the 2×2 matrices, which correspond to these two, differ from each other by a sign:

$U(\vec{\phi})$ and $U(360°)U(\vec{\phi}) = -U(\vec{\phi})$, respectively, for 2×2 matrices.   (3.44)

So to each rotation $\vec{\phi}$, there correspond two 2×2 matrices: $\pm U(\vec{\phi})$. We shall come back to this two-to-one correspondence when we discuss groups.

3.3   ADDITION OF ANGULAR MOMENTA AND CLEBSCH-GORDAN COEFFICIENTS

Given two angular momenta ($\vec{J}_1$ and $\vec{J}_2$), e.g., $\vec{L}$ and $\vec{S}$ or two subsystems with different j's, what is their sum? Classically the result is

$$\vec{J} = \vec{J}_1 + \vec{J}_2. \qquad (3.45)$$

In quantum mechanics this is an operator relationship, and the j's combine to yield many possibilities for the resultant j:

$$|j_1 - j_2| \leq j \leq (j_1 + j_2), \qquad (3.46)$$

where the possible values run in integer steps between the two limits. The J_z values *do* simply add:

$$m = m_1 + m_2. \qquad (3.47)$$

(See Problem 18.)

Let us examine how the quantum mechanical result in Eq. (3.46) comes about for some simple examples. We will use the notation $|j,m\rangle$ in what follows.

We start with wavefunctions which are products of the wavefunctions of the subsystems. The rules concerning addition of angular momentum are most easily obtained by considering such states, which do form a complete set. So we will construct product states for the combined system and will then examine how the total j can be determined.

1. Suppose $j_1 = 0$.

   Then subsystem 1 has only one possible state (with $m = 0$): $|0,0\rangle$, which does not change under rotations. The product states are then of the form $|0,0\rangle|j_2, m_2\rangle$, of which there are still $(2j_2 + 1)$ with $m = 0 + m_2 = m_2$ running from $-j_2$ to $+j_2$. It is apparent that this combination behaves like $j = j_2$.

2. Suppose $j_1 = \frac{1}{2}$ and $j_2 = \frac{1}{2}$.

   Let us introduce a convenient notation, "u" for spin up and "d" for spin down, for the states of these subsystems:

	$|j_1, m_1\rangle$	notation	$|j_2, m_2\rangle$	notation
spin up ↑	$|\frac{1}{2}, \frac{1}{2}\rangle$	$u(1)$	$|\frac{1}{2}, \frac{1}{2}\rangle$	$u(2)$
spin down ↓	$|\frac{1}{2}, -\frac{1}{2}\rangle$	$d(1)$	$|\frac{1}{2}, -\frac{1}{2}\rangle$	$d(2)$

We now construct all possible products:

Product State	m Value
↑↑ $\equiv u(1)u(2)$	1
↑↓ $\equiv u(1)d(2)$	0
↓↑ $\equiv d(1)u(2)$	0
↓↓ $\equiv d(1)d(2)$	−1

These mix only with each other when a rotation is performed, since $u(1)$ mixes only with $d(1)$, etc.

All members of each multiplet (irreducible representation) appearing in our possible states must be present, since (as we have seen in Section 3.1 and Appendix A) any one member of an irreducible representation rotates into all the others. The state with the largest m is ↑↑, with $m = 1$. Since this is the largest m, no multiplet with j greater than 1 can be present (which would require the presence of *all* of its states, including $m = j > 1$). Therefore, ↑↑ must belong to a multiplet with $j = 1$. Similarly, the ↓↓ state with $m = -1$ must belong to a multiplet with $j = 1$. If a $j = 1$ multiplet is present there must be a state for each possible m value ($m = +1, 0, -1$). There is just one state each for $m = \pm 1$, so they must belong to the same $j = 1$ multiplet. However, there are two $m = 0$ states and it is not evident which of them or, more accurately, which linear combination of them belongs to the $j = 1$ multiplet. To find the $m = 0$ state belonging to the $j = 1$ multiplet, we can apply the J_- operator to the $m = 1$ state, as we have learned in Section 3.1.

For the combined system, the operator $\vec{J}$ is the sum of the angular momentum operators of the individual systems:

$$\vec{J} = \vec{J}_1 + \vec{J}_2, \tag{3.48}$$

which implies that $J_- = J_{1-} + J_{2-}$ (since $J_{k-} = J_{kx} - iJ_{ky}$ for $k = 1,2$), where J_{1-} operates only on the factor containing 1-variables; likewise for $k = 2$. This yields

$$J_-|1,1\rangle = J_- u(1)u(2) = [J_{1-}u(1)]u(2) + u(1)[J_{2-}u(2)]. \tag{3.49}$$

From Eq. (3.11), we have

$$J_{1-}u(1) = J_{1-}|\tfrac{1}{2},\tfrac{1}{2}\rangle = |\tfrac{1}{2},-\tfrac{1}{2}\rangle = d(1) \text{ and a similar result for } J_{2-}u(2),$$

so Eq. (3.49) yields

$$J_-|1,1\rangle = d(1)u(2) + u(1)d(2). \tag{3.50}$$

But Eq. (3.11) also tells us that

$$J_-|1,1\rangle = \sqrt{2}|1,0\rangle,$$

therefore, inserting this into Eq. (3.50), we find

$$|1,0\rangle = \tfrac{1}{\sqrt{2}}(d(1)u(2) + u(1)d(2)), \tag{3.51}$$

which is a linear combination of the two product states. (It is obvious, without using Eq. (3.11), that the combination in Eq. (3.50) needs a factor of $1/\sqrt{2}$ to be properly normalized.) Thus, we have found the $j = 1$ multiplet, which contains three states:

$$|1,1\rangle = u(1)u(2)$$

$$|1,0\rangle = \tfrac{1}{\sqrt{2}}(d(1)u(2) + u(1)d(2))$$

$$|1,-1\rangle = d(1)d(2). \tag{3.52}$$

But we had four independent (product) states and should have the same number of degrees of freedom when we are done. This dilemma is easily solved by recalling that there should be two independent linear combinations of the two $m = 0$ product states. The missing one must be orthogonal to the $|1,0\rangle$ state. It is left to the reader to verify that the state below (designated $|0,0\rangle$) is indeed that orthogonal state and that it has $j = 0$ (see Problem 20(b)):

$$|0,0\rangle = \tfrac{1}{\sqrt{2}}(d(1)u(2) - u(1)d(2)). \tag{3.53}$$

For a rotationally invariant interaction, the states in the $j = 1$ multiplet are degenerate, but the separate $j = 0$ multiplet (which has only one state) need not have the same energy.

Notice that the spin 1 states are all symmetric to an interchange of the two subsystems $(1 \leftrightarrow 2)$, whereas the $j = 0$ state is antisymmetric to such an interchange. From Eq. (3.48), we see that the operator J_- is symmetric to the interchange of the subsystems. Therefore, when acting on a state, J_- will not change its behavior under subsystem interchange. Since $\uparrow\uparrow$ is symmetric to an interchange, it is not surprising that the symmetric linear combination appears for the $m = 0$ member of that multiplet. The generalization to the combination of many $j = \tfrac{1}{2}$ subsystems also follows from the structure of J_- as a sum. (See Problem 21.) This has the consequence that **all of the members of a multiplet must behave the same way under subsystem interchange as does the highest state of that multiplet**. Since the highest resulting j always has the state with all spins up, which is totally symmetric, **all the states in the highest resultant j multiplet must be symmetric to an interchange of any two subsystems**.

In order to generalize their structure, let us collect our equations using the notation $|j_1,m_1;j_2,m_2\rangle$ for the product state $|j_1,m_1\rangle|j_2,m_2\rangle$:

$$|1,+1\rangle = |\tfrac{1}{2},+\tfrac{1}{2};\tfrac{1}{2},+\tfrac{1}{2}\rangle$$

$$|1,0\rangle = \tfrac{1}{\sqrt{2}}|\tfrac{1}{2},-\tfrac{1}{2};\tfrac{1}{2},+\tfrac{1}{2}\rangle + \tfrac{1}{\sqrt{2}}|\tfrac{1}{2},+\tfrac{1}{2};\tfrac{1}{2},-\tfrac{1}{2}\rangle$$

$$|1,-1\rangle = |\tfrac{1}{2},-\tfrac{1}{2};\tfrac{1}{2},-\tfrac{1}{2}\rangle$$

$$|0,0\rangle = \tfrac{1}{\sqrt{2}}|\tfrac{1}{2},-\tfrac{1}{2};\tfrac{1}{2},+\tfrac{1}{2}\rangle - \tfrac{1}{\sqrt{2}}|\tfrac{1}{2},+\tfrac{1}{2};\tfrac{1}{2},-\tfrac{1}{2}\rangle. \tag{3.54}$$

The coefficients appearing in these equations are called **Clebsch-Gordan coefficients** and tell us how much of each product state is present in the state of given j,m. (Note that many of the Clebsch-Gordan coefficients are 0; for example, unless $m_1 + m_2 = m$, the corresponding Clebsch-Gordan coefficient is 0.) They are often expressed by the notation: $\langle j,m|j_1,m_1;j_2,m_2\rangle$. Reading Eq. (3.54), we obtain

from the first line: $\langle 1,+1|\tfrac{1}{2},+\tfrac{1}{2};\tfrac{1}{2},+\tfrac{1}{2}\rangle = +1$,

$$\langle 1,+1|\tfrac{1}{2},-\tfrac{1}{2};\tfrac{1}{2},+\tfrac{1}{2}\rangle = 0, \text{ etc., and}$$

from the second line: $\langle 1,0|\tfrac{1}{2},+\tfrac{1}{2};\tfrac{1}{2},+\tfrac{1}{2}\rangle = 0$,

$$\langle 1,0|\tfrac{1}{2},-\tfrac{1}{2};\tfrac{1}{2},+\tfrac{1}{2}\rangle = +\tfrac{1}{\sqrt{2}}, \text{ etc.} \tag{3.55}$$

The same procedure can be used for any j_1,m_1 and j_2,m_2. Clebsch-Gordan coefficients for many combinations have been worked out (using more sophisticated methods) and have been tabulated for us.[6] In Mathematical Table II at the end of the book, we display the Clebsch-Gordan coefficients for small values of j.

The general form of the recoupling equations may be written:

$$|j,m\rangle = \sum_{m_1 m_2} \langle j,m|j_1 m_1;j_2 m_2\rangle|j_1 m_1\rangle|j_2 m_2\rangle = \sum_{m_1 m_2} C^{j\ j_1\ j_2}_{m\,m_1\,m_2}|j_1 m_1\rangle|j_2 m_2\rangle,$$

$$\tag{3.56}$$

where we have also represented the Clebsch-Gordan coefficients by $C^{j\ j_1\ j_2}_{m\,m_1\,m_2}$.

PROBLEMS

1. a. From their definitions in terms of differential operators, verify that $[L_x,L_y] = iL_z$ (and cyclic).
   b. Explain why J_x, J_y, and J_z cannot be diagonalized simultaneously.
2. a. Eq. (3.21) shows the matrix for an active rotation about the z axis (x into y). Write the corresponding matrix for a passive rotation (R_z).

---

6. Actually, since the overall phases of states are arbitrary, different tables may have different choices of phases and, thus, different phases in their coefficients. The reader should carefully check the phase conventions when using more than one table.

   b. Write the analogous equation for a passive rotation about the x axis (y into z) (R_x) and that for a passive rotation about the y axis (z into x) (R_y).

   c. Show that $[R_x(\theta_1), R_y(\theta_2)] \neq 0$ for the matrices found in part a.

3. a. From the appropriate commutation relations show that $[J_i, J^2] = 0$.

   b. Explain how this verifies that J^2 is a rotationally invariant operator.

   c. Show that this commutator implies that operating with J_x, J_y, or J_z, on a state with angular momentum quantum number j, does not change the j value.

4. a. Show that $[J_z, J_+] = J_+$, and from this show that J_+ raises the m value of the state $|j, m\rangle$ by $+1$.

   b. Follow the analogous steps for J_-.

5. a. From the commutation relations find $[J_+, J_-]$.

   b. Show that the constant in Eq. (3.11) is consistent with the answer to part a.

6. Consider a passive rotation about the z axis by ϕ. Find $U^\dagger J_i U$, for $i = 1, 2,$ and 3. Use the Baker-Hausdorff lemma, displayed in Problem 10 of Chapter 2, to sum the series.

7. Verify that U of Eq. (3.22) is unitary.

8. a. Show that the representation spanned by $\begin{pmatrix} V_+ \\ V_z \\ V_- \end{pmatrix}$ is reducible (and in reduced form) for rotations about the z axis. Look at Eq. (3.24).

   b. Find R_x and R_y in this basis. Hint: Perform a similarity transformation on the matrices found in Problem 2, using U of Eq. (3.22).

   c. Show that R_x and R_y mix V_+, V_z, and V_-, verifying that this three-dimensional representation of the full rotation group is not reducible.

9. Consider a rotation about the z axis by ϕ, with L_z of Eq. (A.2) as the generator, operating on the $\ell = 1$ states in Eq. (A.3).

   a. Write the matrix $R_z(\phi)$ for this basis.

   b. For the three matrices L_x, L_y, and L_z for $\ell = 1$ shown in Eq. (A.2), use a similarity transformation (Eq. 2.33), to show that in the rotated coordinate system:

$$^{new}\langle b|L_x|a\rangle^{new} = \langle b|L_x|a\rangle \cos\phi + \langle b|L_y|a\rangle \sin\phi,$$

$$^{new}\langle b|L_y|a\rangle^{new} = -\langle b|L_x|a\rangle \sin\phi + \langle b|L_y|a\rangle \cos\phi,$$

$$^{new}\langle b|L_z|a\rangle^{new} = \langle b|L_z|a\rangle,$$

as expected for the components of a vector operator.

10. Verify the assignments in Eq. (3.25) for the vector $\vec{V} = \vec{r}$ by applying the operator

$$L_z = -i\left(x\frac{\partial}{\partial y} - y\frac{\partial}{\partial x}\right) \text{ directly to } V_\pm = \frac{1}{\sqrt{2}}(x \pm iy) \text{ and } V_z = z.$$

11. Find J_x and J_y for $j = 1$ from the $J_\pm$ matrices shown in Eqs. (3.28) and (3.29); compare the results with the L_x and L_y matrices of Appendix A.

12. Verify that the $J_\pm$ matrices of Eqs. (3.28) and (3.29) acting on

$$|1, +1\rangle \rightarrow \begin{pmatrix} 1 \\ 0 \\ 0 \end{pmatrix}, \quad |1, 0\rangle \rightarrow \begin{pmatrix} 0 \\ 1 \\ 0 \end{pmatrix}, \text{ and } |1, -1\rangle \rightarrow \begin{pmatrix} 0 \\ 0 \\ 1 \end{pmatrix}$$

obey Eq. (3.11).

13. Verify that $\vec{J} = \frac{1}{2}\vec{\sigma}$ for $j = \frac{1}{2}$ follows from Eq. (3.11).

14. a. From Eq. (3.40), explicitly write the 2×2 matrix for a passive rotation about the z axis by ϕ in terms of $e^{\pm i(\phi/2)}$. Verify that $\boldsymbol{u^\dagger \vec{\sigma} u}$ **is a vector**, where the components of $\vec{\sigma}$ are the usual Pauli matrices, as follows:

   b. Show that $u'^\dagger \vec{\sigma} u' = u^\dagger (U^\dagger \vec{\sigma} U) u$, where u' is a rotated state.

   c. Perform the similarity transformation $(\sigma_i)' = U^\dagger \sigma_i U$ for the rotation around the z axis to show that

   $$(\sigma_x)' = \sigma_x \cos \phi + \sigma_y \sin \phi, \ (\sigma_y)' = -\sigma_x \sin \phi + \sigma_y \cos \phi, \ \text{and} \ (\sigma_z)' = \sigma_z.$$

   (σ_x, σ_y, and σ_z are shown in Eq. 3.33.) Use this result to show that $u^\dagger \vec{\sigma} u$ behaves like a vector should for this rotation. This can actually serve as a proof, since we can choose to call the axis of rotation the z axis.

   d. Find $(\sigma_x)'$, $(\sigma_y)'$, and $(\sigma_z)'$ for $\phi = \pi/2$.

15. a. Find $(\sigma_x)^2$ and $(\sigma_y)^2$.

   b. Calculate $\{\sigma_x, \sigma_y\} \equiv \sigma_x \sigma_y + \sigma_y \sigma_x$, $\{\sigma_y, \sigma_z\}$, and $\{\sigma_z, \sigma_x\}$. These are called anticommutators.

   c. Calculate $(\vec{\sigma} \cdot \hat{n})^2$ for $\hat{n}$ in an arbitrary direction ($\hat{n} = n_1 \hat{x} + n_2 \hat{y} + n_3 \hat{z}$ with $n_1^2 + n_2^2 + n_3^2 = 1$).

   d. Derive the 2×2 rotation matrix for a rotation by ϕ about the $\hat{n}$ direction (which is a general rotation).

16. Calculate the anticommutator $\{L_x, L_y\}$ for $\ell = 1$. Note that the answer is not analogous to the result for $\{\sigma_x, \sigma_y\}$.

17. Equation (3.40) shows a two-dimensional matrix corresponding to a rotation about the z axis.

   a. Write the analogous equation for a rotation about the x axis and that for a rotation about the y axis.

   b. Calculate $[R_x(\theta_1), R_y(\theta_2)]$ for the two-dimensional matrices found in part a. The answer should not be zero, which confirms the fact that they do not commute.

18. Show that $J_z = J_{1z} + J_{2z}$ implies that $m = m_1 + m_2$.

19. a. Use Eq. (3.11) to construct the $j = \frac{3}{2}$ combination of $j_1 = 1$ and $j_2 = \frac{1}{2}$.

   b. From the result in part a find all the non-zero Clebsch-Gordan coefficients:

   $$\langle \tfrac{3}{2}, m \,|\, 1, m_1; \tfrac{1}{2}, m_2 \rangle.$$

20. a. Express J^2 in terms of J_+, J_-, and J_z.

   b. Show that the state designated $|0,0\rangle$ in Eq. (3.53) has $j = 0$ by applying the J^2 operator to it.

   c. For the coupling of $j_1 = 1$ and $j_2 = \frac{1}{2}$ studied in Problem 19, construct the resulting state $|j, \frac{1}{2}\rangle$, which is orthogonal to the state $|\frac{3}{2}, \frac{1}{2}\rangle$, and find its j value by applying the J^2 operator to it.

21. a. Prove that the highest j multiplet resulting when N $j = \frac{1}{2}$ spinors are combined is symmetric to the interchange of any two spinors.

   b. Prove that combining N multiplets, all with the same j value (not necessarily $\frac{1}{2}$), produces a state which is symmetric to the interchange of any two subsystems (called totally symmetric) for the largest resultant $j_{total} = Nj$.

Chapter 4

Relativistic Quantum Mechanics and Quantum Field Theory

In this chapter, we will discuss relativistic quantum mechanics and will show how it leads to the existence of spin, antiparticles, and the concept of renormalization. We shall see that the discovery of creation and annihilation of particles requires a theory beyond the limitations of quantum mechanics which, however, must be included within the more extensive theory. That theory is quantum field theory which arises from relativistic quantum mechanics. A particular type of quantum field theory called a *gauge theory* is believed to explain all the interactions we see. In this chapter we will provide only a thumbnail sketch of an introduction to quantum field theory, because a solid background in quantum mechanics at the graduate level is required for such an introduction. Instead, we present a description[1] of those features which are essential to an understanding of the beauty and power of the gauge theories. Much of our later elucidation of the theories indicated in the chart on the inside front cover will be based upon the semiquantitative discussions that we now undertake.

4.1 RELATIVISTIC FORMULATION AND NOTATION

When events are looked at by two observers in relative motion to one another along the x direction, the times (measured from $t = 0$ at the crossing of the two origins) and locations will be related by a Lorentz transformation:

$$x' = \gamma(x - vt), \; y' = y, \; z' = z, \text{ and } t' = \gamma(t - [v/c^2]x), \qquad (4.1)$$

where $\gamma \equiv 1/\sqrt{1 - (v^2/c^2)}$. Let us call

$$x \to x^1, \; y \to x^2, \; z \to x^3, \text{ and } ct \to x^0.$$

---

1. There exists another formulation of quantum field theory, based on the path-integral formulation of quantum mechanics, which will not be discussed in this book.

where $g_{00} = +1$, $g_{11} = g_{22} = g_{33} = -1$, and $g_{\mu\nu} = 0$ for $\mu \neq \nu$. The indices of $g_{\mu\nu}$ are written as subscripts with good reason, as we shall see.

The entity $g_{\mu\nu}$ is called the metric tensor, and it can be written in a matrix array for easy reference:

$$g_{\mu\nu} = \begin{pmatrix} 1 & 0 & 0 & 0 \\ 0 & -1 & 0 & 0 \\ 0 & 0 & -1 & 0 \\ 0 & 0 & 0 & -1 \end{pmatrix} . ^3 \tag{4.6}$$

The location and time of an event is designated (using natural units) by

$$x^\mu = (t; \vec{r}), \ \mu \text{ running from 0 to 3,} \tag{4.7}$$

where the $x^i (i = 1,2,3)$ have been combined into the vector $\vec{r}$ as a notational simplification. We define a four-vector (a^μ) as an object with four components, which undergo the same mixing as do the corresponding components of x^μ under rotations and Lorentz transformations. (The mixing for Lorentz transformations is shown in Eq. 4.2.)

Space-time derivatives form a four-vector *operator*:

$$\partial_\mu = \left(\frac{\partial}{\partial t} ; \vec{\nabla} \right), \tag{4.8}$$

where the $\partial/\partial x^i$ have been combined into the vector operator $\vec{\nabla}$. The generalization of the dot product (in this four-dimensional space) for any two four-vectors a^μ and b^μ is

$$\sum_{\nu=0}^{3} \sum_{\mu=0}^{3} a^\mu g_{\mu\nu} b^\nu = a^\mu g_{\mu\nu} b^\nu = a^\mu b_\mu = a_\nu b^\nu, \tag{4.9}$$

where the second form uses the summation convention in which a sum from 0 to 3 over all repeated indices (one upper and one lower) is *understood*, and the third and fourth forms result from the definitions:

$$b_\mu \equiv g_{\mu\nu} b^\nu \text{ and } a_\nu \equiv a^\mu g_{\mu\nu}. \tag{4.10}$$

Actually, the definitions in this equation are equivalent due to the symmetry of $g_{\mu\nu}$. These summations over repeated indices (one upper and one lower) are referred to as **contractions**. We shall see that these covariant products are invariant to Lorentz transformations, as well as rotationally invariant.

Consider a four-vector $b^\mu = (b^0; \vec{b})$. Let us construct b_μ from its definition in Eq. (4.10). First we will find b_0 by setting $\mu = 0$ in the definition:

$$b_0 = g_{0\nu} b^\nu.$$

---

3. In Einstein's theory of gravitation, where the interaction is due to a curvature in space-time, $g_{\mu\nu}$ is the dynamical field. Its elements then are no longer 0's and 1's, but instead it has 10 independent components (since it is actually a symmetric 4 × 4 matrix).

We write the indices (1, 2, 3, and 0) as superscripts for reasons which will soon become apparent. We have here what looks like a four-component object, whose 1, 2, and 3 components are just those of $\vec{r}$ and whose 0th component is the length ct. Let us write it as x^μ with μ having four possible values (0,1,2,3). Under rotations the three space components mix as usual and the 0th (time) component is unchanged. Under Lorentz transformations, the time component and space components mix with each other, as can be seen by looking at the Lorentz transformations in Eq. (4.1). From that equation, the $x^0 - x^1$ mixing may be written:

$$x'^0 = \gamma x^0 - \gamma\beta x^1 \text{ and } x'^1 = \gamma x^1 - \gamma\beta x^0, \text{ where } \beta \equiv \frac{v}{c}, \tag{4.2}$$

whose form is similar to that for a rotation.[2]

In classical mechanics and electrodynamics, many of the equations are vector equations. This form guarantees that the theory is rotationally invariant, since each side of the equation would behave the same way under rotations (i.e., if $A_i = B_i$, then in a rotated frame $A_i' = B_i'$, since the new components on each side are the *same* mixture of the corresponding old components, which were equal, as is shown in Problem 1). Such a form for the equations is referred to as **manifest covariance**. Recall that for rotation,

$$\sum_{i=1}^{3} (x^i)^2 = r^2 \tag{4.3}$$

is a scalar; therefore, it is (a quadratic) invariant to rotations. If we wish to have an analogous invariant for Lorentz transformations, we might consider constructing it from x^μ, since it has a time component and those transformations mix space and time with one another. However, the Lorentz transformation is not of exactly the same form as a rotation in the $x^0 - x^1$ "plane," as shown in footnote 2, so we do not expect $\sum_{\mu=0}^{3} (x^\mu)^2$ to be an invariant. Instead, the quadratic form:

$$(x^0)^2 - \sum_{i=1}^{3} (x^i)^2 \tag{4.4}$$

is invariant under Lorentz transformations (as well as rotations), and it is left as an exercise for the reader to check this for a Lorentz transformation in the x direction. Note that this differs from the usual dot product of a vector with itself, only because some + signs are replaced by − signs. Let us write this expression in a more elegant form:

$$\sum_{\nu=0}^{3} \sum_{\mu=0}^{3} x^\mu g_{\mu\nu} x^\nu, \tag{4.5}$$

---

2. The *sum* of the squares of the coefficients is not 1 as it is for rotations ($\sin^2\theta + \cos^2\theta = 1$); rather, their difference is 1. Also, both transformations in Eq. (4.2) have negative coefficients in them.

The repeated index ν is to be summed 0 to 3:

$$b_0 = g_{00} b^0 + g_{01} b^1 + g_{02} b^2 + g_{03} b^3.$$

Now, looking back at $g_{\mu\nu}$ in Eq. (4.6), we see that it is diagonal, so that $g_{01} = g_{02} = g_{03} = 0$ and $g_{00} = +1$. This yields

$$b_0 = b^0. \tag{4.11}$$

Now let us find b_1:

$$b_1 = g_{1\nu} b^\nu = g_{10} b^0 + g_{11} b^1 + g_{12} b^2 + g_{13} b^3.$$

Again looking at $g_{\mu\nu}$, we see that $g_{11} = -1$ (and the off-diagonal elements are 0), so that

$$b_1 = -b^1$$

and, similarly,

$$b_2 = -b^2 \text{ and } b_3 = -b^3. \tag{4.12}$$

Now let us consider the scalar product of the four-vector a^μ with b^ν, using Eqs. (4.11) and (4.12). Let us write this as $a^\rho b_\rho$, where we have switched Greek letters to emphasize that any repeated index represents the same sum of terms:

$$a^\rho b_\rho = a^0 b_0 + a^1 b_1 + a^2 b_2 + a^3 b_3$$
$$= a^0 b^0 + a^1 (-b^1) + a^2 (-b^2) + a^3 (-b^3) = a^0 b^0 - \vec{a} \cdot \vec{b}, \tag{4.13}$$

which is an invariant. We have already verified that it is invariant for both a^ν and b^ν being the same four-vector x^ν (Eq. 4.4). Exactly the same algebraic manipulations will show that $a^\rho b_\rho$ is indeed invariant, since the result depends only on the fact that their components mix the same way as do those of x^μ under Lorentz transformations or rotations. (Each term in Eq. (4.13) is obviously rotationally invariant.) The scalar product $a^\rho b_\rho (= a_\rho b^\rho)$ is sometimes referred to as the dot product $a \cdot b$ in four dimensions.

EXAMPLE OF A FOUR-VECTOR: Energy and Momentum of a Free Particle

In elementary physics courses the concept of the rest mass (m_0) of a particle is encountered, and in some courses the "relativistic mass" $= m_0/\sqrt{1 - \beta^2}$ is introduced. Furthermore, the total energy (rest energy + kinetic energy) may be expressed as $E = m_0 c^2/\sqrt{1 - \beta^2}$. Notice that the "relativistic mass" is the same as the energy, except for a multiplicative constant (c^2), which is actually 1 in natural units. Therefore, it is redundant to employ relativistic mass as a separate entity, so we will never do so; the **rest mass will be referred to as the mass,** and we will dispense with the subscript 0 to **write m for mass.**

In this example we will examine how to incorporate the momentum $\vec{p}$ of a particle into a four-vector. Consider a particle of mass m moving along the

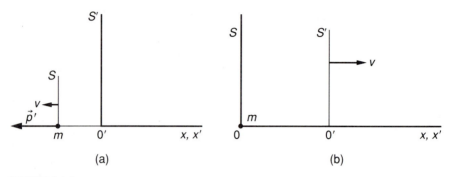

FIGURE 4.1
A particle of mass m (a) moving along the negative x axis (in frame S'), (b) in its rest frame (S).

negative x axis with speed v as shown in Fig. 4.1(a). Let us call the frame of the observer (at O') S'. The momentum and energy of this particle (in S') are

$$p'_x = -\gamma m v \text{ and } E' = \gamma m c^2, \text{ where } \gamma \equiv \frac{1}{\sqrt{1-\beta^2}}. \tag{4.14}$$

Let us now move along with the particle. In our frame of reference, which is the rest frame of the particle (we will call it S), the particle is at rest and the S' frame is moving along the $+x$ axis with speed v as shown in Fig. 4.1(b). Since the particle is at rest in S, we have

$$p_x = 0 \text{ and } E = m c^2 \text{ (in the particle's rest frame).} \tag{4.15}$$

Using Eqs. (4.14) and (4.15), we can express p'_x and E' in terms of p_x and E:

$$p'_x = \gamma \left(p_x - \beta \frac{E}{c} \right)$$

$$\frac{E'}{c} = \gamma \left(\frac{E}{c} - \beta p_x \right), \tag{4.16}$$

where we have "invented" the form of the p_x terms (since $p_x = 0$ in this example). It is left as an exercise (in Problem 3) for the reader to verify that, for a particle moving with speed w in the x direction in S, the momentum and energy in S' are indeed related to those in S by the formula above. These transformations have exactly the same form as those for x^0 and $\vec{r}$, so that the energy and momentum form a four-vector which we shall designate as p^μ:

$$p^\mu = \left(\frac{E}{c}; \vec{p} \right) = (E; \vec{p}) \text{ (in natural units).} \tag{4.17}$$

We have seen that the dot product of a four-vector with itself is an invariant; let us see what we get for $p^2 \equiv p^\mu p_\mu$:

$$p^2 \equiv p^\mu p_\mu = E^2 - \vec{p}^2 = m^2 \text{ (in natural units).} \tag{4.18}$$

It is left to the reader to check this in S' and in S, using Eqs. (4.14) and (4.15), respectively. ∎

We have learned that the laws of electrodynamics are relativistically invariant and should expect that this theory can be formulated in a *manifestly* relativistically covariant form (e.g., with four-vector equations or their generalizations[4]). Let us consider the electric current $\vec{J}$. We can construct a four-current J^μ which has its three-space components identical with those of $\vec{J}$ and its time component equal to ρ:

$$J^\mu = (\rho; \vec{J}), \tag{4.19}$$

as shown in Appendix B. The equation of continuity (Eq. 2.15) may be written in a covariant form as

$$\partial_\mu J^\mu = 0. \tag{4.20}$$

Let us integrate this continuity equation over all of space:

$$\partial_0 \int J^0 \, d^3r + \int \vec{\nabla} \cdot \vec{J} \, d^3r = 0,$$

where the ∂_0 has been pulled out of the integral, which is permissible since we are integrating over all of space (which is independent of time). Using the divergence theorem for the second term we obtain

$$\partial_0 \int J^0 \, d^3r + \int_{S_\infty} \vec{J} \cdot \hat{n} \, dS = 0.$$

For all physical states, $\vec{J} \to 0$ on the boundary (at $r \to \infty$); therefore, the second term vanishes. So the continuity equation implies

$$\partial_0 Q = 0 \text{ \textbf{conservation of charge}},$$

$$\text{where } Q = \int J^0 \, d^3r \text{ is the charge.} \tag{4.21}$$

In our study of the properties of elementary particles, we shall encounter currents corresponding to numbers other than electric charge, which obey a continuity equation. Those numbers (analogous to charge) will be conserved.

4.2 RELATIVISTIC QUANTUM MECHANICS

The Schrödinger equation was based on the classical equation:

$$E = (\vec{p})^2/2m + V.$$

---

4. We will explain the generalizations in Chapter 7.

However, this equation is only correct in the nonrelativistic limit ($v/c \ll 1$, for all v in the problem). We know, from millions of particle scatterings, that the laws of nature are invariant under Lorentz transformations; therefore, we must formulate a relativistically invariant theory. The connection between E and $\vec{p}$ for a free particle, which we obtained in Eq. (4.18), may be written:

$$E^2 = (\vec{p})^2 + m^2. \tag{4.22}$$

Following the same procedure used for the Schrödinger equation to obtain a quantum mechanical theory, we would replace E and $\vec{p}$ by time- and space-derivative operators, respectively, operating on the wavefunction (Φ) to obtain

$$(i\partial/\partial t)^2 \Phi = (-i\vec{\nabla}) \cdot (-i\vec{\nabla}) \Phi + m^2 \Phi.$$

Bringing all the terms to one side we have

$$(\partial^2/\partial t^2 - \nabla^2 + m^2)\Phi = (\partial^\mu \partial_\mu + m^2)\Phi = 0,$$

where $\partial^\mu = (\partial_0; -\vec{\nabla})$, whose relation to ∂_μ is indeed the same as that displayed for four-vectors in Eq. (4.10). Thus the quantum mechanical equation to be solved is

$$(\partial^\mu \partial_\mu + m^2)\Phi = 0, \tag{4.23}$$

which is called the **Klein-Gordon equation**. This equation is second order in the time derivative, and so we must know not only the state at $t = 0$, but also the time derivative of Φ at that time.

Along came Paul Dirac. He pointed out that our theory should need only Φ at $t = 0$ to predict its behavior at later times, thus only a first-order time derivative should be present.[5] This suggests taking the square root of the relativistic equation to obtain

$$p_0 = \pm\sqrt{(\vec{p})^2 + m^2}. \tag{4.24}$$

However, this expression is unsatisfactory, since p_0 and $\vec{p}$ are being treated unsymmetrically and the square root of an operator is very difficult to handle. In fact, it is not understood how to put interactions (with a field) into this type of a theory.

So Dirac proposed a theory of the form:

$$i\partial/\partial t = \vec{\alpha} \cdot \vec{p}_{op} + \beta m, \tag{4.25}$$

which is *linear* in the differential operators, where $\vec{\alpha}$ must rotate like a vector (so that the expression is rotationally invariant) and β is a scalar. Let us consider eigenstates of the energy as we have done in Chapter 2. The left-hand side

---

5. We shall see in Section 4.4 that, in quantum field theory, the Klein-Gordon equation does describe the behavior of (boson) field operators.

of this equation yields the energy of the state and the right-hand side is called the Dirac Hamiltonian (H_{Dirac}):

$$H_{Dirac} = \vec{\alpha} \cdot \vec{p} + \beta m. \tag{4.26}$$

However, the energy must be related to the momentum by the relativistic connection in Eq. (4.22):

$$E^2 = \vec{p}^2 + m^2.$$

Applying H_{Dirac} twice to an energy eigenstate yields E^2, so that we must have

$$H_{Dirac} H_{Dirac} = E^2 = (\vec{p})^2 + m^2.$$

Substituting from Eq. (4.26), yields

$$(\vec{\alpha} \cdot \vec{p} + \beta m)(\vec{\alpha} \cdot \vec{p} + \beta m) = (\vec{p})^2 + m^2,$$

which is

$$(\alpha_1 p^1 + \alpha_2 p^2 + \alpha_3 p^3 + \beta m)(\alpha_1 p^1 + \alpha_2 p^2 + \alpha_3 p^3 + \beta m)$$
$$= (p^1)^2 + (p^2)^2 + (p^3)^2 + m^2. \tag{4.27}$$

Looking at the terms $(p^1)^2$, $(p^2)^2$, $(p^3)^2$, and m^2 in Eq. (4.27), we see that **the square of each component of $\vec{\alpha}$ and β^2 must be 1.** The right-hand side has no term with $(p^1 p^2)$ in it, so that term on the left-hand side must be 0:

$$(\alpha_1 \alpha_2 + \alpha_2 \alpha_1)(p^1 p^2) = 0.^6 \tag{4.28}$$

Since $p^1 p^2$ is an operator, we have the requirement on the α's that

$$\alpha_1 \alpha_2 + \alpha_2 \alpha_1 = 0. \tag{4.29}$$

Looking at the other terms which contain cross-terms in the components of $\vec{p}$, we obtain, in an identical manner,

$$\alpha_2 \alpha_3 + \alpha_3 \alpha_2 = 0 \text{ (from } p^2 p^3) \text{ and } \alpha_3 \alpha_1 + \alpha_1 \alpha_3 = 0 \text{ (from } p^3 p^1). \tag{4.30}$$

These equations cannot be satisfied (for non-zero α's with numbers, because numbers commute so that the two terms on the left-hand sides are identical. The expression $(\alpha_1 \alpha_2 + \alpha_2 \alpha_1)$ looks like a commutator, except that there is a plus sign between the two terms. It is called an **anticommutator** and is abbreviated:

$$\{\alpha_1, \alpha_2\} \equiv \alpha_1 \alpha_2 + \alpha_2 \alpha_1. \tag{4.31}$$

Equations (4.29) and (4.30) show that $\boldsymbol{\alpha_1}$, $\boldsymbol{\alpha_2}$ and $\boldsymbol{\alpha_3}$ **anticommute with one another.**

Noting the absence of terms with $m p^i$, by the same arguments used for the components of $\vec{\alpha}$, we see that $\boldsymbol{\beta}$ **anticommutes with $\vec{\alpha}$.** We may summarize all of these results as follows:

$$\{\alpha_i, \alpha_j\} = 2\delta_{ij}, \; \{\beta, \alpha_i\} = 0, \text{ and } \beta^2 = 1. \tag{4.32}$$

---

6. We have used the fact that $p^1 p^2 = p^2 p^1$, i.e., they commute.

As we have said above ordinary numbers cannot anticommute, so the α_i and β must be more complicated objects. When Dirac was pursuing this study physicists were acquainted with matrices, which do have nontrivial commutation properties, so it seems natural that Dirac searched for a set of matrices which obey these relations. He showed that there did not exist a set of four 2×2 matrices nor a set of 3×3 matrices which satisfy these anticommutation relations, but found a satisfactory set among the 4×4 matrices:

$$\alpha_i = \begin{pmatrix} 0 & \sigma_i \\ \sigma_i & 0 \end{pmatrix} \text{ and } \beta = \begin{pmatrix} 1 & 0 \\ 0 & -1 \end{pmatrix}, \tag{4.33}$$

where the σ_i are the 2×2 Pauli matrices and the $1, 0, -1$ entries are 2×2 matrices as well.[7] An argument analogous to our remarks concerning 2×2 and 3×3 matrices, in our discussion of angular momentum in Chapter 3, shows that when 4×4 matrices represent operators, **four 4-columns** are needed to **span the space of states** and we may choose them to be

$$\begin{pmatrix} 1 \\ 0 \\ 0 \\ 0 \end{pmatrix}, \begin{pmatrix} 0 \\ 1 \\ 0 \\ 0 \end{pmatrix}, \begin{pmatrix} 0 \\ 0 \\ 1 \\ 0 \end{pmatrix}, \text{ and } \begin{pmatrix} 0 \\ 0 \\ 0 \\ 1 \end{pmatrix}. \tag{4.34}$$

Thus, ψ is now a four-column as well as a function of space. $\vec{\alpha}$ and β operate on ψ according to the rules of matrix multiplication, and the components of $\vec{p}(\sim\vec{\nabla})$ take space derivatives of ψ.

Although this work by Dirac was brilliant mathematics, a physicist at the time might have questioned whether it had anything to do with the real world (i.e.,

---

7. This means

$$\pm 1 = \begin{pmatrix} \pm 1 & 0 \\ 0 & \pm 1 \end{pmatrix} \text{ and } 0 = \begin{pmatrix} 0 & 0 \\ 0 & 0 \end{pmatrix},$$

so that $\alpha_1 = \begin{pmatrix} 0 & 0 & 0 & 1 \\ 0 & 0 & 1 & 0 \\ 0 & 1 & 0 & 0 \\ 1 & 0 & 0 & 0 \end{pmatrix}$, $\alpha_2 = \begin{pmatrix} 0 & 0 & 0 & -i \\ 0 & 0 & i & 0 \\ 0 & -i & 0 & 0 \\ i & 0 & 0 & 0 \end{pmatrix}$, $\alpha_3 = \begin{pmatrix} 0 & 0 & 1 & 0 \\ 0 & 0 & 0 & -1 \\ 1 & 0 & 0 & 0 \\ 0 & -1 & 0 & 0 \end{pmatrix}$,

and $\beta = \begin{pmatrix} 1 & 0 & 0 & 0 \\ 0 & 1 & 0 & 0 \\ 0 & 0 & -1 & 0 \\ 0 & 0 & 0 & -1 \end{pmatrix}$.

with physics). Let us continue to analyze this theory. Notice, from Eq. (4.26), that *states with* $\vec{p} = 0$ would have a Hamiltonian:

$$H = \beta m = \begin{pmatrix} m & 0 & 0 & 0 \\ 0 & m & 0 & 0 \\ 0 & 0 & -m & 0 \\ 0 & 0 & 0 & -m \end{pmatrix}. \tag{4.35}$$

Applying $H = \beta m$ to the first of our basis states, we find

$$H \begin{pmatrix} 1 \\ 0 \\ 0 \\ 0 \end{pmatrix} = \begin{pmatrix} m & 0 & 0 & 0 \\ 0 & m & 0 & 0 \\ 0 & 0 & -m & 0 \\ 0 & 0 & 0 & -m \end{pmatrix} \begin{pmatrix} 1 \\ 0 \\ 0 \\ 0 \end{pmatrix} = \begin{pmatrix} m \\ 0 \\ 0 \\ 0 \end{pmatrix} = m \begin{pmatrix} 1 \\ 0 \\ 0 \\ 0 \end{pmatrix}, \text{ for } \vec{p} = 0.$$

Thus, the first state is an eigenstate of the Hamiltonian with energy m (i.e., a particle at rest). Likewise, the three other basis states are eigenstates of H (for $\vec{p} = 0$) with energy m, $-m$, and $-m$, respectively. So the latter two are negative-energy states. It can be shown (see Problem 8) that even when the momentum is nonzero, we will still find two **negative-energy states**. This should not be surprising because Eq. (4.24), for p^0 in terms of $\vec{p}$ and m, has a $\pm$ in front of the square root.

Even for positive energies, we find that there are two states (the first two in Eq. 4.34) instead of one. Is this a dilemma or is it telling us something insightful? If we ignore the lower two entries (which are 0) of these first two states of Eq. (4.34), we see that they look just like the states we had when we discussed $j = \frac{1}{2}$ states, viz., $\begin{pmatrix} 1 \\ 0 \end{pmatrix}$ and $\begin{pmatrix} 0 \\ 1 \end{pmatrix}$. For the negative-energy states, we can similarly ignore the upper two 0 entries to reach the same conclusion. It has been shown (see Problem 7) that indeed these two states correspond to a spin $\frac{1}{2}$ particle with spin up and spin down, respectively. So we learn that when special relativity is incorporated into quantum mechanics, spin $\frac{1}{2}$ particles arise naturally. Actually, the existence of electron spin had been inferred from spectroscopic data slightly earlier but, until this theory was proposed, spin was put into theories in an ad hoc manner. So this is not a defect but a strength of this theory.

The energy dilemma is still present, however. There are twice as many states as expected, since we have negative-energy states. The existence of these negative-energy states also implies that there is no lowest state because, as $|\vec{p}| \rightarrow \infty$, $E \rightarrow -\infty$. So a charged particle can make radiative transitions to states of lower and lower energy (approaching $E = -\infty$). Think of all the energy that would be emitted! Why do positive-energy particles still exist (like those of which we are made)? This dilemma certainly looks like a fatal defect of the theory. However, Dirac suggested a way out of the catastrophe. Suppose that the entire "neg-

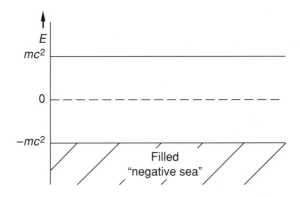

FIGURE 4.2
The vacuum state in Dirac's formulation of quantum mechanics. There are no states with
E between $-mc^2$ and $+mc^2$.

ative sea" of states was already filled, even for the vacuum (no particle) state,
as shown in Fig. 4.2. When particles are placed into the positive-energy states,
the exclusion principle tells us that no transitions to the negative-energy states
can occur. We are saved! However, can physical states, including the vacuum
state, really consist of an infinite number of charged particles? Wouldn't each
physical state's total charge then be infinite? And what about their energies?
With all the negative-energy states filled, even the energy of the "vacuum" state
should be $-\infty$. Dirac's resolution of these difficulties may be stated as follows:
We always tacitly compare our states to the vacuum state when we make measurements, so that we detect only how a state differs from the vacuum state.
Hence, the observed charge is the *difference* of the total charge of the state from
that of the vacuum; this explanation works as well for the energy. (This is
reminiscent of potential energy, where only differences in potential energy or
electrical potential are physically meaningful.) In this theory no fundamental
calculations of charges or masses can be made. Instead, their experimentally
observed values are merely inserted for the corresponding differences (of infinite results). After this substitution has been made in calculations of transition
probabilities, decay rates, or other properties of systems, the results should be
the finite observed values of those probabilities, decay rates, etc.

Are we to proclaim the validity of a theory which at first produces an infinite result, but from which another infinite result must be subtracted to yield
the finite value corresponding to experiment, or is this just Dirac in Wonder-
land? Even if the finite differences which result do agree with experiment, isn't
this just a "sweeping under the rug" of a fundamental flaw? This question has
not yet been successfully answered. However, this feature is not *so* different
from the theories of the nineteenth century, in which charge and mass were also
treated as parameters to be inserted. It is the appearance of infinite results, at
the intermediate stage, that has given this feature "bad press." The use of dif-

Alright Ruth, I about got this one renormalized   COPYRIGHT JFCARTIER 1984

FIGURE 4.3
Renormalization.

ferences as results is present also in all quantum field theories and is referred
to as **renormalization**. (See Fig. 4.3.)

Another interesting feature of the negative sea was discovered by Dirac.
If a photon in a vacuum has enough energy, it might be absorbed by one of the
negative-energy electrons, which would then make a transition to a positive-
energy state as shown in Fig. 4.4. So we start with one photon and end with a
positive-energy electron and a hole in the negative sea. What does a state with
a hole in the negative sea, i.e., with one electron missing from a negative-energy
state, correspond to? Well, we must compare it to the vacuum state (renormalization). Because a negative-energy electron is missing, the state has a charge
which is $+e$ more than the vacuum and an energy which is higher than that of
the vacuum state. So experimentally, **the state with a hole in the negative sea
would be a positively charged particle of (ordinary) positive energy**. The only
known positively charged fundamental particle was the proton, so Dirac hoped
that this hole state would correspond to a proton. However, it soon became
apparent that its mass would have to be identical to the electron mass. This particle is the antiparticle to the electron—the **positron (e^+)**. With this imaginative theory Dirac had predicted the existence of antiparticles, a type of matter
that no one at the time knew existed. About a year after this theory was proposed, the positron was discovered in cosmic rays.

The process in Fig. 4.4 can then be described as a photon changing into
a particle (the positive energy electron) and an antiparticle (a hole in the nega-

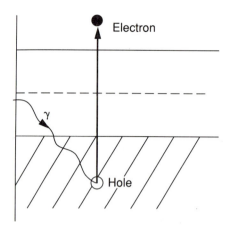

FIGURE 4.4
The absorption of a photon by an electron in a negative energy state causes it to make a transition to a positive energy state.

tive sea). The opposite process can occur as well, viz., beginning with a particle and an antiparticle (a hole in the negative sea), the particle can make a transition down to the empty state (in the negative sea) emitting a photon in this radiative transition (just like those in atoms). So we start with an electron and positron and are left with a photon. The first process (Fig. 4.4) was pair creation, and this reverse process is pair annihilation.

Thus, from the imagination or intuition of Dirac had sprung a theory which

1. had a natural explanation for spin ($j = \frac{1}{2}$),
2. predicted the existence of antiparticles,
3. predicted particle creation and annihilation (albeit only in pairs of particle-antiparticle), and
4. required renormalization.

At any time in this century, we could find several theories being proposed in journals and preprints. They would each have strong points and flaws. This one seemed to have some bizarre features, but we have seen that 1 through 3 above are, in fact, great insights into unknown realms.

4.3   MANIFEST COVARIANCE

The Dirac theory is Lorentz invariant by construction, but its form does not make that fact immediately evident. We have seen that it is desirable and more elegant to write equations in a manifestly covariant form. Consider the Dirac equation:

$$i\frac{\partial}{\partial t}\psi = (\vec{\alpha}\cdot\vec{p} + \beta m)\psi. \tag{4.36}$$

Let us bring all the terms to the left-hand side, remembering that $\vec{p} = -i\vec{\nabla}$, to obtain

$$\left(i\frac{\partial}{\partial t} + i\vec{\alpha}\cdot\vec{\nabla} - \beta m\right)\psi = 0. \tag{4.37}$$

Notice that the time-derivative operator has only an implied unit matrix as a factor, but the space derivatives are multiplied by the matrices α_i; the α_i and 1 cannot form a four-vector, since the 1 remains unchanged by unitary transformations. If we multiply the equation from the left by β, the result might be more suitable to our purpose. The resulting equation can be written (using $\beta^2 = 1$):

$$(i\gamma^\mu\partial_\mu - m)\psi = (\gamma^\mu p_\mu - m)\psi = 0,$$

$$\text{where } \gamma^\mu = (\beta; \beta\alpha^i). \tag{4.38}$$

From the anticommutation relations of the $\vec{\alpha}$ and β, it follows that (see Problem 9)

$$\{\gamma^\mu, \gamma^\nu\} = 2g_{\mu\nu}. \tag{4.39}$$

It is beyond the level of this book to discuss the behavior of the γ^μ matrices under general Lorentz transformations; however, it can be shown that they do indeed behave like a four-vector under Lorentz transformations and rotations. (By solving Problems 8 and 9, you can verify their four-vector behavior for a Lorentz transformation in the z-direction.) Thus Eq. (4.38) is **manifestly covariant, guaranteeing the invariance of the theory.**

We wish to construct Lorentz covariant objects to correspond to observables and, in analogy with $\psi^*O\psi$ of elementary quantum mechanics, we expect them to be of the form $\psi^\dagger A\psi$. Here ψ is a column, A is a square matrix, and $\psi^\dagger$ is a row, so that the product is a number (1×1 matrix). (Try it out for four-dimensional matrices.) However, we expect that $\psi^\dagger\psi$ does not behave like a scalar, since $\psi^*\psi$ was a density, i.e., the zeroth component of a four-vector. Instead,

$$\psi^\dagger\gamma^0\psi = \bar{\psi}\psi \text{ behaves like a scalar,} \tag{4.40}$$

$$\bar{\psi}\gamma^\mu\psi \text{ behaves like a four-vector, etc.,}$$

$$\text{where } \bar{\psi} \equiv \psi^\dagger\gamma^0,$$

which you can verify by doing Problem 9. The use of $\bar{\psi}$ has the nice property that each expression behaves exactly as the indices displayed, under Lorentz transformations. For example, consider the four-vector

$$J^\mu = \bar{\psi}\gamma^\mu\psi. \tag{4.41}$$

(The electric current is qJ^μ.) Notice that

$$J^0 = \bar{\psi}\gamma^0\psi = \psi^\dagger\gamma^0\gamma^0\psi = \psi^\dagger\psi,^8 \tag{4.42}$$

---

8. since $(\gamma^0)^2 = \beta^2 = 1$ (the unit matrix).

which *is* the expression expected for a density. We leave it to the reader (Problem 9) to show, using the Dirac equation, that J^μ is conserved:

$$\partial_\mu J^\mu = 0. \tag{4.43}$$

Space inversion ($x_i \to -x_i$) plays a central role in the structure of the electroweak theory; anticipating that discussion, let us define

$$\gamma^5 \equiv i\gamma^0\gamma^1\gamma^2\gamma^3. \tag{4.44}$$

It can be shown that γ^5 is invariant to rotations, but changes sign under space reflections and inversion of the coordinate system (see Problem 11); such objects are called pseudoscalars. (An example of such an object would be $\vec{a} \cdot (\vec{b} \times \vec{c})$, with $\vec{a}$, $\vec{b}$, and $\vec{c}$ being vectors.) Likewise, $\gamma^\mu\gamma^5$ picks up an extra minus sign under reflections and inversion, and such objects are called pseudovectors or axial vectors. (An example of such an object would be $(\vec{a} \times \vec{b})$, with $\vec{a}$ and $\vec{b}$ being ordinary (or polar) vectors.) So we can construct

$\bar{\psi}\gamma^5\psi$, **a pseudoscalar,** and $\bar{\psi}\gamma^\mu\gamma^5\psi$, **a pseudovector or axial vector.**   (4.45)

4.4   SKETCH OF RELATIVISTIC QUANTUM FIELD THEORY

4.4.1   Underlying Structure

We shall present here enough of the *flavor* of quantum field theory to enable the reader to appreciate the beauty and power of gauge theories. (The actual structure is beyond the level of this book.)

Recall that in *classical mechanics*, the *state of a system of N particles* is specified by their *positions* $\vec{x}^{(i)}$ ($i = 1, \ldots, N$), *a set of numbers*, and their velocities $\dot{\vec{x}}^{(i)}$. Their momenta $\vec{p}^{(i)}$ are defined in general by constructing a Lagrangian:

$$L = \Sigma_i L(\vec{x}^{(i)}, \dot{\vec{x}}^{(i)}), \tag{4.46}$$

which includes all the particles (i) and may have interaction terms containing more than one particle's variables, and setting

$$p_j^{(i)} \equiv \frac{\partial L}{\partial \dot{x}_j^{(i)}}, \tag{4.47}$$

where the (i) refers to the particular particle and the sub-j refers to the direction in space ($j = 1$, 2, or 3).

One can also discuss *classical fields*. Let us consider a single real *scalar field* $\phi(\vec{r})$. The *field* $\phi(\vec{r})$ *is a set of numbers*, one for each point in space ($\vec{r}$). Therefore, ϕ is analogous to the set of particle position variables ($\vec{x}^{(i)}$) labeled, however, by the *continuum* label $\vec{r}$ rather than by the discrete labels i used for particles, i.e., each point in space is analogous to a different particle. When we construct Lagrangians for a collection of particles in classical mechanics, we sum over i, as shown in Eq. (4.46), and analogously for a field, we have to integrate over all points of space. Thus, we have

$$L = \int d^3r \mathcal{L}(\phi(\vec{r}), \dot{\phi}(\vec{r})), \tag{4.48}$$

where $\mathcal{L}$ is the Lagrange density. In Appendix C, we outline the variational method for obtaining the equations of motion from the Lagrangian. (It is the Lagrangian which plays a major role in quantum field theory, and, in fact, we will not need the equations of motion in what follows.) There we see that a Lorentz- and rotationally invariant quantity called the action (S) is the central object of the variational formulation of mechanics. Substitution into its definition yields

$$S \equiv \int L \, dt = \int \mathcal{L} \, d^4x. \tag{4.49}$$

From this equation, we see immediately that $\mathcal{L}$ **must be a four-scalar (Lorentz-and rotationally invariant)**, since d^4x is an invariant.

In *quantum mechanics* for *particles*, the x_j and p_j are "promoted" to play the role of operators obeying the relationship:

$$[p_i, x_j] = -i\delta_{ij}, \tag{4.50}$$

and **a new "space" is used to describe the state of a system**, namely the **function space** made up of wavefunctions. Operators (e.g., derivatives or multiplication by functions of position) acting on one function produce another so that **we do not leave this function space; the operators connect one function to another**.

We have seen that the introduction of relativity into quantum mechanics, via the Dirac equation, led to the prediction of the existence of antiparticles, before their existence was even suspected, as well as the annihilation and creation of fermions (albeit in pairs). (This still has conservation of particle number, because antiparticles are assigned the particle number -1.) Moreover, the antiparticle has a very complicated structure, as does the vacuum state (and all others as well), because of the presence of an infinite number of particles in the negative sea of states. A theoretical formulation has been created in which there are no negative-energy states (the vacuum state is indeed empty) and in which the antiparticle is as fundamental as the particle. The physical states, which have a finite number of particles and antiparticles, are represented by states with that same content, rather than the infinity of particles appearing in the original Dirac theory. This formulation is called **quantum field theory**. We have said previously that in quantum field theory the number of particles is not fixed, and so it must contain annihilation and creation operators. There are, in fact, many quantum field theories, so that we may consider this to be a category of theories. In these theories, the wavefunction (ψ) is "promoted" to the role of an operator which can create and annihilate particles. **The state of a system lies in a *(Hilbert) space*** *of states* (**sometimes called a** *Fock space*[9]), which have different numbers of particles (of each kind). These states are "connected" to one another when acted upon by the ψ operators. For a given kind of particle, examples of these states follow:

State	Description
$\lvert 0 \rangle$	vacuum, i.e., no particles or antiparticles are present
$\lvert p \rangle$	1 particle of momentum p is present
$\lvert p_1, p_2 \rangle$	2 particles, one of momentum p_1 and one of p_2, are present
$\lvert \bar{p} \rangle$	1 antiparticle of momentum p is present

Classical physics provides functions for those entities that are considered fields, and quantum mechanics produces a field-like function, viz., the wavefunction ψ, for "particles." These are treated in a very similar way when they become operators. Each will be an operator with annihilation and creation operators in it. Thus, the classical field is quantized and is seen to be composed of particles similar to those of the quantized theory for "particles." The classical particle and wave have both become entities possessing "wave-particle duality." Furthermore, the particle's wavefunction can be incorporated into a Lagrangian in a manner similar to the treatment used classically for fields.

As an example, let us follow this "quantization" for a real scalar field $\phi(\vec{r})$. In a manner analogous to that used in quantum mechanics, a canonical momentum operator $\pi(\vec{r})$ is defined by

$$\pi(\vec{r}) \equiv \frac{\partial \mathcal{L}(\phi(\vec{r}), \dot{\phi}(\vec{r}))}{\partial \dot{\phi}(\vec{r})}, \tag{4.51}$$

and these operators (on the space of states) must obey

$$[\pi(\vec{r}), \phi(\vec{r}')] = -i\delta^3(\vec{r} - \vec{r}') \qquad \text{at } t = t', \tag{4.52}$$

which is reminiscent of the commutator $[p, x]$ of ordinary quantum mechanics. We will not be concerned with the presence of the δ-function here (although it can be understood by following the analogy with Eq. (4.50) for the quantum mechanics of particles), but notice that we have commutation relations among the field operators.

Let us examine the *wave operator* $\phi(\vec{r})$ in more detail. The usual formulation uses the Fourier analysis of the wavefunction in quantum mechanics, with Fourier coefficients replaced by operators: [10]

$$\phi(\vec{r}, t) = \int \frac{d^3 k}{\sqrt{(2\pi)^3 2\omega}} \, e^{i\vec{k} \cdot \vec{r}} (d^\dagger_{-\vec{k}} e^{+i\omega t} + d_{\vec{k}} e^{-i\omega t}), \tag{4.53}$$

where $d_{\vec{k}}$ annihilates a particle with momentum $\vec{k}$, $d^\dagger_{-\vec{k}}$ creates a particle[11] with momentum $-\vec{k}$, and $\omega(\vec{k}) = \sqrt{(\vec{k})^2 + m^2}$. Notice that there is a positive fre-

---

9. named after the Soviet physicist V. Fock.
10. Here we display explicitly that ϕ is a function of t.
11. In general, it would be the antiparticle, but in this case they are the same.

quency part (the first term) which corresponds to negative energies, since we expect the wavefunction to behave like e^{-iEt}. Negative energy is now in the operators instead of in the states.

Similarly for *the photon field $\vec{A}$ (the vector potential)*, we have

$$\vec{A}(\vec{r},t) = \int \frac{d^3k}{\sqrt{(2\pi)^3 2\omega}} \, e^{i\vec{k}\cdot\vec{r}}(\vec{d}^{\dagger}_{-\vec{k}} e^{+i\omega t} + \vec{d}_{\vec{k}} e^{-i\omega t}), \qquad (4.54)$$

where the x component of $\vec{d}_{\vec{k}}$ **annihilates a photon** of momentum $\vec{k}$, polarized in the x direction, etc., for y and z. Similarly $\vec{d}^{\dagger}_{-\vec{k}}$ **creates a photon** of momentum $-\vec{k}$, and $\omega = |\vec{k}|$.

The *wave operator for an electron* is written:

$$\psi(\vec{r},t) = \Sigma_s \int \frac{d^3k}{\sqrt{(2\pi)^3 2\omega}} \, e^{i\vec{k}\cdot\vec{r}}(b^{\dagger}_{-\vec{k},-s} v_s e^{+i\omega(\vec{k})t} + a_{\vec{k},s} u_s e^{-i\omega(\vec{k})t}), \qquad (4.55)$$

where $a_{\vec{k},s}$ **annihilates an electron** of momentum $\vec{k}$ and z component of spin $= s$, $b^{\dagger}_{-\vec{k},-s}$ **creates a positron** of momentum $-\vec{k}$ and z component of spin $= -s$, the u_s and v_s are two-columns[12] that we saw in Chapter 3 used to describe the spin state, and we sum over the two possible spins ($s = \uparrow$ and $s = \downarrow$). Correspondingly, the operator $\psi^{\dagger}$ (and therefore also $\bar{\psi}$) *creates* an electron or *annihilates* a positron. It is referred to as **the conjugate field** to ψ and vice versa.

When operating on a state, a creation operator produces a state with one more particle (or 0 if that would violate the exclusion principle for electrons or positrons), whereas an annihilation operator gives 0 unless a particle with the quantum numbers of that operator is in the state. If there is such a particle present, the annihilation operator produces a state with that particle gone.

For photons, the commutation relation in Eq. (4.52) manifests itself as the commutation relations between the annihilation and creation operators, as follows (see Problem 12):

$$[d_{\vec{k}}, d^{\dagger}_{\vec{k}'}] = \delta^3(\vec{k} - \vec{k}'), \qquad (4.56)$$

and the commutators of a creation operator with a creation operator, or an annihilation operator with an annihilation operator, vanish.

For electrons (spin $\frac{1}{2}$ particles), the creation and annihilation operators obey anticommutation rules instead:

$$\{a_{\vec{k}',s'}, a^{\dagger}_{\vec{k},s}\} = \delta^3(\vec{k} - \vec{k}')\delta_{s,s'}, \text{ etc.} \qquad (4.57)$$

Equations (4.56) and (4.57) lead to the correct statistics: Bose-Einstein for photons and Fermi-Dirac for electrons.

In Problem 13 the reader can verify that **boson field operators (like ϕ and $\vec{A}$) obey the Klein-Gordon equation, and fermion field operators (like ψ) obey the Dirac equation.**

---

12. We will not use the detailed structure of u_s and v_s in our later discussions; however, see Problem 8.

4.4.2   Interaction of the Electron with the Electromagnetic Field — Feynman Diagrams

The usual formulation of quantum field theory is based on the construction of a Lagrange density $\mathcal{L}$ for the fields involved. The Lagrange density contains energy densities, so that we are interested in the energy of interaction. In the realm of those experiments and observations where the classical theory is valid, the quantum field theory should give the same results. This is referred to as the correspondence principle. Keeping this in mind, we use the same form for the energy of an electromagnetic field as appears classically. In classical electromagnetism, we learn[13] that the interaction of charge with the electromagnetic field appears in $\mathcal{L}$ as $-(\rho V - \vec{J} \cdot \vec{A})$. Recall (from Section 4.1 and Appendix B) that ρ and $\vec{J}$ form a four-vector $J^\mu = (\rho; \vec{J})$. It is also shown in Appendix B that $\vec{A}$ (the vector potential) and V (the scalar potential) form a four-vector, $A^\mu = (V; \vec{A})$, so that the interaction may be written in covariant form:

$$\mathcal{L}_{INT} = -J^\mu A_\mu = e\bar{\psi}\gamma^\mu\psi A_\mu. \tag{4.58}$$

This is what appears as the interaction energy in the Lagrangian. In quantum electrodynamics (the quantum field theory of electrons and photons) $\bar{\psi}$, ψ, and A_μ are operators with both creation and annihilation operators in each, as we have shown in Eqs. (4.54) and (4.55). Therefore, the form of Eq. (4.58) is

$$\underset{\text{(iii)}}{\mathcal{L}_{INT} \sim (\text{create } e^- \text{ or annihilate } e^+)} \underset{\text{(ii)}}{(\text{create } e^+ \text{ or annihilate } e^-)} \underset{\text{(i)}}{(\text{create or annihilate } \gamma)},$$

$$\tag{4.59}$$

which should be read from right to left, as is usual for operators. We shall see shortly how this interaction can mediate pair creation and annihilation, bremsstrahlung, the photoelectric effect, and scattering. It also can, in principle, account for bound states.

In the Lagrangian interaction terms for the standard model, which we shall study in later chapters, there are juxtapositions of annihilation and creation operators of *different types of fermions*. Combinations of such terms can describe the decays of particles (to other types). These decays, along with all other possible processes, must then be incorporated into the S-matrix formalism with suitable interaction matrix elements (M), as described for scattering processes in Chapter 2, Section 2.10.

In general, then, ψ and $\bar{\psi}$ might refer to *different* particles; however, the interaction Lagrangian must be Hermitian so that the Hermitian conjugate of each term appears in it as well. In the Hermitian conjugate term, the conjugate field operators are used, in which the creation operators become annihilation operators and vice versa. The very structure of the field operators, with both annihilation and creation operators (for particles and their antiparticles, respectively), and the identical structure to that of Eq. (4.59) with particle and

---

13. thinking of distributed charge for particles.

antiparticle interchanged that is present in the Hermitian conjugate term, lead to the equality of the matrix elements (amplitudes) of many related processes. For example:

(1) $a + b \rightarrow c + d$, (2) $c + d \rightarrow a + b$, (3) $a + \bar{c} \rightarrow d + \bar{b}$, (4) $a \rightarrow \bar{b} + c + d$, etc.,

are all described by the same interaction. The simple rules of thumb are:

  i. **the reaction can be read backwards** — because (2) is produced by the interaction terms which are the Hermitian conjugates of those which produce (1).

  ii. **any particle can be pulled across the arrow of the reaction if it is also changed to its antiparticle** — because (3), (4), etc., follow since the operator which creates a c particle also annihilates a $\bar{c}$, that which annihilates a b also creates a $\bar{b}$, etc. This property of quantum field theories is referred to as **crossing symmetry**.

Feynman has created a diagrammatic method to represent the processes which $\mathcal{L}_{INT}$ (Eq. 4.58) mediates. This operator ($\mathcal{L}_{INT}$) is represented as a **vertex** — a point at which two fermion lines and one photon line meet. The vertex is drawn in Fig. 4.5 without indicating which are incoming and which are outgoing particles.

This same vertex describes, in a rudimentary way, all those processes that result from different incoming-outgoing choices (subject to the requirement that the vertex conserve fermion number, as will be explained below). Furthermore, actual processes are described by using combinations of vertices as "building blocks." As shown in Eq. (4.58), the charge e appears as a multiplicative factor in the interaction Lagrangian, which has different types of fields (electrons and photons) coupled together. Consequently, the charge is referred to as the **coupling constant** and is included in the diagram. In general, **the size of the coupling constant is a measure of the strength of the interaction**. The elements of the interaction matrix M for scattering and decay will depend on a series in powers of the coupling constant, as we shall soon discuss. Consequently, particles decaying via **stronger interactions** (in general) have greater probabilities of decay, hence **shorter lifetimes**, and decays due to **weaker interactions** have **longer lifetimes**.

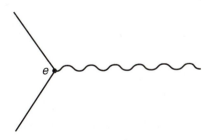

FIGURE 4.5
The vertex.

It is convenient to put an arrow on each of the fermion lines, which represents fermion (electron) number, with the direction of the arrow showing the flow of this number; it is continuous from one line to the other. When the direction of motion of the "particle" is the same as that of the arrow, it represents a particle with fermion (here electron) number = +1, whereas, if the direction of motion of the "particle" is opposite that of the arrow, it represents an antiparticle with fermion number = −1, since fermion number is dropping in the direction of the particle's motion. In practice, if the incident state and the final state have the same kind of particle present, they are thought of as the *same* particle whose properties have changed. This is in contrast to the quantum field theory language which has annihilation or creation of each line joining the vertex.

Let us see how this quantum field theory formulation incorporates pair production and pair annihilation, which were intrinsic to Dirac's theory. Refer to Eq. (4.59). If we have *a state with one photon and no electrons or positrons*, the $\mathcal{L}_{INT}$ operator: (i) can create another photon or **annihilate the photon in the initial state**, (ii) would **create a positron**, because the annihilation part of ψ gives 0 (since there is no electron in the state), (iii) can **create an electron** or annihilate the positron that was created.[14] This sequence can thus account for **pair creation**. With other initial states, the vertex describes other rudimentary processes, some of which are shown in Fig. 4.6. (See Problem 14.)

We have[15] in

Fig. 4.6(a) both fermions out and photon in, which is the rudimentary *pair creation*, described above;

Fig. 4.6(b) both fermions in and photon out, which is a rudimentary *pair annihilation (of a free pair or positronium)*;

Fig. 4.6(c) one fermion in, one out and photon out, which is a rudimentary *bremsstrahlung* (German for "braking radiation");

Fig. 4.6(d) one fermion in, one out and photon in, which is a rudimentary *photoelectric effect*.

For actual processes, **quantum field theory generates an infinite series of terms for the interaction matrix M, which is a power series in the coupling constant (charge).** The **interaction vertex** above is the **building block** from which these terms are constructed.

As a consequence of the construction of combinations of vertices, a new entity arises in quantum field theory which is called a **virtual particle**. Let us

---

14. Actually not all of the operator products that appear here are used in the order that they appear. A so-called *normal* ordering is part of the theory so that the vacuum is correctly defined.

15. In reality, processes (a), (c), and (d) require an interaction with another system as well, so that energy and momentum are conserved in the interaction. A free photon cannot decay into a pair, bremsstrahlung (braking radiation) occurs when the charged particle is accelerated (or decelerated), and the photoelectric effect involves a bound electron. Also, pair annihilation (b) of free particles must produce at least *two* photons in order to conserve both energy and momentum. See Problems 16 and 17.

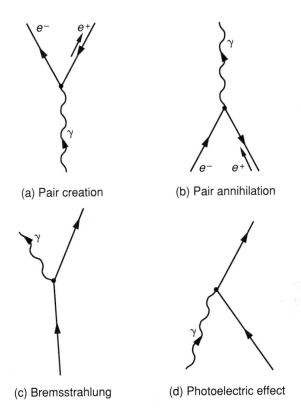

(a) Pair creation   (b) Pair annihilation

(c) Bremsstrahlung   (d) Photoelectric effect

FIGURE 4.6
Rudimentary processes which can be represented by the vertex of $\mathcal{L}_{INT}$. *To make it easier to follow the processes*, we have twisted the vertices somewhat to put all incoming
lines at the bottom and outgoing lines at the top, i.e., **time flows upward in these diagrams.** Diagrams (c) and (d) each represent two distinct processes: One involves electrons
when the direction of the particle's motion follows the arrows, and the other involves
positrons when motion is opposite to the arrows (in which case the fermion lines should
be flipped if we wish to maintain the time-flow direction).

see how the interaction vertex is utilized and how virtual particles appear in an
investigation of *the scattering of two electrons from one another.* In this case,
no photon appears in the initial or final state. However, vertices (of the electron line with a photon) *are* the building blocks of the interaction. Let us examine the relevant Feynman diagrams. The least number of vertices required is two,
since the photon from a single vertex would be free and would, therefore, have
to appear in the initial or final state. The corresponding Feynman diagram is
shown in Fig. 4.7.

The structure of Feynman calculations is such that we need not put arrows
on virtual boson lines since, when the corresponding integrals are done, a particle going one way and its antiparticle going the other are both automatically

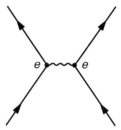

FIGURE 4.7
Feynman diagram of lowest order in e^2 for electron-electron scattering. It has no closed loops and is called a tree-level diagram.

included. Notice that the photon line appears to be exchanged between the two electron lines, and so the photon is only "virtually" present. We can refer to this as a **virtual photon**. It is not present in the initial or final system, but in this diagrammatic approach it appears as the carrier of the interaction from one electron to the other. Furthermore, four-momentum is conserved at each vertex, with the consequence that the square of the four-momentum of the virtual photon in this diagram is not equal to the square of the mass (zero) of a real photon. (See Problem 18). This may be expressed by saying that this virtual photon is "off its mass shell."

It is apparent that the *effective* coupling here is proportional to e^2. The reader might wonder whether there could be more than one virtual particle present, and the answer is yes; the infinite series which we talked about above contains terms described by an infinite set of diagrams, each of which has one more virtual photon than the preceding one. Figure 4.8 shows those diagrams which have an extra virtual photon in the various topologically distinct ways. Notice that all the diagrams in Fig. 4.8 have a closed loop of lines; they are referred to as one-loop diagrams. **Diagrams with no closed loops are called tree-level diagrams.** The contributions to M due to diagrams with loops are referred to as **radiative corrections** to the tree-level result.[16] The effective contribution of the one-loop diagrams to the sum is proportional to $e^2 \cdot e^2$, as shown explicitly in Fig. 4.8(a), since there are four vertices in these one-loop diagrams.

Perhaps you are wondering whether there are virtual fermions (electrons) in the theory as well; the answer is yes. Technically, the electron lines between two vertices in Fig. 4.8 are virtual particles. It is left to the reader to construct the tree-level diagram for photon-electron (Compton) scattering, in which the incident photon is absorbed by the electron, which subsequently emits the final photon; the intermediate electron line is a virtual electron. (See Problem 19.) In Fig. 4.9 we show the insertion of an electron loop in the exchanged photon line of Fig. 4.7, which may be thought of as the combination of two vertices

---

16. Diagrams for processes with extra initial and/or final state radiation are *related* to those of the original processes by the presence of extra vertices, on the incident or emerging charged lines, from which the extra photons emerge. Nevertheless, these are considered *different* tree-level diagrams.

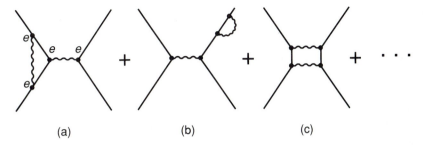

FIGURE 4.8
Distinct types of one-loop diagrams for *e-e* scattering, resulting from the insertion of an extra virtual photon. (a) shows the coupling *e* explicitly. Diagrams like (a) contribute to charge renormalization and those like (b) to mass renormalization. Diagram (c) is not a renormalization type graph; it is a pure second-order scattering diagram.

on a closed electron line or two vertices whose respective electron lines have been joined. In a similar manner, such loops can be inserted into the photon lines of all the diagrams we have considered so far to generate new diagrams.

Every time a virtual photon or an electron loop is inserted into a diagram, we obtain a factor of *e* at each of the two new vertices, yielding an extra factor of e^2. In order to discuss the relative sizes of the resulting terms, it is important to consider the size of that factor. However, a power series in a *dimensionless* quantity may be investigated in a more meaningful way because the relative sizes of successive terms are independent of units. The reader may think that e^2 has dimensions of coulombs squared, but remember that we are working in natural units where $\hbar = c = 1$. To find the dimensionless constant which e^2 represents, we need merely divide by $\hbar$ and c to obtain

$$\alpha \equiv e^2/\hbar c \approx (137)^{-1}, \text{ known as the } fine\text{-}structure \text{ } constant.[17] \qquad (4.60)$$

---

17. We are using Gaussian units. In Heaviside-Lorentz units, α is defined with an extra 4π in the denominator.

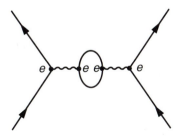

FIGURE 4.9
Feynman diagram for *e-e* scattering with an electron loop insertion in the exchanged (virtual) photon line.

Thus the successive terms generated by the quantum field theory form a power series in α. Notice that $\alpha \ll 1$, so we expect successive terms to be smaller and smaller (after renormalization, which will be discussed shortly). However, it is an open question as to whether the series will *really* converge. An argument[18] exists which indicates that it is only an asymptotic series, a series that appears to converge as more and more terms are included in the sum, but eventually diverges. The argument indicates though that we have to include about 137 terms before it begins to diverge. Notice that each extra power of α diminishes the contribution of successive terms by a factor of 10^{-2}, so we are shifting the result by smaller and smaller amounts. Calculations which include about eight such terms have yielded results that agree with extremely precise experiments to many decimal places. In that respect, **this is the most accurate theory ever proposed**.

Quantum field theory has succeeded in removing the infinite sea containing an infinite number of particles, but it still has the same kinds of problems with infinities that arose in Dirac's original formulation. Feynman diagrams are accompanied by detailed rules for calculating the actual transition probabilities, but when these rules are followed, they yield *infinite results* for diagrams with loops. So the correction terms, which we first considered small, now appear infinite. Remember that in Dirac's formulation of relativistic quantum mechanics infinite charge and energy were renormalized away. About half a century ago, an analogous procedure was discovered for quantum field theory. Notice that in Fig. 4.8, diagram (a) has a vertex modified by the inserted virtual photon; this can be shown to contribute to a shifting of the value of the effective charge, since the sum of all such vertex modifications is equivalent to an unmodified vertex with a shifted value of the coupling constant. Diagram (b) in that figure appears to have a modified outgoing electron line, and this can be shown to contribute to a shift of the effective mass of the electron, based on the occurrence of such modifications even for a single (isolated) electron. (Diagrams with the other vertex or the other electron lines modified must be included as well.) Although each of these terms is infinite, they all must be added to the first term (Fig. 4.7). The resulting sum turns out to be treelike, but with the charge and mass of the electron shifted by an infinite amount. "Shifted" from what? Shifted from the original parameters put into the Lagrangian. However, the finite values of these quantities observed in experiments would include the shift, i.e., we "see" only the renormalized quantities, which are the sums of the original parameters and the corresponding shifts. Thus, these renormalized quantities must be finite. This implies that the original parameters (for mass and charge) must have been infinite also. A detailed rigorous method has been devised to replace the

---

18. The argument, originally due to Dyson, shows that negative values of α lead to an unstable vacuum state. This result follows from the fact that negative α would have like charges attracting and opposite charges repelling, with the inevitable result that a state with many e's in one region of space and many $\bar{e}$'s in another region would have lower energy than the vacuum state. If a series converges around $\alpha = 0$, it must converge for negative values, which would not be the case for the unstable vacuum situation. For more information, see *An Introduction to Relativistic Quantum Field Theory* by Schweber.

resulting values of the renormalized charge and mass by their observed values in the Feynman diagrams of each process. It has been proven for QED that this renormalization procedure removes *all* infinities. The observed values (of transition probabilities, etc.) are, of course, finite, and they match the theoretical (renormalized) predictions to the high accuracy alluded to above. This renormalization was achieved independently by Schwinger, Tomanaga, and Feynman in the 1940s. (The innovative diagrammatic methods of Feynman were shown by Dyson[19] to be equivalent to the other formulations of quantum field theory.) It should be pointed out, however, that the observed coupling constants (here, the charge) and masses of particles are not understood, but must be inserted as parameters; the "sweeping under the rug" criticism of this procedure still persists. (See Fig. 4.3.) However, the amazing accuracy of the predictions of QED is convincing evidence of the power of these techniques.

4.5   RANGE OF THE INTERACTION AND THE MASS OF THE FORCE FIELD PARTICLE

The range of the electromagnetic interaction is infinite, and the mass of the field particle responsible for that interaction (the photon) is zero. These two facts are not unrelated. On the contrary, there is a direct connection between them. Similarly, the short range of the weak interaction is directly related to the very large masses (~90 times the proton mass) of the W's and Z (the "cousins" of the photon).

In order to obtain a qualitative understanding of this connection in a quantum field theory with a massive field particle (like a massive "photon"), we shall utilize a semiclassical argument. For simplicity, let us consider (classically) a massive scalar field. We will let this field obey the Klein-Gordon equation, since the corresponding field quanta (particles) are bosons. Suppose there is a point source of this field at the origin, so that the Klein-Gordon equation is modified to become

$$(\partial^\mu \partial_\mu + m^2)\Phi(\vec{r}, t) = g\delta^3(\vec{r}), \tag{4.61}$$

where $\delta^3(\vec{r})$ is the Dirac delta function, which is 0 everywhere except at the point $\vec{r} = 0$ (the origin), where it is ∞. It has the property that $\int \delta^3(\vec{r})\, dr = 1$ for any region which includes the origin.[20] To find the static[21] solution of this equation, we must solve

$$(\nabla^2 - m^2)\Phi(\vec{r}) = -g\delta^3(\vec{r}). \tag{4.62}$$

---

19. The book *Disturbing the Universe* by Freeman Dyson gives an interesting account of the developments and interactions of the physicists involved.

20. $\delta^3(\vec{r})$ is not a true function, but instead is called a *generalized function* or *distribution*. It is obviously the correct expression to use for the density of a unit point charge.

21. "Static" implies that $\partial \Phi / \partial t = 0$.

This looks like the equation for the electrostatic potential due to a point charge at the origin, except that here we have a mass term. Let us solve this equation by utilizing the Fourier transform. Let

$$\Phi(\vec{r}) = \int \chi(\vec{k}) e^{i\vec{k}\cdot\vec{r}} d^3k. \tag{4.63}$$

We can Fourier analyze $\delta^3(\vec{r})$ to obtain

$$\delta^3(\vec{r}) = (2\pi)^{-3} \int e^{i\vec{k}\cdot\vec{r}} d^3k. \tag{4.64}$$

Substituting this into Eq. (4.62) yields

$$\int (-k^2 - m^2)\chi(\vec{k}) e^{i\vec{k}\cdot\vec{r}} d^3k = -g(2\pi)^{-3} \int e^{i\vec{k}\cdot\vec{r}} d^3k \text{ or}$$

$$\int [(-k^2 - m^2)\chi(\vec{k}) + g(2\pi)^{-3}] e^{i\vec{k}\cdot\vec{r}} d^3k = 0. \tag{4.65}$$

An obvious solution[22] for this is obtained by setting the bracket in the integrand equal to zero, to obtain

$$\chi(\vec{k}) = \frac{g}{(2\pi)^3(k^2 + m^2)}. \tag{4.66}$$

Inserting this into Eq. (4.63), we obtain

$$\Phi(\vec{r}) = \frac{1}{(2\pi)^3} \int \frac{g}{k^2 + m^2} e^{i\vec{k}\cdot\vec{r}} d^3k. \tag{4.67}$$

This integral is done in Appendix D and yields

$$\Phi(\vec{r}) = \frac{g}{4\pi r} e^{-mr} = \frac{g}{4\pi r} e^{-(r/R)}, \text{ with } R = \frac{1}{m} \text{ (in natural units).} \tag{4.68}$$

So the field extends beyond the source (at the origin), falling exponentially with a range which is characterized by R. The correspondence principle tells us that the quantum field theory should also produce a range for the interaction of this order of magnitude. From Eq. (4.68), we have $R = m^{-1}$ in natural units, and to convert to *SI* units we must put in the appropriate factors of $\hbar$ and c. This yields

$$R = \frac{\hbar}{mc}, \tag{4.69}$$

the Compton wavelength of that particle which propagates the force.

Let us see what size mass is necessary to account for the range of the weak interaction, which is $R \sim 10^{-18}$ meters. To get the rest energy (or mass) in GeV, we leave it as an exercise for the reader to solve this for (mc^2) and convert the

---

22. due to the "orthogonality" of the $e^{i\vec{k}\cdot\vec{r}}$'s.

answer from Joules to GeV.[23] This yields 200 GeV, and a range twice as long would yield a mass near 100 GeV.[24] We shall learn in Part B that gauge theories have massless field particles (called gauge particles). However, the standard model contains an ingenious breaking of its gauge theory to produce *massive* gauge particles yet preserve the gauge invariance of the Lagrangian as well as renormalizability. The standard model predictions for the masses of the W's and the Z coincide with the experimentally observed values. The predictions were made about a decade before the experiments were done.

PROBLEMS

1. Verify that an equation of the form: $\vec{A} = \vec{B}$ is rotationally invariant, if $\vec{A}$ and $\vec{B}$ rotate like vectors. In other words, show that if $A_i = B_i$, $i = 1,2,3$ then $A'_j = B'_j$, $j = 1,2,3$.

2. Verify that the expression in Eq. (4.4) is an invariant. Hint: Take the same expression in S':

$$(x'^0)^2 - \sum_{j=1}^{3} (x'^j)^2$$

and express the primed quantities in terms of the unprimed ones (Eq. 4.1) to show that it is equal to the same expression in terms of the unprimed quantities. This is analogous to showing that $x'^2 + y'^2 + z'^2 = x^2 + y^2 + z^2$ under rotations, i.e., the magnitude of the vector is unchanged by rotations.

3. Verify Eq. (4.16) for a particle moving in the $+x$ direction with speed w. Hint: In S we have

$$p_x = \frac{mw}{\sqrt{1 - \dfrac{w^2}{c^2}}} \quad \text{and } E = \frac{mc^2}{\sqrt{1 - \dfrac{w^2}{c^2}}}.$$

In S', we call the particle's speed w', and so we have

$$p'_x = \frac{mw'}{\sqrt{1 - \dfrac{w'^2}{c^2}}} \quad \text{and } E' = \frac{mc^2}{\sqrt{1 - \dfrac{w'^2}{c^2}}}.$$

The velocity addition law of special relativity tells us that w' (which is the particle's speed in S') is related to w by

$$w' = \frac{w - v}{1 - \dfrac{vw}{c^2}}, \quad \text{where } v \text{ is the relative speed between frames.}$$

---

23. $mc^2 = hc/R$ and $1.6 \times 10^{-19} \, J = 1$ eV.

24. Yukawa originally proposed this idea for nuclear (strong) interactions, and predicted the mass of the particle to be ~200 MeV. The pion, which was discovered later, seemed to fit the bill ($m_\pi \approx 140$ MeV) and for a long time was thought to be the particle responsible for the strong interaction.

By substituting for w' in the equations for p_x' and E', show that the transformations of Eq. (4.16), expressing p_x' and E' in terms of p_x and E, are correct.

4. In a system of two equal mass (m) particles:
   a. Calculate the invariant four-scalar $(p_1 + p_2)^2$ in terms of E_1, E_2, m, and the angle (θ) between the directions of motion of the particles, where p_1 and p_2 are the four-momenta of the corresponding particles.
   b. Evaluate $(p_1 + p_2)^2$ in the CM frame, starting from the definition of this expression. Compare that result with the general result found in part a, when the latter is used in the CM frame.

5. a. Show that $\{\sigma_i, \sigma_j\} = 2\delta_{ij}$ and $[\sigma_i, \sigma_j] = 2i\epsilon_{ijk}\sigma_k$, where ϵ_{ijk} is the Levi-Civita symbol whose nonzero components are $\epsilon_{123} = \epsilon_{312} = \epsilon_{231} = +1$ and $\epsilon_{132} = \epsilon_{213} = \epsilon_{321} = -1$.
   b. From the commutation and anti-commutation relations of the σ's, show that $(\vec{\sigma}\cdot\vec{a})(\vec{\sigma}\cdot\vec{b}) = \vec{a}\cdot\vec{b} + i\vec{\sigma}\cdot\vec{a} \times \vec{b}$.

The two dimensional (compressed) notation for four-dimensional matrices is very useful in many calculations involving the Dirac equation. You may multiply these matrices as if they were 2×2 matrices even though each entry is itself a 2×2 matrix; it is important, however, to preserve the order in the product of these sub-matrices — $\sigma_x\sigma_y \neq \sigma_y\sigma_x$, as shown in Problem 5(a), etc. In the following problems, keep the 2×2 forms as much as possible.

6. Check that the $\vec{\alpha}$ and β matrices anticommute and that their squares are 1. You may use $\{\sigma_i, \sigma_j\} = 2\delta_{ij}$, which was shown in Problem 5(a).

7. In the two-dimensional (compressed) notation for four-dimensional matrices, we may write the basis states of Eq. (4.34) as

$$\begin{pmatrix} \chi_s \\ 0 \end{pmatrix} \text{ and } \begin{pmatrix} 0 \\ \chi_s \end{pmatrix},$$

for positive energy and negative energy states, respectively, where $s = +$ or $-$ and $\chi_+ = \begin{pmatrix} 1 \\ 0 \end{pmatrix}$ and $\chi_- = \begin{pmatrix} 0 \\ 1 \end{pmatrix}$.

   a. Write H for a particle at rest in this two-dimensional (compressed) notation and, using this notation, show that the basis states are indeed eigenstates of H with energy $\pm m$, respectively.
   b. In the rest system of a spin $\frac{1}{2}$ particle,

$$S = \frac{1}{2}\vec{\sigma}1 = \begin{pmatrix} \frac{1}{2}\vec{\sigma} & 0 \\ 0 & \frac{1}{2}\vec{\sigma} \end{pmatrix}$$

   is the spin operator. Show that this definition agrees with our designation in the text of the spin up and spin down states, respectively, as

$$\chi_+ = \begin{pmatrix} 1 \\ 0 \end{pmatrix} \text{ and } \chi_- = \begin{pmatrix} 0 \\ 1 \end{pmatrix} \text{ in both } \begin{pmatrix} \chi_s \\ 0 \end{pmatrix} \text{ (positive } E\text{)}$$

$$\text{and } \begin{pmatrix} 0 \\ \chi_s \end{pmatrix} \text{ (negative } E\text{) states,}$$

   by finding the effect of $S_\pm$ and S_z.

8. Here we shall investigate how the Dirac states look in a moving frame. If we move in the $-\hat{v}$ direction, **a positive energy state will have momentum $\vec{p} = \gamma m\vec{v}$**. How-

ever, since a positron, which corresponds to a hole in the negative sea, would also have momentum $\vec{p}$, **a negative energy state will have momentum $-\vec{p}$.**

a. Show that

$$u_s(\vec{p}) = N \begin{pmatrix} \chi_s \\ \dfrac{\vec{\sigma} \cdot \vec{p}}{E + m} \chi_s \end{pmatrix},$$

where χ_s is a two-column normalized spinor and $E > 0$, is an eigenstate of $H_{\text{Dirac}} = \vec{\alpha} \cdot \vec{p} + \beta m$ with eigenvalue E.

b. Show that $N = \sqrt{(|E| + m)/2m}$ yields a ψ normalized so that $\int d^3 r \, \psi^\dagger \psi$ is Lorentz invariant. Hint: $d^3 r$ is not an invariant under Lorentz transformations; instead, $d^3 r' = \gamma^{-1} d^3 r$ (see Appendix B). Also note that $\vec{p} = \gamma m \vec{v}$ and $|E| = \gamma m$.

c. Show that

$$v_s(-\vec{p}) = N \begin{pmatrix} \dfrac{\vec{\sigma} \cdot \vec{p}}{|E| + m} \chi_s \\ \chi_s \end{pmatrix}$$

is an eigenstate of $H_{\text{Dirac}} = \vec{\alpha} \cdot (-\vec{p}) + \beta m$ with eigenvalue $-|E|$.

d. Show that the ψ for rest states (Eq. 4.34) can be changed into $u_s(\vec{p})$ and $v_s(-\vec{p})$, respectively, by application of the transformation:

$$B_{\vec{p}/|E|} = \begin{pmatrix} N & N \dfrac{\sigma \cdot \vec{p}}{|E| + m} \\ N \dfrac{\vec{\sigma} \cdot \vec{p}}{|E| + m} & N \end{pmatrix}.$$

$B_{\vec{v}}$ shows us how the particle appears in a frame moving at velocity $-\vec{v}$ with respect to the particle's rest frame, i.e., it boosts the particle from rest to velocity $\vec{v}$. Consequently, it is called a **boost**.

e. Because $\psi^\dagger \psi$ is not invariant, B is not unitary; instead, $B^{-1} = \gamma^0 B^\dagger \gamma^0$. Verify that this is the inverse matrix.

9. a. Show that the γ's obey the anticommutation relations of Eq. (4.39):

$$\{\gamma^\mu, \gamma^\nu\} = 2g_{\mu\nu}.$$

b. Use the boost matrix B for motion along the $-z$ direction, setting $\vec{p} = -p\hat{z}$ to obtain $(\gamma^\mu)' \equiv B^{-1} \gamma^\mu B$. **This corresponds to a passive Lorentz transformation in the $+z$ direction.** Hint: Use the 2×2 compressed notation. For $\mu = i$, the result should be

$$(\gamma^i)' = N^2 \begin{pmatrix} -\{\sigma_i, \sigma_z\} \dfrac{p}{|E| + m} & \sigma_i + \sigma_z \sigma_i \sigma_z \left(\dfrac{p}{|E| + m} \right)^2 \\ -\left(\sigma_i + \sigma_z \sigma_i \sigma_z \left(\dfrac{p}{|E| + m} \right)^2 \right) & \{\sigma_i, \sigma_z\} \dfrac{p}{|E| + m} \end{pmatrix},$$

in which the result for the anticommutators of the Pauli matrices (shown in Problem 5) can be used.

c. Prove that $J^\mu = \bar{\psi} \gamma^\mu \psi$ behaves like a four-vector under this passive Lorentz transformation in the $+z$ direction, i.e., transform the states but use the standard γ^μ in this expression. Remember that $\bar{\psi} \equiv \psi^\dagger \gamma^0$ and use $B^{-1} = \gamma^0 B^\dagger \gamma^0$ along with the result for part b.

    d. Show that $\partial_\mu J^\mu = 0$ follows from the Dirac equation.

10. Find the spin operator for ψ's of moving states.

11. a. Show that $\vec{a} \cdot \vec{b} \times \vec{c}$ changes sign under reflection, where $\vec{a}$, $\vec{b}$, and $\vec{c}$ are ordinary vectors (like the momenta of particles). For simplicity, call the plane, through which we are reflecting, the x-y plane.

    b. Show that an inversion of coordinate system results if we follow a reflection with a rotation by π about the normal to the plane of reflection.

    c. We may take Inv $= \gamma^0$ as the transformation of ψ to an inverted coordinate system. Show that:

        (i) $\bar{\psi}\gamma^i\psi (i = 1,2,3)$ behave like the components of a vector should for inversion,

        (ii) $\bar{\psi}\gamma^5\psi$ changes sign under this transformation of the states, and

        (iii) $\bar{\psi}\gamma^i\gamma^5\psi (i = 1,2,3)$ behave like the components of an axial vector.

12. Consider the Lagrangian $\mathcal{L} = \frac{1}{2}\partial_\nu\phi\partial^\nu\phi - \frac{1}{2}M^2\phi^2$ for a real scalar field ϕ.

    a. Find the canonical momentum $\pi(\vec{r})$.

    b. Show that the commutation relation between $\phi(\vec{r})$ and $\pi(\vec{r}')$ $[\pi(\vec{r}),\phi(\vec{r}')] = -i\delta^3(\vec{r} - \vec{r}')$ at $t = t'$ (of Eq. 4.52) leads to the commutation relations $[d_{\vec{k}}, d_{\vec{k}'}^\dagger] = \delta^3(\vec{k} - \vec{k}')$ of Eq. (4.56). Hint: Note the time independence of the commutation relations. You may need the result $\int e^{i\vec{K}\cdot\vec{r}} d^3r = (2\pi)^3\delta^3(\vec{K})$, where $\delta^3(\vec{K})$ is the three-dimensional Dirac delta function. The Dirac delta function is 0 everywhere except where its argument vanishes, such that $\int F(\vec{K})\delta^3(\vec{K}) d^3K = F(0)$ as long as the region of integration includes the point $\vec{K} = 0$; otherwise, the integral is 0.

13. a. Verify that the boson field operators ϕ and $\vec{A}$ obey the Klein-Gordon equation.

    b. Verify that the fermion field operator ψ obeys the Dirac equation. The properties of u_s and v_s are shown in Problem 8.

14. Verify that $\mathcal{L}_{INT}$ has the correct annihilation and creation operators to mediate the processes shown in Fig. 4.6 (b) pair annihilation, (c) bremsstrahlung, and (d) the photoelectric effect.

15. In Figs. 4.8 and 4.9, we show the distinct one-loop Feynman diagrams for electron-electron scattering. For the purposes of this problem, treat the two electrons as if they are different kinds of charged particles. (Their identity is taken care of in the next stage of the Feynman rules, when amplitudes are combined with appropriate relative signs.) Put arrows on the external lines and

    a. sketch all the one-loop diagrams.

    b. sketch all the two-loop diagrams resulting from the one-loop diagram of Fig. 4.8(a). Of what order in α are they?

    c. Answer (a) and (b) for electron-positron scattering.

16. a. Show that the rudimentary diagrams for pair production, bremsstrahlung, and the photoelectric effect cannot describe a process in empty space. Hint: Consider conservation of energy and momentum (or of four-momentum).

    b. Assume that there is another charged particle present "to take up the recoil" and draw the tree-level diagrams for these processes.

17. a. Show that, because of conservation of energy and momentum (or of four-momentum), pair annihilation cannot produce just one photon, but that two-photon production is allowed.

    b. Draw the tree-level Feynman diagram for pair annihilation into two photons.

    c. Draw a one-loop diagram for the same process, which has a virtual electron loop.

    d. Draw the distinct one-loop diagrams for the same process, which have virtual photons.

18. Find the square of the four-momentum of the exchanged photon (which is the square of the "mass" of this virtual photon) in the tree-level diagram (Fig. 4.7) for electron-electron scattering, in terms of the momenta of the electrons in their CM system (p^{CM}) and the angle of scattering. Hint: Conservation of energy requires that the outgoing momenta have the same magnitude (p^{CM}) as the incoming momenta. See the treatment of the Mandelstam variable t in Appendix E.

19. Construct the tree-level diagram for photon-electron (Compton) scattering and identify the virtual electron in it.

20. Can a photon scatter from a photon in free space? if not, explain. If so, draw the lowest order corresponding Feynman diagrams. What does your answer imply about the principle of linear superposition for electromagnetic waves?

21. From footnote 23, show that the Yukawa particle has mass $\approx$ 200 MeV.

Chapter 5

Groups and Their Representations

Our understanding of the fundamental interactions is based upon mathematical structures called groups. Their representations correspond to the fundamental particles of our physical theories. Thus, it is crucial to gain a familiarity with groups and their representations in order to appreciate the structure and content of the gauge theories which describe nature. Consequently, we shall present an intuitive glance at these mathematical concepts in this chapter.

5.1 GENERAL PROPERTIES AND DEFINITIONS

We have already discussed sets of operations such as translations and rotations in Chapters 2 and 3. Let us begin by considering the set of all rotations about the z axis, which are rotations in two dimensions. In Chapter 3 (Eq. 3.22), we wrote the corresponding operators for active rotations as matrices:

$$\begin{pmatrix} V'_x \\ V'_y \end{pmatrix} = \begin{pmatrix} \cos \theta_1 & -\sin \theta_1 \\ \sin \theta_1 & \cos \theta_1 \end{pmatrix} \begin{pmatrix} V_x \\ V_y \end{pmatrix}, \qquad (5.1)$$

for a rotation by θ_1 about the z axis, for a vector in two dimensions ($V_z = 0$). If we rotate further by θ_2 about the z axis, we have

$$\begin{pmatrix} V''_x \\ V''_y \end{pmatrix} = \begin{pmatrix} \cos \theta_2 & -\sin \theta_2 \\ \sin \theta_2 & \cos \theta_2 \end{pmatrix} \begin{pmatrix} V'_x \\ V'_y \end{pmatrix}. \qquad (5.2)$$

Furthermore, the result of these two successive rotations is a rotation by $\theta = \theta_1 + \theta_2$, as is evident from Fig. 5.1. We have also seen, in Section 3.1, that if we use $V_\pm$ instead of V_x and V_y, we obtain

$$V'_\pm = e^{\pm i\theta} V_\pm$$

for a rotation by θ, in which the operation is represented by a 1×1 matrix (the numbers $e^{\pm i\theta}$). Let us note the following properties possessed by this set of two-dimensional rotations:

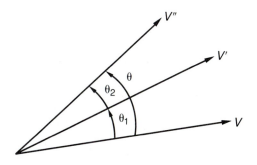

FIGURE 5.1
Two successive (active) rotations about the z axis. Obviously, the net result is a rotation by $\theta = \theta_1 + \theta_2$.

1. Given **any two rotations R_1 and R_2**, their product[1,2] $R_1 R_2$ **is also a rotation** — *closure under group multiplication*.

2. When applying three successive rotations, we can combine (multiply) two successive ones to get the equivalent rotation and then combine (multiply) that by the third one to get the final result, in either of two ways. Both ways give the same result: $(R_1 R_2)R_3 = R_1(R_2 R_3) = R$ — *associative law*.

3. There exists an **identity element I** such that $RI = IR = R$ for all rotations (R) in the set. The identity element is a rotation by $0°$ (which is equivalent to no rotation at all)[3] — *existence of an identity element.*

4. For every rotation R **there exists in the set an inverse rotation R^{-1}** such that $RR^{-1} = R^{-1}R = I$. Here, the inverse to a rotation by θ is a rotation by $-\theta$ — *existence of an inverse*.

Any set of elements (which, in physical contexts, are usually operations) that obey these four rules, under the group multiplication defined for that set, is called a **GROUP**. These rules have been isolated by mathematicians because they have been able to find a rich and diverse set of properties for operations which form a group.[4] Here we shall be interested in only a few of those results.

The two-dimensional rotations we have discussed above have other interesting properties as well:

---

1. Applying two operations in succession is referred to as multiplying them. We do indeed multiply the matrices which represent the two operations to obtain the resulting matrix. See Problem 1.

2. Physicists read this product as corresponding to the application of R_2 first followed by R_1, i.e., from right to left, as if it is operating on a state ket or wavefunction following it.

3. A $360°$ rotation is considered to be the same group element as a $0°$ rotation. In fact, rotations about a given axis, whose angles differ by an integral multiple of $360°$, are considered to be the same rotation.

4. This is a beautiful field of mathematics, which provides insights for a wide range of physical and mathematical concepts.

5. When two successive two-dimensional rotations are performed, the result does not depend on their order: $R_1 R_2 = R_2 R_1$. This may be concluded from Fig. 5.1, from which we infer that $R(\theta_1)R(\theta_2) = R(\theta_1 + \theta_2) = R(\theta_2)R(\theta_1)$ or $[R(\theta_1), R(\theta_2)] = 0$ — *commutativity*, **the group is called an *abelian group*.**

6. The group operations are defined by **a continuous parameter (the angle θ)**, and a **derivative with respect to that parameter** can be performed. More explicitly, we may construct a space of that parameter, each point of which corresponds to an element of the group. (The parameter space (of θ) for the two-dimensional rotation group may be taken as a circle, with θ running from $0°$ to $360°$, the point corresponding to $360°$ coinciding with that for $0°$.) Operations of the group are found to be analytic functions on that parameter space, i.e., all order derivatives are well defined. This is a particular kind of *continuous* or *topological group*, usually called a *Lie group*.

In our consideration of three-dimensional rotations, in Chapters 2 and 3, we saw that there were several (three) Hermitian generators, which did not all commute with one another. However, the commutator of any two of them produced another one of them, viz., $[J_1, J_2] = iJ_3$ (and cyclic). For general Lie groups, the generators are, indeed, closed under commutation, meaning that the commutator of any two of them yields a linear combination of them. This is usually expressed by writing

$$[G_i, G_j] = ic_{ijk} G_k \text{ (sum over } k), \tag{5.4}$$

where G_i are the generators. The c_{ijk} are a set of numbers called the **structure constants** and the Hermiticity of the generators requires that they be **real**. (See Problems 17 through 19.) Mathematicians express the content of Eq. (5.4) by saying that **the generators form a closed Lie algebra**, where commutation replaces the operation of multiplication of ordinary algebra.

It is left to the reader to verify that the structure constants for the rotation group are ± 1 or 0, according to the following rules: the structure constant is $+1$ for ijk being 1 2 3, or obtainable from 1 2 3 by an even number of interchanges; it is -1 for ijk obtainable from 1 2 3 by an odd number of interchanges;[5] and it is 0 if any two indices ijk are the same. Actually, these rules describe the elements of the Levi-Civita symbol ε_{ijk} (where i, j, and k can each be 1, 2, or 3) which are $+1$ when the order of ijk is a cyclic permutation of 1 2 3; are -1 when they are anticyclic;[6] and are zero if any two indices are the same. The non-zero elements are $\varepsilon_{123} = \varepsilon_{312} = \varepsilon_{231} = +1$ and $\varepsilon_{132} = \varepsilon_{213} = \varepsilon_{321} = -1$.

---

5. For example: 132 is obtained from 123 by interchanging the second and third numbers; 312 is obtained from 123 by two interchanges: 123 → 321 if we interchange the first and the third and 321 → 312 when we interchange the second and third entries.

6. Think of a clock face with 1 at the 12 o'clock position, 2 at the 4 o'clock position and 3 at the 8 o'clock position. Starting at any of these three numbers and moving clockwise gives a cyclic order (123, 231 and 312). Moving counterclockwise we obtain the anticyclic orders.

5.2   REPRESENTATIONS

The elements or operations of a group (A_i) can be represented by matrices (M_i) — a matrix corresponding to each element — with the group multiplication of two elements represented by matrix multiplication of the corresponding matrices. (Matrix multiplication obeys the associative property, so that property 2 above is automatically satisfied.) In fact, in group theory mathematicians use the word **"representation"** to mean a **representation by a set of matrices**. For every product of the group elements, $A_i A_j = A_k$, the corresponding matrices must have the analogous relationship, $M_i M_j = M_k$; the matrix multiplication must duplicate the group multiplication. Although each element of the group is represented by a matrix, the matrices corresponding to different elements need not be different. If **each element** of the group corresponds to a **different matrix**, then the representation is called a **faithful representation**.

Every group has the one-dimensional representation (1×1 matrices) $M_i = $ **1, for all i,** which clearly duplicates any group multiplication structure ($1 \cdot 1 = 1$). This is called **the trivial representation** of the group. For physical operations which form a group, we recognize this representation as characterizing the behavior, under group operations, of objects or mathematical forms which are invariant; they are multiplied by 1 by the group operations.

The importance of representations to physical considerations is a result of the fact that they show how our physical constructs are affected by symmetry operations. In this section, we shall consider the representations of the rotations, from which we will learn the properties of representations.

5.2.1   Two-Dimensional Rotations

The abelian group we have used as our example above, rotations in two dimensions, is essentially defined by the 2×2 matrices in Eq. (5.1). Notice that these matrices (R) have the property that

$$R^T R = R R^T = I, \tag{5.5}$$

where R^T is the transpose of R:

$$R = \begin{pmatrix} \cos\theta & -\sin\theta \\ \sin\theta & \cos\theta \end{pmatrix} \text{ and } R^T = \begin{pmatrix} \cos\theta & \sin\theta \\ -\sin\theta & \cos\theta \end{pmatrix},$$

so that $R^{-1} = R^T$. Matrices which obey this property are called **orthogonal matrices**. They leave an orthogonal coordinate system orthogonal and do not change the lengths of vectors, as may be checked by solving Problem 3. Of course, applying any number of such transformations (which is equivalent to multiplying the corresponding matrices) does not change this orthogonality; therefore, the set of **orthogonal 2×2 matrices** form a group which is called (no surprise!) **O(2).** That set of 2×2 matrices is called the **defining representation,** since the group has been determined from and named for this representation. (We shall soon see that groups have many representations.) The 2×2 rotation matrices in Eq. (5.5) are not the only two-dimensional orthogonal matrices; notice that they have the special property that their determinants are $+1$.[7] This "unimodular" property is expressed by calling the matrices "Special" and the two-dimensional rotation group is given the name **SO(2)**.

We have seen in Eq. (3.24) that suitable linear combinations ($V_{\pm}$) of the original components (V_x and V_y) are merely multiplied by $e^{\pm i\theta}$ by rotations, so that, using $V_{\pm}$ as our basis set, we obtain

$$\begin{pmatrix} V'_+ \\ V'_- \end{pmatrix} = \begin{pmatrix} e^{i\theta} & 0 \\ 0 & e^{-i\theta} \end{pmatrix} \begin{pmatrix} V_+ \\ V_- \end{pmatrix}. \tag{5.6}$$

(These matrices are no longer orthogonal, because of the complex nature of the transformation to the new basis.) Notice that we have two 1×1 matrices ($e^{\pm i\theta}$) with 0's in the positions that would mix the V_+ and the V_-. For any θ, this matrix, is in **reduced form**. More generally, a matrix is in reduced form if it is in block diagonal form, i.e., if it has square submatrices along the "diagonal" with 0's in rows and columns which connect any of these submatrices. (For an example see Eq. 5.16.) The fact that, by taking such linear combinations of the components, we obtain a reduced form for all θ is expressed by saying that the two-dimensional representation is **reducible**. We shall see, in later chapters, that particles correspond to **irreducible representations (irreps)** of groups, i.e., **representations whose matrices cannot all be put into block diagonal form by the same choice of basis set**. Therefore, we wish to learn how to find all the irreducible representations of a group.

Consider a reducible representation of a group in reduced form, e.g.,

$$M_i = \begin{pmatrix} a_i & 0 \\ 0 & b_i \end{pmatrix}$$

for all i, where a_i and b_i are smaller submatrices. For the matrix multiplication $M_i M_j = M_k$ (which duplicates the corresponding group multiplication) we find

$$M_i M_j = \begin{pmatrix} a_i a_j & 0 \\ 0 & b_i b_j \end{pmatrix} = \begin{pmatrix} a_k & 0 \\ 0 & b_k \end{pmatrix},$$

so that the products of the submatrices obey $a_i a_j = a_k$; $b_i b_j = b_k$. Consequently each of these submatrix sets is also a representation of the group. The same conclusion can be reached for any number of submatrices strung out along the diagonal; in particular, we can transform to a basis set for which the submatrices appearing along the diagonal cannot be reduced further—they are irreducible representations of the group. Once we have found all of the irreducible representations, it is trivial to build the reducible representations from them. From

---

7. The matrix $\begin{pmatrix} 1 & 0 \\ 0 & -1 \end{pmatrix}$ is an orthogonal matrix (as the reader should verify), which is not included in the set of R (rotation) matrices. Likewise, the product of this matrix with any 2×2 rotation matrix yields an orthogonal matrix whose determinant is -1; hence, the product is not a rotation.

Eq. (5.6), we conclude that $e^{i\theta}$ is a (one-dimensional, hence, irreducible) representation of the two-dimensional rotation group and $e^{-i\theta}$ is another irreducible representation of that group, SO(2). (See Problem 6.)

Let us investigate how we could build other representations of SO(2). In the Lorentz-covariant formulation of classical electrodynamics, which we shall discuss in Chapter 7, the electromagnetic field tensor is defined as $F_{\mu\nu} \equiv \partial_\mu A_\nu - \partial_\nu A_\mu$. This is one example of the construction of an object with two indices from products of components of two vectors and/or vector operators. Such an object is called a second-rank tensor. (If an object were a product of three factors, each a vector, it would be a third-rank tensor, etc.) A second-rank tensor for SO(2) can be built by considering two vectors and taking products of their components, one factor from each vector. Consider vectors $\vec{V}$ and $\vec{W}$ with components V_x, V_y and W_x, W_y respectively. We could construct the second-rank tensor $T_{ij} \equiv V_i W_j$ ($T_{11} \equiv V_x W_x$, etc.; there are four of them). Under rotations, the V_x and V_y mix (in V'_x, V'_y), so that

$$T'_{11} \equiv V'_x W'_x = (\cos\theta V_x - \sin\theta V_y)(\cos\theta W_x - \sin\theta W_y),$$

which is a linear combination of all of the T_{ij}. Since the combinations $V_\pm$ and $W_\pm$ behave more simply under these rotations, let us build our second-rank tensor from them instead. The four components of the tensor are:

$$T_{++} \equiv V_+ W_+, \; T_{+-} \equiv V_+ W_-,$$

$$T_{-+} \equiv V_- W_+, \text{ and } T_{--} \equiv V_- W_-, \tag{5.7}$$

which are linear combinations of the components of T_{ij}. (See Problem 5.) These then form a (basis for a) four-dimensional representation of SO(2):

$$\begin{pmatrix} T'_{++} \\ T'_{+-} \\ T'_{-+} \\ T'_{--} \end{pmatrix} = \begin{pmatrix} e^{2i\theta} & 0 & 0 & 0 \\ 0 & 1 & 0 & 0 \\ 0 & 0 & 1 & 0 \\ 0 & 0 & 0 & e^{-2i\theta} \end{pmatrix} \begin{pmatrix} T_{++} \\ T_{+-} \\ T_{-+} \\ T_{--} \end{pmatrix}, \tag{5.8}$$

since both V_+ and W_+ change by the factor $e^{i\theta}$, and V_- and W_- change by $e^{-i\theta}$. Notice that this is in reduced form. Even if we had used simple products of the x and y components (the T_{ij}), we would call the representation reducible, since suitable linear combinations of the T_{ij} yield a reduced matrix. We can build higher-dimensional representations by using higher-rank tensors, and, by using the $V_\pm$ forms, we obtain matrices in completely reduced form (in fact, containing only 1×1 matrices along the diagonal). We thus generate the representations $e^{in\theta}$, where n is an integer. These are, in fact, *all* the representations of SO(2). (We can imagine changing from θ to an arbitrary parameter α, thus obtaining the representations $e^{ia\alpha}$, where a is any real number.) They are all one-dimensional unitary matrices and if $e^{in\theta}$ are selected as the defining representations, the group would be called **U(1)**. We see here that there is a $1 \leftrightarrow 1$

correspondence between the elements of U(1) and those of SO(2), with identical results for corresponding elements under group multiplication; thus, these two groups correspond to the *same* abstract group. This relationship between two groups is called an **isomorphism** and is expressed by saying that they are **isomorphic**:

$$SO(2) \cong U(1), \tag{5.9}$$

where the symbol $\cong$ stands for "is isomorphic to."

All of the representations we have found here have been reducible to a collection of one-dimensional representations. We have just established that all of the irreducible representations of SO(2) are of the form $e^{in\theta}$. Why are there no irreducible representations of more than one dimension? Can there be irreps of greater dimension for *any* group? The answer to the latter question is yes. When we consider three-dimensional rotations, we shall find many higher-dimensional irreducible representations. However, for SO(2), an abelian group, there are none. In fact it is a property of **abelian groups** that **all** of their **irreducible representations are one-dimensional.**

5.2.2   Three-Dimensional Rotations

Let us now look at another of our favorite groups, namely the rotations in three dimensions. We have already seen in Chapter 3 (using two books, Fig. 3.1), that the order in which successive three-dimensional rotations are performed *is* relevant, so that the group is **nonabelian**, giving us hope of having irreducible representations which are not 1×1 matrices (one-dimensional representations[8]). If we examined the general matrix for an arbitrary rotation of a vector,[9] we would find that it cannot be reduced. It is irreducible! Since it is a 3×3 matrix, we will refer to this irrep as **3**.

Three-dimensional rotations preserve the orthogonality of the coordinate system and the lengths of vectors, and so the corresponding matrices are orthogonal for the same reasons discussed for two-dimensional rotations. (It is straightforward to check that $RR^T = I$ for the general rotation matrix. See Problem 9.) Let us consider Eq. (5.5), for three-dimensional matrices. Taking the determinant of both sides yields

$$\det(RR^T) = \det I = \det \begin{bmatrix} 1 & 0 & 0 \\ 0 & 1 & 0 \\ 0 & 0 & 1 \end{bmatrix} = +1. \tag{5.10}$$

It is a property of determinants that the determinant of a product is equal to the product of the corresponding determinants. Furthermore, it follows from

---

8. The word "dimensional" is being used here to represent the size of the representation. We are still considering rotations in three dimensions.

9. We could use Euler angles to describe the orientation of the new axes in terms of the old.

the definition of the determinant that $\det R^T = \det R$. Consequently, from Eq. (5.10), we find

$$(\det R)^2 = 1, \text{ thus } \det R = \pm 1. \tag{5.11}$$

Starting with

$$I = \begin{pmatrix} 1 & 0 & 0 \\ 0 & 1 & 0 \\ 0 & 0 & 1 \end{pmatrix}, \tag{5.12}$$

noting that $\det I = +1$, let us consider an infinitesimal rotation $d\theta$ about an arbitrarily chosen axis. As $d\theta$ approaches 0, the rotation matrix approaches I continuously, its elements being shifted by terms of order $d\theta\,(O(d\theta))$. Since the determinant is a linear combination of products of those elements, it will be shifted from $+1$ (at most) by terms $O(d\theta)$, which approach 0 as $d\theta \to 0$, i.e., $\det R(d\theta) = 1 + O(d\theta)$. Thus, the determinant for an infinitesimal rotation can differ from $+1$ only by an *infinitesimal* amount. However, we have just seen that the determinant must be ± 1, so that the determinant for an infinitesimal rotation is $+1$. Furthermore, all possible rotations can be reached continuously from I by a succession of infinitesimal rotations, none of which can cause the determinant to jump to -1. Thus **the determinants of the matrices for all rotations are $+1$**, i.e., they are unimodular or "Special". For this reason the **three-dimensional rotation group** is given the name **SO(3). The unimodular 3×3 orthogonal matrices form the defining representation.**

What about those matrices whose determinants are -1? Let us consider a simple example of such a matrix:

$$Inv \equiv \begin{pmatrix} -1 & 0 & 0 \\ 0 & -1 & 0 \\ 0 & 0 & -1 \end{pmatrix}, \tag{5.13}$$

which represents inversion through the origin ($x \to -x$, $y \to -y$, and $z \to -z$). It is equivalent to doing a reflection through the x-y plane ($z \to -z$) followed by a rotation by π about the z axis ($x \to -x$ and $y \to -y$). Since the determinant of the rotation is $+1$, the determinant of the reflection must be -1,[10] so that their product will be equal to the determinant of Inv. Furthermore, Inv treats all directions the same way (Inv $\vec{V} = -\vec{V}$ for all vectors $\vec{V}$), and we could have chosen the reflection through *any* plane followed by the π rotation about its normal to obtain Inv. Thus, **the determinant is -1 for all reflections.** The set of 3×3 matrices, whose determinants are -1, involve reflections and they do not form a group. (Why?)

---

10. This can be checked directly; see Problem 8.

As we did for SO(2), let us now build a second-rank tensor for SO(3) from vectors to see what representation we obtain. We have just learned that the components of a vector **span** a three-dimensional irrep **(3)**. Using $\vec{V}$ and $\vec{W}$, as we did for two-dimensional rotations, we have nine possible products $(V_i W_j)$. Under rotations these nine products mix, and 9×9 matrices are needed to represent the effects of the rotations. We say that these nine components of the second-rank tensor span a nine-dimensional representation of the rotation group. For SO(2), we saw that a suitable choice of linear combinations of the components of the original tensor (T_{ij}) put the representation in reduced form, so that the reducibility of that representation became apparent. Here, we will again choose suitable linear combinations of the products, to see that this 9×9 representation is reducible. First consider $(V_1 W_1 + V_2 W_2 + V_3 W_3)$. How does this behave under rotations? Does it look familiar? It should, for it is $\vec{V} \cdot \vec{W}$, the **scalar** product, which we know does not change under rotations. We will take this product (suitably normalized) as the first component of our nine-component object:

$$T_1 = \tfrac{1}{\sqrt{3}} (V_1 W_1 + V_2 W_2 + V_3 W_3), \tag{5.14}$$

which is invariant to rotations: $T_1' = 1T_1$. Consider next $\vec{V} \times \vec{W}$. It is a **vector** under rotations, whose components mix together. So if we take

$$T_2 = \tfrac{1}{\sqrt{2}} (\vec{V} \times \vec{W})_1 = \tfrac{1}{\sqrt{2}} (V_2 W_3 - V_3 W_2),$$

$$T_3 = \tfrac{1}{\sqrt{2}} (\vec{V} \times \vec{W})_2 = \tfrac{1}{\sqrt{2}} (V_3 W_1 - V_1 W_3), \text{ and}$$

$$T_4 = \tfrac{1}{\sqrt{2}} (\vec{V} \times \vec{W})_3 = \tfrac{1}{\sqrt{2}} (V_1 W_2 - V_2 W_1), \tag{5.15}$$

we know that T_2, T_3, and T_4 mix only with each other under rotations, and the resulting 3×3 matrix is irreducible. Thus the matrices of our 9×9 representation are all reduced, so far, to the form: [11]

$$\begin{bmatrix}
1 & 0 & 0 & 0 & 0 & 0 & 0 & 0 & 0 \\
0 & X & X & X & 0 & 0 & 0 & 0 & 0 \\
0 & X & X & X & 0 & 0 & 0 & 0 & 0 \\
0 & X & X & X & 0 & 0 & 0 & 0 & 0 \\
0 & 0 & 0 & 0 & X & X & X & X & X \\
0 & 0 & 0 & 0 & X & X & X & X & X \\
0 & 0 & 0 & 0 & X & X & X & X & X \\
0 & 0 & 0 & 0 & X & X & X & X & X \\
0 & 0 & 0 & 0 & X & X & X & X & X
\end{bmatrix}, \tag{5.16}$$

---

11. The mixings we have talked about occur when we transform from system S' to S (using the transpose matrix) as well as S to S', so that zeros appear at both row and column locations in Eq. (5.16).

where the X's represent entries, which depend upon the rotation being represented, and are not all equal. The first row and column refer to $\frac{1}{\sqrt{3}} \vec{V} \cdot \vec{W}$ (i.e., T_1) and so show no mixing with the other components; the next three rows and columns refer to the components of $\frac{1}{\sqrt{2}} \vec{V} \times \vec{W}$ (i.e., T_2, T_3, and T_4) and mix only among themselves, as indicated by the X's and 0's; and the last five have not yet been discussed.

Notice that T_2, T_3, and T_4 are antisymmetric to the interchange of $\vec{V}$ and $\vec{W}$, i.e., $\vec{V} \times \vec{W} = -\vec{W} \times \vec{V}$. Let us instead take $+$ signs between the two terms in each line in Eq. (5.15), i.e., consider the symmetric combinations: *Repeated*

Notice that T_2, T_3, and T_4 are antisymmetric to the interchange of $\vec{V}$ and $\vec{W}$, i.e., $\vec{V} \times \vec{W} = -\vec{W} \times \vec{V}$. Let us instead take $+$ signs between the two terms in each line in Eq. (5.15), i.e., consider the symmetric combinations:

$$T_5 = \tfrac{1}{\sqrt{2}}(V_2 W_3 + V_3 W_2),$$

$$T_6 = \tfrac{1}{\sqrt{2}}(V_3 W_1 + V_1 W_3), \text{ and}$$

$$T_7 = \tfrac{1}{\sqrt{2}}(V_1 W_2 + V_2 W_1). \tag{5.17}$$

(They are independent of the first four T_i.) T_5, T_6, and T_7 were constructed by analogy to the corresponding antisymmetric products, but we can write three more symmetric products, namely, $V_1 W_1$, $V_2 W_2$, and $V_3 W_3$. These three products are not independent of T_1 (the scalar product, which is their sum), but any two of them are. We can, for instance, take

$$T_8 = \tfrac{1}{\sqrt{2}}(V_1 W_1 - V_2 W_2) \text{ and}$$

$$T_9 = \tfrac{1}{\sqrt{6}}(V_1 W_1 + V_2 W_2 - 2V_3 W_3), \tag{5.18}$$

where T_8 and T_9 have been constructed to be orthogonal to T_1 (considering $V_1 W_1$, $V_2 W_2$, and $V_3 W_3$ to be mutually orthogonal). Since we have already used T_1 (the scalar product, which does not mix at all), only these five independent symmetric products ($T_5 \rightarrow T_9$) remain. These span the 5×5 sub-matrix shown in our 9×9 matrix above. It can be shown (when this is constructed in detail) that the 5×5 matrix is irreducible. Thus the 9×9 representation is reducible to a **1**, a **3**, and a **5** (all irreducible representations). We already knew about the existence of scalars (the **1**), and vectors (the **3**), but we have found a higher dimensional irrep here (the **5**). By constructing higher- and higher-rank tensors and then examining them (as we have done here for the second-rank tensor), we always find a new irreducible representation of higher dimension. Furthermore, it can be proven that they are always of odd dimension.

Recall, from Chapter 3, that a vector behaves like an $\ell = 1$ object under rotations. We also learned there that rotation operators can be written as $e^{i\vec{L} \cdot \vec{\theta}}$ (ignoring spin for now) and, for any given ℓ, the dimension of the matrix representing this operator is $(2\ell + 1)$. A scalar behaves like an $\ell = 0$ object (which has only one, invariant, member). The reader may have guessed by now that the five (symmetric) components which mix together irreducibly behave like an $\ell = 2$ multiplet. This is in fact true, and we can generate all the possible ℓ's this way using tensors of ever higher rank. The irreducible representations of SO(3) *are* the ℓ-multiplets; thus, ℓ is the appropriate quantum number to label the states of any rotationally invariant system (when spin is not present). Here we have found that combining two three-dimensional irreps produces a one-dimensional irrep plus a three-dimensional irrep plus a five-dimensional irrep, viz.:

$$3 \times 3 = 1 + 3 + 5,$$

which is the same as $(\ell = 1) \times (\ell = 1) = (\ell = 0) + (\ell = 1) + (\ell = 2),$ (5.19)

a familiar result from atomic physics. (See Eq. 3.46.) It has been shown that *all* the representations of SO(3), the three-dimensional rotation group, can be obtained in this way; therefore, they are all odd-dimensional. (Note that for $(2\ell + 1)$ odd, ℓ must be an integer, so spin is not yet incorporated.)

5.2.3   Spin

Now let us consider spin. We have learned that the world has objects which behave like spinors ($j = \frac{1}{2}$). Under rotations, these are operated upon by a set of 2×2 matrices $e^{i\vec{J} \cdot \vec{\theta}}$, where $\vec{J} = \frac{1}{2}\vec{\sigma}$, and the $\vec{\sigma}$ are the Pauli matrices. Furthermore, we have seen in Chapter 3 (Eq. 3.44) that two of these 2×2 matrices correspond to each rotation. These 2×2 matrices, therefore, do not form a representation of the rotation group, but instead they form (or may be taken as the defining representation of) a group with a 2 to 1 correspondence with the rotation group. This is called a **homomorphism**. The matrices $e^{i\frac{1}{2}\vec{\sigma} \cdot \vec{\theta}}$ are two-dimensional and unitary. Let us take the direction of the vector $\vec{\theta}$ as our new z direction (i.e., $\hat{z}' = \hat{\theta}$). In this new coordinate system the matrix reads $e^{iJ'_z\theta}$. We can diagonalize J'_z so that we obtain $e^{im\theta}$ (with $m = \pm\frac{1}{2}$), and the matrix is

$$e^{iJ'_z\theta} = \begin{pmatrix} e^{i(\theta/2)} & 0 \\ 0 & e^{-i(\theta/2)} \end{pmatrix}.$$ (5.20)

The determinant of this matrix is

$$\det e^{iJ'_z\theta} = e^{i(\theta/2)} e^{-i(\theta/2)} = e^{i(\frac{1}{2} + \frac{-1}{2})\theta} = +1,$$ (5.21)

for any angle θ.[12] No new choice of the basis vectors (by a unitary transformation, as discussed in Chapter 2) will change the determinant. Consequently, for any rotation, $e^{i\vec{J} \cdot \vec{\theta}}$ is a unimodular matrix.[13] This **set of 2 $\times$ 2 unitary unimodular matrices** is taken as the defining representation of this group which is then called **SU(2)**. As we have said above, this group is homomorphic with SO(3), and it is a two to one homomorphism. SU(2) has both even- and odd-dimensional irreps. Very often people refer to the even-dimensional irreps (corresponding to $j = \frac{1}{2}, \frac{3}{2}, \dots$) as **spinor representations of the rotation group**. Actually, they are not representations of SO(3) at all, but are representations of SU(2).

---

12. This is true for higher-dimensional representations as well: $\det e^{iJ'_z\theta} = e^{i(\Sigma m)\theta} = +1$ since exponents add and $\Sigma m = 0$.

13. Alternatively, using the matrix for $e^{i(1/2)\vec{\sigma} \cdot \vec{\theta}}$ constructed in Chapter 3, Problem 15, we can check directly that its determinant is $+1$.

The notation for the three Pauli matrices $(\sigma_x, \sigma_y, \sigma_z)$ seems to indicate that they form a vector under rotations. This is indeed the case, as shown in Chapter 3, Problem 14. They undergo a similarity transformation, and when the resulting matrices are written in terms of the Pauli matrices, they are seen to be the correct mixture of those components. In fact, it can be proven (using the commutation relations) that the *generators* of rotations and of SU(2) (J_x, J_y, and J_z) actually rotate like the components of a vector. (See Chapter 3, Problem 6.) Their behavior under rotations is described by the same set of 3×3 matrices that describe the behavior of a vector (as shown in Chapter 3, Problems 6 and 9), so that **the generators span a three-dimensional representation of the rotation group**.

5.3   THE GROUP SU(3)

We shall soon see that groups larger than those considered thus far occur in our physical theories. The group on which quantum chromodynamics is built is **SU(3)**, defined by the unitary unimodular three-dimensional matrices.

We shall use what we have learned about SU(2) to understand the structure of this group. The generators must be Hermitian, as we learned in Chapter 2. We also saw in Eq. (5.21) and the associated footnote that for the determinant to be $+1$, the sum of the diagonal elements of the diagonalized generator must be 0. This sum is called the **trace** of the matrix, and its value is unchanged by the change of basis vectors. Consequently, by choosing different basis sets, we can diagonalize each of the generators in turn to find that **each generator must be "traceless."**[14] The Pauli matrices σ_x, σ_y, and σ_z span the space of all 2×2 Hermitian traceless matrices.[15] That is why using them as generators yields *all* the unitary unimodular 2×2 matrices that make up SU(2). Here we shall present the argument for the completeness of the generators of SU(3) and leave it to the reader (in Problem 14) to show that the Pauli matrices span the space of the generators of SU(2). (Try it after reading through the SU(3) discussion.) For SU(3), we must find a complete basis set for the 3×3 Hermitian traceless matrices. We can use the Pauli matrices for our inspiration. Let us write matrices like them that mix only the 1 and 2 components:

$$\lambda_1 = \begin{bmatrix} 0 & 1 & 0 \\ 1 & 0 & 0 \\ 0 & 0 & 0 \end{bmatrix}, \lambda_2 = \begin{bmatrix} 0 & -i & 0 \\ i & 0 & 0 \\ 0 & 0 & 0 \end{bmatrix}, \text{ and } \lambda_3 = \begin{bmatrix} 1 & 0 & 0 \\ 0 & -1 & 0 \\ 0 & 0 & 0 \end{bmatrix}. \quad (5.22)$$

---

14. The trace is equal to 0.

15. This means that all 2×2 Hermitian traceless matrices can be written as real linear combinations of the Pauli matrices. If this idea seems a bit obscure, consider the set of all 2×2 real matrices. The most general matrix can be written $\begin{pmatrix} a & b \\ c & d \end{pmatrix}$, where a, b, c, and d are real numbers. This general matrix can be written as a sum: $\begin{pmatrix} a & b \\ c & d \end{pmatrix} = a\begin{pmatrix} 1 & 0 \\ 0 & 0 \end{pmatrix} + b\begin{pmatrix} 0 & 1 \\ 0 & 0 \end{pmatrix} + c\begin{pmatrix} 0 & 0 \\ 1 & 0 \end{pmatrix} + d\begin{pmatrix} 0 & 0 \\ 0 & 1 \end{pmatrix}$, so that the independent matrices $\begin{pmatrix} 1 & 0 \\ 0 & 0 \end{pmatrix}$, $\begin{pmatrix} 0 & 1 \\ 0 & 0 \end{pmatrix}$, $\begin{pmatrix} 0 & 0 \\ 1 & 0 \end{pmatrix}$, and $\begin{pmatrix} 0 & 0 \\ 0 & 1 \end{pmatrix}$ span the space of 2×2 real matrices.

These have the Pauli matrices in the upper-left 2×2 matrix, surrounded by 0's. We could, instead, mix the first with the third rows and columns to write

$$\lambda_4 = \begin{pmatrix} 0 & 0 & 1 \\ 0 & 0 & 0 \\ 1 & 0 & 0 \end{pmatrix} \text{ and } \lambda_5 = \begin{pmatrix} 0 & 0 & -i \\ 0 & 0 & 0 \\ i & 0 & 0 \end{pmatrix}, \tag{5.23}$$

for the nondiagonal ones, and mixing the second with the third rows (and columns) we obtain

$$\lambda_6 = \begin{pmatrix} 0 & 0 & 0 \\ 0 & 0 & 1 \\ 0 & 1 & 0 \end{pmatrix} \text{ and } \lambda_7 = \begin{pmatrix} 0 & 0 & 0 \\ 0 & 0 & -i \\ 0 & i & 0 \end{pmatrix}. \tag{5.24}$$

Let us now write the corresponding diagonal matrices for Eqs. (5.23) and (5.24):

$$\lambda_? = \begin{pmatrix} 1 & 0 & 0 \\ 0 & 0 & 0 \\ 0 & 0 & -1 \end{pmatrix} \text{ and } \lambda_{??} = \begin{pmatrix} 0 & 0 & 0 \\ 0 & 1 & 0 \\ 0 & 0 & -1 \end{pmatrix},$$

where we have given them strange names. (That is because these are not the matrices we will use.)

How many such independent matrices do we expect? There are six off-diagonal elements in a 3×3 matrix. However, only three of them are independent in a generator, since it must be Hermitian, so that the transpose of any non-zero element must be its complex conjugate. For a basis set, we can choose one of these three to be 1 and the other two to be 0, with zeros down the diagonal. The three different choices give us λ_1, λ_4, and λ_6, shown above. Suppose instead we had chosen i to be the non-zero element. The transpose of that non-zero element must be $-i$, so these matrices are independent of those just discussed. (They are not just i times the previous ones.) We obtain λ_2, λ_5, and λ_7.

Now let us discuss the basis matrices corresponding to diagonal elements. We might start by putting a 1 in any one of the three possible diagonal positions with 0 everywhere else. This is not satisfactory, however, because the resulting three matrices are not traceless. Tracelessness demands that the sum of the three diagonal elements be 0; thus, only two diagonal elements may be chosen independently, and this yields only two independent diagonal matrices. Furthermore, we cannot use i on the diagonal because such a matrix would not be Hermitian. Thus, altogether there should be two such diagonal matrices. But we have found three: λ_3, $\lambda_?$, and $\lambda_{??}$. There is no inconsistency, however. This merely means that those three diagonal matrices cannot be independent. The reader can check that

$$\lambda_3 - \lambda_? + \lambda_{??} = 0.$$

Instead of $\lambda_?$ and $\lambda_{??}$, we use

$$\lambda_8 = \frac{1}{\sqrt{3}} \begin{bmatrix} 1 & 0 & 0 \\ 0 & 1 & 0 \\ 0 & 0 & -2 \end{bmatrix}, \tag{5.25}$$

which is proportional to $\lambda_? + \lambda_{??}$.[16] As in SU(2), the generators are defined as $G_i = \frac{1}{2}\lambda_i$.

In our discussion of **SU(2)**, we pointed out that **the generators behave like a vector (a 3)**[17] under rotations, i.e., the σ_i mix with each other just like the components of a vector. Thus, they span an irreducible representation of the group. In general, the representation that **the generators form** is called **the adjoint representation**. The adjoint representation of SU(2) is the **3**. We have found *eight* **independent generators for SU(3)** which, it can be shown (see Problem 21), mix among themselves under SU(3) transformations. We conclude that these span an eight-dimensional irreducible representation. **The adjoint representation of SU(3) is an 8.**

So, for SU(3) we have the **3** which is the defining representation and the **8** which is the adjoint representation. For any group, we can imagine objects which are invariant, i.e., none of them mix at all; each of those, of course, span a one-dimensional representation—the **1**. We can build all the irreducible representations of SU(3) by constructing tensors in an analogous manner to that used for SO(3).[18]

Table 5.1 summarizes the groups we have discussed, so far, in this chapter.

5.4   DISCRETE (OR FINITE) GROUPS

The groups we have discussed so far are continuous groups, since we can move within the group continuously from one element (or operation) to another $(R(\vec{\theta}_1) \rightarrow R(\vec{\theta}_2)$ for the rotation group) by the continuous variation of parameters $(\vec{\theta})$ which define the group elements. We now turn our attention briefly to groups with a finite number of elements. The elements of the group are obviously discrete, since a continuous parameter would define a continuum of elements, consequently an infinite number of them.

We shall be interested in the simplest, nontrivial, discrete group, namely **the group with only two elements.**[19] Since one element must be the identity, let

---

16. The proportionality constant $\frac{1}{\sqrt{3}}$ is chosen so that $Tr(\lambda)^2 = 2$ for all the λ's, which is convenient for calculations.

17. Note that the **3** of SU(2) is different from the **3** of SU(3).

18. We use the **3** and a representation conjugate to it (the $\bar{3}$, which we shall discuss in Chapter 6). This method is applicable to a large class of Lie groups.

19. There does exist a trivial group with only one element, and that must be the identity element (I), with the multiplication rule $II = I$.

TABLE 5.1
Summary of Continuous Groups

Group	Irreducible Representations	Representation of Generators — Adjoint Representation
U(1) ≅ SO(2) (two-dimensional rotations)	$e^{ia\alpha}$, for any a, all one-dimensional	one-dimensional (to be discussed in particular examples)
SO(3) (three-dimensional rotations)	$\ell = 0$ (scalar), $\ell = 1$ (vector), $\ell = 2, \ldots, \ell$ **1** **3** **5** **(2ℓ + 1)** D	$3(L_x, L_y, L_z)$
SU(2)	$j = 0, \frac{1}{2}, 1, \frac{3}{2}, 2, \ldots, j$ **1 2 3 4 5 (2j + 1)** D	$3(J_x, J_y, J_z)$
SU(3)	**1 3 8** D	$8(G_i = \frac{1}{2}\lambda_i, i = 1, \ldots, 8)$

Notes: D means "Defining Representation."
    The **1**s and $a = 0$ for U(1) are invariants.

us call the two elements **I and P**. In order to obey the four properties which define a group, the products of these elements must be

$$II = I, \; IP = P, \; PI = P, \text{ and } PP = I.$$

The result of the last product follows from the fact the P^{-1} must be in the group, but there is no other possible inverse besides P itself. A physical symmetry group whose operations follow these rules is the set: reflection in a plane (corresponding to P) and no reflection (corresponding to I), since two successive reflections (PP) is equivalent to no reflection (I). Hence, this group is often referred to as the reflection group. We shall see that other physical groups also correspond to this abstract group.

For discrete groups, it is useful and usual to display their multiplication rules in a **multiplication table**. The entries in Table 5.2 are the result of the product (row element)·(column element). This is the **multiplication table for the reflection group**. In Problems 22 and 23, we will prove and use the fact that for any finite group, **each element of the group appears once in each row (column) of the multiplication table**.

TABLE 5.2
Multiplication Table for the Reflection Group

	I	P
I	I	P
P	P	I

PROBLEMS

1. Show that
$$\begin{pmatrix} V_x'' \\ V_y'' \end{pmatrix} = \begin{pmatrix} \cos(\theta_1 + \theta_2) & -\sin(\theta_1 + \theta_2) \\ \sin(\theta_1 + \theta_2) & \cos(\theta_1 + \theta_2) \end{pmatrix} \begin{pmatrix} V_x \\ V_y \end{pmatrix},$$
   by multiplying the matrices appearing in Eqs. (5.1) and (5.2).

2. Prove that a faithful representation of a nonabelian group cannot be one-dimensional.

3. In order to apply the 3×3 rotation matrix to a vector, we write the vector as a column:
$$V = \begin{bmatrix} V_x \\ V_y \\ V_z \end{bmatrix}.$$

   The dot product of two vectors $\vec{V} \cdot \vec{W}$ becomes the matrix product $V^T W$ or $W^T V$. Consider the unit vectors
$$\hat{x} = \begin{bmatrix} 1 \\ 0 \\ 0 \end{bmatrix}, \hat{y} = \begin{bmatrix} 0 \\ 1 \\ 0 \end{bmatrix}, \text{ and } \hat{z} = \begin{bmatrix} 0 \\ 0 \\ 1 \end{bmatrix}.$$

   Show that the orthogonality property of all rotation matrices, viz., $RR^T = R^T R = 1$, implies that $(\hat{x})' = R\hat{x}$, $(\hat{y})' = R\hat{y}$, and $(\hat{z})' = R\hat{z}$ remain mutually orthogonal and do not change in length.

4. Find the 4×4 matrices which represent the two-dimensional rotations when $T_{ij} \equiv V_i W_j$ ($i, j = 1, 2$) are used as the basis set.

5. a. Express the $T_{\pm\pm}$ in terms of the T_{ij} and show that the transformation from the T_{ij} to the $T_{\pm\pm}$ can be accomplished by a unitary matrix.
   b. Use the unitary matrix obtained in (a) to transform the 4×4 matrices of Problem 4. The result should be the matrices shown in Eq. (5.8).

6. a. Prove directly that $e^{in\theta}$ is a representation of the group of two-dimensional rotations (about a fixed axis).
   b. For what value(s) of n are the representations faithful?

7. a. Prove directly that $e^{i\vec{w} \cdot \vec{a}}$ is a representation of the group of translations (where $\vec{a}$ is the translation) for any real vector $\vec{w}$.
   b. For a particle, what does $\vec{w}$ correspond to?

8. a. Construct the 3×3 matrix for the reflection of a vector through the x-y plane (call it Ref_{xy}) and calculate its determinant.
   b. Explain how the reflection through any plane is related to Ref_{xy} by a matrix of the form $R^{-1}(Ref_{xy})R$ by explaining what R should be. Show, from its form, that the determinant of the matrix for any reflection is -1.
   c. Prove that the set of 3×3 matrices, whose determinants are -1, do not form a group. Which of the four requirements of a group are violated?

9. A general rotation can be obtained by performing three successive rotations: $R = R_z(\gamma) R_y(\beta) R_z(\alpha)$, where α, β and γ are arbitrary angles, usually called Euler angles.

a. Write the 3×3 matrix for $R_y(\beta)$ explicitly and verify that it is an orthogonal, unimodular matrix.

b. From the fact that the 3×3 matrix for R_z is also orthogonal and unimodular, prove that the 3×3 matrix for a general rotation R is an orthogonal, unimodular matrix, without explicitly doing the matrix multiplication to obtain the resultant matrix.

10. Construct the 9×9 matrix corresponding to a rotation about the z axis by angle θ, which represents the behavior of the second-rank tensor $T_{ij} = V_i W_j$ with $i,j = 1,2,3$. Notice that it is not in reduced form. Nevertheless, this nine-dimensional representation of the rotation group is reducible, as shown in the text. (See Problem 11.)

11. a. Find the transformation (U) from $T_{ij} = V_i W_j$ to the set $T_1, \ldots, T_9$ defined in the text:

$$\begin{pmatrix} T_1 \\ \vdots \\ \vdots \\ \vdots \\ T_9 \end{pmatrix} = U \begin{pmatrix} T_{11} \\ T_{12} \\ \vdots \\ \vdots \\ T_{33} \end{pmatrix}.$$

b. Show that U is unitary. Note that the orthonormality of the rows, treated as if they were nine-dimensional vectors, automatically guarantees it.

12. a. From the behavior of $\vec{V}$ and $\vec{W}$ under rotations (by ϕ) about the z axis, find the set of matrices (one for each value of ϕ) which represent these rotations using T_2, T_3 and T_4, as the basis. The latter span a three-dimensional irreducible representation of the rotation group, and the matrices are the 3×3 matrices for rotations about the z direction.

b. Find the matrices which represent these rotations using $T_5, \ldots, T_9$, as the basis. The latter span a five-dimensional irreducible representation of the rotation group, and the matrices are the 5×5 matrices for rotations about the z direction.

c. From the results in parts a and b, construct the matrices (one for each value of ϕ) which represent these rotations using $T_1, \ldots, T_9$ as the basis. These are 9×9 matrices in reduced form, which represent rotations by ϕ about the z axis. They are explicit examples of rotation matrices of the form shown in Eq. (5.16).

13. Show that $\det e^{iG\alpha} = +1$ (for arbitrary α) if and only if $TrG = 0$. You may assume that the eigenstates of G form a complete orthonormal set, so that they may be used as a basis set.

14. Show that σ_x, σ_y and σ_z span the "space" of (or can serve as a basis for) the 2×2 Hermitian traceless matrices.

15. Write out the matrix $e^{i(\frac{1}{2}\vec{\sigma} \cdot \hat{n}\theta)}$ (see Chapter 3, Problem 15) and show directly that its determinant is $+1$.

16. a. Show that the structure constants for the three-dimensional rotation group are the elements of the Levi-Civita symbol ϵ_{ijk}, i.e., show that:

$$[J_i, J_j] = i\varepsilon_{ijk} J_k \text{ (summed over } k).$$

b. Show that ε_{ijk} is antisymmetric to the interchange of any two indices, e.g., $\varepsilon_{231} = -\varepsilon_{213}$.

c. Show that $(\vec{A} \times \vec{B})_i = \varepsilon_{ijk} A_j B_k$ (summed over repeated indices).

17. a. For the generators $G_i = \frac{1}{2}\lambda_i$, $i = 1,2,\ldots,8$ of SU(3), find the commutation relations $[G_1, G_4]$ and $[G_4, G_5]$ and express them as linear combinations of the generators G_i.

  b. Show that the fact that the G_i span the space of Hermitian traceless 3×3 matrices implies that $[G_i, G_j] = ic_{ijk}G_k$ (sum over k), with real coefficients c_{ijk}. Hint: Prove that the commutators are anti-Hermitian ($A^\dagger = -A$), or that i times the commutators are Hermitian, and that they are traceless. The c_{ijk} are the structure constants of the Lie group.

  c. Show that the structure constants c_{ijk} are antisymmetric to the interchange of the first two indices.

18. a. Show that the structure constants for any Lie group are antisymmetric to an interchange of the first two indices. Recall that they are defined by the equation

$$[G_i, G_j] = ic_{ijk}G_k \text{ (sum over } k).$$

  b. For SU(3), show that the λ_i matrices obey $Tr(\lambda_i\lambda_j) = 2\delta_{ij}$. For all our symmetry (Lie) groups, the λ_i can be chosen so that this is true.

  c. Prove that $Tr(ABC) = Tr(CAB) = Tr(BCA)$ for any three $n \times n$ matrices A, B, and C.

  d. Show that the structure constants are antisymmetric to an interchange of the second and third indices.

  Hint: To show the antisymmetry to an interchange of the second and third indices, consider the matrix representations of the generators $G_i = \lambda_i/2$. Substitution into the commutator yields

$$Tr(\lambda_i[\lambda_j, \lambda_k]) = Tr(\lambda_i 2c_{jkl}\lambda_l) = 4c_{jki}.$$

  Use part c to show that $Tr(\lambda_i[\lambda_j, \lambda_k]) = -Tr(\lambda_k[\lambda_j, \lambda_i]) = -4c_{jik}$.

  e. Using the antisymmetries already proven, show that the structure constants are antisymmetric to the interchange of *any* two indices.

19. a. Show that the generators of SU(n) are Hermitian and traceless. You may assume that the eigenstates of any generator form a complete orthonormal set.

  b. From this property, find the number of independent generators. They must span the space of $n \times n$ Hermitian traceless matrices.

  c. Prove that the commutators of the generators may then be written: $[G_i, G_j] = ic_{ijk}G_k$ (sum over k), with real coefficients c_{ijk}. Hint: Prove that the commutators are anti-Hermitian ($A^\dagger = -A$), or that i times the commutators are Hermitian, and that they are traceless. The c_{ijk} are the structure constants of the Lie group.

  d, e, and f. Redo parts a, b, and c for the group SO(n) with real entries. Hint: For a real basis, the generators must be pure imaginary so that the transformations are real. Thus SO(n) is a subgroup of SU(n). For example, the generator for SO(2) is σ_y and

$$e^{i\sigma_y\theta} = \begin{pmatrix} \cos\theta & \sin\theta \\ -\sin\theta & \cos\theta \end{pmatrix}.$$

20. a. Verify the relationship: $[[A,B],C] + [[C,A],B] + [[B,C],A] = 0$, which is known as the **Jacobi identity**.

  b. Show that the Jacobi identity yields $\sum_d (c_{abd}c_{dce} + c_{cad}c_{dbe} + c_{bcd}c_{dae}) = 0$, for the structure constants. Hint: Replace A by G_a, etc., in part a.

21. a. Consider the infinitesimal SU(3) transformation $U = e^{i\lambda_1\alpha} \approx 1 + i\lambda_1\alpha$. Calculate the quantities $U^\dagger\lambda_i U$ for $i = 1, \ldots, 8$, to order α, expressing them as linear combinations of the λ_i. (It is noteworthy that the coefficients are related to structure constants of the group.)

b. Use the $\lambda_i (i = 1, \ldots, 8)$ as the basis to construct the 8×8 matrix (M) which represents this transformation, viz., $(\lambda_i)' = M\lambda_i$. The M is a matrix in the adjoint representation of SU(3), which the λ_i form.

22. The Ath row of the multiplication table of a finite group is made up of the elements AX, where X runs through all of the elements of the group.

   a. Show that no two elements in a row can be the same, i.e., show that $AX \neq AY$ for $X \neq Y$. Hint: Suppose $AX = AY$.

   b. Show that this implies that every element appears once and only once in each row. The Ath column of the multiplication table of a group is made up of the elements XA, where X runs through all of the elements of the group.

   c. Show that no two elements in a column can be the same, i.e., show that $XA \neq YA$ for $X \neq Y$.

   d. Show that this implies that every element appears once and only once in each column.

23. a. Consider a group which contains only three elements. Show that there is only one possibility for the abstract group by constructing its multiplication table.

   b. Consider the set of rotations about a fixed axis by angles 0, $2\pi/3$, and $4\pi/3$ radians. Show that they form a group and construct the corresponding multiplication table. How does it compare to the multiplication table constructed in part a?

   c. Consider the cyclic group I, a, a^2 where $a^3 = I$. How does this group compare with the group in part b?

24. a. Construct the multiplication table for the cyclic four group: I, a, a^2, a^3, where $a^4 = I$.

   b. Find a different group of four elements with a fundamentally different multiplication table, i.e., not one that can be found from the first by reordering the elements.

Chapter 6

"Internal" Symmetries and
Conserved Numbers

In this chapter, we define and discuss internal symmetries and show their connection to conserved numbers. We then proceed to show how these lead to the classification of the observed particles into multiplets as well as an understanding of their quark content. The experimental evidence and their implications for the standard model will be discussed in more detail in Chapter 15. To appreciate the excitement and wonder which accompanied those discoveries, the reader is strongly urged to read Part C of this book.

6.1 (STRONG) ISOSPIN

During the study of nuclear physics, in the 1930s and 1940s, it was realized that the nuclear force between pairs of constituents of the nucleus, viz., protons and neutrons, did not depend on the *identity* of the interacting particles. In other words, the strong interaction between two protons, that between two neutrons, and the interaction between a proton and a neutron were essentially the same. This was referred to as the charge independence of nuclear forces.[1] Considering the near equality of the mass of the proton (938.3 MeV) and the mass of the neutron (939.6 MeV), which differ by less than 0.2 percent, it was thought possible that this difference was an electromagnetic effect (a small finite remainder after mass renormalization) due to their charge difference.

An imaginative analogy with space-time transformations was created, which accounted for the charge independence of nuclear forces and near equality of neutron and proton masses in a natural way. It considered the proton and the neutron to be two components of the *same* particle, just as an electron with spin

---

1. It was detected and verified by the study of mirror nuclei (interchange the number of protons with the number of neutrons, e.g., Li^7 and Be^7) and adjacent isobars such as O^{14}, N^{14}, and C^{14}.

up is the same particle as an electron with spin down. Of course, unless we can say something more definite about this n-p relationship it is an empty one, which merely expresses these observations in another way. Recall that rotations in space mix the spin up and spin down states in a very definite way, as specified by the rotation group. A new concept called "isotopic spin" ($\vec{I}$) was invented, which was to be analogous to spin. It is now called **isospin**. The proton and neutron were thought of as forming a doublet with $I = \frac{1}{2}$ (analogous to $S = \frac{1}{2}$ for spinors). The proton can be considered to be a nucleon with $I_z = +\frac{1}{2}$ and the neutron a nucleon with $I_z = -\frac{1}{2}$, so that the nucleon wave operator ψ is an isospin doublet:

$$\psi = \begin{pmatrix} p \\ n \end{pmatrix}, \text{ the nucleon isodoublet.} \qquad (6.1)$$

Rotations in an invented space, referred to as **isospace**, were assumed to mix these components in exactly the same way as the components of *spin* doublets are mixed under ordinary space rotations. For this isodoublet, the generators of the isorotations were written

$$I_i = \tfrac{1}{2}\tau_i,$$

where the τ_i are identical to the Pauli matrices, but here act on isospinors like that of Eq. (6.1). Isospace is referred to as an "internal" space in order to distinguish it from the three-dimensional space we live in. The charge independence of nuclear forces is then very clearly expressed by demanding that $\vec{I}$ commute with the Hamiltonian.

But consider a state of a proton. After an **isorotation** (a rotation in isospace), it is a state whose wavefunction (or ket) is part proton ($I_z = +\frac{1}{2}$) and part neutron ($I_z = -\frac{1}{2}$). It doesn't even have a definite charge. How can that be? When we set up our detectors, would it not be detected either as a proton or a neutron? If it is sitting in an electric field, it should either feel a force (if it is charged) or not (if it isn't). (This difference of electromagnetic forces on the proton compared to those on the neutron shows that isospin is not conserved by the electromagnetic interactions.) The answer to this dilemma lies in the nature of quantum mechanics. If we have a state of an electron which is a linear combination of spin up and spin down and we set our detectors to measure S_z, the theory provides *probabilities* for the measurement of up or down. If we have a particle localized to a finite region of space, the wavefunction contains many (a distribution of) $\vec{p}$'s in it, and a measurement of momentum can find *any* of those values. The probability of measuring a particular value equals the absolute square of the coefficient of the term with that value. So these "rotated" isostates would produce either a proton or a neutron with the probability obtained from the coefficient of that term in the ket. What kind of an observer would we be after an isospin rotation? There is no clear answer to that question, but the demand that the *strong part* of the Hamiltonian be invariant to such transformations leads to the requirements that **the particles we find fill iso-**

multiplets[2] and that **the interactions must be isoscalars**. These are the *observable* consequences of isospin invariance.

From the definition of isospin in terms of the nucleon doublet, we see a connection between charge (Qe) and the value of I_z, viz.,

$$Q = I_z + \tfrac{1}{2} \tag{6.2}$$

for the nucleon doublet. Notice that the charge operator (Q) does not commute with $\vec{I}$ (the isospin generators).[3] Since Q gives the coupling to the electromagnetic field, **the electromagnetic interaction is not invariant to an isospin rotation.** The suggestion was that **the strong part of H has isospin symmetry.** The usefulness of a symmetry of only a part of the Hamiltonian may seem dubious; however, this strong interaction symmetry is broken only by the much weaker electroweak part of H. For consideration of the strongly interacting particles, it is appropriate to begin by ignoring the electroweak interactions. We expect, therefore, to find isomultiplets whose members have very nearly the same masses, just as was true for the neutron and proton.

In 1948 another strongly interacting particle was discovered called the *pi meson (pion)*.[4] It was a spin zero ($j = 0$) particle, but came in three different charge states: π^-, π^0, and π^+. The masses of these states were within 4 percent of one another.[5] The isospin value that would yield three states is $I = 1$, with $I_z = -1, 0, +1$ corresponding to the three pion states, respectively. Thus **the pion is an isovector.** This implies that:

$$Q = I_z + 0, \tag{6.3}$$

for pions, which is different than what we had for the nucleon. The general form appears to be

$$Q = I_z + \tfrac{1}{2} Y, \text{[6]} \tag{6.4}$$

where $Y = 1$ for the nucleon and $Y = 0$ for the pion. **Y is called the hypercharge.** At this point it appears that Y is no different than baryon number; however, we shall find that other properties, discovered more recently, also contribute to Y.

The **pion** (which is an isovector) has been found to behave like a scalar under space rotations. However, its ket (or wavefunction) changes sign under

---

2. We have learned, when studying rotations, that a complete set of states must contain all the members of a multiplet (an irreducible representation) because, in general, rotations mix them all together.

3. Recall that $[I_z, I_x]$ and $[I_z, I_y]$ are not zero.

4. It was originally believed to be the particle which transmitted the nuclear force between nucleons, as predicted by Yukawa.

5. The fact that the masses were so close, along with the similarity of the production processes, convinced people that they were three charge states of the same particle, called the pion (π).

6. Some authors omit the factor of $\tfrac{1}{2}$ in front of the Y. This is merely a different definition of Y, with Y values half the size of those used here.

space inversion; this property is referred to as **odd parity**. Consequently, the pion is called a **pseudoscalar**. (See Chapter 4, Section 4.3. Further discussion of inversion and parity is in Chapter 12.) We will use that property now to reconstruct the term in the Lagrangian which contains the interaction between the pion and the nucleon, as it was understood in the 1950s. In order to build a vertex with two nucleons and a pion (analogous to that for quantum electrodynamics) invariant to rotations, inversion,[7] and isorotations, we must couple the pion field to a *pseudoscalar* current, as follows:

$$\mathcal{L}_{\pi-N} = g(\bar{\psi}\gamma^5\vec{\tau}\psi)\cdot\vec{\pi} = g\bar{\psi}\gamma^5(\vec{\tau}\cdot\vec{\pi})\psi, \tag{6.5}$$

where the current is an isovector (behaving like $\vec{\tau}$), the dot product is in isospace (so that $\mathcal{L}_{\pi-N}$ is a scalar under isorotations), and g is the coupling constant. Let us examine the isospace structure of this interaction. (The Dirac-matrix structure will not be discussed here.) We can write

$$\vec{\tau}\cdot\vec{\pi} = \tau_1\pi_1 + \tau_2\pi_2 + \tau_3\pi_3 = \begin{pmatrix} 0 & 1 \\ 1 & 0 \end{pmatrix}\pi_1 + \begin{pmatrix} 0 & -i \\ i & 0 \end{pmatrix}\pi_2 + \begin{pmatrix} 1 & 0 \\ 0 & -1 \end{pmatrix}\pi_3$$

$$= \begin{pmatrix} \pi_3 & \pi_1 - i\pi_2 \\ \pi_1 + i\pi_2 & -\pi_3 \end{pmatrix}. \tag{6.6}$$

Recall that we have already discussed structures identical to these isospace structures when considering space rotations. From their respective I_z values, we conclude that $|\pi^+\rangle(I_z = +\frac{1}{2})$ behaves like V^+, $|\pi^-\rangle(I_z = -\frac{1}{2})$ behaves like V^-, and $|\pi^0\rangle(I_z = 0)$ behaves like V^0, but in isospace.

More explicitly, we obtain for the pion, in isospace,

$$|\pi^+\rangle \sim V^+ = \tfrac{1}{\sqrt{2}}(V_x + iV_y), \text{ etc.} \tag{6.7}$$

as shown in Eq. (3.25). Let us write $|\pi^+\rangle = \pi^{+\dagger}|0\rangle$, defining $\pi^{+\dagger}$ to be the field operator which creates a π^+ (or destroys a π^-).[8] In this notation, Eq. (6.7) shows that

$$\pi^{+\dagger} = \tfrac{1}{\sqrt{2}}(\pi_1 + i\pi_2) \text{ creates a } \pi^+, \quad \pi^{-\dagger} = \tfrac{1}{\sqrt{2}}(\pi_1 - i\pi_2) \text{ creates a } \pi^-, \text{ and}$$

$$\pi^{0\dagger} = \pi_3 \text{ creates a } \pi^0.[9] \tag{6.8}$$

---

7. The observed strong interactions have inversion invariance.

8. This is usually written as π^-, but that notation obscures the effect of the field operator on the vacuum. For our purposes, this modified notation will facilitate matters.

9. Actually, since the π^0 is its own antiparticle, π^0 and $\pi^{0\dagger}$ are the same; but, for uniformity, we shall use the dagger here too.

TABLE 6.1

Light Isomultiplets

Spin $\frac{1}{2}$ Baryons				Spin 0 (Pseudoscalar) Mesons			
Baryons	I-Spin	Mass (MeV)	Name	Mesons	I-Spin	Mass (MeV)	Name
p, n	$\frac{1}{2}$	939	nucleon	π^+, π^0, π^-	1	139	pion
Λ^0	0	1116	lambda	K^+, K^0	$\frac{1}{2}$	495	kaon
$\Sigma^+, \Sigma^0, \Sigma^-$	1	1193	sigma	η^0	0	549	eta
Ξ^0, Ξ^-	$\frac{1}{2}$	1318	cascade*	$K^-, \bar{K}^0$	$\frac{1}{2}$	495	(anti)kaon
				η'^0	0	958	eta prime

*We are told by Pais in *Inward Bound*: "After Gell-Mann and I had thought of the symbol Σ, we needed a name for the next higher state . . . the name of the society $\Sigma\Xi$ came to mind."

Inserting Eq. (6.8) into Eq. (6.6), we have

$$\vec{\tau}\cdot\vec{\pi} = \begin{pmatrix} \pi^{0\dagger} & \sqrt{2}\pi^{-\dagger} \\ \sqrt{2}\pi^{+\dagger} & -\pi^{0\dagger} \end{pmatrix}. \tag{6.9}$$

The isospace structure of the isorotationally invariant interaction of Eq. (6.5) is then

$$\mathcal{L}_{\pi-N} \sim gN^{\dagger}\vec{\tau}\cdot\vec{\pi}N = g(p^{\dagger}n^{\dagger})\begin{pmatrix} \pi^{0\dagger} & \sqrt{2}\pi^{-\dagger} \\ \sqrt{2}\pi^{+\dagger} & -\pi^{0\dagger} \end{pmatrix}\begin{pmatrix} p \\ n \end{pmatrix}$$

$$= g[p^{\dagger}p\pi^{0\dagger} + \sqrt{2}p^{\dagger}n\pi^{-\dagger} + \sqrt{2}n^{\dagger}p\pi^{+\dagger} - n^{\dagger}n\pi^{0\dagger}]. \tag{6.10}$$

Notice that the four *different* vertices appearing ($p^{\dagger}p\pi^{0\dagger}, p^{\dagger}n\pi^{-\dagger}$, etc.) all have the *same* coupling constant g, albeit some have an extra factor of $\sqrt{2}$. There is only one independent coupling constant (g); this is one of the powerful features of isospin invariance. Although we now know that neither the pion nor the nucleon are fundamental particles, isospin invariance requires that the effective Lagrangian for pion-nucleon coupling have the form of Eq. (6.10), with the resulting relationship among the effective couplings. These relationships have been verified experimentally, showing that **the strong interaction has isospin invariance**.

   As higher-energy accelerators were built and more experiments were done, many more strongly interacting particles were found. There were baryons, whose baryon number was conserved, and mesons like the pion, whose number was not conserved. They were built into isomultiplets using reasoning similar to that outlined above. Some of the multiplets found are shown in Table 6.1 (roughly in order of their masses). Determining the identifications and properties listed in Table 6.1 required much hard work and ingenuity. In particular, the *I*-spin assignments of the *K* mesons required further insight, as we shall now discuss.

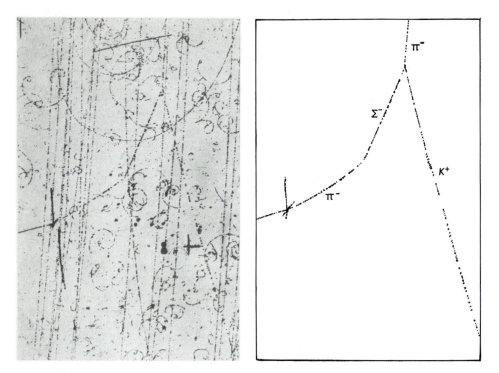

FIGURE 6.1
Bubble chamber picture of $\pi^- + p \rightarrow K^+ + \Sigma^-$. The picture on the left is the actual picture; the one on the right has isolated the tracks of interest.

6.2   STRANGENESS (S)

It was found that whenever a Σ or a Λ (both are baryons) was produced in a scattering of particles in the first line of Table 6.1 (pions and nucleons), a K meson also appeared. (See the bubble chamber picture in Fig. 6.1.) This "associated production" was finally understood as evidence for the existence of another (independent) quantum number **conserved by the strong interaction**, which was dubbed **strangeness**. The K meson carries a unit (+1) of strangeness and the Λ and Σ have strangeness -1; the pion and nucleon have 0 strangeness. Conservation of strangeness would then require that, when a pion beam bombards nucleons, a pion and nucleon could "fuse"[10] to create a Σ or Λ *only* with the associated production of a K.[11]

---

10. In field theory language, they are annihilated and the strange particles are created.

11. We would need a lot more energy than there was in these initial experiments to produce the heavier strange mesons.

In order to understand how the existence of strangeness was ascertained, as well as its implications, let us look at several related reactions.

1. $\pi^- + p \to K^+ + \Sigma^-$
2. $\pi^- + p \to \pi^+ + \Sigma^-$
3. $\pi^- + p \to K^- + \Sigma^+$
4. $\Sigma^- \to n + \pi^-$
5. $\Sigma^+ \to n + \pi^+$ or $p + \pi^0$ (6.11)

Reaction 1 occurs at a high rate; reactions 2 and 3 do not occur; and the decays 4 and 5 occur at a rate suppressed by a factor $\sim 10^{13}$ compared to the production rates of reactions like 1. Reaction 2 is a typical example of a reaction that should occur at a rate comparable to if not greater (because of phase space considerations, which we shall discuss in Chapter 13) than that of reaction 1. How does the concept of strangeness and its conservation provide a rationale for these disparate results? The initial particles in 1 and 3 are not strange particles (they have $S = 0$), the K^+ has $S = +1$, and the Σ's have $S = -1$. Thus, reaction 1 conserves strangeness. The nonoccurrence of reaction 2 is a typical example of the "associated production" mentioned above. In reaction 2 one strange particle (Σ^-) appears, and there is an obvious violation of strangeness conservation. This is no surprise since strangeness was invented to explain the nonoccurrence of just such events.

The nonoccurrence of reaction 3 is yet to be explained here. Many experiments, including reactions 1, 2, 4, and 5, have taught us that the Σ^- and the Σ^+ both have $B = +1$. This means that the Σ^+ cannot be the antiparticle of the Σ^-,[12] but should be in the same isomultiplet. Isospin rotations mix the Σ's together, and if strangeness is a separate independent number all of them should have the same strangeness $S = -1$.[13] So in reaction 3 the final baryon (Σ^+) has $S = -1$. If the K^- were in the same isomultiplet as the K^+, it would have $S = +1$ and the reaction should go. It does not! However, the argument we used for the Σ's concerning baryon number has no counterpart for the mesons.[14] The nonoccurrence of reaction 3 leads to the conjecture that the K^- is the antiparticle of K^+, so it has $S = -1$. With this strangeness assignment, reaction 3 would then have a final state with $S = -2$, which would violate strangeness conservation; hence, it would not occur. Within the context of strangeness conservation of the strong interaction, many other experiments confirm the existence

---

12. The antiparticle of the Σ^- is the $\overline{\Sigma^-}$, whose charge is indeed positive, but whose baryon number is -1. The baryon number of the Σ^+ is $+1$.

13. Starting with any member of a multiplet, we can generate any other member with a suitable isospin rotation. Since it is an independent quantum number, all the members of an isomultiplet have the same strangeness; **isospin rotations change one member of an isomultiplet into another without affecting the strangeness**.

14. There is no conserved meson number as can be seen from the π-N vertices above. The mesons can be produced or annihilated in an analogous way to bremsstrahlung and absorption of photons.

of the $\bar{K}$ isodoublet containing the neutral $\bar{K}^0 (S = -1)$, which is distinct from
the $K^0 (S = +1)$.[15]

The decays, 4 and 5, obviously violate strangeness conservation, but have
rates which tell us that they are probably due to the weak interaction. We shall
see that these decays are indeed due to the weak interaction, which explains why
they do not conserve strangeness. In fact, the lifetimes of the Σ's are very long
compared to particles which decay via the strong interaction. Why don't they
decay strongly? Possible modes would be:

$\quad$ 6. $\Sigma^- \rightarrow \pi^- + p + K^-$
$\quad$ 7. $\Sigma^- \rightarrow n + K^-$
$\quad$ 8. $\Sigma^+ \rightarrow p + \bar{K}^0$ $\hfill$ (6.12)

Reaction 6 is, in fact, related by crossing symmetry to reaction 1. Therefore,
it should occur with the same strength as the scattering 1. Each process of
Eq. (6.12) *does* conserve strangeness, but the sum of the masses of the final particles exceeds the mass of the Σ.[16]

The bubble chamber picture in Fig. 6.1 is not unlike the millions of others
examined. Notice that there is no place where the particle tracks have written
out (like a plane skywriting in the sky) that this track was made by a particle
with strangeness = ___. In fact, strangeness was *invented* to explain the kinds
of "occurs, does not occur" observations discussed here. *Assignments* of strangeness values were made to present a consistent picture of **strangeness conservation for the strong interaction**.

Now let us consider the electric charges of the strange particles. The strange
baryons have baryon number +1, and if we use that for Y in Eq. (6.4), we would
not get the correct values for their charges. (See Table 6.1.) To that end, we need
to amend the definition of Y:

$$Y = B + S. \hfill (6.13)$$

This gives the correct charges for the strange mesons as well. (Try it out!)

6.3   FLAVOR SU(3)

With the inclusion of strangeness as a conserved number, the strong interaction
then had two conserved "internal" quantum numbers (besides baryon number),
viz., I_z and S or Y. The number of light baryons found, all of which had

---

15. The idea that there were two different neutral K mesons and that the K forms an isodoublet (rather than the more "natural" assignment of an isotriplet K^+, K^0, and K^-) was devised to
explain associated production. In his Nobel address, V. L. Fitch said, "Their idea was implausible
and daring in the face of available data."

16. Although mass is not conserved, as pointed out in the Introduction, the reader should
be able to prove that a particle cannot decay into a set of particles whose total mass exceeds that
of the particle. (See Problem 15.)

spin $\frac{1}{2}$, was *eight* in total. (They are the ones listed in Table 6.1.) For the pseudoscalar mesons (π, K, η, and η'), the total number was *eight* as well (with the ninth one at considerably higher mass). There were *nine* vector (spin 1) mesons found as well. Now suppose that the two conserved numbers are proportional to generators of a group. Those generators would both have to be diagonal so as to give a quantum number for each particle. It may seem natural, because we have seen (in Chapter 5) that SU(3) has two diagonal generators, to guess that perhaps a very strong part of the Hamiltonian is invariant to SU(3).[17] Perhaps λ_3 corresponds to I_z and λ_8 to Y. We would then be postulating the existence of *eight* conserved quantities, proportional to the *eight* generators of SU(3) (in analogy to the three generators of SU(2) corresponding to the conserved components of angular momentum, viz., J_x, J_y, and J_z). Two of them, proportional to I_z and Y, form a commuting set, because they are simultaneously diagonal. (None of the others commutes with both.) Each set of particles we have mentioned could then be an **8** (the adjoint representation) of SU(3), and the one extra pseudoscalar meson and the one extra vector meson might each be a **1**. The masses of the spin $\frac{1}{2}$ baryons listed in Table 6.1, which are supposed to be in an **8** of SU(3), are not very close;[18] but, if there were some significant breaking of the SU(3) symmetry, we would not expect them to be. There also was some theoretical indication that the pion's mass *should* be anomalously low,[19] so that the pseudoscalar mesons listed might very well form an **8**. It was noted that all of these strongly interacting particles appeared to form **8**s or **1**s of SU(3).[20]

The members of a given SU(3) multiplet are distinguished by their values of I, I_z, and Y; the I-Y combinations (with I_z running from $-I$ to $+I$) present in the multiplet are completely determined by SU(3). By using I_z and Y as axes, the particles can be placed in multiplet plots as shown in Fig. 6.2, which contains (a) the baryon octet (with $j = \frac{1}{2}$), (b) the pseudoscalar octet, and (c) the vector nonet (**8 + 1**). Besides the particles with $j = 0$ or $\frac{1}{2}$ (which form octets and nonets) discussed above, baryons with $j = \frac{3}{2}$ were also found. Nine such particles were known at the time, and they are listed in Table 6.2. These decayed strongly and were seen only as enhanced scattering events, called *resonances*, in scattering experiments (a topic that will be discussed in Chapter 13).

There are nine particles in Table 6.2, but there is no isosinglet to peel off to leave an **8**, as we did for the sets of particles considered previously; furthermore, there is an $I = \frac{3}{2}$ multiplet which could not possibly be in an **8**. (See Fig. 6.2 showing the **8** multiplets.) For flavor SU(3) to be valid, this collection of

---

17. In fact, it took great insight and a familiarity with groups to realize that SU(3) was the appropriate choice. There are other continuous groups which have exactly two independent commuting generators. (They are referred to as rank 2 groups.)

18. They should be the same for exact flavor SU(3) symmetry, by reasoning analogous to that for the rotation group presented in Chapter 3, Section 3.1.

19. This idea is discussed in Chapter 17, in the section on chiral symmetry.

20. This was referred to by Gell-Mann (one of its discoverers) as "The Eightfold Way."

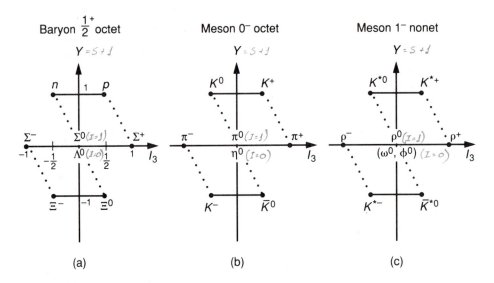

Baryon $\frac{1}{2}^{+}$ octet

Meson 0^{-} octet

Meson 1^{-} nonet

FIGURE 6.2

Flavor octets of the light hadrons: (a) the spin $\frac{1}{2}$ baryons; (b) the pseudoscalar (spin 0, odd parity) mesons (the η' is not included since it is essentially a **1**); (c) the vector (spin 1, odd parity) mesons. (The observed particles ω^{0} and ϕ^{0} are each part **8** and part **1**.) The horizontal lines connect members of the same isotopic multiplets; the negative-slope lines are lines of constant Q.

particles must then be part of a *higher*-dimensional irreducible representation of SU(3). The next highest irrep of SU(3) is a **10**, which indeed has isomultiplets (with Y values) identical to those of the particles listed above plus an iso-singlet with $Y = -2(S = -3)$. We have already argued that all members of a multiplet *must* exist[21] if the symmetry is unbroken. For a broken symmetry to have some validity, we expect that, although the equality of the masses may no longer be valid, all the particles of a multiplet should exist. Therefore, this scheme predicted the existence of a spin $\frac{3}{2}$ baryon with $Y = -2(S = -3)$, which was called Ω^{-}. Furthermore, using group theoretical arguments and the masses of the other particles in this multiplet, the *mass* of the Ω^{-} was predicted. (See Fig. 6.3.) The predicted mass of the Ω^{-} (1672 MeV) was so low that it could not decay into other hadrons via the strong interaction (which conserves strange-

---

21. They must exist because SU(3) operations "rotate" them into one another.

TABLE 6.2
Spin $\frac{3}{2}$ Baryons

Particles	I-Spin	Y (S)	Mass (MeV)	Multiplet Name
$\Delta^{++}, \Delta^{+}, \Delta^{0}, \Delta^{-}$	$\frac{3}{2}$	1 (0)	1232	Δ
$\Sigma^{*+}, \Sigma^{*0}, \Sigma^{*-}$	1	0 (−1)	1385	Σ^{*}
Ξ^{*0}, Ξ^{*-}	$\frac{1}{2}$	−1 (−2)	1530	Ξ^{*}

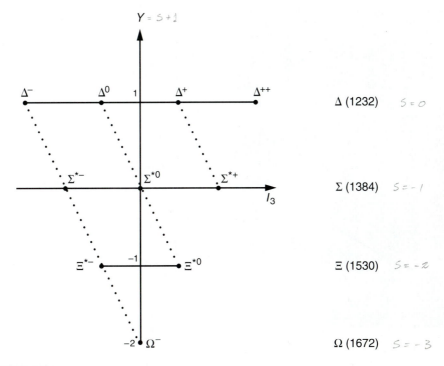

FIGURE 6.3
The $\frac{3}{2}^+$ baryon decuplet. The horizontal lines connect members of the same isotopic multiplets; the negative-slope lines are lines of constant Q.

ness).[22] Thus, it should live long enough after being created to produce a track in a bubble chamber. This (metastable) particle was found at exactly the mass predicted (with the correct properties), as shown in Fig. 6.3, confirming the **SU(3) classification scheme.** Fig. 6.4 shows the actual bubble chamber picture published in the paper which heralded the discovery of the Ω^-. Notice that the Ω^- did indeed leave a visible track in the bubble chamber. Heavier SU(3) multiplets (both **8**s and **10**s) of baryons as well as mesons have subsequently been found.

6.4   QUARK STRUCTURE OF HADRONS

The next question asked was why do the observed particles fit into an SU(3) classification scheme? This association within an SU(3) multiplet of seemingly unrelated particles is reminiscent of the periodic table of elements constructed a century ago. Perhaps the observed particles are made of *more fundamental* par-

---

22. We found the same reason for the nonoccurrence of Σ decay via the strong interaction in Eq. (6.12).

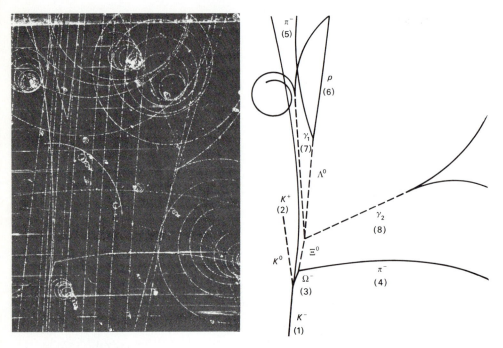

FIGURE 6.4
Bubble chamber picture (and associated line drawing) showing the first published evidence of the existence of the Ω^-. Notice the track of the Ω^-. The event was interpreted as

$$K^- + p \rightarrow \Omega^- + K^+ + K^0$$
$$\hookrightarrow \Xi^0 + \pi^-$$
$$\hookrightarrow \Lambda^0 + \pi^0$$
$$\hookrightarrow \gamma_1 + \gamma_2$$
$$\hookrightarrow e^+ + e^-$$
$$\hookrightarrow e^+ + e^-$$
$$\hookrightarrow \pi^- + p$$

ticles, which form a **3**, the defining representation of SU(3); their antiparticles would form a **3̄**. These conjectured fundamental particles were dubbed "quarks" by Gell-Mann[23] (and "aces" by Zweig). The *observed* particles would presumably be bound states of these fundamental particles. The mathematics of the group would then predict that the observed particles should belong to the higher representations of SU(3), because those representations result from the tensor products of **3**s and/or **3̄**s (as we saw in Chapter 5). Thus, did this "periodic table"

---

23. "Quark" is taken from a line in *Ulysses* by James Joyce.

of elementary particles lead to the next level of "fundamentality." The smallest number of **3** multiplets needed to obtain a **10** is three, strongly suggesting that the **baryons are made of three quarks**. Each quark must then have baryon number $B = \frac{1}{3}$; furthermore, since the strange baryons were assigned $S = -1$ when strangeness was invented, **the strange quark s has $S = -1$.**

The combination (product) of three **3**s has been found, by studying the corresponding Clebsch-Gordan coefficients, to be:

$$3 \times 3 \times 3 = 1 + 8 + 8 + 10, \tag{6.14}$$

which are just those **baryon multiplets** observed.

Likewise, it was found that the mesons are made of a quark and an antiquark ($B = 0$), and the product of the corresponding representations is

$$3 \times \bar{3} = 1 + 8, \tag{6.15}$$

which produces just those **meson multiplets** observed in nature.

6.4.1   Mesons: $q\bar{q}$

The particles discussed so far are thought to be made of three kinds (flavors) of quarks: u, d, and s. Look at Fig. 6.5, showing the **3** of quarks and the $\bar{3}$ of antiquarks. (The $\bar{3}$ is obtained from the **3** by reversing the signs of I_z and Y for each entry.) The **3** and the $\bar{3}$ are referred to as conjugate representations of one another. Notice that the u and d quarks form a doublet under the SU(2) of isospin and, so do $\bar{d}$ and $\bar{u}$:

$$q = \begin{pmatrix} u \\ d \end{pmatrix} \text{ and } \bar{q} = \begin{pmatrix} -\bar{d} \\ \bar{u} \end{pmatrix}, \tag{6.16}$$

where the minus sign in front of $\bar{d}$ is required to get the correct behavior under isorotations. (We shall see in Chapter 10 that it is more fruitful to talk about a generalized isospin, in which the s quark is also part of a doublet.) It is left to the reader to verify the Y assignments (using Eq. 6.13) and to show, from Eq. (6.4), that **the electric charges are $\frac{2}{3}$ and $-\frac{1}{3}$ for the upper and lower components** of these doublets, respectively.

Let us investigate the quark content of the light mesons. The u and d quarks are very light, whereas the s is somewhat heavier (as we shall soon argue). Therefore, we expect that the lightest nonstrange mesons are made of u and d quarks. We use the method we learned in Chapter 3, for combining two spin $\frac{1}{2}$ objects,[24] to construct them; they are listed in Table 6.3.

Strange Mesons   Suppose that we replaced the quark in each state of Table 6.3 by an s (strange quark) or alternatively the $\bar{q}$ by an $\bar{s}$. We have seen that, because

---

24. It yields (upper)(upper) for $I = I_z = +1$; $[\frac{1}{\sqrt{2}}]$[(upper)(lower) + (lower)(upper)] for $I = 1$, $I_z = 0$; etc.

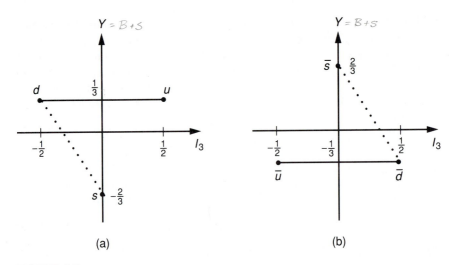

FIGURE 6.5
(a) The quark triplet (**3**); (b) The antiquark antitriplet ($\bar{\mathbf{3}}$). The horizontal lines connect members of the same isotopic multiplets; the negative-slope lines are lines of constant Q.

of the way strangeness was assigned when it was discovered, the s has $S = -1$. The s has strong isospin 0 (as can be seen from Fig. 6.5). The resulting mesons are shown in Table 6.4.

We expect that an $s\bar{s}$ state should exist as well. This is an $I = 0$, $S = 0$ state and there is another $I = 0$, $S = 0$ state (made of u- and d-flavor quarks), viz., $\frac{1}{\sqrt{2}}(u\bar{u} - d(-\bar{d})) = \frac{1}{\sqrt{2}}(u\bar{u} + d\bar{d})$ (where we have written "$\eta^o, \eta'^o(?)$" for NAME in Table 6.3). The linear combination of these two $I = 0$, $S = 0$ states, which behaves like a singlet under SU(3), is

$$\eta_{\text{SU(3)singlet}} \equiv \tfrac{1}{\sqrt{3}}(u\bar{u} + d\bar{d} + s\bar{s}), \tag{6.17}$$

TABLE 6.3
Mesons Made of Nonstrange Quarks

	Pseudoscalars (Spin$^{parity} = 0^-$)*		
		$I = 1$	
	$I_z = 1 : u(-\bar{d})$ $Q = +1$	$I_z = 0 : \frac{1}{\sqrt{2}}(u\bar{u} + d(-\bar{d}))$ $Q = 0$	$I_z = -1 : d\bar{u}$ $Q = -1$
Name:	$\boldsymbol{\pi^+}$	$\boldsymbol{\pi^0}$	$\boldsymbol{\pi^-}$
		$I = 0$	
		$I_z = 0 : \frac{1}{\sqrt{2}}(u\bar{u} - d(-\bar{d}))$** $Q = 0$	
Name:		$\boldsymbol{\eta^o, \eta'^o}$ (?)	

*Parity refers to behavior under space inversion; it will be discussed in Chapter 12.
This state has $I = 0$ and $Y = 0$, but it is not a member of the **8; keep reading to Eq. (6.18).

TABLE 6.4
Mesons Containing One Strange Quark

	$I = \frac{1}{2}; S = -1$		$I = \frac{1}{2}; S = +1$	
	$I_z = +\frac{1}{2} : s(-\bar{d})^*$	$I_z = -\frac{1}{2} : s(\bar{u})$	$I_z = +\frac{1}{2} : \bar{s}u$	$I_z = -\frac{1}{2} : \bar{s}d$
	$Q = 0$	$Q = -1$	$Q = +1$	$Q = 0$
Name:	$\bar{K}^0$	K^-	K^+	K^0

*The minus is irrelevant here and in π^+ of Table 6.3 because it corresponds to an *overall* phase.

which obviously is *unchanged* by any interchange of quark flavors or by *any* flavor SU(3) transformation.[25] The neutral member of the octet must be orthogonal to $\eta_{SU(3)singlet}$, so that we obtain

$$\eta_{SU(3)octet} \equiv \frac{1}{\sqrt{6}}(u\bar{u} + d\bar{d} - 2s\bar{s}).^{26} \qquad (6.18)$$

The s is "heavier" than the u and d quarks, which shows that flavor SU(3) is broken; however, the observed particles η and η' are nearly the pure multiplet states shown above:

$$\eta \approx \eta_{SU(3)octet} \text{ and } \eta' \approx \eta_{SU(3)singlet}. \qquad (6.19)$$

Vectors (Spin^{parity} = 1$^-$) Most vector mesons have the same quark-flavor structure as do the pseudoscalar mesons shown above, but the names are changed: $K \rightarrow K^*$, $\pi \rightarrow \rho$, and $\eta, \eta' \rightarrow \omega, \phi$,[27] as shown in Fig. 6.2. However, the mixing of the $I = 0$, $S = 0$ member of the octet with the SU(3) singlet has been found to be manifested quite dramatically here with the **observed particles**:

$$\omega = \frac{1}{\sqrt{2}}(u\bar{u} + d\bar{d}) \text{ and } \phi = s\bar{s}. \qquad (6.20)$$

The ω is lighter, being made of the light quarks, while ϕ is made solely of the heavier s quark.

Many **8**s and **1**s of heavier mesons have been found since these early discoveries.

6.4.2 Baryons: qqq

The quarks are spin $\frac{1}{2}$ particles and as such should obey Fermi-Dirac statistics. The baryons are made of three quarks, and since the quark flavors are considered different states of *the quark*, we must consider the Pauli exclusion principle when investigating their structure.

Pauli Exclusion Principle Suppose we interchange two identical particles in a multiparticle state. The resulting state will be the same *physically* as the origi-

---

25. It looks like a generalized flavor dot product.
26. Remember that *different* product wavefunctions are orthogonal and the product wavefunctions are normalized as well.
27. The η' does not appear in the **8**, as explained above, but both ω and ϕ are partly in the **8**.

nal state. However, this interchange could change the overall phase of the state, so we write[28]

$$\psi(\vec{r}_2, \vec{\sigma}_2; \vec{r}_1, \vec{\sigma}_1; \dots) = \eta\psi(\vec{r}_1, \vec{\sigma}_1; \vec{r}_2, \vec{\sigma}_2; \dots), \tag{6.21}$$

where η is the phase change ($e^{i\alpha}$, with α unknown so far) due to a particle interchange. Now let us interchange these two particles again (in $\psi(\vec{r}_2, \vec{\sigma}_2; \vec{r}_1, \vec{\sigma}_1; \dots)$); the resulting state should equal $\eta\psi(\vec{r}_2, \vec{\sigma}_2; \vec{r}_1, \vec{\sigma}_1; \dots)$. However, the particles are now in their original states, so that the wavefunction is again $\psi(\vec{r}_1, \vec{\sigma}_1; \vec{r}_2, \vec{\sigma}_2; \dots)$. Thus, we may write

$$\psi(\vec{r}_1, \vec{\sigma}_1; \vec{r}_2\vec{\sigma}_2; \dots) = \eta\psi(\vec{r}_2, \vec{\sigma}_2; \vec{r}_1, \vec{\sigma}_1; \dots) = \eta\eta\psi(\vec{r}_1, \vec{\sigma}_1; \vec{r}_2, \vec{\sigma}_2; \dots). \tag{6.22}$$

This tells us that $\eta^2 = +1$ and $\eta = \pm 1$.

It has been shown (for relativistic quantum field theories) that $\eta = +1$ for particles with integral spin (bosons), and $\eta = -1$ for those with odd-half-integral spin (fermions). This is referred to as **the spin-statistics connection**. The behavior of fermions ($\eta = -1$) leads directly to **the exclusion principle: Two identical fermions cannot be in the same quantum state**. For if we interchange two identical particles which are in the *same* quantum state, then the wavefunction will certainly be unchanged; whereas, for fermions it should change sign. Since the fundamental particles of matter, **leptons and quarks**, are fermions, they **have $\eta = -1$**.

Baryon Structure   The quarks have spin, isospin, and strangeness (so far). We are looking for low-mass combinations of three quarks, i.e., bound states, which would presumably correspond to the light baryons. The **space configuration** which would give the most attraction for a strong, short-range force would be **symmetric** to an interchange of position of any two. This is apparent since anti-symmetric wavefunctions vanish when the two individual wavefunctions are evaluated at the same point: $\frac{1}{\sqrt{2}}(u(\vec{r}_1)v(\vec{r}_2) - u(\vec{r}_2)v(\vec{r}_1))$ vanishes for $\vec{r}_1 = \vec{r}_2$, whereas a symmetric combination is very large.

Let us consider the Δ^{++} particle, shown in Fig. 6.3 and listed in Table 6.2. It has $I_z = +\frac{3}{2}$ and so must have the (flavor) structure uuu. The Δ is also a *spin* $\frac{3}{2}$ combination of three quarks; let us first discuss the component which has $S_z = \frac{3}{2}$. Its wavefunction has the structure:

$$\psi_S(\vec{r}_1, \vec{r}_2, \vec{r}_3)\, u\uparrow\, u\uparrow\, u\uparrow = \psi_S(\vec{r}_1, \vec{r}_2, \vec{r}_3)uuu\uparrow\uparrow\uparrow, \text{ for } \Delta^{++} \text{ with } S_z = +\tfrac{3}{2}, \tag{6.23}$$

where the space part $\psi_S(\vec{r}_1, \vec{r}_2, \vec{r}_3)$ is symmetric to the interchange of any two quarks. Just as for the space part, we have written the spin parts together and the isospin parts together so that the full wavefunction (Ψ) has the form:

$$\Psi = \psi(\text{space})\psi(\text{isospace})\psi(\text{spin}). \tag{6.24}$$

---

28. The particles may have other quantum numbers besides the spin; for this discussion it is unnecessary to display them, and we use the spin as a representative of them all.

Not only do we expect the space part to be symmetric (for these lightest baryons), but Eq. (6.23) shows that the isospin part of the Δ^{++} and the spin part for $S_z = +\frac{3}{2}$ are symmetric as well. When we interchange two quarks, the interchange occurs in ψ(space), ψ(isospace), and ψ(spin), which are each symmetric, so that the full wavefunction of the Δ^{++} with $S_z = +\frac{3}{2}$ is **totally symmetric** to an interchange of quarks. This is a puzzle, since the quarks are fermions. A lot of thought and concern has gone into this problem, as we shall soon discuss. First let us ask about the other S_z states of the Δ^{++}. Recall that we obtain them by lowering the *symmetric* $S_z = \frac{3}{2}$ state with $S_- = S_{1-} + S_{2-} + S_{3-}$. The operator S_- is obviously symmetric to any interchange of particles; consequently, S_- **will not change the behavior of a state under interchange**. Thus, the states of the Δ^{++} with lower S_z values are symmetric under spin interchanges (just like the $S_z = \frac{3}{2}$ state); consequently, their total wavefunctions are symmetric. Identical reasoning applies for the isospin part of the wavefunction, because the mathematics is identical, so the wavefunctions of the entire Δ isomultiplet appear to be symmetric. The other lowering operators of SU(3) have this same symmetric structure $(\mathcal{O}_1 + \mathcal{O}_2 + \mathcal{O}_3)$, and so we have the same symmetry of the wavefunction in the entire **10** multiplet.

We now consider two related questions concerning spin. When we combine three spin $\frac{1}{2}$ states,

1. Can we get a totally antisymmetric state?
2. What is the symmetry associated with the spin $\frac{1}{2}$ composite (which would correspond to the baryon octet of Fig. 6.2)?

Answer 1 is no! If we begin by antisymmetrizing in the first two and then try to antisymmetrize that with the third particle, we will not succeed (Try it! Eq. 6.25 below is the first step.)

Answer 2 results when we pursue the attempt to construct antisymmetric states further. The two states we first obtain, after antisymmetrizing in the first two, are

$$\tfrac{1}{\sqrt{2}}(\uparrow\downarrow - \downarrow\uparrow)\uparrow \text{ and } \tfrac{1}{\sqrt{2}}(\uparrow\downarrow - \downarrow\uparrow)\downarrow, \tag{6.25}$$

where the order indicates which particle (1, 2, or 3) is referred to. Notice that the first two particles are coupled to spin 0, so that the two states in Eq. (6.25) correspond to an $S = \frac{1}{2}$ doublet; the first has $S_z = +\frac{1}{2}$, the second $S_z = -\frac{1}{2}$.

However, the states in Eq. (6.25) are neither symmetric nor antisymmetric to interchanges of quarks 2 and 3. In fact, such an interchange produces states which are not contained in Eq. (6.25):

$$\tfrac{1}{\sqrt{2}}(\uparrow\downarrow - \downarrow\uparrow)\uparrow \equiv \tfrac{1}{\sqrt{2}}(\uparrow\downarrow\uparrow - \downarrow\uparrow\uparrow) \rightarrow \tfrac{1}{\sqrt{2}}(\uparrow\uparrow\downarrow - \downarrow\uparrow\uparrow), \quad \text{for } 2 \leftrightarrow 3. \tag{6.26}$$

This property is referred to as **mixed symmetry**. We shall not study mixed symmetry in detail; however, such a study shows that appropriate combinations of products of these spin states of mixed symmetry with isospin states of mixed symmetry (times a symmetric space part) yield states which are totally symmetric to an interchange. Thus, we can construct spin $\frac{1}{2}$ states which are totally symmetric to an interchange, so that *all* the observed baryons would have this property.

The reader may wonder whether the mixed symmetry spin and isospin parts could be combined to form an antisymmetric state. The answer is yes, but the decuplet (10) has told us already that at least some of the states built in this quark model are symmetric. We shall soon see that the symmetric result *is satisfactory*.

Instead of the coupling in Eq. (6.25), we could couple particles 2 and 3 to a singlet and then couple that singlet to particle 1 to obtain an $S = \frac{1}{2}$ entity, or couple 3 and 1 to $S = 0$ first with the same final outcome. The resulting three doublets are not independent, as the reader can prove by solving Problem 12. This may be understood by considering the rules for combining representations of SU(2); we obtain

$$(j = \tfrac{1}{2}) \times (j = \tfrac{1}{2}) \times (j = \tfrac{1}{2}) = (j = \tfrac{1}{2}) \times [(j = 0) + (j = 1)]$$
$$= (j = \tfrac{1}{2}) \times (j = 0) + [(j = \tfrac{1}{2}) \times (j = 1)]$$
$$= (j = \tfrac{1}{2}) + [(j = \tfrac{1}{2}) + (j = \tfrac{3}{2})], \tag{6.27}$$

showing that only two $(j = \frac{1}{2})$ irreps occur.

In Table 6.5 we list the quark-flavor structure of the baryons. For the $j = \frac{1}{2}$ states, we indicate the quark (flavor) content without trying to keep track of the interchange structure.

Let us now explore possible solutions to the statistics dilemma, which is a consequence of the symmetric (instead of antisymmetric) behavior of the qqq states under particle interchange:

1. We could abandon the spin-statistics connection for quarks.
2. We could reject the quark theory.
3. We could create a new kind of statistics.[29]
4. We could invent a new quantum number and make the wavefunction *antisymmetric* with respect to interchanges of its assignments.

For several years after this problem was realized, alternative 2 was widely subscribed to and the quark theory was not taken seriously. However, alternative 4 turned out to be the correct approach,[30] so let us pursue it further. We have already seen (in our discussion of mixed symmetry) that if the new quantum number has only two possible values, then we cannot build an antisymmetric state of three objects. Let us consider the case where the new number can have *three* possible values; we will call them r, g, and b. Consider the following combination of three quarks:

$$rgb - rbg + gbr - grb + brg - bgr, \tag{6.28}$$

---

29. This was done and named "Parastatistics."

30. Alternative 3, the introduction of more complicated statistics, has been shown to be mathematically equivalent to 4. However, the new quantum number has *physical* significance, as we shall learn in Chapter 9 on QCD.

TABLE 6.5
Quark-Flavor Structure of the Baryons

$j = \frac{3}{2}$ Baryons		
(those with ~ are symmetrized sums)		

Nonstrange

$$\Delta^- = ddd \qquad \Delta^0 \sim ddu \qquad \Delta^+ \sim duu \qquad \Delta^{++} = uuu$$

$$S = -1$$

$$\Sigma^{*-} \sim sdd \qquad \Sigma^{*0} \sim sud \qquad \Sigma^{*+} \sim suu$$

$$S = -2$$

$$\Xi^{*-} \sim ssd \qquad \Xi^{*0} \sim ssu$$

$$S = -3$$

$$\Omega^- = sss$$

$j = \frac{1}{2}$ Baryons
(mixed symmetry combinations occur multiplied by mixed spin symmetry factors)

Nonstrange

$$n \sim udd \qquad p \sim uud$$

$$S = -1$$

$$\Sigma^- \sim dds \qquad \Sigma^0 \sim \frac{1}{\sqrt{2}}(ud + du)s \qquad \Sigma^+ \sim uus$$

$$\Lambda^0 \sim \frac{1}{\sqrt{2}}(ud - du)s$$

$$S = -2$$

$$\Xi^- \sim dss \qquad \Xi^0 \sim uss$$

where, in each term, **the first symbol refers to quark 1, the second to quark 2, etc.** Observe that an interchange of the letters in any two positions, identically in all the terms, yields the negative of the same expression. If we associate r, g, and b with the numbers 1, 2, and 3 respectively, we see that this corresponds to combining all possible permutations of 123, with (-1) as the coefficient for those terms that are obtainable from 123 by an odd number of interchanges and $(+1)$ for those which come from an even number of interchanges. In fact we can use ε_{ijk}, the Levi-Civita symbol (where i, j, and k can each be 1, 2, or 3), defined in Chapter 5, to rewrite Eq. (6.28) as

$$\varepsilon_{ijk}q^i(1)q^j(2)q^k(3), \tag{6.28'}$$

where we are using the summation convention, we have explicitly indicated (in parentheses) which quark the symbol is describing,[31] and q^1, q^2, and q^3 are r, g, and b, respectively. (Try Problem 13(a).) The Levi-Civita symbol is, in fact,

---

31. We have explicitly indicated which quark each symbol represents so that this will not look as if it is identically zero. Actually, this notation is redundant since **the first symbol always refers to particle 1, etc.**

antisymmetric to an interchange of any two indices; this property implies that the expression in Eq. (6.28') is antisymmetric to the interchange of any two quarks. (The interchange of quarks 1 and 2 means that, in each term, the value of i is interchanged with the value of j. See Problem 13(b).) You may have guessed by now that r, g, and b are short for "colors" red, green, and blue. This designation is indeed *colorful*, but may be misleading, for it has nothing to do with actual colors. We are merely dealing with a quantum number whose values have three different possibilities.

The expression in Eq. (6.28') is invariant to any unitary transformation mixing the three colors, i.e., to $SU_c(3)$. Thus, it is a singlet under $SU_c(3)$, and the corresponding states are sometimes referred to as **colorless** or **white**.[32] All the **observed** baryons are such **color singlets**, due to **color confinement**.

So far, we have not shown any relationship between the colors and *physical* properties. We shall see in Chapter 9 that in quantum chromodynamics the colors are the coupling constants and correspond to three different kinds of color "charge."

6.5   QUARK FLOW DIAGRAMS

Quark flow diagrams are an efficient way to keep track of the conserved (in QED and QCD) flavor numbers. We will justify their validity after we study quantum chromodynamics, but will describe and utilize them here. These diagrams become more appropriate than the introduction of higher unitary groups as new flavors of quarks are discovered. This is because the "masses" of the latter are quite far from those of the three we have already discussed. This means that the masses within the hadron multiplets are quite disparate, and we may not even find the entire multiplet in our experiments. Furthermore, the physical states may be *mixtures* of members of *different multiplets* of the larger flavor group (e.g., $SU(n)$ for n quark flavors). However, we do wish to keep track of the quark-flavor content, which is not affected by QCD or QED at all. To this end, we use the quark-flavor content of the initial and final particles and draw lines showing where those flavors come from and where they go. The actual quarks may not follow the simple lines drawn, but this does keep track of the flavors, which are not changed by interactions with gluons or photons.

Look at Fig. 6.6, which shows the quark flow diagram for the reaction: $\pi^- + p \to K^+ + \Sigma^-$. (A bubble chamber picture of this reaction was shown in Fig. 6.1.) Here all incoming and outgoing quarks are shown, with pair annihilation (the u of the p with the $\bar{u}$ of the π^-) and pair creation (the s and $\bar{s}$ which then go into different final particles). No virtual particles are drawn, not even those responsible for pair creation or those resulting from pair annihilation. The

---

32. This choice of colors, viz., red, green, and blue, provides a simple way of remembering that they couple (but in the definite combination shown in Eq. 6.28) to a "white" state, just as the combination of light beams of these primary colors would combine to produce white.

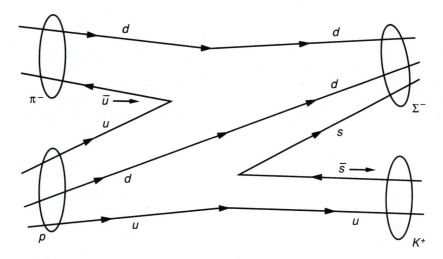

FIGURE 6.6
Quark flow diagram for $\pi^- + p \to K^+ + \Sigma^-$.

ovals cutting across quark lines tell us that a bound state of those quarks is being considered and the hadron name is shown. In this diagram, we can see the conservation of strangeness in an obvious way. (Try to draw a quark flow diagram for $\pi^- + p \to K^- + \Sigma^+$, which violates conservation of strangeness; you will see that it is impossible.)

We have seen that the vector mesons have considerable (multiplet) mixing to yield:

$$\omega = \tfrac{1}{\sqrt{2}}(u\bar{u} + d\bar{d}) \text{ and } \phi = s\bar{s}. \qquad (6.29)$$

Notice that the ϕ is made solely of strange quarks whereas the ω has no strange particles in it. Although the ϕ does have two strange quarks in it, it has $S = 0$. We may say that it has **hidden strangeness**. Let us consider its main decay modes:

$$\phi \to \begin{cases} K^+K^- & 49\% \\ K^0\bar{K}^0 & 34\% \\ \rho\pi & 13\% \\ \pi^+\pi^-\pi^0 & 2\% \end{cases}.$$

It is not evident why the decays to $\rho\pi$ and 3π are suppressed compared to the decays to kaons (a total of about 84 percent of the decays). On the contrary, we would expect the decay to pions (2 percent) to be enhanced because their total mass is considerably less than that of the final kaons. (Such considerations will be studied in Chapter 13, when we discuss phase space.) Let us examine typical quark flow diagrams for these decays, as shown in Fig. 6.7.

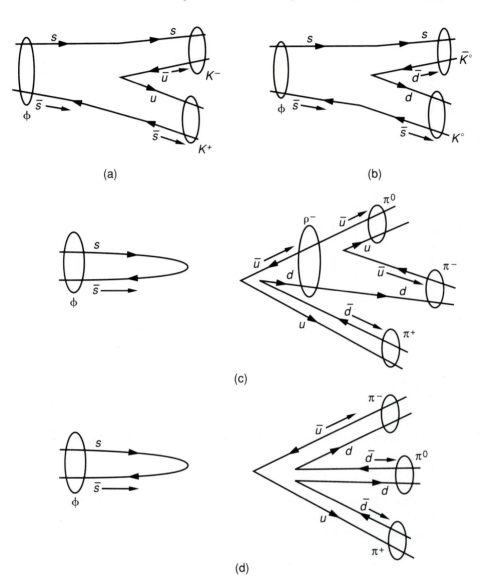

FIGURE 6.7
Typical quark flow diagrams for the decay of the ϕ: (a) $\phi \to K^+ + K^-$; (b) $\phi \to K^0 + \bar{K}^0$; (c) $\phi \to \rho^- + \pi^+$ followed by $\rho^- \to \pi^- + \pi^0$; (d) $\phi \to \pi^+ + \pi^- + \pi^0$.

In Fig. 6.7, we see that the initial quarks appear in the final observed particles for the decay to kaons, (a) and (b), but there is no continuous quark line from initial to final state for the decays to pions, (c) and (d). We shall see (in Chapter 9) that QCD provides us with a rationale for saying that such a disconnectedness yields a suppression of those processes. This is called the OZI rule (named after Okubo, Zweig, and Iizuka) or the Zweig Rule.

TABLE 6.6
Lightest Particles with c Quarks

Particle	Type	Spin	Structure	Mass (MeV)
$D^0, \bar{D}^0$	meson	0	$c\bar{u}, u\bar{c}$	1865
$D^\pm$	meson	0	$c\bar{d}, d\bar{c}$	1869
$D_s^\pm$	meson	0	$c\bar{s}, s\bar{c}$	1971
D^*	meson	1	same as corresponding D	2010
Λ_c^+	baryon	$\frac{1}{2}$	udc	2281
η_c	meson	0	$c\bar{c}$	2981
J/ψ	meson	1	$c\bar{c}$	3097

6.6 HEAVIER FLAVORS

The ϕ meson has been found to be nearly a pure $s\bar{s}$ bound state, as we have just discussed. This structure (particle-antiparticle of the same flavor) of ϕ is repeated, for the heavy quarks, in the heavier vector mesons and has led to the discovery of two more flavors, viz., charm and bottom, as we shall discuss in Chapter 15.

A relatively long-lived[33] massive vector meson, the J/ψ particle,[34] presumably a q-$\bar{q}$ bound state of a new flavor, was discovered in 1974. Its discovery was unexpected, although some theorists had already expressed (in 1964) the notion that there should be as many quarks as leptons and whimsically named the new flavor "charm." (There were four leptons and three quark flavors known at the time.) By 1974 it had been shown that the existence of a new quark flavor would have important ramifications for strangeness-changing decays (see the GIM mechanism, Section 11.3). An understanding of the enhancement of the lifetime of the J/ψ follows from what we have just learned in our discussion of the ϕ. If the J/ψ is a pure $c\bar{c}$ state, it has hidden charm. Its decay into the known lighter hadrons (which do not have c or $\bar{c}$ quarks) would be suppressed by the OZI rule. Suppose that the masses of the charmed particles, into which it should decay, were so large that their sum exceeded that of the J/ψ.[35] Then the only decay modes available would be OZI- (or Zweig-) suppressed and the decay would be slower than the usual strong decays. This is indeed the case. The lightest observed particles, which contain charmed mesons, are shown in Table 6.6. Notice that the lightest charmed meson (the D^0) does have a mass which is more than one half that of the J/ψ, so that the latter cannot decay into a pair of them.

Although the c quark is considerably heavier than the others it might still be useful to define a flavor SU(4) based upon u, d, s, and c quarks. (See Fig. 6.8.) Notice that the D and D_s particles fill out the cuboctahedron (15-plet)[36] of the

---

33. Its lifetime was $\sim 10^3$ times the expected lifetime ($\sim 10^{-23}$ sec.) for a strong decay.

34. It has two names because of its independent discovery by two research groups.

35. Recall that an analogous situation has already been discussed for the strange members of the baryon octet.

36. The SU(4) **15** contains the SU(3) multiplets: **8** + **1** with charm = 0; **3** with charm = +1; $\bar{\mathbf{3}}$ with charm = -1.

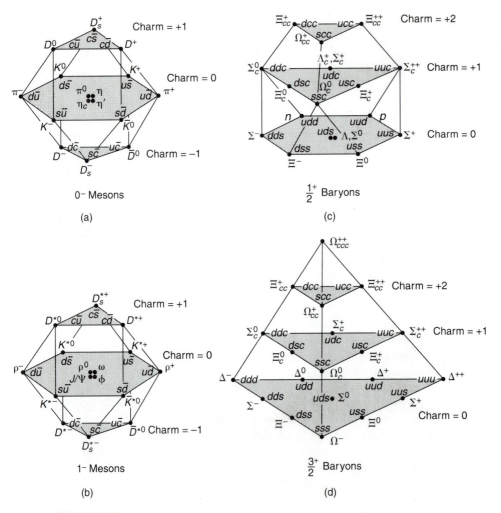

FIGURE 6.8

Some SU(4) multiplets of hadrons. The pseudoscalar (0^-) mesons (a) and the vector (1^-) mesons (b) each form a cuboctahedron (15-plet) plus a **1**. The baryons form two different kinds of **20**-plets ((c) and (d)). The charm = 0 planes contain the SU(3) particle multiplets we have already discussed. Notice that charm ≠ 0 planes are contained in these SU(4) multiplets, which have new, charmed particles (forming their own SU(3) multiplets). These new particles are thereby associated with the uncharmed hadrons (already discovered) in the same SU(4) multiplet. (Source: Review of Particle Properties, Phys. Rev. D, *45* (1992).)

pseudoscalar (0^-) mesons and the D^* and D_s^* do the same for the vector (1^-) mesons. Because of the great mass splittings and mixings, it is not deemed useful to take these multiplets of SU(4) very seriously.

Studies of e^+e^- reactions at even higher energies produced an analogous

TABLE 6.7
Lightest Particles with b Quarks

Particle	Type-Spin	Structure	Mass (MeV)
$B^{\pm}$	meson-0	$u\bar{b}, b\bar{u}$	5271
$B^0, \bar{B}^0$	meson-0	$d\bar{b}, b\bar{d}$	5275
Υ	meson-1	$b\bar{b}$	9460

vector meson Υ (called the *Upsilon*) at a mass of 9460 MeV. It was subsequently shown to behave like a bound state of a quark and an antiquark of another new flavor. As we shall see when we study the electroweak theory in Chapter 10, there is a generalization of the strong isospin at the heart of the theory. The first four quarks we have discussed (u, d, c, and s) form two doublets of that generalized isospin. The new flavor was seen to behave like the bottom part of a new doublet (the third generation of quarks) and is usually called *bottom*. It is conjectured that there must be a partner to form a doublet under the isospin SU(2), and it is called top. (Some like to call b and t *beauty* and *truth*, respectively. So all together, the new "qualities" produced in our accelerators are: strangeness, charm, beauty, and truth. Notice the progression from the superficial to fundamental wisdom.[37]) The lightest particles made of bottom quarks, found so far, are shown in Table 6.7.

The common structure of the $\phi(s\bar{s})$, the $J/\psi(c\bar{c})$, and the $\Upsilon(b\bar{b})$ is reminiscent of positronium ($e^-e^+ = e\bar{e}$). They are therefore referred to as **quarkonium systems** (charmonium, bottomonium, etc.). The quantum field theory for these quark systems is quantum chromodynamics, which is qualitatively more difficult to understand for bound systems than is quantum electrodynamics. Nevertheless, considerable success has been achieved in calculating the energy levels and transitions of these systems. The charmonium system includes the J/ψ, the η_c, and a number of other states, some "bound" (with OZI-suppressed decays) and some "quasibound" (with unsuppressed strong decays). Likewise, bottomonium has been studied and possesses many more bound states than does charmonium. Mixed "-onium" systems, analogous to muonium ($e^-\mu^+$), are also being considered. We will discuss quarkonium systems more fully in our discussions of QCD in Chapters 9 and 15.

The search for the top quark proceeds without success, so far. It appears that, if it exists, its mass will be larger than what present-day accelerators can reveal ($m_{top} > 91$ GeV). Detailed experimental studies of electroweak parameters (e.g., W and Z^0 masses and the Z^0 width) show that radiative corrections (to one loop in QCD calculations) should contain virtual top quarks with a mass of roughly 160 GeV. The search for the top quark will be elaborated upon in Chapter 15. The existence of the top quark is crucial for the standard model, as we shall see in Part B of this book.

.37. If you haven't smiled yet, be assured that this is meant as a joke.

PROBLEMS

1. Show that the nucleon current operator of the form $\psi^\dagger \psi$ (omitting the Dirac structure) is an $I = 0$ operator and the current operator $\psi^\dagger \vec{\tau} \psi$ is an $I = 1$ structure, where
   $\psi \sim \binom{p}{n}$ and $\psi^\dagger \sim (p^\dagger n^\dagger)$ and $\vec{\tau}$ are the Pauli matrices. Hint: Recall, from Eq. (6.16),
   that the isodoublet is $\binom{p}{n}$ and $\binom{-n^\dagger}{p^\dagger}$, where the upper components have $I_z = +\frac{1}{2}$
   and the lower components have $I_z = -\frac{1}{2}$.

2. Use Eq. (6.4) for Q to find the electric charges of the upper and lower components
   of q and of $\bar{q}$.

3. For the pseudoscalar π-N coupling model of Eqs. (6.5) and (6.10), draw the tree-
   level Feynman diagrams for (a) n-p scattering and (b) n-n scattering. In the diagrams
   indicate the coupling constants. Note that the nucleons can switch identities at the
   vertices.

4. Check that using the equation: $Y = B + S$ in the equation for Q (Eq. 6.4), gives the
   correct charge assignments to the (strange) K mesons.

5. The Ξ has $S = -2$. Check that the observed charges for the Ξ doublet result from
   $Q = I_z + Y/2$.

6. Show that if the strangeness operator (S) commutes with isospin ($\vec{I}$), then all the
   members of an isomultiplet have the same strangeness.

7. Which of the following reactions can occur (as a strong interaction)? Which are pro-
   hibited and why? Consider baryon number, charge, and strangeness conservation.
   a. $\pi^- + n \rightarrow \Xi^- + K^0 + K^0$
   b. $K^- + p \rightarrow \Xi^0 + K^0$ or $K^- + p \rightarrow \Xi^0 + \bar{K}^0$
   c. $\pi^- + p \rightarrow \Xi^0 + \pi^0$
   d. $K^- + p \rightarrow \Lambda + \pi^0$
   e. $\pi^- + p \rightarrow \Xi^0 + K^0$
   f. $\pi^- + p \rightarrow \Xi^- + K^+$
   g. $K^- + p \rightarrow \Lambda + n$
   h. $\Sigma^- + p \rightarrow \Xi^- + K^+$

8. Apply $S_- = S_-^{(1)} + S_-^{(2)} + S_-^{(3)}$ to $\uparrow\uparrow\uparrow$ and verify directly that the result is symmetric to an interchange of any two spins in the resulting state.

9. a. Find the effects of $\lambda_1, \ldots, \lambda_8$ (the matrices in Section 5.3 of Chapter 5) on

   $$q^1 = \begin{pmatrix} 1 \\ 0 \\ 0 \end{pmatrix}, q^2 = \begin{pmatrix} 0 \\ 1 \\ 0 \end{pmatrix}, \text{ and } q^3 = \begin{pmatrix} 0 \\ 0 \\ 1 \end{pmatrix},$$

   expressing your answer in terms of the q^i, e.g., $\lambda_1 q^1 = q^2$, etc.

   b. The three-quark state $\varepsilon_{ijk} q^i(1) q^j(2) q^k(3)$ is a color singlet and, therefore, must
   be an eigenstate of the generators of SU(3) with 0 eigenvalue. The generators are
   $G^\alpha = \lambda_\alpha/2$, $\alpha = 1, \ldots, 8$, where the λ_α are the matrices in Section 5.3 and
   $G^\alpha = G^{\alpha(1)} + G^{\alpha(2)} + G^{\alpha(3)}$. (Here $G^{\alpha(1)}$ operates on the quark in position 1,
   etc.) Verify that $\varepsilon_{ijk} q^i(1) q^j(2) q^k(3)$ is an eigenstate of the generators of SU(3)
   with 0 eigenvalue for $\alpha = 1, 2, 3, 4,$ and 8. Hint: λ_1 changes a q^1 to a q^2, etc.,
   as was shown in Part (a).

   c. Show that the fact that G_i operating on $\varepsilon_{ijk} q^i(1) q^j(2) q^k(3)$ yields 0 implies that
   a general SU(3) transformation on $\varepsilon_{ijk} q^i(1) q^j(2) q^k(3)$ leaves it unchanged. This
   means that $\varepsilon_{ijk} q^i(1) q^j(2) q^k(3)$ is a singlet under SU(3) since any SU(3) operator is represented by the 1×1 matrix 1.

10. Fig. 6.5 shows the **3** and $\bar{\mathbf{3}}$ representations of SU(3). They each have an isodoublet and an isosinglet in them. Multiply these representations by combining the isospins in the product terms (i.e., an $I = \frac{1}{2}$ times an $I = \frac{1}{2}$ produces an $I = 0$, and an $I = 1$, etc.) and adding the corresponding Y's. Show that the resulting isomultiplets, with their corresponding Y values, match those that appear in an **8** plus a **1**, thus verifying that $\mathbf{3} \times \bar{\mathbf{3}} = \mathbf{8} + \mathbf{1}$ is plausible (although this is not a complete proof).

11. Do the same multiplications as outlined in Problem 6 for $\mathbf{3} \times \mathbf{3} \times \mathbf{3} = \mathbf{1} + \mathbf{8} + \mathbf{8} + \mathbf{10}$.

12. Prove that the three spin $\frac{1}{2}$ states with $S_z = +\frac{1}{2}$ constructed by coupling two spins to 0: (a) particles 1 and 2 coupled to 0; (b) particles 2 and 3 coupled to 0; (c) particles 3 and 1 coupled to 0, are not independent. Hint: Find a linear combination of them which is 0.

13. a. Show that the expression in Eq. (6.28'), $\varepsilon_{ijk} q^i(1) q^j(2) q^k(3)$, is identical to the sum in Eq. (6.28).
    b. Verify that ε_{ijk} is antisymmetric to the interchange of any two indices, and show that this implies that $\varepsilon_{ijk} q^i(1) q^j(2) q^k(3)$ is antisymmetric to the interchange of any two quarks.

14. Draw quark flow diagrams, or show that it is impossible to do so, for:
    a. $\pi^- + p \to \Lambda + \bar{K}^0$
    b. $\pi^- + p \to \Lambda + K^0$
    c. $\pi^- + n \to \Xi^- + K^0 + K^0$
    d. $K^- + p \to \Xi^0 + K^0$
    e. $p + p \to \Sigma^+ + n + K^+$
    f. $\pi^- + p \to \Xi^0 + K^0$
    g. $\pi^- + p \to \Lambda + \pi^0$

15. a. Prove that if a particle decays into two or more (product) particles, then the sum of the masses of the final particles cannot exceed the mass of the decaying particle.
    b. Construct a decay of the Ω^- to a Ξ^-, i.e., name the other decay product(s), which is allowed by the strong interaction, draw the quark flow diagram, and explain why this decay does not occur.
    c. Construct a decay of the Ω^- to a Ξ^0, which is allowed by the strong interaction, draw the quark flow diagram, and explain why this decay does not occur.
    d. Construct a decay of the Ω^- to a Σ^0, which is allowed by the strong interaction, draw the quark flow diagram, and explain why this decay does not occur.
    e. Construct a decay of the Ω^- to an N, which is allowed by the strong interaction, and explain why this decay does not occur.

Part B

The Standard
Model

Chapter 7

Quantum Electrodynamics
as a Gauge Theory

7.1 INTRODUCTION

Quantum electrodynamics is the simplest of the gauge theories and serves as a prototype for all the others. All the theories listed in the chart on the inside front cover are, in fact, gauge theories, and so it is important and instructive to understand QED from this new point of view. In this chapter we shall consider a manifestly covariant formulation of quantum electrodynamics, utilizing the Lagrange density approach discussed in Chapter 4. We will then proceed to put this theory in the form of a gauge theory, thereby defining gauge theories for the reader. We will see that QED is a gauge theory based upon the group U(1) and will show how to construct other U(1) gauge theories. In the course of this discussion, we will define a covariant derivative, which simplifies the process of constructing gauge theories.

7.2 THE LAGRANGIAN

We have shown in Section 4.4 that the Lagrangian density $\mathcal{L}$ must be Lorentz invariant, i.e., **$\mathcal{L}$ is a space-time (four-)scalar**. In order to construct a manifestly covariant formulation of electrodynamics, we must build covariant objects for all its physical entities. The physical fields in this theory are the electric ($\vec{E}$) and magnetic ($\vec{B}$) fields. We know that they mix under Lorentz transformations, even from the simple fact that a wire moving through a magnetic field has an induced $\mathcal{EMF}$ in it and consequently "sees" an electric field in its frame. There are six components of these fields (three from each) and so they cannot form a four-vector. (We have seen in Section 4.1 and Appendix B that ρ and $\vec{J}$ form a four-vector J^{μ}.)

The electric and magnetic fields may be expressed in terms of a scalar potential (V) and a vector potential ($\vec{A}$) by the following equations:

$$\vec{E} = -\vec{\nabla}V - \partial\vec{A}/\partial t \text{ and } \vec{B} = \vec{\nabla} \times \vec{A}. \tag{7.1}$$

It is shown in electrodynamics (see Appendix B) that the potentials form a four-vector $A^\mu = (V; \vec{A})$, whose components mix appropriately under Lorentz transformations. Since the fields are space and time derivatives of the potentials, we would like to construct a Lorentz-covariant object of that form, using the covariant derivative operator ∂_μ and A^μ; we will call it $F_{\mu\nu}$.

If we used $\partial_\mu A_\nu$, we would have 16 independent quantities (since μ and ν each can have four possible values), which are too many. However, if we also demand symmetry or antisymmetry, the number of independent components drops. Let us consider the antisymmetric form:

$$F_{\mu\nu} = \partial_\mu A_\nu - \partial_\nu A_\mu. \tag{7.2}$$

Since there are four possible values for each subscript and $F_{\mu\nu}$ is antisymmetric to an interchange, it follows that there are only six independent elements.[1] Since $\vec{E}$ and $\vec{B}$ together also have six components perhaps $F_{\mu\nu}$ contains only those fields. Let us check:

$$F_{01} = \partial_0 A_1 - \partial_1 A_0 = \frac{\partial(-A_x)}{\partial t} - \frac{\partial V}{\partial x} = E_x;^2 \tag{7.3}$$

likewise, F_{02} and F_{03} give the other two components of $\vec{E}$. The space-space components are the components of the magnetic field:

$$F_{12} = \partial_1 A_2 - \partial_2 A_1 = \frac{\partial(-A_y)}{\partial x} - \frac{\partial(-A_x)}{\partial y} = -B_z; \tag{7.4}$$

likewise, the other two components of $\vec{B}$ are obtained from the remaining space-space components. These six are the only independent components, since $F_{10} = -F_{01}$, $F_{21} = -F_{12}$, etc., and all diagonal elements are 0. Thus, $F_{\mu\nu}$ contains only $\vec{E}$ and $\vec{B}$.

As we have been reminded before, the Lagrangian involves energies. Recall that the energy density of an electric field $\sim E^2$ and that of a magnetic field $\sim B^2$. So we expect that two factors of $F_{\mu\nu}$ must be in each term of $\mathcal{L}$. Since $\mathcal{L}$ is a scalar, the indices must be paired and summed over (contracted) just as occurs in the scalar product of two four-vectors. This gives us

$$\mathcal{L}_{EM} = -\tfrac{1}{4} F^{\mu\nu} F_{\mu\nu}, \tag{7.5}$$

where the proportionality constant $(-\tfrac{1}{4})$ is appropriate for the units used in this book. ($F^{\mu\nu}$ is related to $F_{\mu\nu}$ the same way b_μ is related to b^μ, viz., lowering a

---

1. Since $F_{\mu\nu} = -F_{\nu\mu}$ implies that $F_{00} = -F_{00}$ (and similarly for the 11, 22, and 33 components), the diagonal elements must be 0. The remaining 12 (off-diagonal) components pair into six independent sets of two (an off-diagonal element and its transpose); only one element in each set is independent, the other being its negative. On the other hand, the symmetric form has 10 independent components, four diagonal and six independent off-diagonal elements.

2. Recall that $A_1 = -A^1 = -A_x$.

space index (1,2,3) gives a factor of -1; lowering a time index leaves the value unchanged. For example, $F^{01} = -F_{01}$, $F^{12} = +F_{12}$, etc.)

Next we must include the electron in our theory. This introduces the terms

$$\mathcal{L}_{DIRAC} + \mathcal{L}_{INT}$$

into $\mathcal{L}$, where

$$\mathcal{L}_{DIRAC} = \bar{\psi} i \gamma^\mu \partial_\mu \psi - m \bar{\psi} \psi,^3 \tag{7.6}$$

and, as we have seen in Eq. (4.58), the interaction term is

$$\mathcal{L}_{INT} = e \bar{\psi} \gamma^\mu \psi A_\mu. \tag{7.7}$$

This interaction is called **minimal coupling**, because more complicated coupling terms do not appear. All together, these individual Lagrange densities add to the total Lagrange density:

$$\mathcal{L} = \mathcal{L}_{EM} + \mathcal{L}_{DIRAC} + \mathcal{L}_{INT}$$
$$= -\tfrac{1}{4} F^{\mu\nu} F_{\mu\nu} + \bar{\psi} i \gamma^\mu \partial_\mu \psi - m \bar{\psi} \psi + e \bar{\psi} \gamma^\mu \psi A_\mu, \tag{7.8}$$

which produces the correct field equations for ψ and A_μ.

7.3   THE GAUGE SYMMETRY

We have learned in electrodynamics that the potentials are not uniquely defined. For example, in electrostatics V can be shifted by a constant and still describe the same electric field ($\vec{E} = -\vec{\nabla} V$); $\vec{A}$ can be shifted by the gradient of any scalar function (χ) and still describe the same magnetic field ($\vec{B} = \vec{\nabla} \times \vec{A}$), since the curl of a gradient is 0. A *time-dependent* shift of $\vec{A}$ of this sort, though, will change $\vec{E}$, as shown in Eq. (7.1), which relates $\vec{E}$ to the potentials. However, if this shift of $\vec{A}$ is accompanied by a suitable shift in V, we may have the same $\vec{E}$ field after all. This can, in fact, be accomplished by the following transformation:

$$A_\mu \rightarrow A'_\mu = A_\mu + \partial_\mu \chi, \text{ a \textbf{gauge transformation}.} \tag{7.9}$$

This means that a shift in $\vec{A}$ by $\vec{\nabla}\chi$ must be accompanied by a shift in V by $\partial_0 \chi$, so that we obtain, from Eq. (7.3),

$$E'_x = \partial_0(A_1 + \partial_1 \chi) - \partial_1(A_0 + \partial_0 \chi) = \partial_0 A_1 - \partial_1 A_0 = E_x,$$

where the extra terms have canceled (because $\partial_0 \partial_1 \chi = \partial_1 \partial_0 \chi$). More generally we see that

$$F_{\mu\nu} \rightarrow F'_{\mu\nu} = \partial_\mu(A_\nu + \partial_\nu \chi) - \partial_\nu(A_\mu + \partial_\mu \chi) = \partial_\mu A_\nu - \partial_\nu A_\mu = F_{\mu\nu}, \tag{7.10}$$

---

3. This gives the Dirac equation, when the Euler-Lagrange equations are used. (See Appendix C.)

because $\partial_\mu \partial_\nu \chi = \partial_\nu \partial_\mu \chi$, so that the extra terms cancel. This shows that the fields are unchanged by the gauge transformation of Eq. (7.9).

Furthermore, notice that if we change ψ by a phase $\psi \to \psi' = e^{-i\alpha}\psi$ (which implies that $\bar{\psi} \to \bar{\psi}e^{i\alpha}$), then $\mathcal{L}(\psi') = \mathcal{L}(\psi)$, i.e., $\mathcal{L}$ is invariant to a change of phase. (This would be true also for the field equations, which are derivable from $\mathcal{L}$). This is referred to as a **global** transformation, since the transformation $(e^{-i\alpha})$ is the *same* at each point in space-time. What about a transformation which changes the phase *differently* at different points of space-time? We have said that overall phase changes should not change results, but this is a change which is **local**, i.e., the change depends on location, so that we must replace α by $\alpha(x^\nu)$. It is obvious that $\mathcal{L}_{EM}$ (Eq. 7.5) and $\mathcal{L}_{INT}$ (Eq. 7.7) are unchanged by a local phase transformation of ψ.

However, let us consider $\mathcal{L}_{DIRAC}$. The first (kinetic energy) term contains a derivative operator, so that we obtain

$$
\begin{aligned}
\mathcal{L}_{DIRAC}(\psi') &= \bar{\psi}e^{i\alpha(x^\nu)}(i\gamma^\mu \partial_\mu - m)e^{-i\alpha(x^\nu)}\psi \\
&= \bar{\psi}e^{i\alpha(x^\nu)}i\gamma^\mu \partial_\mu(e^{-i\alpha(x^\nu)}\psi) - m\bar{\psi}\psi \\
&= \bar{\psi}e^{i\alpha(x^\nu)}\{i\gamma^\mu e^{-i\alpha(x^\nu)}\partial_\mu\psi + i\gamma^\mu \partial_\mu(e^{-i\alpha(x^\nu)})\psi\} - m\bar{\psi}\psi \\
&= \bar{\psi}\{i\gamma^\mu \partial_\mu\psi + i\gamma^\mu e^{i\alpha(x^\nu)}\partial_\mu(e^{-i\alpha(x^\nu)})\psi\} - m\bar{\psi}\psi \\
&= \bar{\psi}(i\gamma^\mu \partial_\mu - m)\psi + \bar{\psi}\gamma^\mu(\partial_\mu\alpha(x^\nu))\psi.
\end{aligned}
$$

Thus we obtain

$$\mathcal{L}_{DIRAC}(\psi') = \mathcal{L}_{DIRAC}(\psi) + \bar{\psi}\gamma^\mu\psi\partial_\mu\alpha(x^\nu), \tag{7.11}$$

so that $\mathcal{L}_{DIRAC}$ is not invariant to a *local* phase transformation. The extra term in Eq. (7.11), proportional to $\partial_\mu\alpha$, has a form like $\mathcal{L}_{INT}$, so that we can write those terms together:

$$\mathcal{L}(\psi',A) = \mathcal{L}_{DIRAC}(\psi) + \mathcal{L}_{EM}(A) + e\bar{\psi}\gamma^\mu\psi\left(A_\mu + \frac{1}{e}\partial_\mu(\alpha(x^\nu))\right) \tag{7.12}$$

under a **local** phase transformation. Now let us shift A_μ by a gauge transformation, setting $\chi = -(1/e)\alpha(x^\nu)$. We obtain

$$A_\mu \to A'_\mu = A_\mu - \frac{1}{e}\partial_\mu\alpha(x^\nu). \tag{7.13}$$

If we replace A_μ by A'_μ in $\mathcal{L}$ (Eq. 7.12), we see that the term generated by the derivative operator is canceled by the gauge shift of A_μ (the $\partial_\mu\alpha$ term in Eq. 7.13). Furthermore, Eq. (7.10) has shown that $F_{\mu\nu}$ is unchanged by gauge transformations of A_μ; consequently, $\mathcal{L}_{EM}$ is invariant to such a (gauge) transformation as well. Therefore, $\mathcal{L}$ is invariant to a combination of local phase and gauge transformations:

$$\mathcal{L}(\psi',A') = \mathcal{L}_{DIRAC}(\psi) + \mathcal{L}_{EM}(A) + \mathcal{L}_{INT}(\psi,A) = \mathcal{L}(\psi,A), \tag{7.14}$$

for $\psi' = e^{-i\alpha(x^\nu)}\psi$ and $A'_\mu = A_\mu - (1/e)\partial_\mu\alpha(x^\nu)$ — **a local phase and gauge transformation**. It is left for the reader to verify that the presence of another field of charge ne in the theory would not spoil the gauge invariance if that second field transformed as $\psi'_2 = e^{in\alpha(x)}\psi_2$. (See Problem 5.)

Recall that in previous chapters we have discussed transformations of ψ of the form $\psi \to e^{i\vec{G}\cdot\vec{\alpha}}\psi$, where $\vec{G}$ were operators which could be represented by matrices (and were referred to as the generators of the transformations). We saw that such transformations of ψ are all unitary. Here we have the same form, but the exponent is merely a number; therefore, $e^{i\alpha}$ is also a number, which may be referred to as a **1 × 1 unitary matrix**. The set of unitary 1×1 matrices of this phase transformation is referred to as U(1). The transformation under which this theory is invariant is **a local U(1) (phase) transformation accompanied by a gauge transformation**. The above structure is therefore called a **U(1) gauge theory. The electromagnetic field is referred to as the gauge field.** Furthermore, the extension to more charged fields, mentioned at the end of the last paragraph, shows that the transformation for a field of charge Qe is $\psi' = e^{iQ\alpha(x^\nu)}\psi$, so that we may consider Q as **the generator.**

The structure of this U(1) gauge theory is the underlying structure of *all* gauge theories. The ∂_μ operator always produces an extra $(\partial_\mu\alpha)$ term, and the gauge field in $\mathcal{L}_{INT}$ shifts by a four-divergence of exactly the correct amount to cancel that term. The form of $\mathcal{L}_{INT}$ is *completely determined* by the requirement that this cancellation occur. Symmetries are ordinarily thought to restrict the *form* of the interaction, so that the *determination* **of the interaction itself by a gauge symmetry** is both extraordinary and intriguing. In Chapter 8, we shall see, in detail, that this noteworthy feature is present in all gauge theories.

7.4   THE CONSTRUCTION OF A U(1) GAUGE THEORY

7.4.1   The Covariant Derivative

In this section we will define the covariant derivative (D_μ) and will see how to employ it to construct U(1) gauge theories. Because of the structure of D_μ, the minimal coupling interaction arises *automatically* in all U(1) gauge theories.

In Section 7.3 we saw how the effect of the derivative operator (in the second term in Eq. 7.8) acting on a local phase transformation is compensated for by the presence of an interaction term of the same Dirac form with $(-ie)A_\mu$ (the fourth term in Eq. 7.8) in place of ∂_μ. If we pull the ∂_μ and $(-ie)A_\mu$ terms together, we can rewrite the Lagrange density for the electron (calling it $\mathcal{L}_\psi$) in a simple and suggestive form:

$$\mathcal{L}_\psi = \bar{\psi}i\gamma^\mu\partial_\mu\psi + \bar{\psi}e\gamma^\mu A_\mu\psi = \bar{\psi}i\gamma^\mu D_\mu\psi, \tag{7.15}$$

where $D_\mu = \partial_\mu - ieA_\mu$ for the electron, and, correspondingly,

$$D_\mu = \partial_\mu + ieQA_\mu \tag{7.16}$$

for any charged particle; here Q is the generator of the gauge transformation. D_μ **is called the covariant derivative.**[4]

We infer that the cancellation of the $\partial_\mu \alpha$ terms would always occur as long as ∂_μ is everywhere accompanied by A_μ (in $\mathcal{L}_\psi$) in the combination D_μ. In fact, the local phase plus gauge transformation affects the covariant derivative of the electron field as follows:

$$\text{for } \psi \to \psi' = e^{-i\alpha(x^\nu)}\psi \text{ and } A_\mu \to A'_\mu = A_\mu - \left(\frac{1}{e}\right)\partial_\mu\alpha(x^\nu),$$

$$D_\mu\psi \to (D_\mu\psi)' = e^{-i\alpha(x^\nu)}D_\mu\psi, \tag{7.17}$$

since the $\partial_\mu\alpha$ terms cancel, so that $D_\mu\psi$ transforms like the terms that do not have derivatives (e.g., ψ itself). Reading Eq. (7.15) from right to left, we may say that the interaction term *automatically* arises when we use the covariant derivative (D_μ) and **it is exactly the minimal coupling interaction of electrodynamics.**

Suppose we consider a free, noninteracting particle, i.e., we start with a theory that contains no interaction. The kinetic energy part of $\mathcal{L}$ has derivative operators, so that it is not invariant to a local phase transformation. Let us modify it so that it becomes a theory which *is* invariant to local phase transformations. As we have seen, this requires the existence of a vector field (call it B_μ), whose "response" to a local phase transformation of the particle field is a shift by a total divergence, so as to cancel the extra term that the derivative operator produces. This is neatly accomplished by replacing ∂_μ by D_μ in $\mathcal{L}$:

$$D_\mu = \partial_\mu + igB_\mu,$$

$$\text{or } D_\mu = \partial_\mu + igGB_\mu \text{ for several fields,} \tag{7.18}$$

where G is the generator, which can have different values for different fields, and the coupling constant g may be arbitrarily chosen. B_μ is referred to as the gauge field. Thus, the demand of local phase invariance, or **local U(1) invariance, requires the introduction of a gauge field, such that the interaction with that gauge field is *completely* determined** (up to a coupling constant).

7.4.2 $\mathcal{L}_B$ and the Masslessness of the Gauge Field

The field B_μ must be a propagating field, so that it requires an $\mathcal{L}_B$ (of its own) analogous to $\mathcal{L}_{EM}$. Of course, $\mathcal{L}_B$ must be invariant to the gauge transformation (the shift by a total divergence) that B_μ was introduced to undergo.

The free part of $\mathcal{L}_B$ must be quadratic in B and should have two derivatives in the kinetic-energy-type term in order to yield the Klein-Gordon equation for B_μ. Thus, to be gauge invariant, the kinetic-energy term can contain the fields only in the combination $F^{\mu\nu}$, i.e., $\mathcal{L}_B \sim F^{\mu\nu}(B)F_{\mu\nu}(B)$.

---

4. It is analogous to that which arises in general relativity when we consider parallel displacement. This kind of invariance has been succinctly described in modern mathematics by the use of fiber bundles.

What about a mass term? How would such a term look? Recall from Chapter 4 that fields which are boson in nature obey the Klein-Gordon equation: [5]

$$(\Box + m_B^2)B_\mu = 0,$$

$$\text{where } \Box \equiv \partial\mu\partial_\mu, [6] \tag{7.19}$$

arising from terms in $\mathcal{L}_B$ which are quadratic in B_μ (as shown in Appendix C). The mass term in $\mathcal{L}_B$ would be proportional to $m_B^2 B^\mu B_\mu$. Under the gauge transformation, we have $m_B^2 B^\mu B_\mu \rightarrow m_B^2 (B^\mu - g^{-1}\partial^\mu\alpha)(B_\mu - g^{-1}\partial_\mu\alpha)$, so that it is obviously not gauge invariant, since the extra terms which depend on $\partial_\mu\alpha$ do not cancel. Thus, we conclude that such a term cannot appear in the Lagrangian, which tells us that the gauge fields must be massless. Nevertheless, we have seen that quantum field theory produces a renormalization of the masses (in general by an infinite amount), so perhaps this result is avoided. The very structure of gauge field theories precludes such a renormalization, however, since the interaction terms do not violate the gauge invariance. Thus, **the mass of the gauge particle is exactly 0.**

7.4.3   The Full Structure

The discussion above shows that, by demanding U(1) gauge invariance, we obtain a theory with the structure of electrodynamics — in electrodynamics $g = e$, $G = Q$, and B_μ is called A_μ.

Summarizing:

1. We start with a free $\mathcal{L}$.
2. We replace ∂_μ by $D_\mu = \partial_\mu + igB_\mu$, or $D_\mu = \partial_\mu + igGB_\mu$ (when several different particles are present), where G is the generator of the U(1) symmetry.
3. We then introduce a gauge invariant $\mathcal{L}_B \sim F^{\mu\nu}(B)F_{\mu\nu}(B)$, to obtain a total $\mathcal{L}$ which is invariant under $\psi \rightarrow e^{i\alpha(x^\nu)}\psi$ or $\psi \rightarrow e^{iG\alpha(x^\nu)}\psi$ (when several different particles are present) together with

$$B_\mu \rightarrow B_\mu - \left(\frac{1}{g}\right)\partial_\mu\alpha(x^\nu). \tag{7.20}$$

In the following chapter, we shall see how this gauge theory serves as a prototype for groups larger than U(1).

PROBLEMS

1. Show that (in a vacuum) Maxwell's inhomogeneous equations (involving ρ and $\vec{J}$) can be written in a manifestly covariant form.

---

5. This came from the generalization of $p^\mu p_\mu - m^2 = 0$ to quantum mechanics, which gave $(\Box + m^2)\Phi = 0$ for a scalar field. For a vector field the same equation is valid: $(\Box + m^2)A_\mu = 0$.

6. $\Box$ is called the D'Alembertian.

2. a. Check that $F^{\mu\nu}F_{\mu\nu} = 2(\vec{B})^2 - 2(\vec{E})^2$, as expected for the Maxwell-Lagrange density.
   b. Use the Euler-Lagrange equations to derive the inhomogeneous Maxwell's equations from $\mathcal{L} = \mathcal{L}_{DIRAC} + \mathcal{L}_{EM} + \mathcal{L}_{INT}$.
3. Derive the Dirac equation (including the electromagnetic interaction) from $\mathcal{L} = \mathcal{L}_{DIRAC} + \mathcal{L}_{EM} + \mathcal{L}_{INT}$.
4. Consider a complex scalar field ϕ whose free Lagrangian is $\mathcal{L}_0 = (\partial^\mu\phi)^*(\partial_\mu\phi) - \mu^2\phi^*\phi$.
   a. Use the requirement of gauge (and local phase) invariance to construct the Lagrangian for the corresponding U(1) gauge theory of electromagnetism. Let q be the charge of the field.
   b. What is the interaction part of $\mathcal{L}$?
   c. Can you see why a real scalar field must be electrically neutral?
   d. How do the number of degrees of freedom differ for a real scalar field and a complex scalar field? How is this related to the fact that a complex scalar field represents a charged particle, but a real scalar field does not?
5. Show that a theory with two charged particles, one of charge e and the second of charge ne, is a U(1) gauge theory, where the transformations are

$$\psi_1 \to e^{i\alpha(x^\nu)}\psi_1, \ \psi_2 \to e^{in\alpha(x^\nu)}\psi_2, \text{ and } A_\mu \to A_\mu - \frac{1}{e}\partial_\mu\alpha.$$

Chapter 8

Yang-Mills Theories

8.1 INTRODUCTION

In the 1950s, long before the quarks or even flavor SU(3) was proposed, Yang and Mills wrote a classic paper generalizing the gauge theory structure of the electromagnetic field. In this generalization, the role played by the (abelian) gauge group U(1) in electrodynamics is played instead by the nonabelian gauge group SU(2) of an "internal" symmetry—Yang and Mills asked what would happen if we demanded that our particle theory have *local SU(2)* "internal" symmetry. Once this generalization to a nonabelian gauge theory was fully understood, the structure of the gauge theory for any nonabelian Lie group became apparent. In fact, there is a field of mathematics (called fiber bundles) that contains a more abstract approach to the same mathematical concepts discovered by Yang and Mills.[1] However, their paper was largely ignored by physicists for about 10 years because the masslessness of the gauge particles was inconsistent with short-range interactions (as we saw in Chapter 4) like the strong and the weak interactions. During that time the concept of spontaneous symmetry breaking in a gauge theory was proposed and developed to overcome that masslessness problem in the weak interaction, as we shall discuss in Chapter 11. (A different mechanism accounts for the apparent short range of the strong interaction.)

The standard model is, in fact, composed of nonabelian gauge theories, and the structure and insights of the original Yang-Mills paper form the basis of our understanding of these theories. In this chapter we will examine the essential features of the original (nonabelian) SU(2) gauge theory studied by Yang and Mills (as $\vec{T}$, isotopic spin), because those features appear for all the larger nonabelian groups as well.

---

1. Indeed, mathematicians refer to a whole class of theories as "Yang-Mills theories."

8.2   (ISOTOPIC) SU(2) GAUGE THEORY

Consider the case where the particle states (and hence the field operators ψ) form a $T = \frac{1}{2}$ irreducible representation of the isotopic SU(2) — ψ is an isospinor and $\vec{T} = \frac{1}{2}\vec{\tau}$ (where the arrow signifies an isovector and $\vec{\tau}$ are the usual Pauli matrices.) Then, under a local SU(2) transformation, analogous to our treatment for a local U(1) transformation, we find that

$$\psi \to \psi' = e^{i(\vec{\tau}/2)\cdot\vec{\theta}}\psi = e^{i\vec{\tau}\cdot\vec{\alpha}}\psi, \text{ where } \vec{\alpha} \equiv \frac{\vec{\theta}}{2} \text{ is an isotopic vector, and}$$

$$\partial_\mu\psi \to \partial_\mu\psi' = \partial_\mu(e^{i\vec{\tau}\cdot\vec{\alpha}}\psi) = e^{i\vec{\tau}\cdot\vec{\alpha}}(\partial_\mu + i\vec{\tau}\cdot\partial_\mu\vec{\alpha})\psi. \tag{8.1}$$

Again we wish to introduce a gauge field to absorb the unwanted $\partial_\mu\vec{\alpha}$ term. We shall see that to cancel the term in question, **the gauge field must be a vector in isospace**. Consequently we introduce an isovector field $\vec{G}_\mu$, which is also a four-vector in space-time (as indicated by the subscript μ). In other words, there are three four-vector fields, $G_{1\mu}$, $G_{2\mu}$, and $G_{3\mu}$, which "rotate" into each other under SU(2) isospin "rotations." Now we replace $\partial_\mu\psi$ with $D_\mu\psi$ everywhere, where the covariant derivative is defined by

$$D_\mu\psi = \left(\partial_\mu + ig\frac{\vec{\tau}}{2}\cdot\vec{G}_\mu\right)\psi, \tag{8.2}$$

using $\frac{1}{2}\vec{\tau}$ for the generators, since ψ is a ($T = \frac{1}{2}$) spinor. (Notice that, in the covariant derivative, the gauge field $\vec{G}_\mu$ has to be an isovector to combine correctly with the generators to form an isoscalar.)

Let us investigate the behavior of this covariant derivative (reversing the procedure we used in Chapter 7). We have learned from our study of electrodynamics (see Eq. 7.17) that we should require

$$D_\mu'\psi' = e^{i\vec{\tau}\cdot\vec{\alpha}}D_\mu\psi. \tag{8.3}$$

From Eq. (8.1) we have $\psi = e^{-i\vec{\tau}\cdot\vec{\alpha}}\psi'$. Inserting this and Eq. (8.2) into the right-hand side of Eq. (8.3), we obtain

$$D_\mu'\psi' = e^{i\vec{\tau}\cdot\vec{\alpha}}D_\mu\psi = e^{i\vec{\tau}\cdot\vec{\alpha}}\left(\partial_\mu + ig\frac{\vec{\tau}}{2}\cdot\vec{G}_\mu\right)e^{-i\vec{\tau}\cdot\vec{\alpha}}\psi'$$

$$= \left(\partial_\mu - i\vec{\tau}\cdot(\partial_\mu\vec{\alpha}) + ige^{i\vec{\tau}\cdot\vec{\alpha}}\left(\frac{\vec{\tau}}{2}\cdot\vec{G}_\mu\right)e^{-i\vec{\tau}\cdot\vec{\alpha}}\right)\psi'. \tag{8.4}$$

Therefore, the resulting expression in parentheses on the right-hand side of Eq. (8.4) should be D_μ'. The form in Eq. (8.2) shows us, therefore, that

$$\partial_\mu + ig\frac{\vec{\tau}}{2}\cdot\vec{G}_\mu' = \partial_\mu - i\vec{\tau}\cdot(\partial_\mu\vec{\alpha}) + ige^{i\vec{\tau}\cdot\vec{\alpha}}\left(\frac{\vec{\tau}}{2}\cdot\vec{G}_\mu\right)e^{-i\vec{\tau}\cdot\vec{\alpha}}, \tag{8.5}$$

so that (after reordering)

$$\frac{\vec{\tau}}{2} \cdot \vec{G}'_\mu = e^{i\vec{\tau}\cdot\vec{\alpha}} \left(\frac{\vec{\tau}}{2} \cdot \vec{G}_\mu \right) e^{-i\vec{\tau}\cdot\vec{\alpha}} - \frac{1}{g}\, \vec{\tau}\cdot\partial_\mu\vec{\alpha}. \tag{8.6}$$

In our investigation of electrodynamics as a gauge theory (in Chapter 7), we did not encounter a term like the first term in Eq. (8.6). However, it has just the structure we should expect, for $\vec{G}$ **is an isovector** and it should rotate like one. In fact, since $\vec{\tau}$ rotates like an isovector, we may write for that term

$$e^{i\vec{\tau}\cdot\vec{\alpha}} \left(\frac{\vec{\tau}}{2} \cdot \vec{G}_\mu \right) e^{-i\vec{\tau}\cdot\vec{\alpha}} = e^{i\vec{\tau}\cdot\vec{\alpha}} \frac{\tau_i}{2} e^{-i\vec{\tau}\cdot\vec{\alpha}} G^i_\mu = R^T_{ij} \frac{\tau_j}{2}\, G^i_\mu, \tag{8.7}$$

where the Latin (i,j) indices refer to isospace, the Greek (μ) subscript refers to space-time, and R^T_{ij} is the transpose of the usual rotation matrix for a vector.[2] Substituting Eq. (8.7) into Eq. (8.6) (and multiplying by 2), we find

$$\tau_j G'^j_\mu = R^T_{ij}\tau_j G^i_\mu - \frac{2}{g}\, \tau_j\partial_\mu\alpha^j. \tag{8.8}$$

The independence of the Pauli matrices allows us to "peel" them from each side of the equation[3] to obtain

$$G'^j_\mu = R_{ji} G^i_\mu - \frac{2}{g}\, \partial_\mu\alpha^j.\text{[4]} \tag{8.9}$$

The gauge field has been rotated (first term), as a vector field should, and shifted by a covariant derivative (second term) as well.

Now we must introduce the kinetic-energy terms for the gauge fields, as we have done in electrodynamics. To this end, we wish to construct a tensor analogous to $F_{\mu\nu}$ but which will also behave like an *isovector* (since $\vec{G}$ does) under rotations in isospace. Let us try

$$F^i_{\mu\nu} \stackrel{?}{=} \partial_\mu G^i_\nu - \partial_\nu G^i_\mu.$$

First consider an isorotation by $\vec{\theta}$. Using Eq. (8.9), we see that

$$\partial_\mu G'^j_\nu - \partial_\nu G'^j_\mu = R_{ji}(\partial_\mu G^i_\nu - \partial_\nu G^i_\mu) + (\partial_\mu R_{ji}(\vec{\theta}))G^i_\nu - (\partial_\nu R_{ji}(\vec{\theta}))G^i_\mu. \tag{8.10}$$

The first term is what we would expect if this (space) four-tensor rotated like an isovector under isotopic rotations. However, the last two terms in Eq. (8.10) (which depend on the space-time derivatives of θ) show that the fact that $\vec{G}$ is a vector in isospace leads to spurious terms for *local* isotopic rotations.

---

2. Equation (8.7) contains the transpose matrix (which is also the inverse) because we have proceeded in this reversed order, so that $U\tau_i U^\dagger$ appears rather than $U^\dagger\tau_i U$. See Problem 1.

3. This is easily accomplished by taking the anticommutator with respect to τ_1, then starting again but with respect to τ_2, and then with respect to τ_3. Try it! Alternatively, we can multiply by τ_1 and take the trace (since $\text{Tr}(\tau_i\tau_j) = 2\delta_{ij}$), etc.

4. $R^T_{ij} = R_{ji}$.

To understand these extra terms in a simple way, we begin from Eq. (8.6). Recall that the rotation group is continuous; because of this we can learn what we need by considering an infinitesimal transformation. For $\vec{\alpha}$ infinitesimal we have

$$e^{\pm i\vec{\tau}\cdot\vec{\alpha}} \approx 1 \pm i\vec{\tau}\cdot\vec{\alpha},$$

$$
\begin{aligned}
e^{i\vec{\tau}\cdot\vec{\alpha}}(\vec{\tau}\cdot\vec{G}_\mu)e^{-i\vec{\tau}\cdot\vec{\alpha}} &= (1 + i\tau_i\alpha^i)(\tau_j G_\mu^j)(1 - i\tau_i\alpha^i) \\
&= \tau_j G_\mu^j + i\tau_i\tau_j\alpha^i G_\mu^j - i\tau_j\tau_i\alpha^i G_\mu^j \\
&= \vec{\tau}\cdot\vec{G}_\mu + i[\tau_i,\tau_j]\alpha^i G_\mu^j \quad \text{to } O(\alpha),
\end{aligned}
\tag{8.11}
$$

so that we obtain, from Eq. (8.6),

$$\vec{\tau}\cdot\vec{G}_\mu' = \vec{\tau}\cdot\vec{G}_\mu + i[\tau_i,\tau_j]\alpha^i G_\mu^j - \frac{2}{g}\vec{\tau}\cdot\partial_\mu\vec{\alpha} \quad \text{to } O(\vec{\alpha}). \tag{8.12}$$

This yields[5]

$$\vec{G}_\mu' = \vec{G}_\mu - 2(\vec{\alpha}\times\vec{G}_\mu) - \frac{2}{g}\partial_\mu\vec{\alpha} \quad \text{to } O(\alpha) \text{ or}$$

$$G_\mu'^i = G_\mu^i - 2\varepsilon_{ijk}\alpha^j G_\mu^k - \frac{2}{g}\partial_\mu\alpha^i \quad \text{to } O(\alpha). \tag{8.13}$$

Following the same procedure for a general Lie group yields

$$G_\mu'^i = G_\mu^i - 2c_{ijk}\alpha^j G_\mu^k - \frac{2}{g}\partial_\mu\alpha^i \quad \text{to } O(\alpha). \tag{8.14}$$

(See Problem 3.) From the form of Eq. (8.13), we infer that the extra terms which arose for *local* transformations on the right-hand side in Eq. (8.10) must come from the derivative acting on $\vec{\alpha}$ in the $\vec{\alpha}\times\vec{G}_\mu$ (second) term. (The third term in Eq. (8.13) produces terms $\sim(\partial_\mu\partial_\nu\vec{\alpha} - \partial_\nu\partial_\mu\vec{\alpha})$, which are obviously 0.) The $\partial_\mu\vec{\alpha}$ terms, arising from Eq. (8.13), are

$$-2\{(\partial_\mu\vec{\alpha})\times\vec{G}_\nu - (\partial_\nu\vec{\alpha})\times\vec{G}_\mu\} \quad \text{to } O(\alpha). \tag{8.15}$$

This form suggests that we need a $\vec{G}_\mu\times\vec{G}_\nu$ term in $\vec{F}_{\mu\nu}$ to cancel such terms. If we modify the definition of the gauge four-tensor to

$$\vec{F}_{\mu\nu} = \partial_\mu\vec{G}_\nu - \partial_\nu\vec{G}_\mu - g\vec{G}_\mu\times\vec{G}_\nu \text{ or}$$

$$F_{\mu\nu}^i = \partial_\mu G_\nu^i - \partial_\nu G_\mu^i - g\varepsilon_{ijk}G_\mu^j G_\nu^k, \tag{8.16}$$

we succeed in constructing a tensor which behaves like an isovector under *local* isorotations. This is easily checked by writing $\vec{F}_{\mu\nu}(\vec{G}')$ and substituting for $\vec{G}'$ in terms of $\vec{G}$ to $O(\alpha)$ from Eq. (8.13); see Problem 2. The second form in

---

5. We have used the fact that the commutators for the Pauli matrices are $[\tau_i,\tau_j] = 2i\varepsilon_{ijk}\tau_k$, where ε_{ijk} is the Levi-Civita symbol defined in Chapter 4 and $\varepsilon_{ijk}\alpha^i G^j\tau_k = \vec{\tau}\cdot\vec{\alpha}\times\vec{G}$. We then "peeled off" the $\vec{\tau}$.

Eq. (8.16) shows a structure which is easily generalized to any nonabelian gauge theory, since ε_{ijk} are the structure constants of SU(2) (and the rotation group O(3) as well):

$$F^i_{\mu\nu} = \partial_\mu G^i_\nu - \partial_\nu G^i_\mu - g c_{ijk} G^j_\mu G^k_\nu, \qquad (8.17)$$

where c_{ijk} are the structure constants and $F^i_{\mu\nu}$ and G^i_μ each form an adjoint representation (e.g., an **8** for SU(3), with $i = 1, \ldots, 8$) of the gauge group. Therefore, i, j, and k run from 1 to n = dimension of the adjoint representation of the group ($n = 3$ for SU(2) and SO(3), and $n = 8$ for SU(3), etc.). (See Problems 2, 3, and 6.)

The SU(2)-invariant Lagrangian for the gauge field is then

$$\mathcal{L}_G \propto \vec{F}^{\mu\nu} \cdot \vec{F}_{\mu\nu} = (\partial^\mu \vec{G}^\nu - \partial^\nu \vec{G}^\mu - g\vec{G}^\mu \times \vec{G}^\nu) \cdot (\partial_\mu \vec{G}_\nu - \partial_\nu \vec{G}_\mu - g\vec{G}_\mu \times \vec{G}_\nu)$$

$$= (\partial^\mu G^{i\nu} - \partial^\nu G^{i\mu} - g\varepsilon_{ijk} G^{j\mu} G^{k\nu})(\partial_\mu G^i_\nu - \partial_\nu G^i_\mu - g\varepsilon_{ijk} G^j_\mu G^k_\nu),$$

$$(8.18)$$

with a sum over all repeated indices. Since $\vec{F}$ has terms which are linear (the first two) and terms which are quadratic (in the last term) in the gauge field ($\vec{G}$), $\mathcal{L}_G$ **has quadratic, cubic, and quartic terms (four factors of $\vec{G}$)**. The quadratic terms are the usual kinetic-energy terms, but the cubic and quartic terms represent interactions (vertices) of the gauge particles among themselves. The cubic terms are three-gauge-boson vertices and the quartic terms are four-gauge-boson vertices. Consequently, the gauge particles are not neutral particles with respect to "isocharge" (the way the photon is neutral with respect to electric charge). For the general case, $\mathcal{L}_G$ has the same form as Eq. (8.18), with the structure constants c_{ijk} replacing ε_{ijk}:

$$\mathcal{L}_G \propto (\partial^\mu G^{i\nu} - \partial^\nu G^{i\mu} - g c_{ijk} G^{j\mu} G^{k\nu})(\partial_\mu G^i_\nu - \partial_\nu G^i_\mu - g c_{ijk} G^j_\mu G^k_\nu), \quad (8.19)$$

consequently, **three-gauge-boson vertices and four-gauge-boson vertices appear in all nonabelian gauge theories.**

This non-neutrality is *expected* since **the gauge fields of a nonabelian gauge theory form a nontrivial (adjoint) representation of the group.** This may be regarded as a consequence of the fact that we must obtain an invariant when the gauge fields are contracted with the generators (which form a *nontrivial* adjoint representation of the group) in the covariant derivative or the fact that the field must absorb a $\partial_\mu \alpha^i$ term (with α^i forming an adjoint representation of the group). In order to produce an invariant, upon such a contraction, the gauge fields must also belong to (form) an adjoint representation of the group. (See Problem 6.) Furthermore, we saw in Eq. (8.7) that the generators form a nontrivial representation of the group because they do not commute; hence, the root cause is that **the group is nonabelian.**[6]

---

6. Remember that the photon was not rotated by the phase rotations of electrodynamics, where the gauge group (U(1)) is *abelian*.

We have found here that more interactions are necessary for *nonabelian* gauge theories than were required for U(1) gauge theories (like electrodynamics). However, it is still true that **all the interactions are completely prescribed by the gauge symmetry**. Thus, the role of the *local* symmetry is predominant, as opposed to the rather mild requirements expected of the usual (global) symmetries.[7]

Summarizing what we have learned **for nonabelian gauge groups**:

1. **The gauge fields must form an adjoint representation (as do the generators) of the gauge group, i.e., they possess "charge."**
2. **The gauge fields have vertices with three gauge particles interacting and vertices with four gauge particles interacting.**

We may again ask about a possible mass term, and again the answer is that any term of the form $(m^2 \vec{G}^\mu \cdot \vec{G}_\mu)$ is not gauge invariant. It is also of crucial concern that the theory would not be renormalizable with a mass term in $\mathcal{L}$ for the (four-vector) gauge particles. So we write:

3. **The gauge particles must be massless.**

These properties are characteristic of any nonabelian gauge group.

Property 3, which is also true for U(1) gauge theories, should make the reader ask: How can the electroweak theory, which is supposedly a gauge theory, have gauge particles (the W's and Z) whose masses are ~90 times a proton mass? The answer to this question took about a decade to emerge from a synthesis of the work of several people. In Chapter 11 we will study the structure of the spontaneous symmetry breaking involved.

PROBLEMS

1. a. Check that $e^{i\vec{\tau}\cdot\vec{\alpha}}\tau_i e^{-i\vec{\tau}\cdot\vec{\alpha}} = R_{ij}^T \tau_j$, for a rotation about the z axis by angle θ ($\alpha = \theta/2$), where R_{ij}^T is the transpose of the correct three-dimensional rotation matrix for θ. This verifies Eq. (8.7).

   Recall that $e^{i\tau_3\alpha} = \cos\alpha + i\tau_3 \sin\alpha$.

   b. Do the same check for a rotation about the y axis.

   Hint: $e^{i\tau_2\alpha} = \cos\alpha + i\tau_2 \sin\alpha$, as shown in Problem 15 of Chapter 3.

   c. Show that the relationship, checked in part a (for a rotation about the z axis) and in part b (for a rotation about the y axis), is true for a general rotation. Use the results of parts a and b and the fact that a general rotation can be written as $R = R_z(\theta_3)R_y(\theta_2)R_z(\theta_1)$. This can be done without explicitly multiplying out the matrices.

   Hint: The theorem we proved in Problem 5 of Chapter 2 may be useful.

---

7. For example, (global) rotational symmetry, in classical or quantum mechanics, restricts the possible forms of the potentials to the rather wide class $V(r)$, independent of θ and ϕ, but does not determine the functional form of V.

2. The first two terms of Eq. (8.13) show that, for infinitesimal rotations, a vector rotates as follows: $\vec{V}' = \vec{V} - 2(\vec{\alpha} \times \vec{V})$, to order α, where $\vec{\alpha} = \frac{1}{2}\vec{\theta}$. Prove that the $\vec{F}_{\mu\nu}$ of Eq. (8.16) rotates like an isovector, i.e., show that

$$\vec{F}'_{\mu\nu} = \vec{F}_{\mu\nu} - 2(\vec{\alpha} \times \vec{F}_{\mu\nu}), \text{ to order } \alpha,$$

when Eq. (8.14) (including the last term) is used for $\vec{G}'_{\mu}$.

3. Write Eq. (8.17) for $F'^{i}_{\mu\nu}$ in terms of G'^{i}_{μ} and substitute Eq. (8.14), $G'^{i}_{\mu} = G^{i}_{\mu} - 2c_{ijk}\alpha^{j}G^{k}_{\mu} - (2/g)\partial_{\mu}\alpha^{i}$ to $O(\alpha)$, to show that $F'^{i}_{\mu\nu} = F^{i}_{\mu\nu} - 2c_{ijk}\alpha^{j}F^{k}_{\mu\nu}$ to $O(\alpha)$. This verifies to first order in α that the $F^{i}_{\mu\nu}$ transform like an adjoint representation of the gauge group; i.e., transform like G^{i}_{μ}, omitting the $\partial_{\mu}\alpha^{i}$ term.
   Hint: Use the Jacobi identity and the antisymmetry of the structure constants, which are shown in the problems of Chapter 5.

4. Follow the steps from Eq. (8.2) on for a general Lie group, with the transformation written as U and generators $\mathcal{G}^{i}$, to derive

$$G'_{\mu} = UG_{\mu}U^{-1} + \frac{i}{g}(\partial_{\mu}U)U^{-1}, \text{ where } G'_{\mu} \equiv \vec{\mathcal{G}} \cdot \vec{G}'_{\mu} \text{ and } G_{\mu} \equiv \vec{\mathcal{G}} \cdot \vec{G}_{\mu}.$$

The dot product is in the adjoint representation, i.e., $\vec{\mathcal{G}} \cdot \vec{G}_{\mu} \equiv \mathcal{G}^{i}G^{i}_{\mu}$ summed $i = 1, \ldots, 3$ for SU(2); $i = 1, \ldots, 8$ for SU(3), etc.

5. a. For the rotation group, show that if $V_{i}W_{i}$ is a scalar and $\vec{V}$ is an arbitrary vector, then $\vec{W}$ also rotates like a vector.
   b. Can the same thing be concluded for the **3** of SU(2)?

6. Verify that the contraction of two objects (V_{i} and W_{i}), which form adjoint representations of one of our symmetry groups, yields an invariant by showing that (for infinitesimal α): $V'^{i}W'^{i} = V^{i}W^{i}$ to order α. As indicated in Problem 3, the transformation is

$$V'^{i} = V^{i} - 2c_{ijk}\alpha^{j}V^{k} \text{ and similarly for } W'^{i}.$$

Hint: Remember that the structure constants are antisymmetric to *any* interchange of indices. (See Chapter 5, Problem 18.)

Chapter 9

Quantum Chromodynamics (QCD)

In this chapter we present that part of QCD generally accepted as part of the standard model. More speculative aspects, such as the nonperturbative considerations which lead to instantons, axions, and the strong CP problem as well as the possible existence of a quark-gluon plasma, are discussed in Chapter 16.

9.1 THE NONABELIAN GAUGE THEORY

We have already seen in Chapter 6 that the existence of a color quantum number alleviates the spin-statistics dilemma for quarks. There was also, in the 1970s, a growing body of *experimental data* which indicated that there should be three quarks of each flavor. The data contained results which were three times, and in some cases nine times, the results predicted without the inclusion of color. It was certainly not a matter of detail and precision, but instead nature was shouting loudly, if not clearly, that a factor of three was missing. The electroweak theory had been the culmination of decades of theoretical developments. It succeeded in incorporating the weak interaction into a nonabelian gauge theory along with QED despite formidable obstacles, as we shall discuss in Chapters 10 and 11. In contrast, quantum chromodynamics seemed to emerge full grown (like Athena) from the heads of its creators.

Quantum chromodynamics is an SU(3) gauge theory. To distinguish it from the **flavor SU(3)** which **is *not* a gauge theory**, it is often written $SU_c(3)$. As we have already seen in Chapter 5, **the generators** of this group **form an 8 — the adjoint representation of SU(3)**. We have learned in Chapter 8 that the gauge particles must therefore also form an **8**; they are called **gluons**.

We shall use r, g, and b as the three color states, as we have already done in Chapter 6. They are analogous to the u, d, and s of flavor SU(3). They form a **3** of $SU_c(3)$, and there is a corresponding $\bar{\mathbf{3}}$ (remember the $\bar{d}$, $\bar{u}$, and $\bar{s}$) whose members are $\bar{g}$, $\bar{r}$, and $\bar{b}$. The gluons behave just like the generators which (we have seen in Chapter 5) behave like the λ_i. Look back at the structure of the λ_i. Notice from Eq. (5.22) that λ_1 changes an r into a g and vice versa as does

144

λ_2 (with different relative coefficients). Analogously, we may symbolically write, for **the color content of the gluons**,

$$G^1 \sim \frac{1}{\sqrt{2}} \, (r\bar{g} + g\bar{r}) \text{ and } G^2 \sim \frac{-i}{\sqrt{2}} \, (r\bar{g} - g\bar{r}) \text{ or}$$

$$\frac{1}{\sqrt{2}} \, (G^1 + iG^2) \sim r\bar{g} \text{ and } \frac{1}{\sqrt{2}} \, (G^1 - iG^2) \sim g\bar{r}. \tag{9.1}$$

These structures are analogous to those of the pseudoscalar meson **8** (also made of $3 \times \bar{3}$): $\pi^+ = u\bar{d}$ and $\pi^- = d\bar{u}$. The interpretation of a combination like $r\bar{g}$ is that the absorption of such a gluon will change a g quark into an r quark (whereas its emission changes an r quark into a g quark). The colors carried by the other six gluons may be determined in a similar manner from the λ_i or from the structure of the pseudoscalar octet. (See Problem 2.) Thus, we see that the gluons carry a color and an anticolor and, **when a quark interacts with a (nondiagonal) gluon, its color will be** *changed*.

As we have learned in Chapter 8, the fact that the gluons themselves carry color (due to the nonabelian nature of QCD) leads to three-gluon and four-gluon vertices. Let us write the gauge-field part of the Lagrangian explicitly; using Eq. (8.19) (also see Problem 3 in Chapter 8), we obtain

$$\mathcal{L}_G \sim \sum_{i=1}^{8} F^{i\mu\nu} F^i_{\mu\nu}$$

$$= \sum_{i=1}^{8} (\partial^\mu G^{i\nu} - \partial^\nu G^{i\mu} - g_s c_{ijk} G^{j\mu} G^{k\nu})(\partial_\mu G^i_\nu - \partial_\nu G^i_\mu - g_s c_{ijk} G^j_\mu G^k_\nu), \tag{9.2}$$

(where g_s is the QCD coupling constant and sums over j and k are implied) from which we can see that the three-gluon vertex has coupling $\sim g_s$, and the four-gluon vertex has coupling $\sim g_s^2$. Diagrams with these vertices must be included in the calculations of amplitudes, which makes them considerably more complicated than the corresponding calculations in an abelian theory like QED. As a simple example, let us compare the tree-level $O(\alpha_s)$ amplitude for quark-gluon scattering in QCD to the $O(\alpha)$ amplitude for electron-photon scattering (Compton scattering) in QED. In Fig. 9.1 we see the (tree-level) $O(\alpha)$ Feynman diagrams for these processes, which give the lowest-order contributions to the amplitudes. Notice the extra diagram, containing a three-gluon vertex, which must be included in the QCD calculation. Its presence leads to a qualitatively different result, with a different angular dependence. (See Problem 3.) Higher-order contributions to this process would also require the inclusion of diagrams with four-gluon vertices; thus, the calculational complications due to the nonabelian nature of the theory proliferate.

Compton (e-γ) scattering has been experimentally studied for the better part of the twentieth century, yielding results that verify the QED predictions; however, we have not directly observed even one quark-gluon scattering event. This is a consequence of the fact that the quarks and the gluons are colored par-

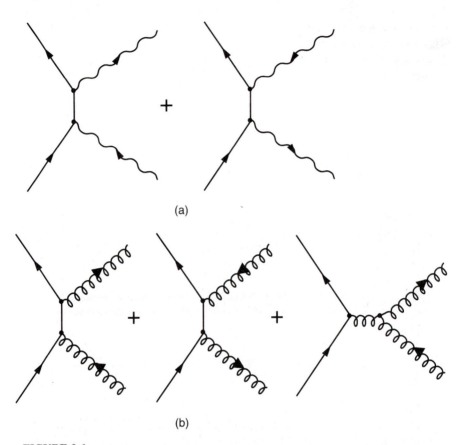

(a)

(b)

FIGURE 9.1
Tree-level $O(\alpha)$ contributions to the amplitudes for: (a) e-γ (Compton) scattering, where the solid lines are electrons and the wavy lines are photons; the second diagram has the identities of the incoming and outgoing photons interchanged and (b) quark-gluon scattering, where the solid lines are quarks and the spiral lines are gluons. Note the extra diagram containing a three-gluon vertex.

ticles and we have not succeeded in producing even one free colored particle. We have already seen the reason in Chapter 6, namely **confinement**. Therefore, quark-gluon scattering occurs in our experiments only as a subprocess of hadron-hadron scattering.

9.2   LONG-DISTANCE BEHAVIOR – INTERACTION RANGE

The gluons are massless as expected for gauge particles. So why is the range of the force between the observed particles so short? The answer is **confinement**.

What is confinement and how does it arise in this theory? The question has not yet been fully answered because of the difficulty of doing calculations in

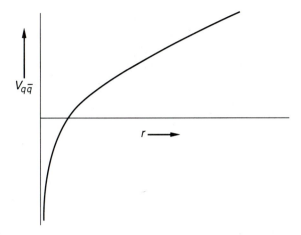

FIGURE 9.2
Potential between a quark and (the anticolor) antiquark. For large distances it is linear
and for short distances it drops as $1/(r \ln r)$ for r greater than the Compton wavelength
of the quark. It is believed that the full quantum field theory must be used for closer
approach, so that an effective potential will not give a realistic representation; therefore,
our V stops there.

this theory. Calculations for bound states, although successfully performed
in relativistic quantum mechanics for electromagnetic interactions, were never
even accomplished in quantum electrodynamics (an *abelian* gauge theory), since
they appear to require nonperturbative techniques. As we have just seen, this
nonabelian theory (QCD) has many *new* Feynman diagrams, such that the resulting calculations are much more difficult than the corresponding QED calculations. The observed hadrons are *low energy* bound states, for which the effective
strong coupling strength (α_s) is very large, as we shall soon see. Therefore, it
is not surprising that QCD bound-state calculations have not been performed.
However, *an SU(3) gauge theory in two dimensions* (with only one space direction) has solved exactly. It has just the confinement we are talking about.
For example, when we consider a system of a red quark (q_r) and an antired
antiquark ($\bar{q}_{\bar{r}}$), the potential energy grows *linearly* with their separation at
large distances, as shown in Fig. 9.2. Thus, to get the quark and antiquark further and further apart we must supply more and more energy, and to break them
free of each other would require *infinite* energy; therefore, it cannot happen.
Ta-dah — CONFINEMENT!

The mesons are $q\bar{q}$ systems and, in analogy to what we have seen in flavor
SU(3), they form a singlet under $SU_c(3)$ group operations if we use

$$\tfrac{1}{\sqrt{3}}(r\bar{r} + g\bar{g} + b\bar{b}), \tag{9.3}$$

where we have done the translation from u, d, and s to r, g, and b, respectively,
in Eq. (6.17). Each part of this sum leads to confinement, so we find that

$$V_{\text{LONG RANGE}} \approx kx. \tag{9.4}$$

In the world of three space dimensions, where we actually live, the size of a 1 GeV particle (like the proton) is about 1 fermi (10^{-15} meters), so we expect that k is about 1 GeV/fermi $\approx$ 10 metric tons.

This behavior is contrary to that of all the interactions we have previously encountered. In fact, our belief that we can discuss isolated systems relies on our belief that interactions do die off with distance. In other words, after pulling our system far enough away from everything else (sometimes by taking a mathematical limit), we can consider it isolated. Here we have a situation where the strength of the force remains constant[1] no matter how far we separate the particles. Is this believable?

9.3 SHIELDING, ANTISHIELDING, AND CONFINEMENT

9.3.1 Running Coupling Constants

In the actual theory, with all four dimensions, there are a number of indications that the confinement property is also present. One such indication concerns the virtual particles which are present in any field theory.

Let us first consider **QED** for electrons. In that theory virtual electron-positron pairs are produced around any given charge,[2] so that the effective charge seen depends on how close to the charge we probe. The net effect of these virtual pairs is a **shielding** of the charge at large distances. This shielding effect may be seen more explicitly as follows.

The interaction between charges requires the calculation of diagrams with all possible insertions of virtual particles, as we have seen when we discussed quantum field theory in Chapter 4. Let us consider e-e scattering. In Fig. 9.3 we show a series of diagrams with 1, 2, . . . virtual electron loops in the virtual photon line (which carries four-momentum p_γ) in a chainlike fashion. The four-momentum exchanged (p_γ) is equal to the change of momenta of the scattered particles. For **a real photon, $p_\gamma^2 = m_\gamma^2 = 0$**; however, **in general, $p_\gamma^2 \neq 0$ for virtual photons**. For $p_\gamma^2 \neq 0$, the exchanged photon has the same qualitative effect as the massive particles we discussed in Section 4.5, namely, it probes regions closer and closer to the charge as p_γ^2 approaches ∞.[3] Since there is less effective shielding the closer we get, we expect that **the effective coupling constant "runs" with p^2**. (We drop the sub-γ from here on.)

---

1. Recall that $F = -(\partial V/\partial r)$, so that, at large distances, the force is the slope of the linear portion of the potential shown in Fig. 9.1.

2. It is as if the charge has polarized the vacuum; virtual pairs are even produced in the pure vacuum state. These are often referred to as vacuum polarization effects. The virtual pairs may also be thought of as a plasma which produces Debye screening.

3. The fact that the momentum (p) increases as r decreases may also be understood in several related ways: (1) large p corresponds to small λ (viz., the DeBroglie wavelength), which is needed to probe small regions; (2) from the uncertainty principle we obtain $\Delta k \Delta r \sim h$, so that, for Δr small, Δk must be large, hence large k components must be present (and $p = \hbar k$); (3) because of the Coulomb barrier (for like-charged particles) and the centrifugal barrier, a larger p is required for the particles to get closer.

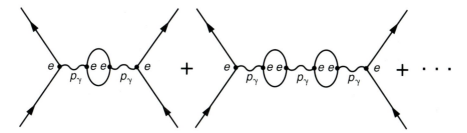

FIGURE 9.3
QED: The series of loops in the exchanged photon (of momentum p_γ), which are repetitions of the single photon loop. Diagrams with more complicated loops (containing more virtual particle insertions) are also repeated in a chainlike fashion.

Although we must consider more complicated loops (containing more virtual particle insertions) that are also repeated in a chainlike fashion, those in Fig. 9.3 dominate at large $|p^2|$. Furthermore, their effect becomes significant only for $|p^2| \geq (2m_e)^2$, which corresponds to probing at distances $\leq \frac{1}{2} m_e^{-1} = \frac{1}{2}\lambda_{e\ Compton}$. This may be qualitatively understood by looking at the related process $e^+ + e^- \to e^+ + e^-$,[4] for which the diagrams in Fig. 9.3 are also relevant. In that case p^2 (the effective mass of the virtual photon) is the total energy in the CM system, and when $p^2 \geq (2m_e)^2$, the effective mass of the virtual photon is large enough to produce a *real* electron-positron pair in the loop insertion.

The fine-structure constant[5] (α) is what we measure very far (compared to the electron Compton wavelength) from the charge, which is therefore the effective coupling strength at $p^2 = 0$; consequently, we may write: $\alpha \equiv \alpha(0)$. Below the threshold $p^2 \leq (2m_e)^2$, the sum of the chainlike diagrams of Fig. 9.3 is negligible so that for $|p^2| < 4m_e^2$, $\alpha(p^2) = \alpha(0) = \alpha = \frac{1}{137}$.

The contribution of the one-loop insertion to the effective coupling strength, above that threshold, has been found (using Feynman rules) to be

$$M_{1\ loop+extra\ virtual\ \gamma} = X(p^2),$$

$$\text{with } X(p^2) = \frac{\alpha}{3\pi} \ln\left(\frac{|p^2|}{m_e^2}\right) \quad \text{for } \frac{|p^2|}{m_e^2} \gg 1. \tag{9.5}$$

The sum of the corresponding chainlike diagrams to the effective α, which we shall call $\alpha(p^2)$ because they depend on p^2, is then

$$\alpha(p^2) = \alpha\left(1 + X(p^2) + \left(X(p^2)\right)^2 + \dots\right) = \alpha\,\frac{1}{1 - X(p^2)}, \tag{9.6}$$

---

4. This is sometimes referred to as *the t-channel*, as discussed in Appendix E.

5. Recall that the effective strength of the interaction between two charges depends on the square of their effective charge, and we have found it convenient to talk about the dimensionless "fine-structure constant" $\alpha = e^2/\hbar c$.

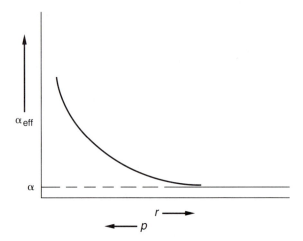

FIGURE 9.4
The behavior of α_{eff} as a function of r (or p) for **quantum electrodynamics (QED)**. Standard measurements of e are made far enough away that the asymptotic value α shown on the graph is the value used in atomic physics.

where the closed form is valid for $|X| < 1$. These terms are sometimes referred to as the *leading-log terms* (since X has a single "ln" in it). At an extremely high energy, X will cross 1 but, in fact, nonperturbative treatments of renormalization have shown that Eq. (9.6) is correct for any value of $|X|$. Although that enormous energy will not interest us here, the fact that $\alpha(p^2)$ blows up and then becomes negative certainly raises apprehension about the handling and/or the validity of QED at very close distances. Perhaps the theory has a fundamental problem, or perhaps the expansion is at best an asymptotic series (as discussed in Section 4.4.2).

In the range of energies accessible to observation, Eq. (9.6) shows that $\alpha(p^2)$ grows as p^2 becomes larger, i.e., as we probe closer and closer to the charged particle. In Fig. 9.4 we show this behavior of α_{eff} as a function of r. The quantum field theory calculations for any process, which are done using Feynman diagrams, are expansions in powers of α; however, we can see that we have a choice of the value of p^2 at which to define the coupling strength. If we **use that value of α which corresponds to a $|p^2|$ of the same order of magnitude as the square of the momentum transferred in the scattering**, the first few terms will give a much more precise result than they would have if we had used $\alpha = \frac{1}{137}$. The result is equivalent to that of a much higher order calculation using $\alpha = \frac{1}{137}$, which would include the important contributions of the chains we have used to find the effective coupling strength.

Actually, we cannot use Eqs. (9.5) and (9.6) (which contain only the effect of *electron* loops) when $\sqrt{|p^2|}$ exceeds twice the mass of the next heavier fundamental fermion (the up quark, with $m_u \approx 3.5$ MeV). Above that energy, loops of that next particle (the up quark) also become significant. The coupling strength then runs (with p^2) at a new rate. Correspondingly, when $\sqrt{|p^2|}$ exceeds twice the mass of the next heavier (than the up) fundamental fermion, loops of *that* particle must be included as well. Consequently, we must work in steps, starting from $\alpha(0)$ and then including electron loops in $X(p^2)$ for $\frac{1}{2}\sqrt{|p^2|} > m_e$. When $\frac{1}{2}\sqrt{|p^2|} > m_u$ we start from the value of α at $|p^2| = 4m_u^2$, obtained at the top of the previous (lower energy) run, and now include up quark loops to obtain the new $X(p^2)$, which determines the new run rate. Thus, we move up in energy, including in $X(p^2)$ loops for all the charged fermions whose masses are below $\frac{1}{2}\sqrt{|p^2|}$ and running to the top of that energy region. We then start with the $\alpha(p^2)$ so obtained and run it with a new $X(p^2)$ in the next energy region. Thus, Eqs. (9.5) and (9.6) must be replaced by a relationship expressing the coupling strength at p^2 in terms of one (already known) at μ^2 (which is presumably at the top of the previous energy region):

$$\alpha(p^2) = \frac{\alpha(\mu^2)}{1 - X(p^2)}$$

$$\text{with } X(p^2) = \left(\sum_{i=1}^{N_f} \left(\frac{q_i}{e}\right)^2\right) \frac{\alpha(\mu^2)}{3\pi} \ln\left(\frac{|p^2|}{\mu^2}\right), \tag{9.7}$$

where N_f is number of fundamental fermions with masses below $\frac{1}{2}\sqrt{|p^2|}$. The sum (in parentheses in Eq. 9.7) arises because the coupling strength is proportional to the squares of the fermion charges (q_i^2) instead of the factor e^2 which is in $\alpha(\mu^2)$.[6] Equation (9.7) is valid below $2M_W$, but above that energy W loops become important and the separate running of the electroweak coupling strengths is studied instead. The quark masses to which $\frac{1}{2}\sqrt{|p^2|}$ should be compared are[7]

$$m_u \approx 3.5 \text{ MeV}, m_d \approx 6.1 \text{ MeV}, m_s \approx 120 \text{ MeV}, m_c \approx 1.35 \text{ GeV, and}$$

$$m_b \approx 5.3 \text{ GeV}.$$

The charged leptons have masses:

$$m_e = 0.5 \text{ MeV}, m_\mu = 105.7 \text{ MeV, and } m_\tau \approx 1784 \text{ MeV}.$$

The change of the effective charge with distance, which we have examined here, is analogous to the classical situation of a charged particle in a dielectric. There we have

$$V \propto \frac{q}{\varepsilon r}, \text{ so that } q_{eff} = \frac{q}{\varepsilon}, \tag{9.8}$$

---

6. Since the sizes of the charges of quarks are $\frac{2}{3}$ and $\frac{1}{3}$ (for up and down types) and each flavor comes in three colors, in X (Eq. 9.7) the sum contains $(3)(\frac{2}{3})^2 = \frac{4}{3}$ and $(3)(\frac{1}{3})^2 = \frac{1}{3}$, respectively.

7. These are sometimes referred to as *current* masses, as opposed to *constituent* masses. The current masses appear as the mass parameters in the currents in the Lagrangian, whereas the constituent masses are effective mass values, which account for the masses of the hadrons, viz., the constituent masses of the u and d quarks are approximately 300 MeV ($\frac{1}{3}$ the nucleon mass).

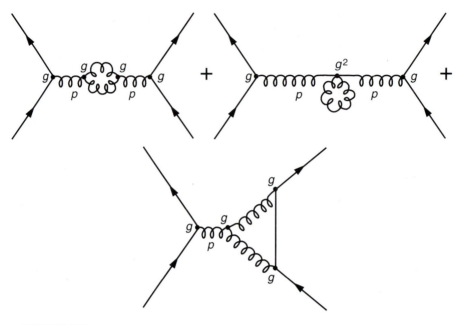

FIGURE 9.5
QCD: The gluon loops which contribute to the leading-log chains.

where ε is the dielectric constant (which is frequency dependent). The calculation in QED of the effective dielectric constant shows that it grows with distance from the charge to give the drop in effective charge that we have noted.

In QCD, the nonabelian nature of the theory has resulted in *colored* gluons. When virtual pairs of these spin 1 particles are produced in the empty space around a colored object, they enhance the color charge; this is referred to as **antishielding**.[8] So we have shielding due to quark-antiquark pairs (in a manner similar to that for e^--e^+ pairs in QED) and antishielding due to gluon pairs. The number of quark flavors and the number of gluons (which is eight for SU(3)) together determine whether shielding or antishielding dominates. For QCD we define $\alpha_s \equiv g_s^2/\hbar c$, where g_s is the SU$_c$(3) coupling constant; however, we cannot think of this as the value of the coupling strength at $p^2 = 0$, since that is just the limit in which the perturbative results do not hold, viz., the (large distance) confinement limit. Instead, the measured value at some convenient momentum transfer squared (μ^2) can be used.

Due to the gluon-gluon vertices, the chainlike diagrams, analogous to those in QED, must now include gluon loops as well; these are shown in Fig. 9.5. Their contributions to $\alpha_s(p^2)$ have the opposite sign to those of the fermion (quark)

---

8. Whether we get shielding or antishielding from virtual colored pairs has been seen to depend on the spin of the particles in the pair; spin $\frac{1}{2}$ particles produce shielding and spin 1 particles produce antishielding.

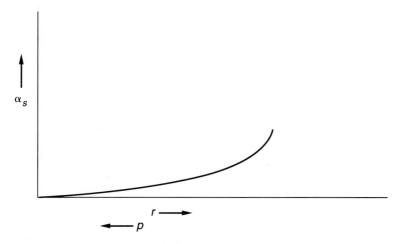

FIGURE 9.6
The behavior of α_s as a function of r (and p) for **quantum chromodynamics (QCD)**.

loops. The running (with p^2) coupling strength, resulting from the series of chainlike repetitions of the one-loop insertions, becomes

$$\alpha_s(p^2) = \alpha_s(\mu^2)\left(1 + Y(p^2) + (Y(p^2))^2 + \cdots\right) = \alpha_s(\mu^2)\frac{1}{1 - Y(p^2)},$$

$$(9.9)$$

$$\text{with } Y(p^2) = \frac{\alpha_s(\mu^2)}{12\pi}(2n_f - 11n_c)\ln\left(\frac{|p^2|}{\mu^2}\right) \quad \text{for } \frac{|p^2|}{\mu^2} \gg 1,$$

where n_f is the number of flavors of quarks with masses less than $\frac{1}{2}\sqrt{|p^2|}$[9] and n_c is the number of colors. The world average of the measured values of the strong coupling constant, at 34 GeV, is

$$\alpha_s((34\text{ GeV})^2) = .148 \pm .018, \text{ for which } n_f = 5.$$

For six (or less) flavors and three colors (corresponding to eight gluons), $(2n_f - 11n_c)$ and, consequently, $Y(p^2)$ are negative. Therefore, the denominator of the expression for $\alpha_s(p^2)$ (Eq. 9.9) grows with p^2. This results in a drop of $\alpha_s(p^2)$ as we probe closer and closer to the (colored) quark; **the color charge appears** *larger, the further away* **we probe**—**antishielding!** In fact, we see that as $p^2 \to \infty$, **which corresponds to** $r \to 0$, $\alpha_s(p^2) \to 0$; this is referred to as **asymptotic freedom**. Figure 9.6 shows the behavior of $\alpha_s(p^2)$. Notice that α_s starts from 0 at $r = 0$ ($p \to \infty$) and grows precipitously at large r (low p or low energy). These features have quite sweeping consequences:

---

9. For quarks with masses larger than $\frac{1}{2}\sqrt{|p^2|}$, the contributions of the corresponding diagrams are greatly suppressed, just as in QED.

1. During high-energy collisions, a quark in one of the hadrons gets very close to a quark in the other and the effective (running) coupling strength is very small. It is approaching 0 asymptotically, so that the particles behave more and more like free particles the closer they get. (This is, indeed, the property referred to above as asymptotic freedom.) Consequently, we expect that the (perturbative) expansion in Feynman diagrams would be valid for these subprocesses.

2. For bound states, which exist at low p, there is an indication that the coupling strength is very large. The calculational procedures known at present for quantum field theories are perturbative in nature (i.e., yield a power series in the coupling), so that no one knows how to produce a definitive answer from QCD about bound states and the confinement of their constituents. With the usual whimsy, the confinement at low energies, if it is really due to a blowing up of the running coupling constant as the energy approaches 0, has come to be called **infrared slavery**.[10]

Finally let us note that, for quantum field theories, it has been shown that **only *nonabelian* gauge theories possess asymptotic freedom**.

9.3.2   Meissner-like Effect in Vacuum Regions

Another analogy to other phenomena recognizes that an empty (vacuum-like) region behaves like a color superconductor, in that, in the presence of color charges, it "expels" (color) field lines. (Recall the Meissner effect in superconductors.) Presumably, this happens because regions with field lines penetrating them effectively do not have (virtual) gluon-antigluon pairs, whose presence would correspond to a much lower energy density than that of field lines in a QED-like configuration.

Consider, in particular, those cases where *the total color charge is 0*. For a given set of color charges, the equilibrium configuration has most of space devoid of field lines, so that virtual gluon-antigluon pairs can exist there; consequently, the field lines are concentrated in a small region of space. The large energy density in the region containing the concentrated field lines is compensated for by the very low energy density in most of space. This is in contrast to electrodynamics, where the field lines extend throughout all of space. (Remember that photons do not carry charge, so that there are no corresponding photon-antiphoton pairs.)

Let us consider two color-charged particles r and $\bar{r}$. Due to the Meissner-like effect in vacuum regions, the field lines run from one particle to the other

---

10. Since the infrared region of the electromagnetic spectrum is below that of the visible, people have come to use the word "infrared" when they mean low energy, even in contexts that have nothing to do with electromagnetic phenomena.

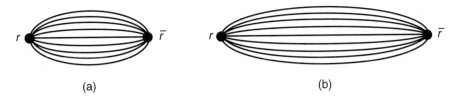

FIGURE 9.7
(a) Field lines between particles r and $\bar{r}$ when they are close to one another; (b) field lines between particles r and $\bar{r}$ when they are farther apart.

in a very compact region, as shown in Fig. 9.7(a). As we pull the particles apart, the lateral size of the region containing the field lines remains approximately the same, but the length of that flux tube is equal to the separation between the particles. The energy contained in that field has the same structure as that for ordinary electrodynamics, viz.,

$$\text{Field Energy} \sim \int |\text{Field}|^2 \, dV. \tag{9.10}$$

The integrand in Eq. (9.10) is roughly constant in the flux tube (and 0 outside), but the volume of that tube is proportional to the separation (x) between the particles, as is evident in Fig. 9.7. Thus the energy we put in, which goes into the field, is proportional to x. This is an energy of position, so that it is the potential energy:

$$V \propto x. \tag{9.11}$$

Furthermore, if we consider *a state which is not color neutral*, some field lines would extend to ∞. In the regions where those field lines exist, the gluon-antigluon pairs would have been dispossessed. Those regions would, therefore, be at higher energy with the result that the state would be an infinite energy above the color-neutral states, since those regions are infinite in extent. Thus, such a (colorful) state is unattainable – *confinement*!

You may be wondering why there is such a complete cancellation of the forces due to the r and $\bar{r}$ quarks on a test (color) charge[11] outside the region of the flux tube, since the interaction strength α_s increases with distance. Why should the cancellation exceed that which occurs in ordinary electrodynamics, where α_{eff} drops off with distance? The answer lies in the "Q" of QCD and in the nonabelian nature of this gauge quantum field theory; the gluon-antigluon virtual pairs, which can be produced in a region effectively devoid of field lines, produce a state of lower energy than that resulting when the field lines are spread out as in QED.

---

11. We imagine using test (color) charges to define the force fields, analogous to the definition of electric field in electrodynamics.

9.3.3   Gluon Masslessness and Dia-electric Behavior

Another argument which accounts for confinement is related to the massless-ness of the gluons. In the vacuum, made up of gluon-antigluon pairs, we expect that the speed of a gluonic wave should equal the inverse of the square root of the product of the color-dielectric constant and the color-magnetic susceptibil-ity. Since the gluons are massless, the speed of the wave should be 1 in natural units:

$$1 = \frac{1}{\sqrt{\varepsilon_c \mu_c}}, \text{ so that}$$

$$\varepsilon_c \mu_c = 1. \tag{9.12}$$

Color-magnetic calculations for virtual gluon-antigluon pairs have shown that $\mu_c > 1$, i.e., paramagnetic behavior. Equation (9.12) then requires that $\varepsilon_c < 1$ in the "empty" regions of space; this is **dia-electric** behavior. Since $V \sim$ (color "charge")/ε_c where the field lines penetrate, V is larger in the flux line region and a lower energy state results if a bag[12] exists, around the $q\bar{q}$ system, outside of which the field lines do not penetrate.

9.4   HIGH-ENERGY SCATTERING AND JETS

We may ask what happens when we "hit" a meson, in an attempt to pull apart the two colored particles (e.g., r and $\bar{r}$) of which it is made? As they separate, more and more energy goes into the fields, and at some point there is enough energy to *create a new pair* so that the field lines are still confined, as shown in Fig. 9.8. It can be shown that the probability of pair creation rises with energy, so that a state with two bound pairs, as shown in Fig. 9.8, becomes highly prob-able. A meson has been hit and two mesons have resulted.

In a high-energy (~GeV) scattering of two (colorless) hadrons, the most likely interaction will be that a quark in one (hard-)scatters from a quark in

---

12. This is reminiscent of a phenomenological bag model created at MIT to describe the prop-erties of hadrons in terms of bound quarks.

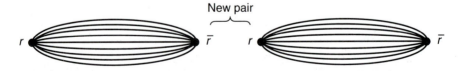

FIGURE 9.8
Two colorless mesons emerge as a result of an attempt to knock the r and $\bar{r}$ of the origi-nal (colorless) meson apart. The r on the far left and the $\bar{r}$ on the far right were the orig-inal quarks, and those in the middle were produced as a new pair.

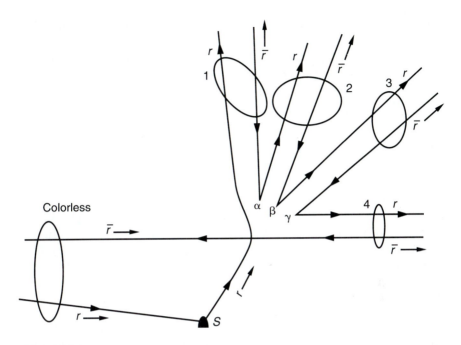

FIGURE 9.9
A schematic example of **hadronization** and the production of jets. The r(ed) quark
(hard-)scatters, from a quark (not shown) in another hadron, in region S. At points α,
β, and γ, virtual gluons produce pairs. The emerging hadrons (1, 2, 3, and 4) are colorless. The kinematics are such that 1, 2, and 3 each have a substantial fraction of the large
transverse momentum of the scattered quark, so that they form a jet whose total momentum is approximately the same as that of the scattered quark. Meson 4, on the other hand,
has very low transverse momentum and is not part of the jet.

the other and the hadron structures are destroyed. The quarks not involved
in the hard-scattering, sometimes referred to as **spectators**, continue roughly in
the beam direction with very little transverse momenta. In contrast, **the hard-
scattered quarks obtain very large transverse momenta**. Since colored objects
cannot break free, pairs will be created by virtual gluons in such a way as to
make each emerging particle a colorless (bound) hadron. This is called **hadronization**, and a simple, schematic example is shown in Fig. 9.9. High-energy **jets**
of hadrons emerge from the collision in transverse directions, one jet moving
in the direction of each one of the (fast-moving) scattered quarks, so that we
expect to see two high-energy jets of "scattered" particles. Their existence is a
verification of the existence of the quarks themselves, even though the quarks
cannot be produced as free particles. We will discuss the experimental evidence
for such jets in Chapter 15.

In QED particle-particle scatterings are always accompanied by scatterings
which produce an extra photon, and that effect is referred to as *bremsstrahlung*.

Its rate and angular distribution, as compared to photonless scattering, is successfully calculated in QED. A similar situation exists in QCD. Gluon bremsstrahlung, at the expected rate and with the correct angular distribution, has been seen as three-jet events (since gluons are also colored and, therefore, cannot be produced as free particles). This is a direct verification of the existence of the *gauge particles* (gluons) of QCD.[13]

9.5   GLUON MATTER

The existence of gluons also implies the existence of some hadrons which do *not* contain quarks. Gluons form an **8** of $SU_c(3)$ and so cannot be produced as free particles. However, it can be shown, in a manner similar to our explanation in Chapter 3 for the products of representations, that $\mathbf{8} \times \mathbf{8} = \mathbf{1} + \cdots$. In fact, we have seen in Chapter 8 how two **8**s can couple to an invariant, i.e., a **1**. (There it was the mathematical form which was the sum of the products of the generators with their corresponding gauge fields.) Presumably, these colorless combinations of gluons *can* be produced as free systems and are referred to as **glueballs** or **gluonium**. Glueball states can sometimes mix with $q\bar{q}$ states to give more states than the $q\bar{q}$ system alone would have. Decay patterns of such states would be modified by the glueball part of their wavefunctions. There does appear to be some evidence for the existence of glueballs, although it is not definitive.

9.6   THE STRONG INTERACTION

Have you been wondering about the strong interaction between the (color-neutral) hadrons at the usual low energies of everyday life? Why should there be a force between "neutral" objects? Consider the electromagnetic interaction of ordinary (*charge* neutral) molecules; when they are brought near one another there *is* a force between them. It is called the *Van der Waals force* and results from the polarization of each molecule by the electric field of the other. It is, in a sense, a *residual force* due to the slight displacement of the negative and positive charges with respect to one another in each molecule. That force decreases with distance much more rapidly than the Coulomb force. Likewise, **the strong interaction we see between** *hadrons* **is a Van der Waals type of force with a very short range.** This "left-over" attractive force is what is responsible for the binding of nucleons to form nuclei, the fusion that occurs in stars, etc. Unfortunately, because the effective α_s is so large for low energies, we have no

---

13. Earlier experiments (to be discussed in Chapter 15) showed the existence of constituents in the hadrons, which were called *partons*, but these ***three-jet*** results verified the existence of the **gauge particles**, validating the gauge theory.

reliable techniques to calculate the bound states of quarks (the hadrons), let alone the residual forces between them.

9.7   THE OZI RULE

The OZI rule (or Zweig suppression), which we discussed in Chapter 6, can be understood to be qualitatively correct if we consider asymptotic freedom (or antishielding) and color confinement combined with some mathematical details of $SU_c(3)$. Recall, from Chapter 6, that this rule predicts the suppression of processes in which the incident quark lines are not connected to the outgoing quark lines, as shown in Fig. 6.7. Color confinement tells us that the incoming and outgoing hadrons must be colorless; consequently, a color singlet of intermediate particles must couple to each in order to preserve $SU_c(3)$. Therefore, at least two[14] virtual gluons must propagate between incident and emerging particles, giving rise to two extra factors of α_s in the amplitude. (Considerations of charge conjugation, to be discussed in Chapter 12, show that *at least three* gluons are needed to couple to the vector mesons $\phi(s\bar{s})$; $\psi(c\bar{c})$, $\psi'(c\bar{c})$, $\psi''(c\bar{c})$, etc.; and $\Upsilon(b\bar{b})$, $\Upsilon'(b\bar{b})$, etc.) Since α_s is very small, due to asymptotic freedom, at the high energies of our experiments, the extra factors of α_s will yield the conjectured suppression.

9.8   LATTICE GAUGE CALCULATIONS

Imagine a space-time which is not continuous, but instead is made up of the discrete points of a four-dimensional lattice. A quantum field theory set up on such a lattice can be solved *nonperturbatively* by using powerful supercomputers. Such studies are being done for QCD. Much effort has gone into the solutions for bound states, with encouraging results. These calculations have also produced the prediction, after extrapolation to the continuum limit,[15] of a **phase transition at high temperatures to a regime where free quarks can exist**. This state of matter is referred to as a **quark-gluon plasma**. It is suspected that such was the situation during the first microsecond after the Big Bang and that it may be occurring in the center of neutron stars. Some experimenters hope to produce such a state of matter in high-energy, heavy-ion scattering experiments. We shall discuss the quark-gluon plasma and the proposed experiments in more detail in Chapter 16.

There are, however, many questions about the validity of these lattice gauge calculations:

---

14. Two gluons can indeed accomplished this, because two **8**s *can* couple to a **1**.

15. It is well known that phase transitions do not occur in finite systems.

1. The sizes of the lattices that can be handled, even by the most powerful supercomputers, are very small (of the order of 8 points in the time direction and 16 points in each space direction, giving about 8×16^3 space-time lattice points).

2. The representation on the lattice of virtual fermions has been done only in a very crude, approximate way.

3. The results on the lattice must be extrapolated to the limit of zero lattice spacing to give results that are valid for the (continuum) world in which we live. This can only be done in an approximate way on a computer.

9.9   SEARCH FOR FREE QUARKS

If quarks could exist as free particles, i.e., if there were no confinement, quantum-field-theoretic considerations tell us that the large energy densities present in the early universe would certainly lead to the production of quark-antiquark pairs.[16] Most would have annihilated with corresponding antiparticles as they met them; however, we would certainly expect that *some* should still exist. Because of their fractional charge, they could never form neutral atoms or molecules with the particles of ordinary matter; **there would always be a net fractional charge on any atom or molecule formed by a free quark with ordinary matter**. There have been many searches for fractionally charged particles, including analysis of tons of sea water and of soil as well as of meteorites, dust from the upper atmosphere, and matter from the moon. Some spurious reports of fractional charge have surfaced from time to time, but none have survived close scrutiny. There was one investigation repeating the Millikan oil drop experiment, but with niobium spheres, which supposedly found evidence of fractional charge.[17] However, a number of similar experiments have found no evidence for fractional charge.[18] The experimental status, at present, is that no fractionally charged particles have been detected in verifiable experiments. We shall see in later chapters that this situation has not attenuated the firm belief in the existence of quarks as *bound* particles. In Chapter 15 we will see some of the suggestive experimental evidence of their existence, in inelastic scattering experiments and more directly in the form of jets, which we have already mentioned in Section 9.4.

---

16. Before the creation of QCD and the understanding of confinement, the search for quarks and the subsequent failure to find any was convincing evidence for many that they did not exist.

17. This was performed by a highly respected experimental physicist (W. Fairbanks) and has come to be known as the Fairbanks experiment.

18. In Millikan's original data sheets, one entry has been found which indicated a fundamental charge of roughly $\frac{2}{3}e$; however, Millikan had crossed it out. It is presumed that that run was defective in some way, and we should rely on the judgement of that renowned experimenter as to the reliability of his data. His experiment has been repeated in many different ways for many years with no indication of fractional charge.

PROBLEMS

1. a. Write the term(s) in the Lagrangian which correspond to the quark-gluon vertices.
    b. Draw the tree-level diagrams for the scattering of a "green" quark (corresponding to the second row and column in the λ_i's) and a "blue" quark (corresponding to the third row and column in the λ_i's). Show which gluon is involved in each diagram and the corresponding coupling constants.
2. a. Consider the gluons G_μ^i. For each value of i ($i = 1, \ldots, 8$), show what colors the gluon carries.
    b. Draw the lowest-order Feynman diagrams for gluon-gluon scattering, showing the power of g_s appearing at each vertex.
    c. Draw the lowest-order Feynman diagrams for gluon-gluon scattering in which a virtual quark appears.
    d. If two gluons form a glueball, what color combinations (or i-value combinations, where $i = 1, \ldots, 8$) of gluons appear in the state.
3. After removing the complications of spin by averaging, it has been found that the absolute squares of the amplitudes for quark-gluon scattering compared to electron-photon scattering have the forms:

$$\overline{|S_{QCD}|^2} \sim \frac{\hat{u}^2 + \hat{s}^2}{-\hat{u}\hat{s}} + \frac{9}{4}\frac{\hat{u}^2 + \hat{s}^2}{\hat{t}^2} \quad \text{compared to} \quad \overline{|S_{QED}|^2} \sim \frac{u^2 + s^2}{-us},$$

where the lines over the absolute squares mean spin averaged, s and $\hat{s}$ are the squares of the total energy in the CM system, and t (and $\hat{t}$) and u (and $\hat{u}$) involve the angle of scattering, as elaborated in Appendix E. The Carets appear over the kinematic variables s, t, and u to remind us that these variables correspond to a subprocess and are not those of the full hadron-hadron scattering.
    a. Show that, for the scattering of equal-mass particles, the kinematic invariant u (defined in Appendix E) may be written: $u = -2(p^{CM})^2(1 + \cos\theta^{CM})$.
    b. Write the terms for $\overline{|S_{QCD}|^2}$ explicitly in terms of CM quantities (neglecting the quark mass) and verify that the first and second terms are different functions of θ^{CM}.
4. a. Find the lowest order in α_s for quark-gluon scattering which contains four-gluon vertices by displaying the corresponding Feynman diagrams.
    b. Draw at least three more diagrams to that order which do not have corresponding diagrams for Compton (e-γ) scattering.
5. a. Show that for pure QED (with only one kind of charged particle, viz., electrons and positrons): $\alpha^{-1}(p^2) = \alpha^{-1} + f(p^2)$, where f does not depend on α.
    b. Show that Eqs. (9.7) and (9.9) yield equations of the form:

$$\alpha^{-1}(p^2) = \alpha^{-1}(\mu^2) + f\left(\frac{p^2}{\mu^2}\right),$$

where f does not depend on α.
6. Calculate the value of $\alpha^{-1}(p^2)$, the inverse of the running coupling strength in QED, at the Z mass ($|p^2| = (90 \text{ GeV})^2$), using the fact that $\alpha^{-1}(0) = 137$:
    a. if the only charged fermions were the electrons and
    b. by including the appropriate charged leptons and quarks.

7. a. Show that the equation for $\alpha_s(p^2)$ can be written in the form

$$\alpha_s(p^2) = \frac{12\pi}{(11n_c - 2n_f)\ln\left(\dfrac{|p^2|}{\Lambda^2}\right)},$$

where Λ is a constant which depends on n_f, by finding an expression for Λ. Note that, since Λ depends on n_f, it makes discrete jumps as $|p^2|$ moves through the value corresponding to four times the square of the masses of particular quark flavors.

   b. For the energy range around the Z mass, find Λ from the expression determined for it in part a. Hint: α_s at 34 GeV is given in the text.

   c. Using the appropriate Λ, find α_s at the Z mass.

8. For pure QED (with only one kind of charged particle, viz., electrons and positrons), find the value of $\sqrt{|p^2|}$ for which the denominator in the expression for $\alpha(p^2)$ vanishes.

9. Find the limit on the number of generations if asymptotic freedom is demanded.

10. Suppose that one person uses μ_1^2 in Eq. (9.9) for α_s, and another uses $\mu_2^2 \neq \mu_1^2$. Show that if they agree on the value of α_s for μ_3^2, then they obtain the same value for all $\sqrt{|p^2|}$.

11. Consider the quark flow diagrams for ϕ decay shown in Fig. 6.7. Show that more virtual gluons are needed for the OZI-suppressed decays by sketching in the minimal number required for each. From this determine the lowest order in α_s required.

12. a. Draw the tree-level diagrams for gluon-bremsstrahlung in quark-quark scattering, which produce three jets.

    b. Draw the tree-level diagrams for other types of events, which produce three jets.

Chapter 10

The Electroweak Theory —
I. The Gauge Theory

10.1   WEAK INTERACTIONS

At the beginning of the twentieth century nuclear decays were observed, in which electrons or positrons emerged from nuclei; these decay products were referred to as β-rays (before their identities were established). The nuclei changed charge accordingly becoming other elements. Typical β decays, as *observed* then, were

$$_5B^{12} \rightarrow {_6}C^{12} + \beta^- \text{ and}$$

$$_7N^{12} \rightarrow {_6}C^{12} + \beta^+. \tag{10.1}$$

It was found, however, that energy, momentum, and angular momentum were not conserved in these reactions. After about two decades of confusion about the validity of these conservation laws, Pauli proposed that a neutral, massless[1] particle, called the **neutrino** (Italian for "little neutral one"), must have been emitted along with the electron in these reactions. Conservation of angular momentum required that the neutrino be a spin $\frac{1}{2}$ particle. With this assumption energy, momentum, and angular momentum were found to be conserved in all such reactions. Our belief in these conservation laws was such that the neutrino was considered a *real* particle long before it was ever directly observed. We have already seen that gaps in bubble chamber pictures are routinely taken as evidence of the appropriate neutral particles, but here we are talking about a *supposed* particle which was not seen to interact or decay in our detectors after its creation. However, this is understandable because it only interacts via the weak interaction. In the 1950s, suitable experiments were set up to directly detect neutrinos, and their existence was indeed confirmed.

The processes in Eq. (10.1) are examples of what is known as β **decay**, and the simplest hadronic β decay (of ordinary matter) is the decay of the neutron:

$$n \rightarrow p + e^- + \bar{\nu}_e, \tag{10.2}$$

---

1. It was found that $m_\nu^2 = E_{missing}^2 - p_{missing}^2 \approx 0$. The most recent upper limit on the electron neutrino's mass is $m_\nu < 8$ eV.

where we have included the antineutrino ($\bar{\nu}_e$) that is emitted along with the electron. Meson decays can also produce leptons; a simple example of such a decay is

$$\pi^- \rightarrow \mu^- + \bar{\nu}_\mu. \tag{10.3}$$

These are referred to as **semileptonic** decays, because only some of the particles involved are leptons.[2]

When a heavier charged lepton was discovered, it was seen to decay into an electron:

$$\mu^- \rightarrow e^- + \bar{\nu}_e + \nu_\mu. \tag{10.4}$$

Decays like these are called **leptonic** decays.

It has been found from experiment that **electron number, muon number, and tau number are each conserved**, with the negatively charged lepton and the neutrino counting as +1, and the positively charged lepton and the antineutrino counting as −1. (Check this for the leptonic and semileptonic decays above). No continuous symmetry has been found to account for this, so far, hence there is no (associated) gauge theory.

When more hadrons were found they were seen to undergo β decays as well. Also, some of their decays, which did not involve leptons (electrons, neutrinos, etc.), were so slow that they were considered to be due to the same interaction, which came to be known as the *weak interaction*. We have discussed such decays in Chapter 6 (Eq. 6.11):

$$\Sigma^- \rightarrow n + \pi^-$$
$$\Sigma^+ \rightarrow n + \pi^+ \text{ or } p + \pi^0. \tag{10.5}$$

These are **nonleptonic** decays.

In Section 10.3 we shall study the gauge theory which accounts for all of these weak decays, be they purely leptonic, semileptonic, or nonleptonic. We shall see that this theory contains QED, in a nontrivial way, as well. This electroweak theory was originally suggested by Glashow and completed by Weinberg and (independently) by Salam.

10.2   HANDEDNESS, HELICITY, AND PARITY

The weak interaction is different for different handednesses of the particles; in this section we will define handedness or helicity and explore its connection to the operation of space inversion, which is called *parity*.

---

2. We shall soon see that a virtual W or Z mediates all the observed weak decays. For semileptonic decays, one end of the virtual W or Z couples to leptons and the other end to quarks; we can also understand the names leptonic and nonleptonic in this more fundamental way.

Consider a spin $\frac{1}{2}$ particle moving with momentum $\vec{p}$; let us call the direction of its momentum the z direction. If we were to measure S_z, we know that the result must be $\pm\frac{1}{2}$. When the result is $+\frac{1}{2}$ the particle is called *right-handed*, and when $-\frac{1}{2}$ it is called *left-handed*.[3] We have discussed the Dirac equation:

$$(\gamma^\mu p_\mu - m)\psi = 0,$$

where $\gamma^0 = \beta = \begin{pmatrix} 1 & 0 \\ 0 & -1 \end{pmatrix}$ and $\gamma^i = \beta\alpha_i = \begin{pmatrix} 0 & \sigma_i \\ -\sigma_i & 0 \end{pmatrix}.$ (10.6)

We also introduced the matrix $\gamma^5 = i\gamma^0\gamma^1\gamma^2\gamma^3$, which is

$$\gamma^5 = \begin{pmatrix} 0 & 1 \\ 1 & 0 \end{pmatrix}.$$ (10.7)

(Recall that each entry in the above matrices is itself a 2×2 matrix.) Notice that γ^0, which is diagonal, multiplies p_0 (the energy) in the Dirac equation. Consider a particle at rest ($\vec{p} = 0$). With this representation of the γ matrices we have seen that the upper two components of ψ (a four-column) correspond to positive energy and the lower two to negative energy. We may write

$$(\gamma^0 p_0 - m)\psi = \begin{pmatrix} (p_0 - m)1 & 0 \\ 0 & (-p_0 - m)1 \end{pmatrix}\psi = 0.$$

Substituting

$$\psi = \begin{pmatrix} 1 \\ 0 \\ 0 \\ 0 \end{pmatrix},$$

the equation yields $\boldsymbol{p_0 = m}$ and likewise for

$$\psi = \begin{pmatrix} 0 \\ 1 \\ 0 \\ 0 \end{pmatrix};$$

whereas for $\psi = \begin{pmatrix} 0 \\ 0 \\ 1 \\ 0 \end{pmatrix}$ and for $\psi = \begin{pmatrix} 0 \\ 0 \\ 0 \\ 1 \end{pmatrix}$, we obtain $\boldsymbol{p_0 = -m}$. (10.8)

---

3. This refers to the fact that a classical object which is spinning has the direction of its angular momentum determined by the "right-hand rule." When S_z is in the direction of the momentum, the particle is considered right-handed.

The same is qualitatively true for a moving particle, viz., a positive energy state
has large upper components, whereas a negative energy state has large lower
components. (See Problem 8 in Chapter 4.) However, the requirement on the
γ's, to make the Dirac equation valid, is that they anticommute;[4] it is not
required that they be the ones constructed by Dirac. For the purposes of separation by *handedness*, it is more convenient to use

$$\gamma^0 = \begin{pmatrix} 0 & 1 \\ 1 & 0 \end{pmatrix}, \gamma^i = \begin{pmatrix} 0 & -\sigma_i \\ \sigma_i & 0 \end{pmatrix} \text{ (the negative of the } \gamma^i \text{ in the original representation),}$$

$$\text{and } \gamma^5 = \begin{pmatrix} 1 & 0 \\ 0 & -1 \end{pmatrix}, \tag{10.9}$$

which can be obtained by the similarity transformation resulting from a particular change of basis. (See Problem 5.) It is significant that γ^5 is diagonal
now, rather than γ^0. We shall see that **the new basis vectors correspond to
helicity states.**

Consider a *massless particle (m = 0)* first. The Dirac equation is

$$(\gamma^0 p_0 + \gamma^i p_i)\psi = 0, \tag{10.10}$$

for a massless particle. Let us multiply Eq. (10.10) from the left by $(-i\gamma^1\gamma^2\gamma^3)$.
This yields

$$(-i\gamma^1\gamma^2\gamma^3\gamma^0 p_0 - i\gamma^1\gamma^2\gamma^3\gamma^i p_i)\psi = 0. \tag{10.11}$$

The first term in the parentheses can be transformed to $+i\gamma^0\gamma^1\gamma^2\gamma^3 p_0 = \gamma^5 p_0$
by moving the γ^0 across the other γ's, using the anticommutativity of the γ^i's.
The second term is $-i\gamma^1\gamma^2\gamma^3\gamma^1 p_1 - i\gamma^1\gamma^2\gamma^3\gamma^2 p_2 - i\gamma^1\gamma^2\gamma^3\gamma^3 p_3$, since the
repeated index i stands for a sum with i running from 1 to 3. Using the repre-
sentation of the γ^i's of Eq. (10.9), we find the second term in Eq. (10.11) to be[5]

$$+\sigma_i p_i 1 = -\vec{\sigma}\cdot\vec{p}1, \text{ since } p_i = -(\vec{p})^i.$$

Putting these results together, we obtain

$$[\gamma^5 p_0 - (\vec{\sigma}\cdot\vec{p})1]\psi = 0. \tag{10.12}$$

We know that

$$E = |\vec{p}|c = |\vec{p}| \text{ in natural units, for a massless particle,}$$

$$\text{therefore, } p_0 = |\vec{p}|. \tag{10.13}$$

---

4. They merely need to obey $\{\gamma^\mu,\gamma^\nu\} = 2g^{\mu\nu}$.

5. You might find it simplest to use the fact that $(\gamma^i)^2 = -1$ for each i ($i = 1, 2,$ or 3), along
with the anticommutation of different γ's to do the calculation.

So Eq. (10.12) becomes $(\gamma^5 - \vec{\sigma}\cdot\hat{p})\psi = 0$, and (using Eq. 10.9 for γ^5)

for $\psi = \begin{pmatrix} 1 \\ 0 \\ 0 \\ 0 \end{pmatrix}$ or $\psi = \begin{pmatrix} 0 \\ 1 \\ 0 \\ 0 \end{pmatrix}$, which have eigenvalue $+1$ for γ^5,

we obtain $\vec{\sigma}\cdot\hat{p} = +1$—called **right-handed**,

whereas for $\psi = \begin{pmatrix} 0 \\ 0 \\ 1 \\ 0 \end{pmatrix}$ or $\psi = \begin{pmatrix} 0 \\ 0 \\ 0 \\ 1 \end{pmatrix}$, which have eigenvalue -1 for γ^5,

we find $\vec{\sigma}\cdot\hat{p} = -1$—called **left-handed.**   (10.14)

The operator $\vec{\sigma}\cdot\hat{p}$ is called the **helicity**; using this terminology we say that **a right-handed particle has positive helicity** and **a left-handed particle has negative helicity**.

We see that the basis vectors break into two sets (of two), one set corresponding to right-handed particles and one to left-handed particles. A general state ψ can be expanded in this basis set to obtain

$$\psi = \psi_R + \psi_L,$$ (10.15)

where ψ_R is the **right-handed part**, which has eigenvalue $+1$ for γ^5, and ψ_L is the **left-handed part**, which has eigenvalue -1 for γ^5. In detail:

$$\psi_R = a_R \begin{pmatrix} 1 \\ 0 \\ 0 \\ 0 \end{pmatrix} + b_R \begin{pmatrix} 0 \\ 1 \\ 0 \\ 0 \end{pmatrix} \text{ and } \psi_L = a_L \begin{pmatrix} 0 \\ 0 \\ 1 \\ 0 \end{pmatrix} + b_L \begin{pmatrix} 0 \\ 0 \\ 0 \\ 1 \end{pmatrix},$$ (10.16)

where a_R, b_R, a_L, and b_L are arbitrary numbers.

We can define **right-handed and left-handed projection operators, P_R and P_L**, as

$$P_R = \frac{1 + \gamma^5}{2} \text{ and } P_L = \frac{1 - \gamma^5}{2},$$

since

$$P_R\psi_R = \frac{1 + \gamma^5}{2}\psi_R = \frac{1 + 1}{2}\psi_R = \psi_R \text{ and } P_R\psi_L = \frac{1 + \gamma^5}{2}\psi_L = \frac{1 - 1}{2}\psi_L = 0,$$

and

$$P_L \psi_L = \frac{1 - \gamma^5}{2} \, \psi_L = \frac{1 - (-1)}{2} \, \psi_L = \psi_L \text{ and } P_L \psi_R = \frac{1 - \gamma^5}{2} \, \psi_R = \frac{1 - 1}{2} \, \psi_R = 0.$$

(10.17)

For a general state, which has a left-handed part and a right-handed part (as shown in Eq. 10.15), we obtain

$$\boldsymbol{P_R \psi = \psi_R} \text{ and } \boldsymbol{P_L \psi = \psi_L},$$
(10.18)

which shows the *projection* behavior of these operators. Notice also that

$$P_R + P_L = 1,$$

$$P_R P_R \psi = P_R \psi_R = \psi_R,$$

which is the same as $P_R \psi$, so that

$$\boldsymbol{P_R P_R = P_R},$$

and likewise we can obtain

$$\boldsymbol{P_L P_L = P_L} \text{ and } \boldsymbol{P_R P_L = P_L P_R = 0.}$$
(10.19)

Equation (10.19) shows that successively performing two handedness projections of the same kind gives a result no different from that of projecting once; whereas, projecting the left-handed part and then trying to project the right-handed part of that (or vice versa) necessarily gives 0.[6] (These projection operators can also be used in the original basis where γ^5 is not diagonal and the basis states do not have definite handednesses.) Note that our discussion of handedness and helicity, so far, has considered only *massless* particles.

Let us now consider a *massive particle*. If it were moving very rapidly (we will call the direction of its motion the $+z$ direction), we expect that the mass term in the Dirac equation would be negligible, so that the above analysis would still be valid. Let us suppose that the particle has its spin aligned along its momentum direction— *it is a positive helicity object*. Now let us view the particle as we move in the $+z$ direction with a speed greater than that of the particle, so that the particle is moving backwards relative to us, i.e., we will do a Lorentz transformation in the $+z$ direction to a frame in which the particle is moving in the $-z$ direction. For a classical particle spinning in the *x-y* plane (whose spin vector is, therefore, in the $+z$ direction), the *direction* of the spin is unchanged by a Lorentz transformation in the $+z$ direction, since transverse components of velocity do not change direction. The quantum mechanical spin behaves the same way under Lorentz transformations. Thus, in the new frame its spin is still in the $+z$ direction but its momentum has reversed direction, so that $\vec{\sigma} \cdot \hat{p}$ is negative, i.e., *its helicity is negative in the new Lorentz frame*. Thus, **helicity is not Lorentz invariant** and **we cannot assign a definite helicity to a mas-**

---

6. The fact that these two projections add up to 1 and are orthogonal tells us that they are the complements of one another.

sive particle. (For a massless particle, $v = c$; we cannot move faster than c, so we *cannot* perform a Lorentz transformation which reverses its helicity.)[7] Although this is true, we can still utilize the P_R and P_L (which we defined above) as operators; they create definite helicity states for massless particles and, for particles with mass, they create states which have the helicity indicated in a frame where the particle is moving very rapidly. Indeed, they yield well-defined field operators because these projection operators are well-defined Dirac matrices:

$$P_R = \frac{1 + \gamma^5}{2} = \frac{1 + \begin{pmatrix} 1 & 0 \\ 0 & -1 \end{pmatrix}}{2} = \begin{pmatrix} 1 & 0 \\ 0 & 0 \end{pmatrix} \text{ and, similarly, } P_L = \begin{pmatrix} 0 & 0 \\ 0 & 1 \end{pmatrix}$$

(10.20)

where each entry is a 2×2 matrix.

In the following we shall use for our field operators:

$\psi_L = P_L \psi$ for the operator which **annihilates a left-handed particle or creates a right-handed antiparticle** and

$\psi_R = P_R \psi$ for the operator which **annihilates a right-handed particle or creates a left-handed antiparticle**.

We will also use their **Hermitian conjugates** for the operators which **create the particle or annihilate the antiparticle**:

$$\bar{\psi}_L = \psi_L^\dagger \gamma^0 = (P_L \psi)^\dagger \gamma^0 = \psi^\dagger P_L \gamma^0$$

$$\text{and } \bar{\psi}_R = (P_R \psi)^\dagger \gamma^0 = \psi^\dagger P_R \gamma^0,$$

since P_R and P_L are Hermitian.

It has been observed that **neutrinos occur only in the left-handed (negative helicity) state and antineutrinos only in the right-handed state**.[8] This implies that only ψ_L and $\psi_L^\dagger$ can be present in the field theory for the neutrinos. Notice from Eq. (10.16), that there are only two left-handed basis states. (They each have 0's for the upper two components for a particle at rest.) It can be shown that, as a consequence, only two two-component states $\begin{pmatrix} 1 \\ 0 \end{pmatrix}$ and $\begin{pmatrix} 0 \\ 1 \end{pmatrix}$ are needed to span the space of left-handed particles, so that this theory may be formulated with two-spinors. Therefore, this theory is referred to as **the two-component theory of neutrinos**. (See Problem 11.)

Parity is the name used for the **operator** that inverts the ordinary three-dimensional space through the origin, i.e., $\vec{r} \to -\vec{r}$. We have learned in Chapter 4 that the γ^5 matrix behaves like a pseudoscalar so that $\gamma^5 \to -\gamma^5$.

---

7. There is an exception to these considerations; a spin $\frac{1}{2}$ particle can have a definite handedness and non-zero mass if it is identical to its antiparticle (Majorana). We shall discuss this *Majorana mass* in our discussion of possible neutrino mixing in Chapter 18.

8. This is consistent with the *masslessness* of the neutrinos, as we have seen that helicity is not well defined for massive particles.

Therefore, **under inversion** we find that $P_L \to P_R$ and $P_R \to P_L$, as we might expect just from looking in the mirror.[9] We shall now proceed to build the gauge theory for the weak interaction. Since neutrinos only occur with one handedness, **parity will not be conserved by the electroweak theory.** This is obvious because inversion invariance would imply that for every reaction involving a left-handed neutrino, an identical reaction involving a right-handed neutrino should occur at exactly the same rate; however, right-handed neutrinos do not even exist. We shall discuss parity further when we consider discrete symmetries in Chapter 12.

10.3   THE ELECTROWEAK GAUGE THEORY $SU_L(2) \times U(1)$

It has been found that the weak interaction can be described by a Yang-Mills type of gauge theory; however, we shall see that quantum electrodynamics is intimately intertwined with the weak interaction. We have described the interaction term in QED with the form:

$$J^\mu A_\mu, \tag{10.21}$$

where J^μ is the electromagnetic current operator and A_μ is the gauge (photon) field operator. The weak interaction is described in a similar manner by an interaction term (vertex) of the form:

$$J^\mu_{WEAK} W_\mu,$$

$$\text{with } J^\mu_{WEAK} = \bar{\psi}_{aL} \gamma^\mu \psi_{bL} = \psi_a^\dagger P_L \gamma^0 \gamma^\mu P_L \psi_b = \tfrac{1}{2} \bar{\psi}_a (\gamma^\mu - \gamma^\mu \gamma^5) \psi_b. \tag{10.22}$$

We see that J^μ_{WEAK} contains only left-handed particles and, from the last expression (which you can derive by solving Problem 6), has the form (vector − axial vector). This $V - A$ **coupling** is a **parity-violating interaction**, since the *relative* sign between the V and the A will change under inversion. (The verification of this is also left (pun intended) as a problem (no. 12) for the reader.) The experimental discovery of this $V - A$ coupling is a fascinating story, which will be discussed in Chapter 12.

Notice that J^μ_{WEAK} is of the form $\bar{\psi}_a$ (operator)$^\mu$ ψ_b, which is similar to that of the electromagnetic current operator, and W_μ is the new gauge field. However, we shall see that state a and state b can be different particles with different values of electric charge[10] for which the corresponding W must carry electric charge (since electric charge is conserved), as shown in Fig. 10.1. J^μ_{WEAK} has a lepton part and a quark part, as we shall elucidate below. These account for the weak interactions of the leptons and hadrons, respectively. In each of the leptonic and semileptonic reactions of Section 10.1, we have two such cur-

---

9. Recall that an inversion is equivalent to a reflection plus a rotation, and reflections change left-handed to right-handed and vice versa.

10. This is a further generalization of the extension of the classical concept of current (flow of charges) to a current *operator* which annihilates and creates particles.

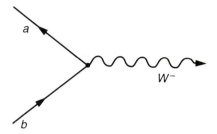

FIGURE 10.1
Weak interaction vertex for $q_a - q_b = +1$. A W^+ can be entering the vertex instead. The charge of the W is reversed for $q_a - q_b = -1$.

rents exchanging a W particle. In Fig. 10.2, we show the tree-level Feynman diagrams for representative leptonic and semileptonic decays. The charge-changing current operators of Eq. (10.22) are sometimes written $J_{WEAK}^{\mu\pm}$, where the superscript $\pm$ stands for the change of charge of the fermion (± 1), and are called the **charged currents**. By the late 1950s it was already conjectured that these vector current operators are the charged members of an $I = 1$ operator, whose neutral member is related to the electromagnetic current. Thus, we expected a theory for the weak interactions which contained QED as well. The relationships among the current operators seemed to indicate that an $SU(2)$ Yang-Mills type of theory might account for both electromagnetic and weak interactions.

In contrast to the colored quarks in QCD, the leptons are certainly not confined, so that the very short range of the weak interactions requires another explanation. Furthermore, the mechanism responsible for that short range must not affect the range of the QED forces. We will discuss that mechanism in Chapter 11. Here we consider the *structure* of the gauge theory. In an $SU(2)$ gauge theory of Ws there is a neutral W (W_3), which appears in the covariant derivative in the term igI_3W_3. The lepton doublets each consist of a charged lepton with $I_3 = -\frac{1}{2}$ and a neutrino with $I_3 = +\frac{1}{2}$; consequently, W_3 couples to the neutrino with the same strength but opposite sign as it does to the charged lepton. Clearly W_3 cannot be the photon, since the photon should not couple at all to neutrinos because they are uncharged. A larger group is required, and we shall now show how the gauge group which succeeds in this task is $SU_L(2) \times U(1)$, which has two independent sets of transformations, viz., isorotations and independent $U(1)$ transformations. As we saw in Eq. (10.22), the weak interaction has $V - A$ coupling — only the left-handed particle operators should be affected by the $SU(2)$ transformations, hence we insert the sub-L on the $SU(2)$ above. The right-handed fermions are (invariant) singlets under this $SU_L(2)$. Despite this handedness structure, the lack of handedness of the electromagnetic interaction is realized.

We have learned that the quarks and leptons occur as isodoublets, but now we see that this is true only for the left-handed particles. In Table 10.1, we write the particle operators again, indicating their behavior under isospin rotations according to their handedness. Note that there are no right-handed neutrinos,

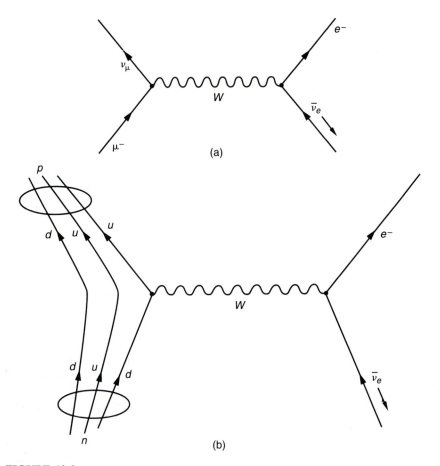

FIGURE 10.2
(a) Tree-level Feynman diagram for the leptonic decay $\mu^- \to e^- + \bar{\nu}_e + \nu_\mu$. (b) Quark flow diagram for the semileptonic decay $n \to p + e^- + \bar{\nu}_e$. In this extension of the quark flow diagrams to the weak interaction the virtual W *is* displayed.

a fact that we have learned from experimental observations, as will be discussed in Chapters 12 and 15.

Recall that when we discussed isotopic spin, we came upon another quantum number called hypercharge (Y) which was related to the charge of the particle by

$$Q = I_z + \frac{Y}{2}. \tag{10.23}$$

Let us use this generalized isospin and the respective charges of the particles to find the associated hypercharges.

For leptons $Y_L = -1$ and $Y_R = -2$; for quarks $Y_L = \frac{1}{3}$ and $Y_R = 2Q$.

($Q = \frac{2}{3}$ for u, c, and t and $Q = -\frac{1}{3}$ for d, s, and b.)

TABLE 10.1

Quark and Lepton $SU_L(2)$ Multiplets

Generation:	1	2	3
Quark doublets:	$\begin{pmatrix} u \\ d \end{pmatrix}_L$	$\begin{pmatrix} c \\ s \end{pmatrix}_L$	$\begin{pmatrix} t \\ b \end{pmatrix}_L$
Quark singlets:	u_R, d_R	c_R, s_R	t_R, b_R
Lepton doublets:	$\begin{pmatrix} \nu_e \\ e^- \end{pmatrix}_L$	$\begin{pmatrix} \nu_\mu \\ \mu^- \end{pmatrix}_L$	$\begin{pmatrix} \nu_\tau \\ \tau^- \end{pmatrix}_L$
Lepton singlets:	e_R^-	μ_R^-	τ_R^-

The Y quantum numbers can be thought of as the values along the diagonal of a diagonalized Y operator; $Y/2$ is taken to be the generator of a new U(1) symmetry, sometimes designated $U_Y(1)$. The elements of the group are $e^{i(Y/2)\theta}$. The Y value gives the representation of $U_Y(1)$ to which the particle belongs.

We employ the covariant derivative construction (of Eq. 8.2) to obtain the interactions involved: [11]

$$\partial_\mu \to D_\mu = \partial_\mu - ig_1 \frac{Y}{2} B_\mu - ig_2 T_i W_{i\mu}, \tag{10.24}$$

with $T_i = \tau_i/2$ for left-handed particles (doublets), as in the Yang-Mills theory, and $T_i = 0$ for right-handed particles (singlets), which do not undergo isorotations. We have introduced a gauge field B_μ for the $U_Y(1)$ part of the gauge group and the $W_{i\mu}$ is an isovector, as in the original Yang-Mills theory. The SU(2) part has one coupling constant, called g_2, and the U(1) group has a separate constant, called g_1, since the "$\times$" between them in $SU_L(2) \times U_Y(1)$ means that they are completely independent.

Let us now look at the $J_{WEAK}^\mu W_\mu$ and $J_{WEAK}^\mu B_\mu$ interactions which arise from the replacement indicated in Eq. (10.24). Let us first look at the *first generation of leptons*. We have

$$\sum_f \bar{\psi}_f i\gamma^\mu D_\mu \psi_f = \sum_f \bar{\psi}_f i\gamma^\mu \left(\partial_\mu - ig_1 \frac{Y}{2} B_\mu - ig_2 T_i W_{i\mu} \right) \psi_f$$

$$= (\nu_L^\dagger \, e_L^{-\dagger}) \gamma^0 \gamma^\mu i \left(\partial_\mu - ig_1 \frac{Y_L}{2} B_\mu - ig_2 \frac{\tau_i}{2} W_{i\mu} \right) \begin{pmatrix} \nu_L \\ e_L^- \end{pmatrix}$$

$$+ (e_R^{-\dagger}) \gamma^0 \gamma^\mu i \left(\partial_\mu - ig_1 \frac{Y_R}{2} B_\mu \right) (e_R^-) + \cdots, \tag{10.25}$$

where f runs through the fundamental fermions and we have explicitly written only the terms for the first generation of leptons.

---

11. The signs of the gauge field terms may be chosen to be negative, which is the usual convention.

From here on we shall omit the γ matrices ($\gamma^0 \gamma^\mu$) which appear in every term. Let us calculate the interaction of the *left-handed leptons* first. When we discussed isotopic spin in Chapter 6, we had a $\vec{\pi} \cdot \vec{\tau}$ interaction with the nucleon doublet (Eqs. 6.5 through 6.10) which was of the same form as the $\vec{W} \cdot \vec{\tau}$ interaction with the left-handed lepton doublet that appears here. Therefore, for the SU(2) part we can copy that result replacing the fields there with the corresponding fields here to obtain

$$\frac{g_1}{2} Y_L \left\{ (v_L^\dagger e_L^\dagger) 1 \begin{pmatrix} v_L \\ e_L \end{pmatrix} \right\} B_\mu + \frac{g_2}{2} \left\{ (v_L^\dagger e_L^\dagger) \begin{pmatrix} W_{3\mu} & W_{1\mu} - i W_{2\mu} \\ W_{1\mu} + i W_{2\mu} & -W_{3\mu} \end{pmatrix} \begin{pmatrix} v_L \\ e_L \end{pmatrix} \right\}$$

$$= \frac{g_1}{2} Y_L \{ v_L^\dagger v_L B_\mu + e_L^\dagger e_L B_\mu \}$$

$$+ \frac{g_2}{2} \{ v_L^\dagger v_L W_{3\mu} + \sqrt{2} v_L^\dagger e_L W_\mu^{-\dagger} + \sqrt{2} e_L^\dagger v_L W_\mu^{+\dagger} - e_L^\dagger e_L W_{3\mu} \}, \quad (10.26)$$

where $W^{\pm\dagger} = \frac{1}{\sqrt{2}} (W_1 \pm i W_2)$ and the daggers create the indicated particles.

In Eq. (10.26) the second and third terms in the second bracket (which is multiplied by $g_2/2$ and came from SU(2)) are the charged currents of the weak interaction (which couple to $W^\pm$). Their effects have been seen in β-decay experiments for almost a century, as we have already discussed. Notice that there is also a $v_L^\dagger v_L$ current operator which is a **neutral current** interacting with the neutral gauge fields:

$$v_L^\dagger v_L \left(\frac{g_1}{2} Y_L B_\mu^0 + \frac{g_2}{2} W_{3\mu} \right). \quad (10.27)$$

The neutrino is not electrically charged, so that this is definitely not an electromagnetic interaction; it must be a part of the weak interaction, but this neutral-current interaction had never been observed (before the theory was proposed). Nevertheless, in this theory the linear combination of the gauge fields shown in Eq. (10.27) couples to the neutral (neutrino) current. That linear combination of gauge fields is called Z_μ^0, so that

$$Z_\mu^0 \propto (g_1 Y_L B_\mu^0 + g_2 W_{3\mu}),$$

which yields

$$Z_\mu^0 = \frac{(-g_1 B_\mu^0 + g_2 W_{3\mu})}{\sqrt{g_1^2 + g_2^2}}$$

for the properly normalized field operator, where we have inserted $Y_L = -1$. Since the sum of the squares of the coefficients in this expression is 1, we can think of them as the sine and cosine of an angle of rotation of the W_3 into the B^0 to form the Z^0:

$$Z_\mu^0 = -\sin \theta_W B_\mu^0 + \cos \theta_W W_{3\mu},$$

where $\cos \theta_W = \dfrac{g_2}{\sqrt{g_1^2 + g_2^2}}$ and $\sin \theta_W = \dfrac{g_1}{\sqrt{g_1^2 + g_2^2}}$. $\quad (10.28)$

As we have said earlier, this theory should include the electromagnetic interaction, so that a linear combination of the neutral gauge particles should correspond to the photon field operator A_μ. Since A_μ cannot couple to the
(electrically neutral) neutrino current, it must be *orthogonal* to the field (Z_μ^0)
that does. That orthogonal linear combination is

$$A_\mu = \cos\theta_W B_\mu^0 + \sin\theta_W W_{3\mu} = \frac{(g_2 B_\mu^0 + g_1 W_{3\mu})}{\sqrt{g_1^2 + g_2^2}}.^{12} \tag{10.29}$$

The neutral gauge fields occurring as distinct entities in experiments are the A_μ
and the Z_μ^0. The A field produces the electromagnetic interactions, and the theory predicts the existence of neutral-current weak interactions mediated by the
Z^0.

It is understandable that the neutral-current weak interactions had not been
observed before their prediction, for the following reasons:

1. To observe the (neutral) neutrino-current interaction with the Z^0 requires the study of neutrino scattering, which necessitates the (extremely
   difficult) detection of neutrinos.
2. To discern the electron-current interaction with the Z^0 is difficult
   because the interaction of the electromagnetic field with the electron current is much larger than the (neutral) weak interaction.

Neutral-current events *were* observed in the 1970s corroborating the theory.

Just as for ordinary rotations, we can invert Eqs. (10.28) and (10.29) and
solve for the B and W fields to obtain

$$B_\mu^0 = \cos\theta_W A_\mu - \sin\theta_W Z_\mu^0$$
$$W_\mu^0 = \sin\theta_W A_\mu + \cos\theta_W Z_\mu^0, \tag{10.30}$$

where we have written W_μ^0 for $W_{3\mu}$.

Let us now look at the coupling of the electron current to the gauge particles. From Eqs. (10.25) and (10.26) above we obtain

$$e_L^\dagger e_L \left(-\frac{g_1}{2} B^0 - \frac{g_2}{2} W_3 \right) + e_R^\dagger e_R (-g_1 B^0)$$

$$= e_L^\dagger e_L \left(A \frac{-g_1 g_2}{\sqrt{g_1^2 + g_2^2}} + Z^0 \frac{g_1^2 - g_2^2}{\sqrt{g_1^2 + g_2^2}} \right)$$

$$+ e_R^\dagger e_R \left(A \frac{-g_1 g_2}{\sqrt{g_1^2 + g_2^2}} + Z^0 \frac{g_1^2}{\sqrt{g_1^2 + g_2^2}} \right). \tag{10.31}$$

(We have inserted $Y_L = -1$ and $Y_R = -2$ in the above and omitted the sub-μ
on the gauge fields.) The coupling to the electromagnetic field has been known

---

12. The field operators $W_{3\mu}$ and B_μ^0 are orthogonal, meaning that they produce orthogonal
states.

for about 100 years to be $-e$ ($= -1.6 \times 10^{-19}$ Coulombs). Equation (10.31) shows us that, in this theory, each handedness of the electron does couple to the electromagnetic field (A_μ) with the same strength (as expected). Furthermore, the combination of g_1 and g_2, which occurs as the effective electromagnetic coupling, must equal $-e$ in order to agree with experiment:

$$\frac{-g_1 g_2}{\sqrt{g_1^2 + g_2^2}} = -e. \tag{10.32}$$

Combining this with the definition of the θ_W in Eq. (10.28), we obtain

$$g_1 = \frac{e}{\cos \theta_W} \text{ and } g_2 = \frac{e}{\sin \theta_W}. \tag{10.33}$$

In order to determine g_1 and g_2 we must know θ_W. The ratio of the rates of neutral current interactions to charged current interactions depends upon the angle θ_W and many different weak interaction measurements give the same result (within 3 percent):

$$\sin^2(\theta_W) = 0.2325 \pm .0008, \theta_W \approx 28°, \text{ at } p \approx M_Z.^{13} \tag{10.34}$$

This then yields: $g_1 \approx 1.14e$ and $g_2 \approx 2.07e$. We see that the weak interactions have larger coupling constants than that of electrodynamics, which is a serious dilemma since the observed weak interactions are much weaker than the electromagnetic interactions.

Let us switch from the Gaussian units for charge, which we have been using until now, to the Heaviside-Lorentz units in which the fine-structure constant is defined as $\alpha = e^2/4\pi\hbar c \approx \frac{1}{137}$, so that our numerical values of the coupling constants will agree with those usually reported. Analogously, we define $\alpha_1 = g_1^2/4\pi\hbar c$ and $\alpha_2 = g_2^2/4\pi\hbar c$. From Eq. (10.34), we find $\alpha_1 : \alpha_2 : \alpha = (1.14)^2 : (2.07)^2 : 1$, with $\alpha \approx \frac{1}{137}$, which yields

$$\alpha_1 = \frac{g_1^2}{4\pi} \approx \frac{1}{100} \text{ and } \alpha_2 = \frac{g_2^2}{4\pi} \approx \frac{1}{30}, \text{ in natural units.}$$

At this point, the theory appears to have two major discrepancies with experiment:

1. **The masslessness of the gauge particles yields long-range interactions, but the observed range of the weak interaction is very short.**
2. **The coupling constants are greater than that of electrodynamics, but the observed weak interactions are much weaker than those of electrodynamics.**

We shall soon learn, in Chapter 11, how the gauge symmetry can be broken in a spontaneous manner to solve both of these dilemmas at once.

---

13. We have learned in Section 9.3 that the g_i vary with energy; consequently, so does θ_W.

The same structure that we have found for the first generation of leptons occurs for the second and third generations.

In order to pursue the structure for the quarks, let us now look back at Eq. (10.24), where we had

$$g_1 \frac{Y}{2} B_\mu + g_2 T_i W_{i\mu}$$

as the form of the coupling in the covariant derivative. Consider the neutral gauge particles contained in this sum:

$$g_1 \frac{Y}{2} B_\mu + g_2 T_3 W_{3\mu}. \tag{10.35}$$

Using Eq. (10.30) and replacing $Y/2$ with $(Q - T_3)$, we find that this yields, for **the effective coupling to the Z^0**:

$$\frac{e}{\cos\theta_W \sin\theta_W} (T_3 - Q \sin^2\theta_W), \tag{10.36}$$

which is a convenient form for the corresponding discussion of quark-Z coupling. In Problem 10, you are asked to perform the analysis for the first quark generation. (The other two generations are heavier "copies.")

PROBLEMS

1. a. Suppose a τ^- decays to one charged particle, which is an electron. What uncharged particles must emerge?
   b. Show the tree-level Feynman diagram(s) for the simplest such process.
2. a. Show that H_{DIRAC} for a massless particle commutes with the handedness projections, P_L and P_R. It may help to replace α^i by $\gamma^0\gamma^i$ in H_{DIRAC} here.
   b. Show that H_{DIRAC}, for particles with mass, does not commute with P_L and P_R.
   c. Find the four-column for a left-handed massless particle of energy $E > 0$ moving in the $+z$ direction. Hint: P_L is a projection operator. Start with the form of $u_s(\vec{p})$ given in Problem 8 of Chapter 4 and put in a general form:

$$\begin{pmatrix} a \\ b \end{pmatrix}, \text{ with } \sqrt{|a^2| + |b^2|} = 1, \text{ for } \chi_s.$$

3. Consider the new representation of the γ matrices, introduced in Section 10.2:

$$\gamma^i = \begin{pmatrix} 0 & -\sigma_i \\ \sigma_i & 0 \end{pmatrix}, \quad \gamma^0 = \begin{pmatrix} 0 & 1 \\ 1 & 0 \end{pmatrix}.$$

   a. Show that they obey the correct anticommutation relations.
   b. Calculate $\gamma^5 = i\gamma^0\gamma^1\gamma^2\gamma^3$ in this new representation.
4. For a particle at rest, find the eigenstates of $H_{\text{DIRAC}} = m\gamma^0$ in the representation of the γ's in which γ^5 is diagonal. (To help you start, we have written γ^0 in place of β; they are equal.) From the result show that each state has both left-handed and right-handed parts in equal amounts.

5. a. Show that a similarity transformation $(U\gamma^\mu U^\dagger)$ of the original representation of the γ^μ's with

$$U = \begin{pmatrix} \frac{1}{\sqrt{2}} & \frac{1}{\sqrt{2}} \\ \frac{1}{\sqrt{2}} & -\frac{1}{\sqrt{2}} \end{pmatrix}$$

   yields the new set.
   b. Write the four-column of a left-handed massless particle in the new representation of the γ^μ's, and then transform it to the old representation.
6. Derive the $V - A$ form of J^μ_{WEAK}, shown in the last expression of Eq. (10.22).
7. Calculate g_1 and g_2 (in terms of e) and determine the values of α_1^{-1} and α_2^{-1}.
8. a. Apply P_L (in the Dirac representation of the γ's) to $u_s(\vec{p})$, given in Problem 8 of Chapter 4, to get the form of the four-column of ψ_L.
   b. Show that for $E \gg m$, the helicity is negative.
9. Calculate the effective coupling constant (in terms of e) for the Z^0 vertex with
   a. ν
   b. e_L, and
   c. e_R.
10. a. Derive Eq. (10.36), the Z^0 coupling to quarks, from Eq. (10.35). Find the effective coupling constant of the Z^0 to
    b. the up quark and
    c. the down quark.
11. a. For massless particles let us start as Dirac did by writing H_{DIRAC} in terms of unknown matrices. Show that only three anticommuting matrices are necessary.
    b. Find three 2×2 matrices which obey the requisite anticommutation relations.
    c. Write the two-column which correspond to a left-handed particle; to a right-handed particle.
    This is the basic structure of the two-component theory of neutrinos.
12. a. Show that γ^0 may be used as the parity operator by examining its effect on γ^μ.
    b. Show that the relative sign in the $V - A$ interaction $\sim \bar{\psi}\gamma^\mu\psi - \bar{\psi}\gamma^\mu\gamma^5\psi$ changes under space inversion.
13. Check the Y_L and the Y_R assignments for leptons and quarks.
14. Draw the quark flow diagram for the decay $\pi^- \to \mu^- + \bar{\nu}_\mu$.

Chapter 11

The Electroweak Theory —
II. Breaking the Symmetry

11.1 RANGE, MASS, RENORMALIZATION, AND SPONTANEOUS SYMMETRY BREAKING

As we have pointed out in our study of gauge theories, no mass term for the gauge particles can occur in the Lagrangian; they must be massless. We have also learned from Yukawa's work (in Chapter 4) that the range of the force in a quantum field theory is proportional to the inverse of the mass of the exchanged particle. Thus, a gauge theory built on an exact symmetry implies infinite-range forces as in QED. However, the range of the weak interaction is very short and, as we saw in Chapter 4, it corresponds to a mass for the exchanged particle $M_G \sim 100$ GeV. Furthermore, there is no confinement of charge, so that the mechanism which led to a short range in QCD cannot be invoked. Our conclusion: $SU_L(2) \times U_Y(1)$ is a *broken* symmetry.

You might guess that we should put a mass term into the Lagrangian and be done with it. However, we must consider the question of the infinities which occur in the perturbative Feynman-diagram calculations of *any* quantum field theory. We have swept under the rug any consideration of the sweeping under the rug of these infinities, which is called *renormalization*. In fact, renormalizability is *not* automatic but is the exception. In general, one cannot reorganize the calculations which yield infinities, so that they all "pack" into the observed charges and masses, producing finite results in terms of those observed values. Instead, most theories will produce *unrenormalizable* infinities when treated as an expansion in Feynman diagrams. At this stage in our understanding of quantum field theories, no one has succeeded in finding a way to make sense out of these **nonrenormalizable** theories. What does this have to do with the masses of gauge particles? A lot! The nonabelian gauge theories (which have massless gauge particles) have been proven to be renormalizable; gauge invariance is crucial to the proof. Putting a mass term for the vector bosons into the Lagrang-

179

ian *destroys* the gauge invariance and, thus, the renormalizability of the theory. However, there is a way that the gauge invariance can *spontaneously* break to produce mass for the gauge particles, without destroying the symmetry of the Lagrangian or the renormalizability. In fact, the proof of the renormalizability of spontaneously broken gauge theories aroused keen attention and intense interest in the entire elementary particle physics community.[1]

The failure of the unbroken $SU_L(2) \times U_Y(1)$ gauge theory to describe the range and strength of the weak interaction was an impasse, whose resolution arose from physics at the opposite end of the spectrum of complexity. In particle physics we attempt to describe the fundamental particles interacting one with another, whereas in solid state physics the collective behavior of a very large number of particles is considered. The mechanism which will make the electroweak theory viable is called **spontaneous symmetry breaking**, and it is well known to anyone who is acquainted with ferromagnetism, a phenomenon studied in solid state physics. (It has been said that one who knows two "things" can find a way around an impasse which appears insurmountable to those who know only one "thing.")

11.1.1   Ferromagnetism

At high temperatures, a ferromagnetic material has zero net magnetism; the magnetic moments of the atoms are randomly oriented. As we lower the temperature below the Curie temperature, the material spontaneously magnetizes. The theory which describes electromagnetism is rotationally invariant (look at Maxwell's equations on all those T-shirts around you). The original system (above the Curie temperature) was also rotationally invariant. However, below the Curie temperature the magnetic moments of the atoms are (nearly) all aligned with one another because of the interaction between neighboring atoms. Clearly, this characterization does not determine one unique state, since the aligned magnetic moments can point in *any* direction. As we lower the temperature and the system goes into a magnetized state, one direction is singled out. If we do this experiment a million times, we will have a *spherically symmetric distribution of that direction*. However, any one of these experiments results in a *breaking* of the spherical symmetry, since a particular direction (in which the magnetization finally points) has been selected. It is analogous to the case of an object at the top of a hill; it is in unstable equilibrium and the slightest disturbance will send it careening down the hill in some direction. There are always fluctuations in the ferromagnetic material, so that the magnetic system will choose one direction along which to magnetize. There is no preferred direction in space, however, so that the direction is "chosen" *randomly*. The result is a spontaneous breaking of the spherical symmetry.

---

1. In 1971 G. 't Hooft, while still a graduate student, constructed a manifestly renormalizable version of spontaneously broken gauge theories (and also independently rediscovered the Higgs mechanism, which we shall discuss in Section 11.1.3).

There exists a phenomenological theory for magnetism which shows this behavior quantitatively. It is the **Landau-Ginsburg theory** which presents an expression for the free energy ($\mathcal{F}$) of a system in terms of its magnetization $\vec{M}$:

$$\mathcal{F} = A|\vec{M}|^2 + B(|\vec{M}|^2)^2 \quad \text{with } B > 0 \tag{11.1}$$

(assuming the sample is small enough so that $\vec{M}$ is uniform). **The system will be in the state that minimizes the free energy.** Let us find the extrema:

$$\frac{d\mathcal{F}}{dM} = 2AM + 4BM^3 = 2M(A + 2BM^2) = 0,$$

$$\text{where } M = |\vec{M}|,$$

$$\text{and } \frac{d^2\mathcal{F}}{dM^2} = 2A + 12BM^2. \tag{11.2}$$

Above the Curie temperature: $A > 0$, so that the expression in parentheses $(A + 2BM^2)$ in $d\mathcal{F}/dM$ is positive, and the extremum occurs at $M = 0$.

$$\text{At } M = 0, \frac{d^2\mathcal{F}}{dM^2} = 2A > 0;$$

thus, this extremum is a minimum and the free energy looks like Fig. 11.1. Thus, the system is spherically symmetric, with $M = 0$.

Below the Curie temperature: $A < 0$, so that either $M = 0$ or $(A + 2BM^2) = 0$ at the extrema.

$$\text{At } M = 0, \frac{d^2\mathcal{F}}{dM^2} = 2A < 0, \text{ so that this is now a maximum;}$$

$$\text{at } M^2 = \frac{-A}{2B} \text{ (for which } A + 2BM^2 = 0),$$

$$\frac{d^2\mathcal{F}}{dM^2} = 2A + 12BM^2 = -4A > 0,$$

so that these are minima. In two dimensions the free energy now looks like Fig. 11.2. Potentials of this shape are often referred to as "Mexican hat" potentials. In three dimensions, we have a whole sphere of minima with radius $|\vec{M}| = \sqrt{-A/2B}$. As we lower the temperature below the Curie temperature, $M = 0$ becomes an *unstable* equilibrium point (a maximum), and the system will move off into one of these (minima) states, with the concomitant breaking of rotational invariance. Thus, we have **spontaneous symmetry breaking**.

11.1.2   Scalar Field Theory – Global Symmetry Transformations

Let us now consider the field theory of a complex scalar field (ϕ). We shall seek the minimum state by looking for the minimum of the potential in the Lagrangian. We will then quantize this theory by expanding about this minimum, which will be considered the vacuum state. This is, in fact, what is tacitly done in the

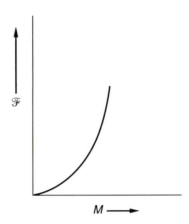

FIGURE 11.1
The free energy $\mathcal{F}$ above the Curie temperature.

usual quantum field theory; however, these words are not used because the minimum is at $\phi = 0$. The Lagrangian will be taken as

$$\mathcal{L} = T - V = (\partial^\mu\phi)^*(\partial_\mu\phi) - \mu^2\phi^*\phi - \lambda(\phi^*\phi)^2, \text{ with } \lambda > 0, \qquad (11.3)$$

$$\text{where } V = \mu^2\phi^*\phi + \lambda(\phi^*\phi)^2.$$

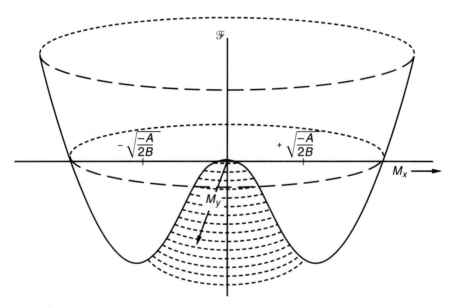

FIGURE 11.2
The free energy $\mathcal{F}$ below the Curie temperature, in two dimensions, as a function of M_x and M_y. The figure is rotated about the $\mathcal{F}$ axis and looks like a "Mexican hat."

This is invariant under the **global** phase transformations $\phi \to e^{i\alpha}\phi$, so that $\mathcal{L}$ possesses **U(1)** symmetry (as we have discussed). It is easier to follow the analogy with the aforegoing if we change to real fields ϕ_1 and ϕ_2 related to ϕ by

$$\phi = \frac{\phi_1 + i\phi_2}{\sqrt{2}}. \tag{11.4}$$

Substituting, we obtain

$$\mathcal{L} = \frac{1}{2}(\partial_\mu \phi_1)^2 + \frac{1}{2}(\partial_\mu \phi_2)^2 - \frac{1}{2}\mu^2(\phi_1^2 + \phi_2^2) - \frac{\lambda}{4}(\phi_1^2 + \phi_2^2)^2. \tag{11.5}$$

This Lagrangian is obviously invariant to a rotation[2] in the ϕ_1-ϕ_2 plane, which is an **SO(2) global invariance**. (In Chapter 5 we learned that SO(2) and U(1) are isomorphic, so this is the same symmetry that was present in the complex form of the theory.) We seek the minima of the potential (for ϕ_1 and ϕ_2 constant in space):

$$V = \frac{1}{2}\mu^2 \zeta + \frac{\lambda}{4}\zeta^2, \tag{11.6}$$

where $\zeta = (\phi_1^2 + \phi_2^2)$. We must require that

$$\frac{\partial V}{\partial \phi_1} = \frac{\partial V}{\partial \zeta}\frac{\partial \zeta}{\partial \phi_1} = \left(\frac{1}{2}\mu^2 + \frac{\lambda}{2}\zeta\right)2\phi_1 = (\mu^2 + \lambda\zeta)\phi_1 = 0$$

$$\text{and } \frac{\partial V}{\partial \phi_2} = (\mu^2 + \lambda\zeta)\phi_2 = 0. \tag{11.7}$$

Similar to our result for the Landau-Ginsburg theory, for $\mu^2 > 0$ the minimum is at $\phi_1 = \phi_2 = 0$, which does not break the SO(2) symmetry. However, if somehow this "mass" term has the wrong sign, i.e., $\mu^2 < 0$, then the minimum occurs at

$$\zeta = \frac{-\mu^2}{\lambda}. \tag{11.8}$$

Equation (11.6) shows that ζ is the square of the "distance" from the origin in ϕ_1-ϕ_2 space; therefore, the minima form a circle of radius

$$v = \sqrt{\frac{-\mu^2}{\lambda}} \tag{11.9}$$

in that space. (See Fig. 11.3.) The system will "choose" one of these states as its lowest or vacuum state resulting in a spontaneous breaking of the SO(2) symmetry.

---

2. This is true as long as the rotation is independent of x^ν, so that the ∂_μ operators do not operate on the rotation angle.

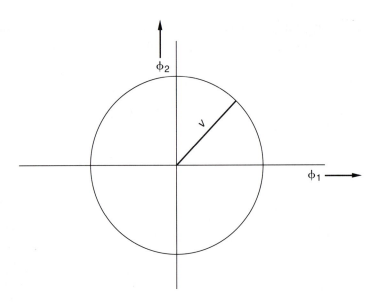

FIGURE 11.3
Circle of minima in ϕ_1-ϕ_2 space.

Let us now expand our fields about the lowest state, which has

$$(\phi_1^2 + \phi_2^2)_{vac.} = v^2.$$

So far, because of the U(1) symmetry, the "directions" of the ϕ_1 and ϕ_2 axes are not prescribed. We can choose them so that the lowest state is along the ϕ_1 direction to obtain

$$\phi_1 = v + \eta(x^\mu) \text{ and } \phi_2 = \rho(x^\mu). \tag{11.10}$$

Substituting into the Lagrangian (Eq. 11.5), we obtain

$$\mathcal{L} = \frac{1}{2}\,(\partial_\mu \rho)^2 + \frac{1}{2}\,(\partial_\mu \eta)^2 - \frac{1}{2}\,(2\lambda v^2)\eta^2 - \lambda v \eta (\eta^2 + \rho^2)$$

$$- \frac{\lambda}{4}\,(\eta^2 + \rho^2)^2 + \frac{\lambda v^4}{4}, \tag{11.11}$$

in which the terms quadratic in ρ canceled because of the relationship between v, μ, and λ. **The term quadratic in a particular field,** having a **coefficient whose sign is opposite that of the kinetic term,** is the **mass term** for that field. We see that the ρ field (the imaginary part of ϕ) is massless, since no term quadratic in ρ appears, and the η field has mass (from the third term in Eq. 11.11):

$$m_\rho = 0 \text{ and } m_\eta = \sqrt{2\lambda v^2}. \tag{11.12}$$

The cancellation of the quadratic terms for the ρ field is not an accident. In fact, there exists a general theorem which states that **the spontaneous break-**

ing of a continuous global symmetry always produces a spin 0 massless state.
This is called the Goldstone theorem, and the massless particle is referred to as
the Goldstone boson.

We can understand the generality of this result by looking at our simple
SO(2) model in Fig. 11.3. Notice that there is a circle on which the potential
is a minimum. Expanding the potential in a Taylor series about any one point
(ϕ_{min}) on the circle, in any direction gives

$$V(\phi) = V(\phi_{min}) + \frac{V''(\phi_{min})}{2!} \Delta^2 + O(\Delta^3), \qquad (11.13)$$

since $V'(\phi_{min}) = 0$, where $\Delta = \phi - \phi_{min}$ (in a particular direction in ϕ-space)
and the prime stands for a derivative in the Δ direction in ϕ-space.

So the square of the mass of the field Δ, which is proportional to the coefficient of the quadratic term, is proportional to V'' in the direction of Δ. When
we take the field Δ **in the direction tangent to the circle of minima,** which is the
ρ direction for our choice of the minimum state above, **the potential is flat,** so
that $V'' = 0$ and that field will be massless. Furthermore, this qualitative argument is quite general; for a continuous "path" of minima, V'' will be zero along
that path and the field corresponding to that *flat* direction will be massless.

11.1.3   Gauge (Local) Transformations and the Higgs Phenomenon

Let us now modify our Lagrangian for the complex scalar field to make it invariant under local phase transformations, i.e., **local U(1).** Recall that we need
merely replace space-time derivatives by covariant derivatives and add a term
quadratic in $F_{\mu\nu}$. The resulting Lagrangian is

$$\mathcal{L} = (D^\mu \phi)^*(D_\mu \phi) - \mu^2 \phi^* \phi - \lambda(\phi^* \phi)^2 - \tfrac{1}{4} F^{\mu\nu} F_{\mu\nu}, \qquad (11.14)$$

where $D_\mu = \partial_\mu - igG_\mu$ and $F_{\mu\nu} = \partial_\mu G_\nu - \partial_\nu G_\mu$ for gauge field G_μ. Let us write
the complex field in polar form: $\phi(x) = H(x)e^{i\chi(x)}$, where x stands for x^μ
(space and time). We can now do a gauge transformation to obtain

$$\phi(x) \to e^{i\alpha(x)} \phi(x) = H(x), \qquad (11.15)$$

where H is a real field, by choosing $\alpha(x) = -\chi(x)$. Since the field is now real,
we have removed the massless Goldstone boson. (Here it is $\chi(x)$; in the global
theory it was ρ, the imaginary part of ϕ.)

Of course, the gauge field also shifts under this gauge transformation, as
we have already learned; we will use G' for it in this new gauge (Eq. 11.15).
Proceeding as we did for the global case, we will expand about the minimum
state ($H = v$):

$$H(x) = \frac{v + h(x)}{\sqrt{2}}, \qquad (11.16)$$

where **h(x)** is called **the Higgs field**. Inserting this into the Lagrangian (Eq. 11.14), we obtain

$$\mathcal{L} = \frac{1}{2} \left[(\partial^\mu - igG'^\mu)(v + h) \right] \left[(\partial_\mu + igG'_\mu)(v + h) \right]$$

$$- \frac{\mu^2}{2} (v + h)^2 - \frac{\lambda}{4} (v + h)^4 - \frac{1}{4} F^{\mu\nu} F_{\mu\nu}. \tag{11.17}$$

Expanding and replacing μ^2 by $(-\lambda v^2)$ (see Eq. 11.9), we obtain

$$\mathcal{L} = \frac{1}{2} \partial^\mu h \partial_\mu h - \frac{1}{4} F(G')^{\mu\nu} F(G')_{\mu\nu} + \frac{1}{2} [g^2 v^2] G'^\mu G'_\mu - \frac{1}{2} [2\lambda v^2] h^2$$

$$- [\lambda v] h^3 - \frac{[\lambda]}{4} h^4 + [g^2 v] h G'^\mu G'_\mu + \frac{1}{2} [g^2] h^2 G'^\mu G'_\mu + \frac{1}{4} [\lambda v^4], \tag{11.18}$$

where we have put the constant parameters in square brackets. The first line contains the kinetic energy parts and the **mass terms generated by the spontaneous symmetry breaking**, viz., the terms quadratic in the fields; the second line has three-particle and four-particle vertices as well as a constant (the last) term. Notice that **the vertices are completely prescribed by the gauge-symmetry requirement. The gauge field G'_μ has a mass** whose square is the coefficient of the $(\frac{1}{2} G'^\mu G'_\mu)$ term:

$$M_G = gv. \tag{11.19}$$

Similarly, the mass of the Higgs field h is

$$m_h = \sqrt{2\lambda v^2}. \tag{11.20}$$

Not only has the gauge field become a massive field, but we have been able to gauge away the field (ρ) that was the Goldstone boson in the global case. This is an important feature, since massless Goldstone bosons resulting from the spontaneous symmetry breaking in the electroweak theory have never been observed experimentally. But in so doing, have we lost a degree of freedom? The final set of fields must contain the same number of independent fields as did the initial set.

The answer is no! The original massless, spin 1 gauge field has two independent helicities, with $S_z = \pm 1$. (Since they are massless they move at speed c and helicity is a Lorentz-invariant property, as we learned in Chapter 10.) This is just like the two possible transverse polarizations of the photon. A *massive* spin 1 particle has three possible states $S_z = 0, \pm 1$, which we know rotate into one another.[3] Thus, just as we would have 2 vector-particle degrees of freedom plus 1 Goldstone-boson degree of freedom in a corresponding broken global theory with a massless vector field, we still have 3 degrees of freedom in the broken local theory which are, however, all in the *massive* gauge field.

---

3. Consider rotations in the rest frame of the particle.

Just as for the photon field, for any massless gauge field (which has only two (transverse) polarizations) there is a gauge in which

$$\vec{\nabla} \cdot \vec{G} = 0,^4 \text{ that in momentum space is } \vec{k} \cdot \vec{G}(\vec{k}) = 0, \tag{11.21}$$

i.e., $\vec{G}$ is polarized transverse (perpendicular) to the momentum $(\vec{k})$. The gauge transformation we have made in Eq. (11.15), which removed the Goldstone boson, shifted the new gauge field to $\vec{G}'$:

$$G'_\mu = G_\mu + \frac{1}{g} \partial_\mu \alpha = G_\mu - \frac{1}{g} \partial_\mu \chi, \tag{11.22}$$

and χ is directly related to ρ of the global theory $(\rho = \sqrt{2} H \sin \chi)$ which was the Goldstone boson. Thus, the gauge field now has a longitudinal component which is essentially the field that was the Goldstone boson in the global theory:

$$\vec{\nabla} \cdot \vec{G}' = \partial_i G'^i = \vec{\nabla} \cdot \vec{G} + \frac{1}{g} \nabla^2 \chi = 0 + \frac{1}{g} \nabla^2 \chi \neq 0. \tag{11.23}$$

In momentum space we now have $\vec{k} \cdot \vec{G}'(\vec{k}) \neq 0$, and a longitudinal polarization has arisen from the Goldstone boson. This is sometimes expressed by saying, "The gauge field has eaten the Goldstone boson and gained weight."

The procedure we have described (for a U(1) gauge theory) is referred to as the **Higgs mechanism**.

11.2   THE BREAKING $SU_L(2) \times U_Y(1) \to U_{EM}(1)$ IN THE ELECTROWEAK THEORY

11.2.1   Ranges of the Interactions and the Masses of the Gauge Particles

We now modify $SU_L(2) \times U_Y(1)$ of the electroweak theory by the introduction of a scalar field which implements the Higgs mechanism. This modification will result in the breaking of the gauge symmetry in a manner that leaves the $U_{EM}(1)$ of electromagnetism unbroken, as required for this exact symmetry of nature. It will produce mass for the W's and Z, yielding a finite range for the weak interactions.

In order to break $SU_L(2)$ the Higgs field is taken to be **an isodoublet scalar field** ϕ_H:

$$\phi_H = \begin{pmatrix} \phi^+ \\ \phi^0 \end{pmatrix} \text{ with } Y_H = +1. \tag{11.24}$$

(The superscripts indicate the electric charge.) We will suppress the sub-H in the following. Let us proceed as before by constructing the Lagrangian for ϕ:

$$\mathcal{L}_\phi = (\partial^\mu \phi)^\dagger (\partial_\mu \phi) - \mu^2 \phi^\dagger \phi - \lambda (\phi^\dagger \phi)^2, \text{ with } \mu^2 < 0, \tag{11.25}$$

---

4. In electrodynamics the gauge in which $\vec{\nabla} \cdot \vec{A} = 0$ is called the *Coulomb gauge*.

where we now use a dagger instead of a star, since the field is a doublet (column) in SU(2). Expressing the two components of ϕ in terms of real fields $\phi^+ = (\phi_1 + i\phi_2)/\sqrt{2}$ and $\phi^0 = (\phi_3 + i\phi_4)/\sqrt{2}$, we obtain

$$\mathcal{L}_\phi = \frac{1}{2} \sum_{i=1}^{4} (\partial^\mu \phi_i)(\partial_\mu \phi_i) - \frac{1}{2} \mu^2 \sum_{i=1}^{4} \phi_i \phi_i - \frac{1}{4} \lambda \left(\sum_{i=1}^{4} \phi_i \phi_i \right)^2, \qquad (11.26)$$

which is **invariant to an SO(4) rotation** of the four fields ϕ_i. In Eq. (11.25) let us replace ∂_μ with D_μ:

$$D_\mu = \partial_\mu - ig_1 \frac{Y}{2} B_\mu - ig_2 \left(\frac{\vec{\tau}}{2} \cdot \vec{W}_\mu \right), \qquad (11.27)$$

as defined in Eq. (10.24). The vacuum (lowest) state of the system is taken to have $\phi^+ = 0$, and can consequently be shown to have $\phi^0 = v/\sqrt{2}$ with $v = \sqrt{-\mu^2/\lambda}$. (See Problem 3.) These minimum values appear in quantum field theory as

$$\langle 0|\phi^0|0 \rangle = \frac{v}{\sqrt{2}} \text{ and } \langle 0|\phi^+|0 \rangle = 0, \qquad (11.28)$$

each of which is called a **vacuum expectation value (vev)**. In what follows, we will use the unconventional notation which takes ϕ^0 and ϕ^+ as fields that *create* the corresponding states.

The vacuum $|0\rangle$ is not invariant to an SU$_L$(2) transformation, since an SU$_L$(2) transformation would mix the ϕ^0 and ϕ^+, producing a new set of vevs. (For example, an SU(2)-transformed vacuum $|0\rangle' = U|0\rangle$ would produce the vev $'\langle 0|\phi^+|0\rangle' = \langle 0|U^\dagger\phi^+ U|0\rangle = \langle 0|\alpha\phi^+ + \beta\phi^0|0\rangle = \beta v/\sqrt{2} \neq 0$.) It is also not invariant under $U_Y(1)$. If $|0\rangle$ had $Y = 0$, then $\phi^0|0\rangle$ would have $Y = +1$, so that they would be orthogonal $-\langle 0|\phi^0|0\rangle = 0-$ which contradicts Eq. (11.28). Therefore, the vacuum must have a mixture of Y values, as would *all* the states in Fock space (since they are built from the vacuum state). Consequently, there would be no apparent conservation of hypercharge (Y). Furthermore, transforming the basis states with a $U_Y(1)$ transformation would change the $Y \neq 0$ parts of the vacuum state and consequently change the vev of ϕ^0. (See Problem 10.)

Just as we did for SO(2) in Section 11.1.3, we can perform a local SO(4) transformation in ϕ-space (i.e., choose a gauge), so that $\phi_1 = \phi_2 = \phi_4 = 0$. This choice (of gauge) then puts the Higgs doublet (of Eq. 11.24) in the form:

$$\phi(x) = \frac{1}{\sqrt{2}} \begin{pmatrix} 0 \\ v + H(x) \end{pmatrix}, \qquad (11.29)$$

where ϕ_3 has been set equal to $(v + H(x))$ and x stands for space-time.

Although electromagnetic theory is included in this theory, the breaking of SU$_L$(2) × U$_Y$(1) is still consistent with the conservation of electric charge. This can be seen as follows: We have defined the vacuum to have 0 charge; applying ϕ^+ to it ($\phi^+|0\rangle$) creates a ϕ particle of charge +1. Since that state has no part of its ket corresponding to the (charge 0) vacuum, it must be orthogonal to the vacuum state yielding the second equality of Eq. (11.28). (If the vacuum expectation value of ϕ^+ had been non-zero, it would have meant that the vacuum was *not* a state of definite charge, and there would be no way to keep track of the charge of states — an apparent breakdown of charge conservation.) Since ϕ^0 does not change the charge of the vacuum state, charge conservation does not produce any such restriction on ϕ^0's vev which can be non-zero, as in Eq. (11.28).

The mass-producing interaction of ϕ with the gauge fields arises from the extra terms in the covariant derivative operators, as in our previous examples, which are

$$\left\{ \left(ig_1 \frac{Y}{2} B^\mu + ig_2 \left(\frac{\vec{\tau}}{2} \cdot \vec{W}^\mu \right) \right) \phi \right\}^\dagger \left\{ \left(ig_1 \frac{Y}{2} B_\mu + ig_2 \left(\frac{\vec{\tau}}{2} \cdot \vec{W}_\mu \right) \right) \phi \right\} \equiv |O\phi|^2.$$

$$(11.30)$$

We wish to examine the mass terms for the gauge fields, which can be extracted from Eq. (11.30) by recalling that such terms arose from the v appearing in ϕ and $\phi^\dagger$. From Eq. (11.28) we see that we need only substitute

$$\phi_{min} = \frac{1}{\sqrt{2}} \begin{pmatrix} 0 \\ v \end{pmatrix}$$

in the equation above for this purpose. Setting $Y = 1$ (see Eq. 11.24), we obtain

$$\frac{1}{8} \left| \begin{pmatrix} g_1 B_\mu + g_2 W_{3\mu} & g_2(W_{1\mu} - iW_{2\mu}) \\ g_2(W_{1\mu} + iW_{2\mu}) & g_1 B_\mu - g_2 W_{3\mu} \end{pmatrix} \begin{pmatrix} 0 \\ v \end{pmatrix} \right|^2$$

$$= \frac{1}{4} v^2 g_2^2 W^{+\dagger\mu} W_\mu^{-\dagger} + \frac{1}{8} v^2 (g_1 B^\mu - g_2 W_3^\mu)(g_1 B_\mu - g_2 W_{3\mu}), \quad (11.31)$$

since the off-diagonal combinations of gauge fields are proportional to the charged W's. Looking back at our definitions of the neutral Z^0 and A fields (Eqs. 10.28 and 10.29), we find that the expression in Eq. (11.31) is equal to

$$[(\tfrac{1}{2}vg_2)^2] W^{+\dagger\mu} W_\mu^{-\dagger} + \tfrac{1}{2} Z^\mu Z_\mu [\tfrac{1}{4} v^2 (g_1^2 + g_2^2)], \quad (11.32)$$

with no term for the A field.[5] Thus, mass has been generated for the gauge particles:[6]

$$M_W = \tfrac{1}{2} vg_2, \quad M_Z = \tfrac{1}{2} v\sqrt{g_1^2 + g_2^2}, \text{ and } M_A = 0. \quad (11.33)$$

The masslessness of A (the photon field) has resulted directly from the fact that the upper component of ϕ_{min} (which is the vev of ϕ^+) is 0, so that the combination of gauge fields which contains a part proportional to A (in the upper left

---

5. The combination of neutral fields that appears in Eq. (11.31) is proportional to the Z field:

$$Z_\mu = \frac{-g_1 B_\mu + g_2 W_{3\mu}}{\sqrt{g_1^2 + g_2^2}}.$$

6. For charged particles the coefficient of the quadratic term is the square of the mass, whereas for neutral particles the coefficient has an extra factor of $\tfrac{1}{2}$.

of the gauge field matrix in Eq. 11.31) does not appear. The vanishing of the vev of ϕ^+ is consistent with charge conservation, as we have argued above. Here we see an intimate connection between charge conservation and the mass-lessness of the photon. Furthermore, this massless gauge field (A_μ) is a linear combination of the gauge field associated with U(1), viz., the B, and one from SU(2), viz., the W^0, as shown in Eq. (10.29); consequently, $U_{EM}(1)$ is not the U(1) of the unbroken theory.

Let us define the ratio

$$\rho \equiv \frac{M_W}{M_z \cos \theta_W}. \tag{11.34}$$

We see from Eqs. (11.33) and (10.28) that the standard model predicts 1 for this ρ parameter.[7] The observed value of ρ is 1.0, within experimental error. Recent one-loop calculations of the masses appearing in ρ, which include contributions from virtual top quarks, indicate that the mass of the top quark should be of the order of 160 GeV. The small experimental uncertainty in the value of ρ does not determine the top-quark mass with precision, nor can this result be thought of as definitive evidence of the existence of the top quark.

11.2.2   Masses of the Fundamental Fermions

We have seen that the left-handed fermions belong to a different multiplet of SU(2) than do the right-handed fermions. To consider mass for the fermions we must analyze the handednesses which would appear in a mass term for a lepton or quark. A mass term in the Dirac Lagrangian has the structure $m\bar{\psi}\psi$, and is called a **Dirac mass term**. Recall from Chapter 10 that the projection operators for handedness are complements of one another, so that $P_R + P_L = 1$. Also recall that projection operators have the property that $P^2 = P$. Using these properties we write

$$m\bar{\psi}\psi = m\bar{\psi}(P_R + P_L)\psi = m\bar{\psi}P_R^2\psi + m\bar{\psi}P_L^2\psi. \tag{11.35}$$

Let us further analyze the first term on the right-hand side:

$$m\bar{\psi}P_R^2\psi = m\bar{\psi}P_R P_R\psi = m\bar{\psi}P_R(P_R\psi) = m\bar{\psi}P_R(\psi_R). \tag{11.36}$$

Using the definitions of P_R and P_L in Eq. (10.17) and the fact that the γ matrices (including γ^5) all anticommute with one another, we obtain

$$m\bar{\psi}P_R = m(\psi^\dagger\gamma^0)P_R = m\psi^\dagger\gamma^0\frac{1+\gamma^5}{2} = m\psi^\dagger\frac{1-\gamma^5}{2}\gamma^0 = m\psi^\dagger P_L\gamma^0,$$

$$\tag{11.37}$$

---

7. More elaborate symmetry-breaking schemes will still predict 1 for the ρ parameter if all the Higgs multiplets are doublets.

bringing the γ^0 across the 1 (no change) and the γ^5 (yielding a change of sign).
The operator P_L is Hermitian

$$\left(P_L = \begin{pmatrix} 0 & 0 \\ 0 & 1 \end{pmatrix}\right) \text{ so that } \psi^\dagger P_L = (P_L \psi)^\dagger = \psi_L^\dagger.$$

Inserting this into Eq. (11.37) yields

$$m\bar{\psi} P_R = m\psi_L^\dagger \gamma^0 = m\overline{\psi_L},$$

so that (after analyzing the second term in the same way) the **Dirac mass term,**
of Eq. (11.35) may be written:

$$m\overline{\psi_L}\psi_R + m\overline{\psi_R}\psi_L.^8 \tag{11.38}$$

(It is left to the reader to show, in Problem 11, that a term of the form $\overline{\psi_L}\psi_L$
or $\overline{\psi_R}\psi_R$ is identically zero.)

Can such a mass term (Eq. 11.38) appear in the SU(2) × U(1)-invariant
Lagrangian of this theory? No, it cannot because ψ_L is a doublet, whereas ψ_R
is a singlet under SU(2), and they cannot combine to produce an invariant. (They
produce a $T = \frac{1}{2}$ object.) However, after the introduction of the Higgs doublet,
we *can* construct a term which will provide mass for the fermions, as follows:

$$\mathcal{L}_{int} = g_\psi[(\overline{\psi_L}\phi)\psi_R + \overline{\psi_R}(\phi^\dagger \psi_L)], \tag{11.39}$$

where g_ψ is an arbitrary coupling constant and the inner parentheses represent
the product of two isospinors. (We have $T = \frac{1}{2}$ coupling to $T = \frac{1}{2}$ to give $T = 0$, an invariant.) This type of term, which has a boson coupled to a fermion line
in a three-particle vertex but has not arisen from any gauge-symmetry require-
ment, is called **Yukawa coupling.**

Let us consider the leptons of the first generation (with $g_\psi \equiv g_e$). Writing
the particle content of the two isodoublets and doing the spinor multiplication,
we find that the first term of Eq. (11.39) is

$$(\overline{\nu_L}\, \overline{e_L}) \frac{1}{\sqrt{2}} \begin{pmatrix} 0 \\ v + H(x) \end{pmatrix} e_R = \frac{1}{\sqrt{2}}\, \overline{e_L}(v + H)e_R. \tag{11.40}$$

After doing the same for the second term, we obtain

$$\mathcal{L}_{int} = g_e \frac{v}{\sqrt{2}} [\overline{e_L}e_R + \overline{e_R}e_L] + g_e \frac{H}{\sqrt{2}} [\overline{e_L}e_R + \overline{e_R}e_L], \tag{11.41}$$

where the first term is quadratic in the fields. Upon comparison of the first term
with those in Eq. (11.38), we see that

$$m_e = \frac{g_e v}{\sqrt{2}}. \tag{11.42}$$

---

8. For fermions which are identical to their antiparticles, a (Majorana) mass term can be constructed which does not mix handedness. See footnote 7 in Chapter 10 and Chapter 18 (neutrino
mixing).

The second term in Eq. (11.41) has vertices of interaction of the Higgs particle with the electron, with a coupling strength equal to $g_{H\text{-}e} = g_e/\sqrt{2}$. Solving Eq. (11.42) for g_e in terms of m_e, we obtain

$$g_{H\text{-}e} = \frac{g_e}{\sqrt{2}} = \frac{m_e}{v}. \tag{11.43}$$

Therefore, we can write the interaction Lagrangian for the first generation of leptons, of Eq. (11.41), as follows:

$$\mathcal{L}_{int} = m_e \bar{e}e + \frac{m_e}{v} \bar{e}eH. \tag{11.44}$$

The form of the interaction term tells us that the Higgs particle couples to fermions with strengths proportional to their respective masses, so that experimental searches for the Higgs particle would be more fruitful if we studied the heaviest fermions available. Notice that no mass term has arisen for the neutrino (nor does it have a vertex with the Higgs). This is as expected, since no ν_R exists in the standard model so that no term $\sim(\bar{\nu}_R\nu_L + \bar{\nu}_L\nu_R)$ can occur.

For quarks the same procedure can be followed but, since there exist right-handed up quarks as well as right-handed down quarks, we should generate a mass term for each. The procedure above will generate a mass for the down quark, analogous to that for the electron. To generate a mass for the up quark we recall that corresponding to any doublet, there exists an antidoublet. (In Section 6.4.1 we learned how to construct the $\bar{q}$ antidoublet.) Corresponding to the Higgs doublet ϕ, there exists an antidoublet ϕ_c:

$$\phi = \frac{1}{\sqrt{2}} \begin{pmatrix} 0 \\ v + H(x) \end{pmatrix} \text{ and } \phi_c = \frac{1}{\sqrt{2}} \begin{pmatrix} -(v + H(x)) \\ 0 \end{pmatrix}. \tag{11.45}$$

Consequently, we can construct a second set of terms in our Lagrangian for up quarks, using the ϕ_c in place of ϕ:

$$\mathcal{L}_{int} = g_d[(\overline{q_L}\phi)d_R + \overline{d_R}(\phi^\dagger q_L)] + g_u[(\overline{q_L}\phi_c)u_R + \overline{u_R}(\phi_c^\dagger q_L)], \tag{11.46}$$

where $q_L = \begin{pmatrix} u_L \\ d_L \end{pmatrix}$. This will give (an independent) mass to the upper quark component because of the structure of ϕ_c; the same conclusion, that the interaction with the Higgs is proportional to the mass of the fermion, ensues. (It is left to the reader to verify these statements in Problem 6.)

11.2.3   The Higgs Field and the Cosmological Constant

The Higgs field is a background field pervading all of space, ever present, even in the vacuum state (since its vev is non-zero). The presence of this field yields an energy density in the vacuum which would curve space-time (via the gravitational interaction) into a region whose space part is roughly the size of a football. The existence of this energy density is most easily seen by noting that the shifted Lagrangian density in Eq. (11.18) includes a constant term ($[\lambda v^4]/4$).[9] Although a constant term never affects the field equations (or the calculated stress-energy tensor), it would appear as a (constant) part of the energy density which couples to the gravitational field. Using Eq. (11.33), we find

$$v = \frac{2M_W}{g_2} \approx \frac{2 \times 80 \text{ GeV}}{2.1e} \approx 250 \text{ GeV,}^{10}$$

for the value of ϕ in the vacuum, yielding

$$\text{Energy Density of the Vacuum} = \frac{[\lambda v^4]}{4} \approx 10^9 \lambda \text{ (GeV)}^4.$$

This does not look like an energy density, but remember that energy and inverse length are equivalent in natural units and we worked out the conversion from GeV to (meters)$^{-1}$ in Chapter 1. Changing to more traditional units yields

$$\text{Energy Density of the Vacuum} \approx 10^{56} \lambda \text{ GeV/m}^3,$$

which depends on the coupling constant λ. For any reasonable value for λ, this energy density is about 10^{54} times the experimental upper limit ($\sim 10^2$ GeV/m^3). Einstein's theory of gravity (general relativity) has room for a term which could cancel these curving effects, but until now that term was believed to be absent. The coefficient of that term is called the **cosmological constant**, and it would have to be adjusted (by theorists?) to one part in 10^{54}, to a value which is exactly that needed (to cancel the curvature). This **fine tuning** makes people skeptical[11] about the fundamentality of the standard model; it is referred to as the **cosmological problem.**

The Higgs field was introduced as a necessary ingredient to achieve the spontaneous symmetry breaking required for the weak interaction. However, its role in the Yukawa-coupled terms of Eqs. (11.39) and (11.46), which produce mass for the fermions, is by no means required, although it is the most efficient way to accomplish the task. The enormous range of fermion masses is a puzzle, and the proposed Yukawa couplings to the Higgs field, in fact, do not *explain* their values at all. These Yukawa couplings merely replace the question of the values of the fermion masses with the question of the sizes of the corresponding Yukawa coupling constants.

---

9. Also, note that the minimum of the potential after spontaneous symmetry breaking (Eqs. 11.6 and 11.8) is of the same order of magnitude.

10. Recall that in Heaviside-Lorentz units (see the discussion below Eq. 10.34) $e \approx \sqrt{4\pi/137}$.

11. Two seemingly unrelated quantities, which are experimentally found to be equal, lead us to wonder whether there is some unknown concept which demands their equality. The experimental equality of inertial and gravitational mass (to one part in 10^{11}) was cited by Einstein as a justification for the equivalence principle, which gives a rationale for them to be exactly equal. The equivalence principle leads directly to the general theory of relativity.

There have been attempts to replace the Higgs field with a composite or condensate of other fields, which would accomplish the spontaneous symmetry breaking in a similar way, but those theories have fallen into disrepute.[12]

At this time, there is no explicit experimental evidence of the existence of the Higgs particle.

11.2.4   Apparent Strength of the Weak Interactions

We are still left with the problem of explaining why the weak interactions appear weak, although their couplings are not smaller than that of electrodynamics (g_1 and g_2 are of the order of e).

Let us consider the exchange of a Z, as shown in Fig. 11.4, as a typical interaction. In order to understand this phenomenon, we must look back to our calculation of the potential due to a massive boson. Before the final k space integration was done, we had an integrand $\sim 1/(k^2 + m^2)$ (Eq. 4.66) where k^2 is $|\vec{k}|^2$. In quantum field theory, an analogous quantity occurs when a virtual particle is exchanged, so that the same type of potential *does* arise in the classical limit in accord with the correspondence principle. It is called the **propagator** for the exchanged particle, which is, after all, *propagating* the interaction or force. **In a relativistically invariant field theory the propagator is $1/(k^2 - m^2)$, where k^2 is the square of the four-momentum.** Thus, the factor arising for Z exchange is

$$g_Z \frac{1}{k^2 - M_Z^2} g_Z, \qquad (11.47)$$

where we have written g_Z for the combination of g_1 and g_2 which occurs in the Z coupling to the particular fermion being considered; it is proportional to e.[13] For most of the twentieth century, the weak interactions were experimentally observed at energies less than or of the order of a few GeV, where the effective value of k^2 in the denominator of the propagator is much smaller than the square of M_Z (91 GeV). k^2 is, therefore, negligible compared to M_Z^2, so that Eq. (11.47) yields an effective interaction strength:

$$`\alpha_Z` \approx \frac{g_Z^2}{M_Z^2} m_p^2, \text{ when hadrons are involved,} \qquad (11.48)$$

where we have put a typical mass size (for hadrons) $m_p \sim 1$ GeV into the expression to produce the typical dimensionless constant which would arise in the calculation. (For leptonic interactions m_{lepton} would arise instead of m_p and produce a considerably smaller result for m_e or m_μ.) Since $m_p^2/M_Z^2 = (\frac{1}{91})^2$, we see a factor of 10^{-4} resulting from the enormous size of the Z mass. Thus,

---

12. They are called *technicolor theories*, and we shall discuss them briefly in Chapter 19.

13. From Eq. (10.31) we see that for the left-handed electron $g_Z^L = (g_1^2 - g_2^2)/\sqrt{g_1^2 + g_2^2} = -1.26e$ and for the right-handed electron $g_Z^R = g_1^2/\sqrt{g_1^2 + g_2^2} = 0.55e$, as we have calculated in Problem 9 in Chapter 10.

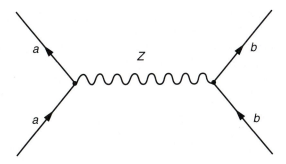

FIGURE 11.4
Neutral current interaction $a + b \rightarrow a + b$.

there is a strong connection between the *apparent* weakness of the interaction and its short range, both of which result from the largeness of M_Z.

11.3   FLAVOR MIXING

Let us ask how the strange particles decay into nonstrange products. Some copious decays of this type are:

$$\text{(a)}\quad K^+ \rightarrow \mu^+ + \nu_\mu \ (64\%),$$

$$\text{(b)}\quad \Sigma^+ \rightarrow p + \pi^0 \ (52\%),$$

$$\text{(c)}\quad \Sigma^- \rightarrow n + \pi^- \ (99.8\%),$$

$$\text{(d)}\quad \Lambda \rightarrow p + \pi^- \ (64\%), \tag{11.49}$$

where we show in parentheses the percentages of the decays to that final state. In Fig. 11.7, we will see the quark flow diagrams for Eqs. (11.49a) and (11.49b), with the appropriate intermediate vector boson inserted, but some discussion is required before we can understand these processes.

Let us consider the coupling of the leptons to the W's first. We learned in Eqs. (10.22) and (10.25) that the form of the SU(2) part of the weak interaction is $\vec{J}^\mu_{WEAK} \cdot \vec{W}_\mu$. In detail (before symmetry breaking), it may be written:

$$\vec{J}^\mu_{WEAK} \cdot \vec{W}_\mu = (\overline{\nu_\ell}\ \overline{\ell^-})_L \gamma^\mu \vec{I} \begin{pmatrix} \nu_\ell \\ \ell^- \end{pmatrix}_L \cdot \vec{W}_\mu, \tag{11.50}$$

one such term for each lepton generation, where ℓ^- is a negatively charged lepton and ν_ℓ is the corresponding neutrino. The *isospin form* of these currents is

$$J^+ = (\overline{\nu_\ell}\ \overline{\ell^-})_L I^+ \begin{pmatrix} \nu_\ell \\ \ell^- \end{pmatrix}_L = \overline{\nu_\ell}\ell^-, \ J^- = (\overline{\nu_\ell}\ \overline{\ell^-})_L I^- \begin{pmatrix} \nu_\ell \\ \ell^- \end{pmatrix}_L = \overline{\ell^-}\nu_\ell$$

$$\text{and } J^0 = (\overline{\nu_\ell}\ \overline{\ell^-})_L I_z \begin{pmatrix} \nu_\ell \\ \ell^- \end{pmatrix}_L = \frac{1}{2} (\overline{\nu_\ell}\nu_\ell - \overline{\ell^-}\ell^-), \tag{11.51}$$

where $I^{\pm} \equiv I_x \pm iI_y$, so that

$$I^+ = \frac{1}{2}(\tau_x + i\tau_y) = \begin{pmatrix} 0 & 1 \\ 0 & 0 \end{pmatrix}, \quad I^- = \frac{1}{2}(\tau_x - i\tau_y) = \begin{pmatrix} 0 & 0 \\ 1 & 0 \end{pmatrix},$$

$$\text{and } I_z = \frac{1}{2}\tau_z = \frac{1}{2}\begin{pmatrix} 1 & 0 \\ 0 & -1 \end{pmatrix}.^{14}$$

Their coupling to the $\vec{W}$ is easily obtained from the identity:

$$\vec{I} \cdot \vec{W} = \frac{1}{\sqrt{2}}[I^+ W^{-\dagger} + I^- W^{+\dagger}] + I_z W_z, \text{ with } W^{\pm\dagger} \equiv \frac{1}{\sqrt{2}}(W_X \pm iW_Y),$$

$$(11.52)$$

so that J^+ couples to $W^{-\dagger}$ and J^- couples to $W^{+\dagger}$.[15] Notice that J^{μ}_{WEAK} has the *same-generation* fermion (doublet) appearing barred and unbarred, since Eq. (11.50) has **a separate term for each lepton generation**, with **no interaction across lepton generations**.

We might expect the same structure for the *quark doublets*. However, if we use the quark doublets corresponding to the generation classification, viz., $\binom{u}{d}$, $\binom{c}{s}$, and $\binom{t}{b}$, then the (second-generation) s quark could not change into a first-generation quark; that would require generation mixing in J^{μ}_{WEAK}. Therefore, the strangeness-changing decays like those in Eq. (11.49) (studied since the 1950s), could not be explained. This problem was investigated (by Cabibbo in 1963) before the standard model or the existence of quarks was even conjectured. In modern terms what Cabibbo found was that **the quark doublets appearing in the weak current operator are not those whose members belong to definite generations (and have definite mass)**.

The simplest way to display **the mixed-generation form of the weak current operator for the quarks** is to start with a particular $I_Z = +\frac{1}{2}$ quark and ask what $I_Z = -\frac{1}{2}$ partner belongs in the same weak-current doublet.[16] Charmed particles had not yet been discovered, and so Cabibbo's result was expressed in terms of isotopic spin for which $\binom{u}{d}$ is a doublet and s is a singlet. Translated into quark language, he showed that all the strangeness-changing decays known up to that time could be understood if the partner to the u was the result of a rotation of the s into the d:

$$d_c = d\cos\theta_c + s\sin\theta_c, \tag{11.53}$$

---

14. Remember that the τ's are the Pauli matrices.

15. The charged field operators, with daggers, are defined in analogy with our discussion of the pion in Chapter 6, e.g., $W^{-\dagger}$ **is meant to be the operator which creates the** W^-; we could have written W^+ for it since the W^+ field operator annihilates a W^+ and creates its antiparticle.

16. The convention of taking the upper quarks as unmixed is universally used.

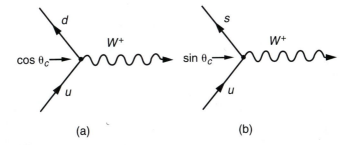

FIGURE 11.5
(a) u-d-W same-generation vertex and (b) u-s-W generation-mixing vertex.

where $\theta_c \approx 13°$; this is known as the *Cabibbo angle*. The doublet appearing in J_{WEAK} is then

$$\begin{pmatrix} u \\ d_c \end{pmatrix} = \begin{pmatrix} u \\ d\cos\theta_c + s\sin\theta_c \end{pmatrix}. \tag{11.54}$$

The replacement of d with d_c alters the coupling strength of the u-d-W vertex by an extra factor of $\cos\theta_c$, and yields a u-s-W (generation-changing) vertex with a similar coupling strength containing $\sin\theta_c$ instead, as shown in Fig. 11.5. The existence of the strangeness-changing decays (from an s quark to a first-generation quark) as well as their strengths relative to the ordinary beta decays were neatly accounted for by this **flavor mixing**. From Eq. (11.54) and Fig. 11.5 we find

$$\frac{\sigma_{\Delta S=1}}{\sigma_{\Delta S=0}} = \tan^2\theta_c, \text{ modulo (phase-space}^{17}\text{) factors due to mass differences.} \tag{11.55}$$

A problem still existed, however, concerning the strangeness-changing decays. In all those decays the charged W's (coupled to charged currents) were the virtual particles involved, whereas no strangeness-changing decays involving the Z^0 (coupled to neutral currents) were found in experiment. Neutral currents were experimentally discovered in the early 1970s; however, **no strangeness-changing neutral currents** have ever been seen. This fact is most easily discerned by looking at the semileptonic decays of hadrons, i.e., the decays which produce lepton pairs, since the charge of the intermediate vector boson in those decays is the same as the total charge of the lepton pair that it produces. For example, the semileptonic decays of the K^+

$$K^+ \rightarrow \pi^0 + \ell^+ + \nu_\ell \; (8\%),$$

---

17. See Chapter 13.

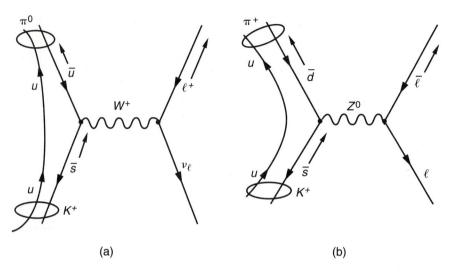

(a)   (b)

FIGURE 11.6
Quark flow diagrams for semileptonic decays: (a) $K^+ \to \pi^0 + \ell^+ + \nu_\ell$ (8%) and (b) $K^+ \to \pi^+ + \ell + \bar{\ell}$ ($<10^{-4}$%).

which result from a virtual *charged W*, can be compared to the (rare) decays

$$K^+ \to \pi^+ + \ell + \bar{\ell} \ (<10^{-4}\%),$$

which result from a virtual *neutral Z*. The corresponding quark flow diagrams are shown in Fig. 11.6. How can there be an **absence of strangeness-changing neutral currents**, in the presence of Cabibbo mixing between generations in the weak quark currents?

Analogous to Eq. (11.51), the isospin form of the weak currents for quarks is

$$J^\pm = (\bar{u}\,\overline{d_c})I^\pm \begin{pmatrix} u \\ d_c \end{pmatrix} \text{ and } J^0 = (\bar{u}\,\overline{d_c})I_z \begin{pmatrix} u \\ d_c \end{pmatrix}. \tag{11.56}$$

J^+ couples to $W^{-\dagger}$ and J^- couples to $W^{+\dagger}$, and these terms correspond to the charged-current weak decays, including the strangeness-changing ones. Let us re-examine the neutral current, with the inclusion of Cabibbo mixing (cf., Eq. 11.51):

$$J^0 \propto (\bar{u}\,\overline{d_c}) \begin{pmatrix} 1 & 0 \\ 0 & -1 \end{pmatrix} \begin{pmatrix} u \\ d_c \end{pmatrix} = \bar{u}u - \overline{d_c}d_c$$

$$= \bar{u}u - \bar{d}d\cos^2\theta_c - \bar{s}s\sin^2\theta_c$$

$$- (\sin\theta_c\cos\theta_c)(\bar{d}s + \bar{s}d). \tag{11.57}$$

The last two terms, which are proportional to $(\sin\theta_c\cos\theta_c)$, are strangeness-changing neutral currents.

In 1970 a theory was proposed which introduced a new quark (the c quark) for arcane field theory reasons, four years before any experimental evidence for the c quark existed. In that theory the s quark is the lower member of a (second generation) doublet of isospin;[18] the upper member was called the **charmed quark**. (This is the flavor doublet structure already discussed.) This doublet should take part in the weak interactions in the same way that the doublet $\binom{u}{d_c}$ does. In the weak current, there would necessarily be generation mixing in the $I_Z = -\frac{1}{2}$ state that goes with the c, since its lower component (s_c) **must be orthogonal to $d_c = d \cos \theta_c + s \sin \theta_c$** (used in the u weak-interaction doublet). Therefore, the second-generation weak doublet is

$$\binom{c}{s_c}, \text{ where } s_c = -d \sin \theta_c + s \cos \theta_c. \tag{11.58}$$

Notice that we can consider the Cabibbo mixing to be a rotation of the lower components of the doublets among themselves (see Eqs. 11.53 and 11.58):

$$\binom{d_c}{s_c} = \begin{pmatrix} \cos \theta_c & \sin \theta_c \\ -\sin \theta_c & \cos \theta_c \end{pmatrix} \binom{d}{s}. \tag{11.59}$$

Including this new doublet (Eq. 11.58) in J^0 (Eq. 11.57), we obtain

$$J^0 \propto \bar{u}u + \bar{c}c - (\overline{d_c}d_c + \overline{s_c}s_c) = \bar{u}u + \bar{c}c - (\bar{d}d + \bar{s}s)$$

by substitution of Eqs. (11.53) and (11.58). Thus, no strangeness-mixing terms survive in J^0. In fact, the absence of strangeness (or, for more general mixing, any other flavor-mixing terms) in the J^0 is easily understood from the fact that the Cabibbo-mixed lower components are related to those unmixed by a rotation, which implies that

$$\overline{d_c}d_c + \overline{s_c}\,s_c = \bar{d}d + \bar{s}s,$$

since this form is identical to that of a dot product. The strangeness-changing neutral currents have canceled,[19] so that the model is consistent with experiment. This structure (and its generalizations) which contains **no flavor-changing neutral current** is called the **GIM mechanism** after the creators of this concept.[20]

Now that we have the correct form for the weak currents, let us look at the quark flow diagrams (Fig. 11.7) which account for the strangeness-changing processes of Eq. (11.49).

---

18. In effect, the old isotopic spin has been generalized to a new isospin by this and subsequent assignments.

19. Actually, for Feynman diagrams in which some of the quarks are virtual particles, the cancellation of the strangeness-changing neutral current is not complete due to the difference in mass of the corresponding members of the two generations. Experimental results verify the calculations.

20. They are Glashow, Iliopoulis, and Maiani.

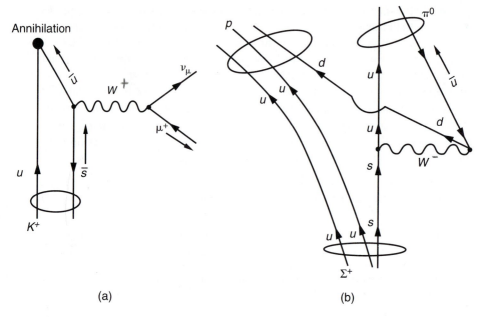

FIGURE 11.7
Quark flow diagrams for (a) $K^+ \to \mu^+ + \nu_\mu$ (64%) and (b) $\Sigma^+ \to p + \pi^0$ (52%). For process (b), the topologically simpler diagram with an $s \to d + Z^0$ vertex does not occur because of the GIM mechanism.

The mixing of the $I_Z = -\frac{1}{2}$ quarks was introduced above as a (Cabibbo) rotation, but we have learned (in Chapter 2) that *any unitary* transformation produces a set of acceptable states. So instead of

$$R = \begin{pmatrix} \cos\theta & \sin\theta \\ -\sin\theta & \cos\theta \end{pmatrix}$$

(the orthonormality of whose rows (columns), when considered as vectors, produces orthonormal states), we may use

$$U = \begin{pmatrix} e^{i\alpha}\cos\theta & e^{i\beta}\sin\theta \\ -e^{i\gamma}\sin\theta & e^{i\delta}\cos\theta \end{pmatrix} \quad \text{with } \delta = \beta + \gamma - \alpha, \qquad (11.60)$$

which is the most general 2×2 unitary matrix. (The normality of the rows (and columns) allows us to write the magnitudes of the elements as sines and cosines of an angle, and the constraint on the phases is required to produce orthogonality of the rows (and columns).) We have learned, however, that the overall phase of a wavefunction (or ket) is undetectable, so that states which differ by an overall phase are considered identical. Let us start with Eq. (11.60), which yields

$$\begin{pmatrix} d' \\ s' \end{pmatrix} = \begin{pmatrix} e^{i\alpha}\cos\theta & e^{i\beta}\sin\theta \\ -e^{i\gamma}\sin\theta & e^{i\delta}\cos\theta \end{pmatrix} \begin{pmatrix} d \\ s \end{pmatrix}.$$

We may remove the phases α and β by redefining our fields: $d_r \equiv e^{i\alpha}d$ and $s_r \equiv e^{i\beta}s$ to obtain

$$\begin{pmatrix} d' \\ s' \end{pmatrix} = \begin{pmatrix} \cos\theta & \sin\theta \\ -e^{i\varepsilon}\sin\theta & e^{i\varepsilon}\cos\theta \end{pmatrix} \begin{pmatrix} d_r \\ s_r \end{pmatrix}, \text{ where } \varepsilon \equiv \gamma - \alpha,$$

and we have used the constraint on the phases from Eq. (11.60) to obtain the same phase (ε) for the 2,1 and the 2,2 elements. If we now redefine $s'' \equiv e^{-i\varepsilon}s'$ and $d'' \equiv d$, we obtain

$$\begin{pmatrix} d'' \\ s'' \end{pmatrix} = \begin{pmatrix} \cos\theta & \sin\theta \\ -\sin\theta & \cos\theta \end{pmatrix} \begin{pmatrix} d_r \\ s_r \end{pmatrix}, \tag{11.61}$$

so that **quantum mechanics has allowed us to define away the extra phases as unobservable**. The Cabibbo form, with all real mixing coefficients, is then the most general for a physical theory with **two generations**.

We now know that there are **three generations**; for each $I_Z = +\frac{1}{2}$ quark, we expect that a linear combination of the three lower ($I_Z = -\frac{1}{2}$) generation-states will appear in the weak current. Thus, a 3×3 unitary matrix arises, and the same kinds of redefinitions can be implemented as were used to obtain Eq. (11.61). However, it has been shown[21] that not all the phases can be removed by such redefinitions; **one *essential* phase remains**. The form of the resulting Cabibbo-Kobayashi-Maskawa mixing matrix may be written:

$$\begin{pmatrix} d' \\ s' \\ b' \end{pmatrix} = \begin{pmatrix} c_1 & -s_1 c_3 & -s_1 s_3 \\ s_1 c_2 & c_1 c_2 c_3 - s_2 s_3 e^{i\delta} & c_1 c_2 c_3 + s_2 c_3 e^{i\delta} \\ s_1 s_2 & c_1 s_2 c_3 + c_2 s_3 e^{i\delta} & c_1 s_2 s_3 - c_2 c_3 e^{i\delta} \end{pmatrix} \begin{pmatrix} d \\ s \\ b \end{pmatrix}, \tag{11.62}$$

where $c_i = \cos\theta_i$ and $s_i = \sin\theta_i$ ($i = 1,2,3$), so that there are three angles and one "essential" phase (δ).[22]

The coupling constants resulting from this transformation will contain $e^{i\delta}$, making them complex. We are very interested in the possibility that the value of this unremovable phase is not 0 because **the existence of a non-zero phase would cause a violation of time-reversal invariance (T)**. We shall discuss this in detail in Chapter 12 (Discrete Symmetries). A short hand-waving argument relies on the realization that, under time reversal, initial and final states are interchanged. Applying T (the time reversal operator) to a stationary state ($\sim e^{-i\omega t}$) therefore produces a state $\sim e^{+i\omega t}$; this is equivalent to including complex conjugation as part of the time-reversal operator. Therefore, we expect that T will

---

21. by Kobayashi and Maskawa.

22. This is the original form presented by Kobayashi and Maskawa. Other forms were subsequently presented by others but they are all equivalent. Here θ_1 is the Cabibbo angle. So far, no experimental determination of δ has been made, and the magnitudes of the CKM matrix elements are approximately

$$\begin{pmatrix} .975 & .22 & .005 \\ .22 & .97 & .045 \\ .01 & .045 & .999 \end{pmatrix}.$$

change the $e^{i\delta}$ (appearing in the resulting weak interaction coupling constants) to $e^{-i\delta}$, with experimentally observable consequences.

We have already noted that conservation of the individual lepton numbers (electron, muon, and tau numbers) has been observed in all experiments to date. These conserved numbers are not known to be related to any gauge theory, but their conservation tells us that **there is no generation mixing in the leptonic weak current.**[23]

11.4   ANOMALIES AND QUARK AND LEPTON GENERATIONS

Another worrisome feature of nonabelian gauge theories is the fact that an anomalous term, which is not present in (pure) QED, appears during renormalization[24] for each fermion in the quantum field theory and spoils the renormalizability. In the standard model the coefficient of the resulting anomaly is proportional to the sum of the charges of the left-handed fundamental fermions in the theory minus the sum of the charges of the right-handed fundamental fermions in the theory, not counting their antiparticles. Since all the fundamental fermions in the theory are left-handed, we see that renormalizability ensues only for

$$\sum_{fermions} Q_f = 0, \tag{11.63}$$

where the sum is over all fermion types (but not their antiparticles). In the sum each generation of quarks (in three colors) produces $3(\frac{2}{3} - \frac{1}{3}) = 1$, and each generation of leptons produces $(0 - 1) = -1$.

Since the unit of charge of quarks and leptons is the same,[25] the required **cancellation of anomalies implies that the number of quark generations must be the same as the number of lepton generations.**

11.5   FUNDAMENTALITY OF THE STANDARD MODEL OR STANDARD THEORY

The standard model appears to be mathematically self-consistent, and so far, there is no experimental evidence which conflicts with its predictions. People have begun referring to it as *the standard theory*. However, one wonders whether it is the fundamental theory of nature or just an approximation of an even deeper theory. We shall explore extensions of the standard model in a later chapter, but here we wish to examine the standard model itself.

---

23. Recent observations of neutrinos from the sun may indicate a mixing of generations. This will be discussed in Chapter 18.
24. It first appears at the one-loop level.
25. The charge of the proton and the charge of the electron are the same size.

The standard model contains many parameters (masses, coupling constants, and mixing angles) which must be determined experimentally. In general (renormalized) quantum field theories do not predict the masses or coupling constants of particles but instead produce (often divergent) integrals which are "absorbed" into the observed values. (Recall our discussions of this renormalization procedure.) The unknown parameters in the standard model are: m_{quark} for each of **six flavors,** $m_e, m_\mu, m_\tau, g_1, g_2, g_3$ (another name for g_s, the QCD coupling constant), λ (the Higgs self-coupling constant, which determines its mass), v (vev of the Higgs which, together with $\theta_W(g_1, g_2)$, gives the W and Z masses), e_{proton}/e (which we know to be $+1$, from the neutrality of atoms, etc., but which could have been different in the standard model),[26] and **the three mixing angles and one phase in the Cabibbo-Kobayashi-Maskawa matrix.** This amounts to **19 free parameters!** Can a theory with 19 undetermined parameters be a fundamental theory of nature? On the other hand, can a theory predict *all* of them without any input, perhaps through some kind of self-consistency requirement? Would it not be like a mathematics with no axioms; is that a philosophical impossibility? If it is impossible, how many parameters are acceptable for the theory to *qualify* as a *fundamental* theory; how many are not? This sounds more like metaphysics, which is beyond the scope of this book (and some would say is beyond the realm of science).

PROBLEMS

1. For the scalar field theory, discussed in Section 11.1.2, verify that the field (ρ) in the unbroken direction remains massless. Hint: Substitute Eq. (11.10) into $\mathcal{L}$.
2. Draw Feynman diagrams for all the vertices in the $\mathcal{L}$ of Eq. (11.18).
3. Verify that for the electroweak $\mathcal{L}_\phi$ of Eq. (11.26), we can choose $\phi^0 = v/\sqrt{2}$ with $v = \sqrt{-\mu^2/\lambda}$ for the minimum of the potential when ϕ^+ is chosen to be 0.
4. Show that the upper-left-hand component of the gauge field matrix in Eq. (11.31) contains the A field.
5. Calculate the value of the ρ parameter (Eq. 11.34) for the standard model.
6. Show that the $\mathcal{L}_{int}$ of the Higgs coupling to the first generation of quarks (Eq. 11.46) can be written in a form analogous to Eqs. (11.41) and (11.44) for electrons. What is the ratio m_u/m_d in terms of g_u and g_d?
7. Draw the lowest-order mass-renormalization diagrams, which involve the top quark, for the W and the Z.
8. Verify that $\lambda v^4/4 \approx 10^{56}\lambda$ GeV/m^3.
9. Suppose we had chosen a Higgs scalar, which was an isovector (like the pion), instead of an isospinor:

$$\phi = \begin{pmatrix} \phi^+ \\ \phi^0 \\ \phi^- \end{pmatrix},$$

26. This is sometimes not mentioned as an unknown, since anomaly cancellation requires it to be $+1$.

which may be expressed in terms of real fields (ϕ_i) as $\phi^{\pm} = \frac{1}{\sqrt{2}}(\phi_1 \pm i\phi_2)$ and $\phi^0 = \phi_3$, where the superscripts are the electric charges and the $\vec{T}$ are now the same 3×3 matrices shown in Appendix A for $\vec{L}$.

    a. What is Y for this Higgs field? How does ϕ behave under the operations of the group $U_Y(1)$?

    b. Write the Lagrangian density for ϕ analogous to Eq. (11.26). Note that it is invariant under **global SU(2) $\times$ U$_Y$(1)**.

    c. Suppose that $\mu^2 < 0$ and find the minimum of the potential (as we did in Section 11.1.2 for a global symmetry).

    d. Select the component(s) of ϕ, which should be non-zero at the minimum of the potential. Remember conservation of charge.

    e. Find the effective masses of the resulting particles. Which are Goldstone bosons? How does the number of Goldstone bosons compare with the number of orthogonal directions in space along which V remains a minimum?

    f. Now consider the effects of demanding invariance under **the (local) SU(2) $\times$ U$_Y$(1) gauge theory** by showing how Eqs. (11.30) through (11.32) are modified and by finding the resulting masses of the W's and Z. Show that your choice of non-zero minimum component(s) in part d has left the photon massless. What does this tell you about the gauge symmetry which remains unbroken?

    g. Does any other gauge particle remain massless? What can you conclude about the unbroken gauge symmetry?

    h. How does the number of massive gauge particles compare with the number of Goldstone bosons, you found in part e, for the global theory? Explain what you expect here.

    i. Calculate the ρ parameter for this model.

10. Show that, for $|0\rangle = a|0\rangle_{Y=0} + b|0\rangle_{Y=1}$, applying a $U_Y(1)$ transformation $e^{i\frac{1}{2}Y\theta}$ to obtain new basis states changes the vacuum state to $|0\rangle' = a|0\rangle_{Y=0} + be^{i\frac{1}{2}\theta}|0\rangle_{Y=1}$; find the new vev for ϕ^0 in terms of the old one.

11. Show that $\overline{\psi_L}\psi_L$ and $\overline{\psi_R}\psi_R$ are identically 0.

12. Verify that $\vec{I} \cdot \vec{W} = \frac{1}{\sqrt{2}}[I^+W^{-\dagger} + I^-W^{+\dagger}] + I_Z W_Z$ (Eq. 11.52).

13. a. Show that $J^+W^{-\dagger}$ and $J^-W^{+\dagger}$ interactions do not violate charge conservation.

    b. Verify that Cabibbo mixing produces a u-s-W vertex and the value of the effective coupling constant.

14. Draw quark flow diagrams for $\Sigma^- \to n + e^- + \bar{\nu}$, $\Sigma^- \to n + \pi^-$, and $\Lambda \to p + \pi^-$, indicating the appropriate Cabibbo factors at the vertices.

15. Verify directly that $\bar{d}_c d_c + \bar{s}_c s_c = \bar{d}d + \bar{s}s$.

16. Show how the GIM mechanism operates to suppress $K^0 \to \mu^+ + \mu^-$. For the long-lived K^0, to be discussed in Chapter 12, it occurs at a fractional rate of 7×10^{-9}, which is many orders of magnitude smaller than the rates of rare (even CP-violating) decay modes.

Chapter 12

The Discrete Symmetries:
P, C, and *T*

12.1   INTRODUCTION

You may wonder why a separate chapter is being devoted to these symmetries, since we have already studied *P* in detail in the construction of the electroweak theory and the others have been discussed as well. Furthermore, since we have already constructed a consistent theory, believed to be *the* theory of the fundamental particles (valid below ~310 GeV), what further can be gained by focusing on these symmetries, whose conservation or lack thereof should already be contained in the theory? The justifications for this closer look at these symmetries are manifold. *First*, the space-time symmetries, reflections (or space inversion *P*) and time reversal (*T*), have intrigued people for centuries; questions concerning the status of antiparticles compared to particles (*C*) have excited the imagination from the time Dirac made us aware of their possible existence. *Second*, the standard model has a phase as a free parameter, so that *T* noninvariance may be incorporated in it (as we have mentioned in Chapter 11). Because of the *CPT* invariance of general quantum field theories, noninvariance under *T* implies noninvariance under *CP*. *Third*, *CP* violations have been observed in K^0 decay (as we shall discuss), but at present, there is no clear theoretical understanding of the cause. Perhaps this is our first hint of physics beyond the standard model—perhaps not.

12.2   PARITY (*P*)

12.2.1   Definition and Properties

If we watch everyday occurrences in a plane mirror, it becomes apparent quite soon that the reflection we see is different than what is seen directly. Right and left are interchanged[1] and, since more of us are right-handed, we will notice (in

---

1. Raise your right hand while looking in a mirror. Which hand is raised in your image?

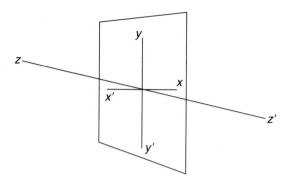

FIGURE 12.1
Behavior of a *right-handed* coordinate system under space inversion ($xyz \rightarrow x'y'z'$).
Notice that the inverted coordinate system is *left-handed*.

the mirror) that more things are done with the *left* hand. On a macroscopic scale, nature is not invariant to reflections! Even on a biological-microscopic scale there are preferred handednesses, e.g., the proteins in living structures are made of left-handed amino acids, which wind into right-handed chains. However, these circumstances could be the result of a chance breaking of an underlying symmetry by the first protein, etc., from which we all (every living cell) evolved. If only one or a small number of *ab initio* creation(s) occurred, then one choice of handedness would prevail. (If that were the case, we would expect that there may be extraterrestrial places where the opposite handedness of biological systems would occur. This is reminiscent of spontaneous symmetry breaking.)

When confronted with reflection symmetry, it is natural to ask "reflection through which plane?" However, the answer would select one particular direction, the normal to the reflection plane, as special. But we believe that the laws of nature are rotationally invariant; therefore, there should be reflection symmetry with respect to planes of any orientation. This is an awkward way to express this property, and it is replaced by the operation (*P*) of inversion through the origin. Inversion invariance is equivalent to invariance under reflection with respect to *any* plane when we have rotational invariance as well. This equivalence may be seen by noting that a reflection with respect to a plane, which takes all things on one side of the plane to the other and vice versa, accompanied by a π rotation about the reflection axis yields an **inversion**. Indeed, if we call the normal to the plane the *z* direction (so that the plane contains the *x* and *y* axes), the reflection causes $z \rightarrow z' = -z$ and the π rotation about *z* axis changes $x \rightarrow x' = -x$ and $y \rightarrow y' = -y$. (See Fig. 12.1.) Inversion treats all directions in an equivalent manner, i.e., there is no preferred direction. The letter *P* is universally used to represent inversion. Obviously, doing two successive inversions brings us back to the original coordinate system, so that

$$P^2 = I. \tag{12.1}$$

Thus, P and I together form the smallest nontrivial group, the mathematically unique group with two elements, sometimes called the reflection group (which we have discussed in Chapter 5). We shall see that T and C define groups isomorphic to this group (although the effect of T is much more subtle than that of P or C). From Eq. (12.1) we learn that the eigenvalues of P are ± 1, which are sometimes referred to as even and odd parity or positive and negative parity, respectively. If parity is conserved, then P commutes with the Hamiltonian so that we can diagonalize both simultaneously. Then we can take all our (stationary-state) wavefunctions to be eigenstates of P with eigenvalues ± 1, and we may write $P\psi = {\tt p}\psi$, where ${\tt p} = \pm 1$ is the parity of ψ.

When two interacting systems are far apart, the total wavefunction becomes a product of those for each part (or a sum of such products):

$$\psi(\vec{r}_1, \vec{r}_2) \rightarrow \psi_1(\vec{r}_1)\psi_2(\vec{r}_2) \text{ for } |\vec{r}_2 - \vec{r}_1| \rightarrow \infty.$$

Applying P to this expression, we find that

$$P\psi \rightarrow P(\psi_1\psi_2) = {\tt p}_1{\tt p}_2(\psi_1\psi_2) \text{ for } |\vec{r}_2 - \vec{r}_1| \rightarrow \infty,$$

where we have used the fact that the product of two wavefunctions becomes the product of the inverted wavefunctions in an inverted coordinate system. If the interaction conserves parity, then the parity (${\tt p}$) of the composite system obeys ${\tt p} = {\tt p}_1 {\tt p}_2$ for any separation (i.e., even when the systems are interacting), so that **parity is a multiplicative quantum number** (as opposed to charge, strangeness, baryon number, etc., which are additive). The same follows for a system with any number of parts.

Let us consider a two-body system with (orbital) angular momentum ℓ. The wavefunction is proportional to a spherical harmonic characterized by ℓ, which is a function of the angular coordinates. A point located at θ, ϕ moves to $(\pi - \theta), (\phi + \pi)$ under inversion, as shown in Fig. 12.2, and (remarkably) the spherical harmonic is changed only by the factor $(-1)^\ell$.[2] Thus, orbital angular momentum must be taken into account in parity considerations, and we may write

$$ {\tt p} = (-1)^\ell {\tt p}_1^{(intrin.)} {\tt p}_2^{(intrin.)}, \tag{12.2} $$

where ${\tt p}_j^{(intrin.)}$ is the intrinsic parity of particle j.

Now we turn to the appearance of physical quantities in the inverted coordinate system:

1. The location of a particle is unchanged by our use of a different coordinate system to measure it, but the new axes point in directions oppo-

---

2. $Y_\ell^m(\pi - \theta, \phi + \pi) \rightarrow (-1)^\ell Y_\ell^m(\theta, \phi)$.

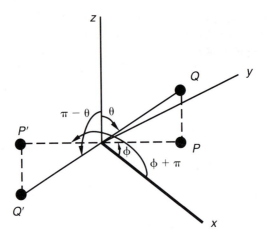

FIGURE 12.2

Behavior of a point under inversion. Point Q, with projection *P* on the *x-y* plane, becomes *Q'* with projection *P'*. The angles change accordingly as shown.

site to those of the old, so that the readings will be the negatives of those in the old system:

$$\vec{r}' = -\vec{r}, \text{ under } P \text{ (space inversion)}, \tag{12.3}$$

as we have already stated.

2. A particle with velocity $\vec{v}$, is moving along impervious to the fact that we are looking at it in an inverted coordinate system. In fact, we can set up two coordinate systems, a usual right-handed system and another inverted with respect to the first, and measure the components of $\vec{v}$ in both. If we measure v_x, v_y, and v_z for its components in the original coordinate system, the components in the inverted system would have to be

$$v'_x = -v_x, \; v'_y = -v_y, \; v'_z = -v_z$$

so that $\vec{v} \to -\vec{v}$—a **vector** under P (space inversion),   (12.4)

because all the axes are in the opposite direction in the inverted system. Both $\vec{r}$ and $\vec{v}$ are vectors, and Eqs. (12.3) and (12.4) show the behavior of ordinary (polar) vectors.

3. Of course, scalar quantities are unchanged by the change of coordinate system:

$$s \to +s - \text{a \textbf{scalar} under } P \text{ (space inversion)}. \tag{12.5}$$

In our studies of nature we also consider quantities which are derived from vectors (and scalars), such as dot- and cross-products as well as tensors.

4. Consider, for example, the angular momentum $\vec{L} = \vec{r} \times \vec{p}$. In the inverted coordinate system the components of both $\vec{r}$ and $\vec{p}$ change sign (e.g., $x'p'_y = (-x)(-p_y) = +xp_y$), so that $\vec{L}' = +\vec{L}$. In general, we may write

$$\vec{a} \rightarrow +\vec{a} - \text{a } \textbf{pseudovector} \text{ or } \textbf{axial vector} \text{ under } P \text{ (space inversion).}$$

(12.6)

5. Consider the dot product $(\vec{v} \cdot \vec{a})$ of a vector $(\vec{v})$ and an axial vector $(\vec{a})$ or the dot and cross between three vectors $(\vec{v}_1 \cdot (\vec{v}_2 \times \vec{v}_3))$, which is a special case of it. The vector changes sign and the axial vector does not, so that this "scalar" changes sign in the inverted coordinate system (or the three vectors change sign giving the same conclusion):

$$(ps) \rightarrow -(ps) - \text{a } \textbf{pseudoscalar} \text{ under } P \text{ (space inversion).} \quad (12.7)$$

6. In general, **we pick up a minus sign for each vector and pseudoscalar in the product,** regardless of the type of product.[3]

Now, let us consider the behavior of the four-vector $\bar{\psi}\gamma^\mu\psi$, which arises when we consider spin $\frac{1}{2}$ particles. Under inversion we expect the space part to change sign but the time part to be unchanged. This can be accomplished by the transformation of the wavefunction:

$$\psi(\vec{r},t) \underset{P}{\rightarrow} \gamma^0\psi(-\vec{r},t), \quad (12.8)$$

which will be verified in Problem 1. The Dirac equation, with its accompanying interpretation, introduced antiparticles associated with the negative energy states; this association persists in quantum field theory, where the spinors for antiparticles are those of the negative energy states of the Dirac equation. (See Eq. 4.55.) The transformation in Eq. (12.8) leaves the positive energy, $\vec{p} = 0$ spinor unchanged, but changes the sign of the negative energy, $\vec{p} = 0$ spinor:

$$\gamma^0 \begin{pmatrix} \chi_s \\ 0 \end{pmatrix} = + \begin{pmatrix} \chi_s \\ 0 \end{pmatrix} \text{ and } \gamma^0 \begin{pmatrix} 0 \\ \chi_s \end{pmatrix} = - \begin{pmatrix} 0 \\ \chi_s \end{pmatrix}. \quad (12.9)$$

Therefore, a state composed of a particle and its antiparticle, each at rest, would have negative parity. The generalization to more complicated states of such a system shows that **a state composed of a spin $\frac{1}{2}$ particle and its antiparticle has odd intrinsic parity,** i.e., under inversion an extra minus sign appears. (This is not true for bosons.) This property explains the observed parity of the lowest-mass mesons, if we accept the spirit of flavor SU(3), which treats all flavors of quarks as the same particle but with different quantum numbers:[4]

---

3. Here we have used a passive (coordinate system) transformation, as described in Chapter 2. If we had used an active transformation, the description of the rules would be different but the conclusions would be the same. However, the realization of an active inversion is hard to imagine.

4. Since spin is also involved, an enlarged group, viz. SU(6), has been considered as well.

Such a $q\bar{q}$ **system**, with orbital angular momentum ℓ, will have

$$\text{parity} = (-1)^{\ell+1}.$$

The space part of the wavefunction of the lowest-lying state of a system is expected to be an $\ell = 0$ (S) state, since the overlap of the two constituent wavefunctions is the greatest, as we have learned from atomic physics. Therefore, we expect the parity to be odd ($(-1)^{\ell+1} = -1$). For two spin $\frac{1}{2}$ particles the total spin can be 0 (a singlet) or 1 (a triplet). Thus, we expect the lowest-lying $q\bar{q}$ states to be:

$^{1}S_0$, a $J = 0$ state (spin $= 0$ and $\ell = 0$) with odd parity, therefore, a **pseudoscalar** meson; and

$^{3}S_1$, a $J = 1$ state (spin $= 1$ and $\ell = 0$) with odd parity, therefore, a **vector** meson.

This extra minus sign in the parity has long been understood and verified in studies of the positronium spectrum and its decays, and it has been found to apply to quarkonium systems as well. (Check the parity assignments in Fig. 15.18.)

12.2.2   Electrodynamics and Photons

In electrodynamics we have the scalar quantity ρ (the charge density) and the vector $\vec{J} = \rho\vec{v}$ (the current density).[5] We have the fields set up by these, which produce forces on charges. Remembering that $\vec{F} = d\vec{p}/dt$ and $\vec{p} = m(d\vec{r}/dt)$, we conclude that $\vec{F}$ is a vector under inversion. The electrodynamic force on a charge is $\vec{F} = q(\vec{E} + \vec{v} \times \vec{B})$, so that $\vec{E}$ must be a vector and $\vec{B}$ an axial vector. Since $\vec{B} = \vec{\nabla} \times \vec{A}$, $\vec{A}$ **is a vector** (the gradient operator, an ordinary vector, changes sign under inversion, as you may verify in Problem 2).[6] It is $\vec{A}$ which is quantized to produce the quanta of the electromagnetic field, viz., the photons. This behavior is expressed by saying that the **photons have odd (intrinsic) parity**, or **the photon is a vector particle**. Since the electromagnetic interaction $\sim \vec{J}\cdot\vec{A}$, it is invariant under inversion, so that **the electromagnetic interaction conserves parity**.

12.2.3   Pions

The π^0 was found to be a spin 0 particle; it decays into two photons, so that this (parity-conserving) electromagnetic decay can be used to determine its parity. Each final photon will have a polarization vector $\hat{\epsilon}_{1,2}$ and the amplitude

---

5. The letter J is used for angular momentum and spin, which are axial vectors, as well as for currents, which are vectors. Do not confuse them; be aware of the context.

6. The vector (as opposed to axial vector) nature of $\vec{A}$ is also easily seen by recalling that for steady currents $\vec{J}$: $\nabla^2\vec{A} \propto \vec{J}$, so that the parity of $\vec{A}$ is the same as that of $\vec{J}$. More generally $\Box\vec{A} \propto \vec{J}$, and the same result ensues.

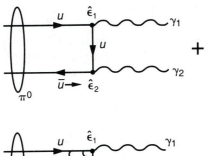

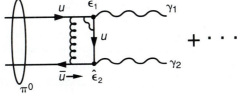

FIGURE 12.3
Feynman diagrams for $\pi^0 \rightarrow 2\gamma$, showing the tree-level diagram and a typical radiative correction, with a virtual gluon (spiral) and a virtual photon. The three dots represent all the other diagrams. Each diagram has one interaction vertex for each final γ.

must be linear in $\hat{\varepsilon}_1$ and $\hat{\varepsilon}_2$, since the Feynman diagrams have one interaction vertex with each final photon. (See Fig. 12.3.)

In the rest frame of the pion, the two types of terms which can appear in the (rotationally invariant) amplitude are

$$\hat{\varepsilon}_1 \cdot \hat{\varepsilon}_2 \ \text{(even parity) and} \ \hat{k} \cdot (\hat{\varepsilon}_1 \times \hat{\varepsilon}_2) \ \text{(odd parity)}, \qquad (12.10)$$

where $\hat{k}$ is in the direction of the momentum of one of the photons.[7] The first expression is a scalar and would result if the π^0 has even parity, and the second is a pseudoscalar that results from odd parity. No decay has ever been seen to have $\hat{\varepsilon}_1 \parallel \hat{\varepsilon}_2$, so that the scalar term is not present, and careful measurements have verified that the π^0 is a pseudoscalar particle.

Detailed studies of reactions involving the charged pions have shown them to be pseudoscalar as well; thus, **the π is a pseudoscalar particle**. In fact, we have seen that the pion is part of a pseudoscalar multiplet of flavor SU(3).

12.2.4 *K* Decays and the *τ-θ* Puzzle

In the early 1950s, when parity was believed to be a symmetry of nature on the elementary-particle level, tracks of charged particles were seen to decay into two

---

7. There can be no $\hat{\varepsilon} \cdot \vec{k}$ terms because of the transverse nature of photons. This follows since, in the rest frame of the π^0, $\vec{k}_2 = -\vec{k}_1 \equiv \vec{k}$ and $\hat{\varepsilon}_1 \cdot \vec{k} = -\hat{\varepsilon}_1 \cdot \vec{k}_1 = 0$ (photons have only transverse polarization); likewise, $\hat{\varepsilon}_2 \cdot \vec{k} = 0$.

pions and, in similar events, nearly identical particles decayed into three pions. Those particles which decayed into two pions had the same mass and lifetime as those which decayed into three. You might think that they were the same unstable particle, for we expect unstable particles to have many different decay channels. However, there was a "catch": If *P* were conserved, the parity of the particles which decayed into three pions would be opposite to the parity of those particles which decayed into two pions. This can be understood from the following discussion.

Let us work in the rest frame of the decaying particle. For simplicity, we will construct the argument assuming that the decaying particle has spin 0[8] (which is indeed the case). Consider first the particle which decayed to two pions. Since pions have no spin, the two-pion system must have $\ell = 0$ due to conservation of angular momentum. The parity of the two-pion system (e.g., $\pi^+ + \pi^0$) is then the product of the intrinsic parities of the pions, so that $p_{2\pi} = (-1)^2 = +1$. The parity of the three-pion system is more complicated to analyze, but careful studies showed that no orbital angular momentum was present if the particle from which they came had spin 0; therefore its parity had to be $p_{3\pi} = (-1)^3 = -1$.

Thus, a particular spin 0 particle could decay to two π's if it had even parity, and three π's (with no orbital angular momentum) if it had odd parity, but not vice versa. That is why it appeared that there were two particles with identical mass and lifetime but with opposite parity. This remarkable coincidence was widely discussed; it was referred to as the τ-θ puzzle, which were the preliminary names given to the two hypothetical particles whose decays were being observed.[9] It was recognized that if, instead, these were decays of the *same* particle (now called the K), then this would be the first evidence that the weak interaction violates parity conservation. But when it had been suggested, a quarter of a century earlier, that the weak interaction violated *energy* conservation, an alternative explanation (the neutrino) was proposed and clutched to the bosom of physics without direct evidence because of our deep-seated belief in that venerable principle. In the next subsection we shall consider the bases of the belief in parity conservation at the time.

12.2.5   Parity Violation in the Weak Interactions

It has long been believed that the laws of physics should not depend upon the particular (Euclidean) coordinate system in which we choose to work.[10] However, notice that there is no way we can reorient ourselves to be in a reflected or inverted coordinate system, as we can for rotations. So perhaps we should

---

8. The argument was originally made for arbitrary spin.

9. The particle decaying into two pions was called θ, and that into three pions τ.

10. Several centuries ago, this concept was generalized to include moving coordinate systems and was known as Galilean relativity. It was generalized by Einstein (to include electrodynamics, etc.) and ultimately yielded general relativity.

not expect nature to be reflection or inversion invariant. Reflection symmetry seemed to be an unexplored question[11] into the mid-1950s. It was tacitly assumed to be true on the elementary-particle level without any experimental exploration of the question. In 1956 the τ-θ puzzle led Lee and Yang to a systematic search for *actual* experimental evidence of this conservation law. They concluded that compelling evidence existed for parity conservation in the strong and electromagnetic interactions, but that it was "only an extrapolated hypothesis unsupported by experimental evidence" in the weak interactions. They proposed several experiments which would directly reveal violations of this conservation law by the weak interaction, some of which were performed within the year. These experiments showed, without any doubt, that the long-standing belief in this symmetry was misguided and that the weak interaction violated parity conservation.

One of the most celebrated of these first experiments was a study of the beta decay of polarized Co^{60} nuclei, performed by C. S. Wu in collaboration with E. Ambler et al., at the National Bureau of Standards.[12] They looked for behavior that would show the existence of a term $\sim \vec{J}_{nucleus} \cdot \vec{p}_{electron}$ in the effective interaction. This term, which is dependent on the angle between the electron momentum and the nuclear spin $\vec{J}_{nucleus}$, changes sign under inversion (since $\vec{J}_{nucleus}$ is a pseudovector, just like $\vec{L}$), so that it violates parity conservation. They found that the electrons were indeed emitted preferentially in the direction opposite to that of the nuclear spins. In Fig. 12.4 we see how such decays look after inversion.

When news of this state-of-the-art achievement was received (before publication of the results), other groups set up further tests of parity violation. Meanwhile, independently of the Co^{60} experiment, Friedman and Telegdi had been studying the decay chain, π-μ-e[13] (with the appropriate neutrino(s) being emitted at each decay) for parity violation. Garwin, Lederman, and Weinrich studied and analyzed the same decay chain, following news of the Co^{60} result. Both groups found a parity violation. See Fig. 12.5 for a description and depiction of this decay chain.

It was also determined in the Co^{60} beta-decay experiments (and confirmed by later experiments) that parity was violated in a maximal way, such that the interaction is purely left-handed ($V - A$), as we have exhaustively discussed already. (The left-handedness of the neutrino and the right-handedness of the

---

11. Dirac had questioned the validity of both reflection symmetry and time-reversal invariance as early as 1949.

12. At the NBS very low temperatures could be achieved (by adiabatic demagnetization), so that the orientation of the nuclear spins would not be random. The nuclei were oriented by the use of a magnetic field.

13. Despite the smaller phase space available, the π's decay almost always to μ's rather than to faster electrons. This results from the fact that the handedness, required by angular momentum conservation, of that charged lepton is opposite to the projection operator appearing. Since the handedness of a fast-moving particle is the same as that of the projection operator (to a good approximation), the decay to a faster-moving electron is highly suppressed. See Fig. 12.5.

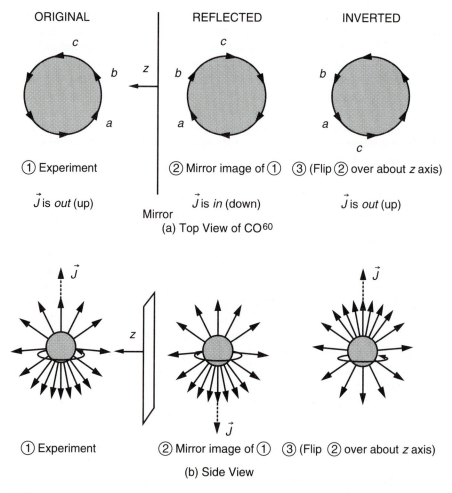

FIGURE 12.4

A visual representation of inversion for Co^{60} beta decay; inversion is obtained by a reflection ② followed by a π rotation ③. (a) A semiclassical picture of the spinning nucleus, with arrows showing the direction of spin and labels a, b, and c to aid the eye in verifying the reflected and inverted images. The direction of the spin $\vec{J}$ follows from the right-hand rule. (b) A side view of (a superposition of) many decays, where the small arrows represent electrons, less of which emerge in the direction of $\vec{J}$ than in the opposite direction. Obviously, the only way that the inverted image ③ *could* be identical to that observed experimentally ① would be if the angular distribution had as many decays parallel to $\vec{J}$ as are antiparallel to it.

antineutrino were determined, even though the neutrinos were not experimentally detected. This was accomplished by determining the spins of the other particles and using conservation of angular momentum, as shown in Fig. 12.5 for the π-μ-e decay chain.)

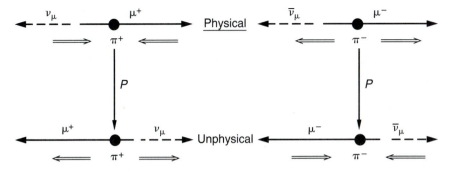

FIGURE 12.5

The decays $\pi^+ \to \mu^+ + \nu_\mu$ and $\pi^- \to \mu^- + \bar{\nu}_\mu$, in the pion rest frames, as observed (physical) and their respective P-images (unphysical). The double arrows show the spin directions; the ν *must be left-handed* and the $\bar{\nu}$ *right-handed*, so that the P-images are not physical. Since the pion has 0 spin, conservation of angular momentum requires that the μ^+ be produced with left-handed spin (the μ^- with right-handed), so that they are polarized. *Their* decays are therefore analogous to the Co^{60} β decays and a correlation of the directions of emergence of the resulting electrons with the muons' directions (and thus their spins) was found, showing that P was violated in the first (π) decay as well as in the subsequent (μ) decay.

Another kind of search for P (and T) violation is being undertaken in experiments to detect electric dipole moments of particles. In Section 12.5 on T invariance we shall discuss the search for the electric dipole moment of the neutron and its significance for these discrete symmetries.

12.3 CHARGE CONJUGATION (C)

12.3.1 Definition and Properties

As soon as the Dirac equation was understood, it was apparent that the interactions among antiparticles were the same as those among particles, so that if each particle in an interaction is replaced with its antiparticle, reaction rates should be unchanged. (Even in classical electrodynamics, Maxwell's equations are invariant to a change of sign of ρ, $\vec{E}$, and $\vec{B}$; a ρ of opposite sign produces the same fields but in opposite directions and the forces between charges are unchanged.) This operation was named *charge conjugation* (represented by C), since, in the original Dirac formulation, the antiparticle and particle differed only in the signs of their charges. With the subsequent discovery of other charge-like additive quantum numbers (baryon number, lepton number, strangeness, etc.), it became apparent that this replacement reverses them all. Nevertheless, the name *charge conjugation* persists even though it is actually a misnomer. So we write

$$C|a,b,\dots\rangle = |\bar{a},\bar{b},\dots\rangle, \tag{12.11}$$

where $a, b, \ldots$ denote particle type and $\bar{a}, \bar{b}, \ldots$ are the corresponding antiparticles. Every particle is the antiparticle of its own antiparticle, and the only reason we refer to one set as antimatter is that those particles are the antiparticles of the ones of which we (and everything around us) are made. Thus, applying C twice brings us back to the original state, so that

$$C^2 = I,$$

and the group (C, I) is isomorphic to the group (P, I) for parity. In general, we do not see eigenstates of C around us; rather, our world is made of "matter" (as opposed to antimatter).[14] There are particles which are their own antiparticles, however, as we have seen even in Table 1.1. Since C reverses the signs of the (additive) quantum numbers — charge, baryon number, lepton number, strangeness, charm, topness, bottomness, etc. — only particles which have 0 for all of these quantum numbers will be eigenstates of C; the eigenvalue can be ± 1 (just as for parity).

Let us consider the photon first. The sources of the A field are charges and currents, which change sign under charge conjugation. The same argument used for parity in footnote 6 shows that **the photon is odd under charge conjugation**. We cannot make the analogous argument for gluons (or the gauge particles of any nonabelian theory) because they carry the group "charge," which is color for QCD, so that they are not eigenstates of C. However, colorless combinations of gluons *can* be eigenstates of C. These considerations arise for (colorless) gluonium states as well as in the discussion of OZI-suppressed strong decays, where the incoming state is colorless and is attached to the colorless final state solely by virtual gluons. Those virtual gluons must be in a colorless combination, since the strong interaction conserves color "charges."

The gluons form an octet of color $SU(3)$ analogous to the octets of flavor $SU(3)$ discussed in Chapter 6, where we showed that $\pi^{\pm\dagger} = \frac{1}{\sqrt{2}}(\pi_1 \pm i\pi_2)$.[15] For C to change a $|\pi^+\rangle$ into a $|\pi^-\rangle$ and vice versa, we see that $C|\pi_1\rangle = +|\pi_1\rangle$ and $C|\pi_2\rangle = -|\pi_2\rangle$ is required. (Other phase choices, viz., $C|\pi^+\rangle = \eta|\pi^-\rangle$, can be made but the extra complication does not affect the conclusion. See Problem 7.) We also had $|\pi^0\rangle = |\pi_3\rangle$, which is a $C = +1$ state, as we shall see shortly; consequently, $C|\pi_3\rangle = +|\pi_3\rangle$. Although we did not go any further for the flavor-$SU(3)$ octets, π_i with i running from 1 to 8 can be used, with $K^{\pm\dagger} = \frac{1}{\sqrt{2}}(\pi_4 \pm i\pi_5)$, $K^{0\dagger} = \frac{1}{\sqrt{2}}(\pi_6 + i\pi_7)$, $\bar{K}^{0\dagger} = \frac{1}{\sqrt{2}}(\pi_6 - i\pi_7)$, and $\eta = \pi_8$. The same argument that we have just used for π_1 and π_2 will show that each

---

14. The states we measure are eigenstates of the charge operator (Q); in fact some people believe that all physical states are eigenstates of Q (which is consequently called a *superselected quantity*). Since C changes the sign of the charge, they are not eigenstates of C. In Problem 6, we will see that $QC = -CQ$, so that Q and C do not commute, which implies that no state of definite non-zero charge can be an eigenstate of C.

15. As in Chapter 6, the dagger means that this is the field which creates the indicated charged particle (or annihilates its antiparticle).

π_i is an eigenstate of $C - C|\pi_i\rangle = c|\pi_i\rangle$ or $C\pi_i C^{-1} = c\pi_i$,[16] with $c = +1$ for π_4, π_6, and π_8 and $c = -1$ for π_5 and π_7, so that $c^2 = +1$ for all π_i. The analogy to the gluon octet of color SU(3) is: $CG^{(i)}C^{-1} = cG^{(i)}$, with $c^2 = +1$. The smallest number of gluons required for a colorless gluon combination is two, in the invariant combination $\sum_{i=1}^{8} G^{(i)}G^{(i)}$. Under charge conjugation it becomes

$$C\sum_{i=1}^{8} G^{(i)}G^{(i)}C^{-1} = \sum_{i=1}^{8} CG^{(i)}C^{-1}CG^{(i)}C^{-1}$$

$$= \sum_{i=1}^{8} c^2 G^{(i)}G^{(i)} = +\sum_{i=1}^{8} G^{(i)}G^{(i)}, \qquad (12.12)$$

where we have inserted $C^{-1}C(=I)$ between the field operators; consequently,

$$C\sum_{i=1}^{8} |g^{(i)}g^{(i)}\rangle = C\sum_{i=1}^{8} G^{(i)}G^{(i)}|0\rangle = C\sum_{i=1}^{8} G^{(i)}G^{(i)}C^{-1}|0\rangle$$

$$= +1\sum_{i=1}^{8} G^{(i)}G^{(i)}|0\rangle = +1\sum_{i=1}^{8} |g^{(i)}g^{(i)}\rangle, \qquad (12.13)$$

since the vacuum is invariant, so that $C^{-1}|0\rangle = |0\rangle$. Equation (12.13) shows that **the two-gluon colorless state has $C = +1$**. The strong interaction (QCD) is invariant to C conjugation, as we shall see below; consequently, states (particles) with $C = -1$ (like the neutral vector mesons ω and ϕ or any other particle-antiparticle states that can arise from a virtual photon) must have at least three gluons connecting the incoming and outgoing (colorless) quark systems in OZI-suppressed strong decays. (The three-gluon state *can* connect them.)

Let us consider *positronium-like systems*, which are bound states of a fermion and its antiparticle; the state has the form: $a_f^\dagger a_{\bar{f}}^\dagger |0\rangle$. Applying the C operator changes the f to an $\bar{f}$ and vice versa. Let us now interchange the space positions associated with the creation operators; for a state of angular momentum ℓ this yields a factor of $(-1)^\ell$. If we now interchange the spin states, we will get $+1$ if the spin were 1 (*even* to an interchange) or -1 if the spin were 0; we can summarize this result as $(-1)^{s+1}$. These three operations have produced an interchange of the two fermion creation operators, which must anticommute, so that $(-1)^{\ell+s+1}C = -1$ for positronium-like systems. This yields

$$C = (-1)^{\ell+s} \text{ for positronium-like systems.} \qquad (12.14)$$

Let us consider the electromagnetic decay of the π^0. Its ket is a sum of $d\bar{d}$ and $u\bar{u}$ positronium-like parts with $\ell = 0 = s$, so that Eq. (12.14) predicts

---

16. Remember (from Chapter 2) that operators undergo similarity transformations when states are transformed and note that $C|\pi_i\rangle = C\pi_i|0\rangle = C\pi_i C^{-1}|0\rangle$. Here we have used the fact that the vacuum is invariant, so that $C^{-1}|0\rangle = |0\rangle$. Although $C^{-1} = C$, we have chosen to write C^{-1} explicitly to retain the familiar form of the similarity transformation.

$C = +1$. It readily decays into two photons, but never into three photons. Hence, even before its substructure was known, it was concluded that **the π^0 is even under charge conjugation**. Other electromagnetic decays, such as positronium decay and η decay, also validate Eq. (12.14), verifying that C is conserved in electromagnetic interactions. (See Problem 5.) The C assignments in Fig. 15.18 are consistent with the expression in Eq. (12.14) for the quarkonium systems appearing.

The strong interaction appears to be invariant under C, the theory (QCD) *is* invariant by its very structure, and those reactions studied experimentally involving mesons seem to possess this invariance.[17] Not many antibaryons have been made so far, but some experimental evidence for C invariance of the strong interaction has been found in $p\bar{p}$ experiments. Study of the energy and angular distributions of the mesons produced has found no violation of C invariance, to an accuracy $\ll 0.1\%$. Also, use of the CPT invariance theorem implies that checks of PT invariance in experiments involving particles (and not necessarily their antiparticles) can serve as tests of C invariance.[18] I have never heard any-one doubt the validity of C invariance for the strong interaction. Of course, that does not preclude such a surprise!

12.3.2   *C* Violation in the Weak Interactions

The weak interaction is seen to violate C invariance in the same experiments which showed a P violation. Simply put, there are only *left-handed* neutrinos and *right-handed* antineutrinos. Applying C to any state with these particles changes it into the C-conjugate system with the left-handed particles now being antineutrinos and the right-handed ones being neutrinos, neither of which exist in nature.[19] C invariance would require all interactions in the C-conjugate system to be identical to those in the original, whereas in nature there is no such (C-conjugate) system. Thus, this is a maximal violation.

12.4   *CP*

12.4.1   Invariance

The maximal violations of both charge conjugation (C) and parity (P) by the weak interaction, in which particles are left-handed and antiparticles are right-handed, has the ramification that the weak interaction *appears to be invariant* under the operation CP. This is apparent from the fact that P changes the left-

---

17. A discrete symmetry called G-parity, which combines C with an isospin rotation, is often used in place of C to verify the effects of the C invariance of the strong interaction. See Appendix G.

18. See Section 12.6.

19. Corresponding statements can be made for the weak interactions of massive particles of the theory if we refer to the projection operators, as explained previously.

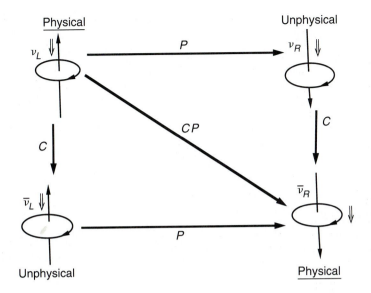

FIGURE 12.6
The behavior of the (physical) left-handed neutrino under P (which reverses the momenta), C (which changes particle to antiparticle), and CP operations. The spin direction is shown pictorially by an arrow on a circle and by the double arrow. The arrows through the centers of the circles are the momenta. Notice that only CP produces a physical state, since only left-handed ν's and right-handed $\bar{\nu}$'s exist. (The arrows can also be followed in reverse order from the $\bar{\nu}_R$.)

handed projection operators occurring for particles into right-handed ones, which are the correct projection operators for antiparticles, and vice versa.

Looking again at neutrinos for simplicity, we see in Fig. 12.6 the effects of C, P, and CP on a *physical state*; notice that only CP produces a physical state. This modified inversion symmetry perhaps mollified those who felt that coordinate systems should be irrelevant to the laws of nature. They were in for a further jolt!

12.4.2   *CP* Eigenstates and K^0 Decay

An interesting example of the application of CP to observable situations arises in the study of the decay of the neutral K mesons. (This study unexpectedly led to a new discovery as well, which we shall discuss in Section 12.4.4.) The weak interaction can change a K^0 into a $\bar{K}^0$, as shown in Fig. 12.7, so that after production by the strong interaction, the particle becomes a mixture with some part being antiparticle.

Alternatively, we can consider the eigenstates of CP. If CP commutes with the weak (and strong) Hamiltonian, we can diagonalize both CP and H simultaneously, yielding states of definite CP value *and* definite energy or mass (and

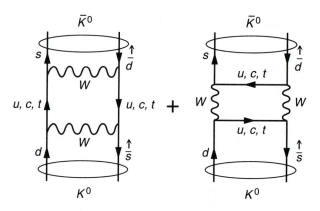

FIGURE 12.7
Two diagrams for the change of a K^0 into a $\bar{K}^0$. A *CP* violation can occur at the
W-vertices via the *CKM* generation-mixing matrix.

lifetime). Let us construct such states: We know that the K is a pseudoscalar particle so that

$$P|K^0\rangle = -|K^0\rangle \text{ and } P|\bar{K}^0\rangle = -|\bar{K}^0\rangle, \tag{12.15}$$

and by definition:

$$C|K^0\rangle = |\bar{K}^0\rangle \text{ and } C|\bar{K}^0\rangle = |K^0\rangle.$$

Thus, under *CP* we have

$$CP|\bar{K}^0\rangle = -|K^0\rangle \text{ and } CP|K^0\rangle = -|\bar{K}^0\rangle, \tag{12.16}$$

so that neither of them is an eigenstate of *CP.* It is left to the reader to verify
that the eigenstates of *CP* are the linear combinations:

$$|K_1\rangle = \tfrac{1}{\sqrt{2}}(|K^0\rangle - |\bar{K}^0\rangle), \text{ which has } CP = +1 \text{ and}$$
$$|K_2\rangle = \tfrac{1}{\sqrt{2}}(|K^0\rangle + |\bar{K}^0\rangle), \text{ which has } CP = -1.\text{[20]} \tag{12.17}$$

Inverting Eqs. (12.17), we obtain

$$|K^0\rangle = \tfrac{1}{\sqrt{2}}(|K_2\rangle + |K_1\rangle) \text{ and } |\bar{K}^0\rangle = \tfrac{1}{\sqrt{2}}(|K_2\rangle - |K_1\rangle). \tag{12.18}$$

We shall see that for a free K^0 beam it is more suitable to use as our basis states
the eigenstates of *CP,* which are states of definite mass (and lifetime),[21] so the
derivation of Eqs. (12.18) is not just an algebraic exercise. We see that the $|K^0\rangle$

---

20. Some authors use the phase freedom of kets to define $C|K^0\rangle = -|\bar{K}^0\rangle$, with the con-
sequence that the two combinations are reversed.
21. This will be slightly modified in Section 12.4.4. Some physicists feel that only states of
definite mass (and lifetime) should be considered physical particles.

and the $|\bar{K}^0\rangle$ are simple quantum mechanical coherent superpositions of states of definite mass and lifetime ($|K_1\rangle$ and $|K_2\rangle$).

Let us consider a strong interaction which produces a K^0 ($d\bar{s}$), which is the neutral K usually produced from nonstrange particles.[22] We wish to follow the particle's time evolution after production. We express the initial condition as

$$|K(0)\rangle = |K^0\rangle = \tfrac{1}{\sqrt{2}}(|K_2\rangle + |K_1\rangle). \tag{12.19}$$

At a later time (t) the state no longer has relative coefficient 1 between its K_1 and K_2 parts, since they evolve differently from one another with time:

$$|K(t)\rangle = \tfrac{1}{\sqrt{2}}(e^{-iE_2t-(\Gamma_2/2)t}|K_2\rangle + e^{-iE_1t-(\Gamma_1/2)t}|K_1\rangle), \tag{12.20}$$

where we have used Γ_1 and Γ_2 for the decay constants.[23] Γ_i^{-1} is the time required for the corresponding *intensity* to drop to $1/e$ of its starting value; that time is commonly called the *lifetime* (τ_i). (We use $\Gamma/2$ in the wavefunction, which yields $e^{-\Gamma t}$ for $|\psi|^2$.) Substituting Eq. (12.17) into Eq. (12.20) yields

$$|K(t)\rangle = \tfrac{1}{2}((e^{-iE_2t-(\Gamma_2/2)t} + e^{-iE_1t-(\Gamma_1/2)t})|K^0\rangle$$
$$+ (e^{-iE_2t-(\Gamma_2/2)t} - e^{-iE_1t-(\Gamma_1/2)t})|\bar{K}^0\rangle),$$

in which we see a non-zero amplitude for $\bar{K}^0$ in this state, which was pure K^0 an instant ago (at its creation); it is proportional to the difference of the time evolution factors of K_1 and K_2. If we convert all quantities to the rest frame of the K, then

$$E \to m, t \to \tau = \sqrt{1 - \beta^2}\, t \quad (\tau \text{ is the proper time, which is the shortest time)}$$

and Γ becomes the usual decay constant. Thus $|K(t)\rangle$ becomes

$$|K(\tau)\rangle = \tfrac{1}{2}((e^{-im_2\tau-(\Gamma_2/2)\tau} + e^{-im_1\tau-(\Gamma_1/2)\tau})|K^0\rangle$$
$$+ (e^{-im_2\tau-(\Gamma_2/2)\tau} - e^{-im_1\tau-(\Gamma_1/2)\tau})|\bar{K}^0\rangle). \tag{12.21}$$

As we have studied in Chapter 2, the probability of detecting (and therefore the relative intensity of) either K^0 or $\bar{K}^0$ is equal to the absolute square of the corresponding coefficient (in Eq. 12.21):

$$I_{K^0}(\tau) = \tfrac{1}{4}\{e^{-\Gamma_2\tau} + e^{-\Gamma_1\tau} + 2e^{-(\Gamma_2+\Gamma_1)\tau}\cos(m_1 - m_2)\tau\} \text{ and}$$

$$I_{\bar{K}^0}(\tau) = \tfrac{1}{4}\{e^{-\Gamma_2\tau} + e^{-\Gamma_1\tau} - 2e^{-(\Gamma_2+\Gamma_1)\tau}\cos(m_1 - m_2)\tau\}. \tag{12.22}$$

The proper time (τ) from production is easily determined in the laboratory from the distance moved (L), since

$$L = vt = v\,\frac{\tau}{\sqrt{1 - \beta^2}}. \tag{12.23}$$

---

22. A virtual gluon produces an $s\bar{s}$ pair, the s going into the baryon.
23. A discussion of the role of Γ as the resonance width appears in Section 13.3.3.

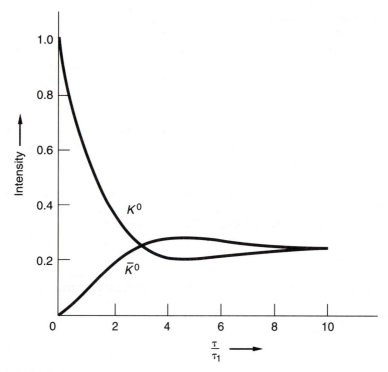

FIGURE 12.8

The relative intensities $I_{K^0}(\tau)$ and $I_{\bar{K}^0}(\tau)$ for a beam which has pure K^0 particles at $\tau = 0$. (Notice that $I_{K^0}(0) = 1$ and $I_{\bar{K}^0}(0) = 0$.) The graph is plotted for $\Delta m \tau_1 = 0.5$.

We shall soon see that $\Gamma_1 \gg \Gamma_2$ or, for the lifetimes: $\tau_1 (= 1/\Gamma_1) \ll \tau_2 (= 1/\Gamma_2)$, and so we can study these intensities for times (τ) greater than or of the order of τ_1, but short compared to the lifetime of the K_2: $\tau \ll \tau_2 = 1/\Gamma_2$.[24]

In Fig. 12.8 we see pronounced oscillations in the intensities (Eq. 12.22), from which we can determine $\Delta m \equiv |m_1 - m_2|$, as will be shown in Problem 9. Figure 12.8 is drawn for $\Delta m \tau_1 = 0.5$, which is very close to the experimental value:

$$\Delta m \tau_1 = 0.477 \pm 0.002.[25]$$

The measured value of the lifetime of K_1 (this identification will need a slight modification — later) is $\tau_1 = .89 \times 10^{-10}$ s, yielding

$$\Delta m = 3.52 \times 10^{-6} \text{ eV.} \tag{12.24}$$

---

24. Measuring the $\bar{K}^0$ intensity as a function of L is easily accomplished by looking for the strange hyperon yield, since the $\bar{K}^0$ has an s quark, which finally is contained in a strange baryon.

25. In regeneration experiments (which we shall learn about in Section 12.4.3), it has been determined that $m_2 > m_1$.

Yes, the exponent really is *minus* 6. A quantum mechanical interference term has enabled us to measure a very tiny quantity: $\Delta m/m \approx 7 \times 10^{-15}$.

Let us now look at the decays of the neutral kaons. In Section 13.3 we shall learn that the probability of decay to a final state depends strongly on the amount of kinetic energy (in the rest frame of the decaying particle) available to the final particles—the more kinetic energy, the larger the probability. We shall see that this dependence is a consequence of the existence of a larger number of final states; effectively, a larger region of momentum space is available for the final particles. This condition is usually expressed by saying that *there is more phase space available*. Thus, **the more phase space available, the greater the decay rate to that final state.** The neutral K's usually decay into pions and, as a result of their respective masses, a two-pion final state has about $2\frac{1}{2}$ times the kinetic energy, therefore much more phase space, of a three-pion final state. (What about a four-pion final state?) In fact, the K^0's decay to two pions in 99% of their decays. The expansion in *CP* eigenstates (Eq. 12.18) and our subsequent discussion suggests that we examine the behavior under *CP* of these modes of decay. If the weak interaction conserves *CP*, then it is important to determine the behavior of these final states under that symmetry transformation:

1. *two-pion final state.* First we consider a $\pi^+\pi^-$ state. The wave operator, which creates this state, is effectively $\psi = \pi^{+\dagger}(\vec{r}_1)\pi^{-\dagger}(\vec{r}_2)$ (when they have moved far enough apart). Working in the rest frame of the K, $r_2 = -\vec{r}_1$, so that P (which reverses both vectors) interchanges the positions of the two particles to yield $P\psi P^{-1} = \pi^{+\dagger}(\vec{r}_2)\pi^{-\dagger}(\vec{r}_1)$. Now applying C, changes the π^+ to a π^- and vice versa; therefore, $CP\psi(CP)^{-1} = \pi^{-\dagger}(\vec{r}_2)\pi^{+\dagger}(\vec{r}_1)$, which looks like the original wave operator but with the individual particle operators interchanged. Since the pions are bosons, the individual particle operators commute (and the state is symmetric to an interchange of bosons); therefore, the resulting wave operator (and state) is identical to the original. Thus, we conclude that $CP = +1$. For a $\pi^0\pi^0$ state the same argument follows. (Since π^0 is an eigenstate of C, the $\pi^0\pi^0$ state is even under P alone. How can the K, which has odd parity, decay into that state?) *This will be one of the (main) decay modes of the K_1, since it is also CP-even.* Because of the **large phase space** available, **the rate of this decay is that typical for the weak interactions.** *The K_2 cannot decay into two pions because it is CP-odd.*

2. *three-pion final state.* First we consider a $\pi^+\pi^-\pi^0$ state. Since $m_{K^0} \approx$ 498 MeV and the masses of pions add to $3m_\pi \approx 415$ MeV, there is very little energy available in the final state, i.e., it has **small phase space.** Thus, **the rate of this decay mode is suppressed.** We have just seen that the $\pi^+\pi^-$ subsystem has $CP = +1$. It can be shown that, with such a small amount of energy, the π^0 will **most likely** have 0 orbital angular momentum with respect to the $\pi^+\pi^-$ subsystem, i.e., $\ell_{\pi^0} = 0$. We know that C has no effect on a π^0 because it is even under C, and P produces

a minus sign because it is a pseudoscalar particle. Thus, for the entire state, we have $CP = -1$ (for $\ell_{\pi^0} = 0$). This will be the main decay mode of the K_2, since it is CP-odd, and so cannot decay into two π's. Higher odd ℓ_{π^0} states are CP-even, and so there can be some three-pion decays of the K_1; however, they are suppressed compared to its two-pion decay mode because they have $\ell_{\pi^0} \neq 0$, they are higher order in the weak interaction, and there is very little phase space available.

For a state of *three π^0s*, $CP = -1$ because only even ℓ-values can occur (as you may show in Problem 10), and each π^0 is even under C.

Thus, the neutral-K decays with $CP = +1$ and those with $CP = -1$ are qualitatively very different:

Initial State	CP	Final State	Phase Space	Lifetime	Dist $c\tau$
K_1	$+1$	$\pi^+\pi^-$ or $2\pi^0$	large	$\tau_1 = 8.9 \times 10^{-11}$ s	2.7 cm
K_2	-1	$\pi^+\pi^-\pi^0$ or $3\pi^0$	small	$\tau_2 = 5.2 \times 10^{-8}$ s	15.5 m

In the last column a measure of the distances traveled before decay (if produced at high energies) is shown. From Eq. (12.23) it is evident that the distances (L) in the laboratory will be longer due to the time-dilation effect, which produced an extra factor of γ $(= 1/\sqrt{1 - \beta^2})$. (For energies near 100 GeV, the short-lived K^0's will have decayed significantly (by a factor of e^{-1}) after traveling about $5\frac{1}{2}$ m, while the long-lived K^0 travels about 3 km before significant decay.)[26,27]

12.4.3   Coherent Regeneration Experiments

Another set of phenomena was predicted from the simple quantum-mechanical superposition principle used above. Although the strong interaction is invariant under C, the matter around us is just that "matter," as opposed to antimatter. Consequently, if we allow a beam which is pure K_2 to interact with that matter, the K^0 part will have a different interaction than will the $\bar{K}^0$ part. In particular, the $\bar{K}^0(s\bar{d})$ can readily supply an s quark to produce strange baryons, whereas the K^0 cannot.[28] Since $\tau_2 \approx 600\,\tau_1$, a time (or distance in the laboratory) can be chosen after which the K^0 beam will be essentially pure K_2.

---

26. Originally, the lifetime of the K^0 was measured by looking at the rapid decay into two pions occurring a few centimeters from production, which we now know is due to the decay of the short-lived component. The long-lived K^0 was found in 1956, shortly after this beautiful quantum-mechanical analysis (to the astonishment of some). At the lower energies available then, the particles traveled 6 m, which was equivalent to about 100 lifetimes of the short-lived component. (Can you find the energy of the K^0 from this information?)

27. The original analysis and the experiment confirming the existence of a long-lived component of the K^0 was done before P and C violation had been discovered, so that these early papers talked about C invariance rather than CP.

28. It has no valence s quark, but it does have sea s quarks. However, "breaking one of them loose" would require high energy and such an event would be highly suppressed.

Suppose that a slab of material is located in the beam, far enough from the point of origin so that a "pure" K_2 beam is entering that slab, as shown in Fig. 12.9.

The material in the slab interacts differently with the K^0 and the $\bar{K}^0$ component, so that the *emerging beam (K_f)* no longer has the same ratio of coefficients of these components:

$$|K_f\rangle = \tfrac{1}{\sqrt{2}}(f_0|K^0\rangle + f_{\bar{0}}|\bar{K}^0\rangle), \qquad \text{with } |f_{\bar{0}}| < |f_0| < 1, \qquad (12.25)$$

because the $\bar{K}^0$ (with strangeness -1) interacts more readily with matter. Substituting Eq. (12.18), we obtain

$$|K_f\rangle = \tfrac{1}{2}((f_0 + f_{\bar{0}})|K_2\rangle + (f_0 - f_{\bar{0}})|K_1\rangle). \qquad (12.26)$$

Some amount of K_1 amplitude has been regenerated by the interactions in the slab because $f_{\bar{0}} \neq f_0$. This is immediately apparent from the copious two-π decay modes appearing in the emerging beam. Here the forward-scattered beam *is* distinguishable from the unscattered beam, which is not true for optical (or other usual) phenomena. This beautiful example of quantum mechanical superposition, is reminiscent of superposition phenomena involving polarizations in optics.[29,30]

12.4.4 *CP* Violation

"Voyages of discovery can be made in new uncharted waters, but also in familiar bays close to port, provided one has observing apparatus that can see familiar objects with detail greater than that previously possible." Thus, V. L. Fitch attributes the role of spark chambers (as the newest particle detectors in 1964) in his discovery (with J. W. Cronin et al.) of the violation of *CP* invariance. There was no theoretical suspicion that such a violation existed, and all experimental evidence up to that time pointed to the validity of this generalized reflection symmetry. That symmetry satisfied the aesthetic sentiments of many. It was while doing coherent regeneration experiments that a test was run with no regenerator which showed a small number of two-π decays of the "pure" K_2 beam.

---

29. Right and left circularly polarized light are linear combinations of equal parts x- and y-linearly polarized amplitudes, with different relative phases. If we pass *right*circularly polarized light, for example, through a calcite crystal, that doubly refractive medium will have a different speed for x and y polarizations. Consequently, the beam that emerges will have the relative phases of the linear amplitudes shifted, which is equivalent to the presence of *some left* circular polarization.

30. Another such example, often cited, is the use of successive Stern-Gerlach experiments to regenerate a spin direction excluded in the first stage: Start with a beam of electrons and take the first magnetic field in the z direction. Use an inhomogeneous field to separate the spin up and spin down particles into separate beams, as in the usual Stern-Gerlach experiment. Send one, say the spin up $S_z = +\tfrac{1}{2}$, beam through a region where the *magnetic field is in the x direction*. Then use an inhomogeneous field, to split the beam into two beams, one with $S_x = +\tfrac{1}{2}$ and one with $S_x = -\tfrac{1}{2}$. If we examine either of these beams, it will have equal parts of $S_z = +\tfrac{1}{2}$ and $S_z = -\tfrac{1}{2}$ in it because each eigenstate of S_x has equal amounts of them (but with different relative phases) as we have learned in Chapter 2.

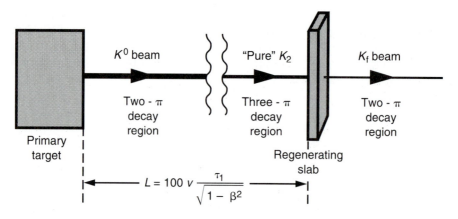

FIGURE 12.9

K^0 regeneration experiment. The strong interaction produces a K^0 beam in the primary target, whose K_1 part decays away in the two-π decay region. After 100 K_1 lifetimes, we have a "pure" K_2 beam, which can only decay via three π's. The K_f beam emerging from the regenerating slab is part K_1, so that two-π decays are again seen.

You may wonder how pure it *was*; well, it was very many K_1 lifetimes (τ_1) beyond the creation point, so that the possible K_1 content was down by many orders of magnitude. (Do Problem 12.) With their improved technology, which had been automatically calibrated by the regeneration experiments being done, they found that the long-lived neutral K decayed to two π's with a branching ratio of 2×10^{-3}. How inelegant of nature to spoil a beautiful idea and provide no alternative beauty in its place![31] This result has wide implications. For if we believe in quantum field theories (of which the standard model is one), then **CPT (all together) is an invariance**. We can **no longer have invariance under *T*** if *CP* is violated, since applying *T* (no violation) followed by *CP* (violation) yields a net violation for *CPT* which is "verboten"! In the next sections we shall discuss *T* and *CPT* more fully. Let us now see how *CP* violation may be understood.

We can imagine two qualitatively distinct reasons for the long-lived state's decay into a $CP = +1$ (two-π) state. The long-lived state's ket may contain a small part with $CP = +1$ and/or the weak decay itself may violate *CP*. A first guess was made, in 1964 (before the invention of the standard model), that there existed a superweak interaction which violated *CP*, so that states of definite mass and lifetime would no longer be eigenstates of *CP*. The beautiful *CP* symmetry of the weak interaction itself was left intact. Thus, the short-lived and long-lived

---

31. The experimenters did not initially believe it themselves and spent about half a year trying to explain it away. That was in 1964, and the result is still with us.

states would not be K_1 and K_2, respectively, but in particular, the long-lived state should be written:

$$|K_L^0\rangle = \frac{1}{\sqrt{1 + \varepsilon^2}} \, (|K_2\rangle + \varepsilon|K_1\rangle), \text{ with } \varepsilon \approx 2 \times 10^{-3}. \quad (12.27)$$

$|K_L^0\rangle$ is called K-long; the state which decays rapidly is designated **$|K_S^0\rangle$ and is called K-short**.

However, we have already learned (in Section 11.3) that the weak interaction contains a mixing[32] of the quark generations, which can have an essential (unremovable) phase. As we have pointed out there and shall elaborate upon soon, this phase would violate T invariance. The CPT theorem implies that a T violation must be accompanied by a CP violation. The essential phase, in fact, leads to a direct violation of CP in the decay itself as well as to states which are not eigenstates of CP. The direct violation is characterized by another parameter called ε', which together with ε characterizes the CP-violating decays. In the "box" diagram of Fig. 12.7, the Cabibbo-Kobayashi-Maskawa (CKM) generation-mixing matrix produces ε, i.e., it produces a CP-even part (K_1) of K_L. In Fig. 12.10 we show the "penguin" diagram of the decay, in which the CKM matrix generates the direct violation of CP in the decay itself, characterized by ε'. Unfortunately, as we have seen in our discussion of QCD, no one knows how to calculate the actual bound states of the initial and final quarks, so that only rough estimates can be made at this time. However, some of the uncertainty is eliminated by taking ratios of decay amplitudes. It has been shown that the standard model predicts a magnitude of ε' which is three orders of magnitude lower than that of ε.[33] (Notice that the top quark enters the calculations for both the penguin and box diagrams; the smallness of ε' is particularly sensitive to and consistent with a large top quark mass.) It is, thus, hard to distinguish experimentally between a superweak theory ($\varepsilon' \equiv 0$) and the standard model. In 1988, after two decades of experiments trying to detect ε', a CERN-Orsay-Pisa-Siegen collaboration and a competing Fermilab group succeeded in measuring $|\varepsilon'/\varepsilon|$. The presently accepted value (1992 Particle Properties Data Booklet) is $|\varepsilon'/\varepsilon| = (2.2 \pm 1.1) \times 10^{-3}$, which is consistent with the prediction of the standard model with a heavy top quark. Thus, after a quarter of a century of wondering whether this CP violation was a signal of a superweak interaction, it appears that the weak interaction, as incorporated in the standard model, can account for the effect. (This conclusion may change when other details of the CP-violating K^0 decays are carefully measured.)

So far, CP violation has been seen only in the K^0-$\bar{K}^0$ system. However, calculations have shown that, for a non-zero phase (δ) in the CKM generation-mixing matrix, **a large CP violation should occur in decays of the charged B**

32. This mixing occurs via the Cabibbo-Kobayashi-Maskawa (3×3) matrix.
33. Their phases are known to be nearly the same, so that they are relatively real.

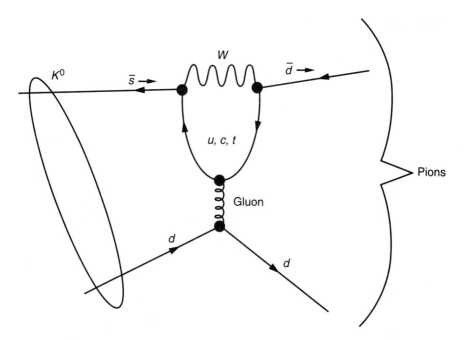

FIGURE 12.10
Diagram for the decay of K^0 to pions (the so-called "penguin" diagram), containing
direct violation of *CP*. The violation occurs at the vertices containing the *W*, through
the CKM mixing matrix. The final quarks hadronize producing pions by combining with
d and $\bar{d}$ or *u* and $\bar{u}$ quarks. (Omit the oval which indicates that the incoming quarks are
in the K^0; does this diagram now look like a penguin?)

mesons. This dramatic prediction is the consequence of Feynman diagrams
which combine QCD and electroweak vertices. The *B* mesons contain one *b*
quark or antiquark and one *u* (or *c*) antiquark or quark: $B_u^- = b\bar{u}$, $B_c^- = b\bar{c}$ and
their positively charged antiparticles. Reactions like $B_u^\pm \rightarrow \pi^\pm \pi^0$, $B_u^\pm \rightarrow K^\pm \pi^0$,
etc., and inclusive reactions (see Section 13.2.2) of the form $B \rightarrow \pi + X$, where
X can be anything, have been considered. In these types of decays the *b* quark
decays as if it were free and the companion *u* or *c* is essentially a spectator, so
that the processes studied are actually $b \rightarrow f + q + \bar{q}$, where *f* is a *d* or *s* quark.
The calculated asymmetry parameter

$$a \equiv \frac{\Gamma_{b \rightarrow fq\bar{q}} - \Gamma_{\bar{b} \rightarrow \bar{f}\bar{q}q}}{\Gamma_{b \rightarrow fq\bar{q}} + \Gamma_{\bar{b} \rightarrow \bar{f}\bar{q}q}} \tag{12.28}$$

has been found to be more than 20δ in some cases. A similar analysis for *s*, *c*,
or *t* quarks shows that the CKM matrix yields a very small asymmetry parameter (*a*). Although large asymmetries are expected in these *B* decays, the branching ratios to them are small. It has been suggested that *B*-meson "factories" be
built in which this intriguing manifestation of *CP* violation could be explored.

12.5   TIME REVERSAL (*T*)

12.5.1   Definition and Properties

The microscopic laws of mechanics appear to be invariant to time reversal, as can be verified by letting $t \to -t$ in Newton's laws. The effect upon the description of a state is

$$\vec{r} \to \vec{r}, \; \vec{v} \to -\vec{v}, \; \vec{p} \to -\vec{p}, \; \vec{L} = \vec{r} \times \vec{p} \to \vec{r} \times (-\vec{p}) = -\vec{L}$$

and the generalization to $\vec{J} \to -\vec{J}$ (for any type of angular momentum, including spin).   (12.29)

Electrodynamics also contains this microscopic invariance, if we remember that $\vec{v}$, $\vec{J}$ (current), and $\vec{B}$ (which we consider to be generated by moving charges) change sign under time reversal. However, wouldn't it be apparent that we were running a film backwards if in the movie all the molecules in a jar were to rush to one corner and go out an open valve (into a room at ordinary atmospheric pressure)? If we follow the collisions involved, however, they do obey the known laws of physics. So how do we *really* know that the film is running backwards? Well, we can only say that it is *overwhelmingly unlikely* that the molecules would have the correct velocities, such that after successive collisions they would have evacuated the jar. Just about any set of velocities at the beginning of the sequence would lead to the same or more *randomness*. This is the entropy increase discussed in thermodynamics, and we would be watching a gross violation of the second law of thermodynamics if the molecules really did rush to one corner (and out the open valve). The second law of thermodynamics is understood as being overwhelmingly probable (by more than twenty factors of 10) in the kinetic (or molecular) theory of matter, which is certainly no longer in dispute. Thus, the usual "arrow" of time is expected on a *probability* basis from a **microscopic theory** which is **time-reversal invariant**.

It is apparent that reversing the time twice in succession, which amounts to running the "movie" forward again, is equivalent to no time reversal at all:

$$T^2 = I.^{34}$$

Thus, *T* together with *I* (the operation of leaving things as they are) forms a group isomorphic to those for *P* and *C*.

However, time reversal is a completely different affair from any symmetry operation we have considered to this point. You might think that the role of time, in the (relativistic) four-dimensional space-time, would imply that time reversal is similar to space inversion. However, it becomes clear that time is com-

---

34. In quantum mechanics physical states can actually belong to "ray representations" such that $T^2|\psi\rangle = \eta|\psi\rangle$ (with $|\eta| = 1$), since the overall phase of a state is irrelevant. The (odd) behavior of the operator $\vec{J}$ under time reversal ($T\vec{J}T^{-1} = -\vec{J}$) leads to $T^2|\psi\rangle = -|\psi\rangle$ **for a spin** $\frac{1}{2}$ **particle** (and, consequently, for any state of an odd number of them). The reader can verify all of this and explore some of its ramifications by solving Problems 19 through 21.

pletely different from space when we consider how experiments are performed. We *start with an initial set of particles* and watch them interact to produce a final set. (Nature flows that way as well.) Thus, time reversal does not merely involve looking at the reactions among time-reversed particles, but also necessitates the transposition of what we consider the initial and final systems. This transposition requires that we re-examine our identification of symmetries with operations on the kets (states) of the system. Let us introduce the notation

$T|\alpha\rangle \equiv |\alpha\rangle^T$ and represent the corresponding bra by $^T\langle\alpha|$.

Time-reversal invariance requires that

$$^T\langle\alpha|\beta\rangle^T = \langle\beta|\alpha\rangle = (\langle\alpha|\beta\rangle)^*, \tag{12.30}$$

where the first equality has the correct transposition of initial and final states and the second equality is merely a property of products of states (as we have learned in Chapter 2). An operator which obeys this property is called **antiunitary**. The requirement in Eq. (12.30) implies that

$$\text{if } |\beta\rangle = b_1|\beta_1\rangle + b_2|\beta_2\rangle, \text{ then } T|\beta\rangle = b_1^* T|\beta_1\rangle + b_2^* T|\beta_2\rangle, \tag{12.31}$$

(as you may verify in Problem 17). It can be shown (see Problem 18) that any antiunitary operator can be written UK, where K is the operation of taking the complex conjugate and U is an ordinary unitary operator. The K (complex conjugating) operation is the part which changes the coefficients (b_1 and b_2) to their complex conjugates in Eq. (12.31). Here we see the reason for the claim that a complex number (or phase) in the mixing matrix will produce a violation of time-reversal invariance. The time-reversed situation will have a different coupling to a state mixed in with a non-zero phase, so that reaction rates involving cross-terms will differ:

$$\text{Rate} \propto |M|^2 = |A_1 + e^{i\delta}A_2|^2 \underset{T}{\to} |A_1 + e^{-i\delta}A_2|^2 \neq |M|^2, \tag{12.32}$$

where A_1 and A_2 are T-invariant (complex) amplitudes.

Another way to see the antiunitary character of T is to utilize time-reversal invariance in the following scenario: Consider a system at time $t = 0$, $|\alpha(0)\rangle$, and watch it evolve for a time δt to become

$$|\alpha(\delta t)\rangle = e^{-iH\delta t}|\alpha(0)\rangle. \tag{12.33}$$

Let us now time-reverse this evolved state to obtain $Te^{-iH\delta t}|\alpha(0)\rangle$. The effect of T is to reverse all momenta that appear in the description of the state. We now watch this time-reversed state (with all the momenta reversed) evolve for a time δt, to become:

$$e^{-iH\delta t}Te^{-iH\delta t}|\alpha(0)\rangle \tag{12.34}$$

(because the time evolution still proceeds with H as the generator). Since all the momenta have been reversed, the configuration returns to the one we had at $t = 0$, but the momenta are still reversed. Now consider applying T to the original state $|\alpha(0)\rangle$. This new state $T|\alpha(0)\rangle$ has the original configuration with reversed momenta, hence, it should be the same as the state we have just obtained (Eq. 12.34):

$$e^{-iH\delta t}Te^{-iH\delta t}|\alpha(0)\rangle = T|\alpha(0)\rangle.$$ (12.35)

Equation (12.35) implies that

$$e^{-iH\delta t}Te^{-iH\delta t} = T \text{ or } e^{-iH\delta t}T = Te^{+iH\delta t}.$$ (12.36)

However, for a T-invariant system, T commutes with H; consequently Eq. (12.36) requires that

$$Ti = -iT,$$ (12.37)

showing that **T must contain a complex conjugation.**

12.5.2   Experimental Status

In nuclear physics experiments, reactions of the form $a + A \leftrightarrow b + B$ have been studied in both directions (where the small letters represent projectiles, such as α, p, e, n, and γ, and the capital letters represent nuclei). Of course the phase space factors are different in the two directions because corresponding particles do not have the same mass, but we know what they are, so that they can be divided away. The absolute square of the matrix elements can then be experimentally compared; their equality is referred to as *detailed balancing* and is required if T is an invariance. These direct tests of the **T invariance of the strong interaction** have found no violation, down to a fraction of a percent.[35] The **electromagnetic interaction has** been well studied over many years, and **no direct violation of T** invariance has been found either.

For the *weak interaction* you would be hard put to run a reaction backwards. The strong interaction overwhelms the situation, so that if the decay products could be brought together, a strong interaction process would be observed. This is why we elaborated on the antiunitary character of T, which allows us to detect a violation of T invariance from the presence of a complex coupling. Also, the experimental detection of a CP violation is usually considered evidence of T violation, as implied by the CPT theorem. (In any quantum field theory, such as the standard model, a complex coupling (or mixing) will yield *both* T violation and CP violation, since CPT invariance is automatic because of the structure of the theory.)

Searches for the *electric dipole moment of the neutron* test both T invariance and P invariance. Let us recall the definition of an electric dipole moment; we will restrict ourselves to the consideration of a charge distribution which has 0 total charge. The most elementary example is that of two point charges $\pm Q$ separated by a distance d. This charge distribution has an electric dipole moment

---

35. The predictions of assumed forms of T-violating amplitudes are found to be undetectable, so that their magnitudes must be less than fractions of a percent of the magnitude of the T-conserving amplitude we know.

of size $\vec{p} = Qd$ pointing from the negative to the positive charge. The energy of the dipole in an electric field, in the limit as $d \to 0$, is

$$H_{ED} = -\vec{p} \cdot \vec{E} \tag{12.38}$$

(which is analogous to the energy of a magnetic dipole in a magnetic field). For a continuous charge distribution, the electric dipole moment is the generalization

$$\vec{p} = \int \rho(\vec{r}) \vec{r} \, d^3r. \tag{12.39}$$

For a distribution containing both positive and negative charges, $\vec{p}$ would be 0 if the positive and negative charge densities had the same centroid, i.e.,

$$\vec{p} = \int \rho_+(\vec{r}) \vec{r} \, d^3r - \int |\rho_-(\vec{r})| \vec{r} \, d^3r = 0,$$

$$\text{for } \int \rho_+(\vec{r}) \vec{r} \, d^3r = \int |\rho_-(\vec{r})| \vec{r} \, d^3r, \tag{12.40}$$

where we have explicitly displayed the sign of the negative part. Let us now consider a neutron[36] at rest, semiclassically. Equation (12.40) tells us that the neutron can have an electric dipole moment only if the center of the positive charge distribution is displaced from that of the negative charge distribution; the direction of this displacement will be the direction of $\vec{p}$. When we consider the full quantum mechanical calculation of $\vec{p}$, the spin of the particle ($\frac{1}{2}\vec{\sigma}$) will be included. Since there is no other vector associated with the neutron (at rest), the displacement and, consequently, $\vec{p}$ itself will point along that direction. Thus, we may write

$$\vec{p}_n = p_n \hat{\sigma} \quad (n \text{ for neutron}), \tag{12.41}$$

which shows that $\vec{p}_n$ behaves like $\vec{S}$ ($= \frac{1}{2}\vec{\sigma}$). We learned (in Eq. 12.29 and Section 12.2) that $\vec{S}$ changes sign under T and is unchanged under P; whereas it is evident from Maxwell's equations that $\vec{E}$ does not change sign under T but does so under P:

$$\vec{p}_n \xrightarrow[T]{} -\vec{p}_n \text{ and } \vec{E} \xrightarrow[T]{} \vec{E} \text{ under } T; \quad \vec{p}_n \xrightarrow[P]{} \vec{p}_n \text{ and } \vec{E} \xrightarrow[P]{} -\vec{E} \text{ under } P. \tag{12.42}$$

Therefore, we see that

$$H_{ED} = -\vec{p}_n \cdot \vec{E} \xrightarrow[T]{} -(-\vec{p}_n) \cdot \vec{E} = -H_{ED}, \text{ and}$$

$$H_{ED} = -\vec{p}_n \cdot \vec{E} \xrightarrow[P]{} -\vec{p}_n \cdot (-\vec{E}) = -H_{ED}. \tag{12.43}$$

In the full quantum theory, the interaction of the neutron's electric dipole moment with the electromagnetic field is identical to H_{ED}. Since H_{ED} changes

---

36. The electric dipole moment of the *neutron* is the most easily and precisely measured electric dipole moment because of its electric neutrality (so that the much larger qV interaction is absent), its accessibility, and its stability ($\tau_{n,mean} \approx 15$ min.).

sign under T or P, it **will yield a violation of both T and P**, in cross-terms with the T- and P-conserving part of the Hamiltonian (by the same reasoning used in Eq. 12.32).

We know that P is violated (by the weak interaction), and violations of CP have been found as well. Furthermore, the CP violation will be accompanied by a violation of T in any quantum field theory because of the CPT theorem. Thus, nature appears to violate both P and T, so that a non-zero value for the electric dipole moments of fermions is possible. In fact, the standard model possesses both P and T violations (the latter, via a possible non-zero phase in the generation-mixing matrix) and predicts $\mathbf{p}_n \approx 10^{-31} e$ cm. So far, it has been found in experiment that $\mathbf{p}_n < 1.2 \times 10^{-25} e$ **cm**. More accuracy will be required to determine whether or not the standard model predicts the correct value.

12.6   *CPT*

It has been rigorously proven that a quantum field theory (in a Hilbert space) which (i) satisfies Lorentz invariance (including rotations) and space-time translation invariance, (ii) possesses a lowest state in its energy spectrum (usually called the vacuum), and (iii) obeys microcausality[37] will be CPT invariant. This means that if we make all three changes — time reversal, charge conjugation, and space inversion — then the theory looks unchanged and all calculations are identical to the corresponding calculations of the original. It says nothing about these discrete symmetries individually or in pairs.

This invariance property is directly related to the observation by Feynman, in the formulation of his diagrams and rules of calculation, that a negative-energy particle moving backwards in time is equivalent to a positive-energy antiparticle moving in the opposite direction (in space) and forward in time [all four components of the momentum vectors p_μ and position vectors r_μ are changing sign (PT) and the particle is changing to an antiparticle (C)].

It is apparent that CPT invariance $([CPT, H] = 0)$ predicts that for every particle there must be a corresponding antiparticle with the same mass and lifetime: Since $(CPT)^2 = 1$, we have

$$\langle a|H|a \rangle = \langle a|H(CPT)^2|a \rangle = \langle a|CPT\, H\, CPT|a \rangle = \langle \bar{a}|H|\bar{a} \rangle,^{38} \quad (12.44)$$

where $|a\rangle$ is a particle at rest whose eigenvalue for the Hamiltonian (H) is its mass (which, for unstable particles, has a complex part corresponding to its width and lifetime).[39] Many detailed measurements, to accuracies of a few

---

37. The detailed statement requires either commutation or anticommutation relations between all distinct fields and automatically produces the correct statistics according to whether the spin is integer or odd half-integer.

38. The interchange of initial and final states due to T is trivial here, since they are the same state.

39. The connection of the complex part of the mass (or energy) to resonance width and lifetime will be discussed in Section 13.3.3.

parts in 10^5, have found mass and lifetime differences between particles and corresponding antiparticles to be 0, as predicted.[40] It can also be shown that the magnetic moment of a particle should be equal in size (in the opposite direction) to that of its antiparticle. Magnetic moment measurements of electrons and positrons as well as muons yield a 0 difference to an even greater accuracy than the mass and lifetime measurements ($\sim 10^{-12}$ for e and 10^{-8} for μ).

PROBLEMS

1. Show that $\gamma^0 \psi(-\vec{r}, t)$ obeys the Dirac equation in the inverted coordinate system if $\psi(\vec{r}, t)$ obeys the Dirac equation in the original system. It may be useful to start by writing the manifestly covariant Dirac equation for ψ and then to apply γ^0 on the left.

2. Show that the gradient operator $\vec{\nabla}$ changes sign under inversion (P). Hint: consider $\vec{\nabla}\phi$, where ϕ is a scalar function of $\vec{r}$. (The reader may choose to Fourier analyze ϕ.)

3. Verify Eq. (12.9).

4. The pseudoscalar η meson is seen to decay into two γ's with a branching ratio of $\sim 39\%$.
   a. What does this tell you about its behavior under C?
   b. Do you expect it to decay, via the electromagnetic interaction, into three γ's? Explain.
   c. Sketch the tree-level Feynman diagram for its electromagnetic decay into $\pi^0 e^+ e^-$. Explain why this decay is forbidden.

5. Use the C invariance of the electromagnetic interaction to find the smallest number of photons produced in the decay of the ground state of ortho-positronium ($S = 1$). Do the same for para-positronium ($S = 0$).

6. a. Assuming that states of definite charge form a complete set, prove that $\{Q, C\} = 0$.
   b. From the two states $|e^-\rangle$ and $|e^+\rangle$, construct an even and an odd eigenstate of C.
   c. Show explicitly that they are not eigenstates of Q.

7. Suppose that $C\pi^{+\dagger}C^{-1} = \eta\pi^{-\dagger}$ and $C\pi^{-\dagger}C^{-1} = \eta'\pi^{+\dagger}$, where $\pi^{+\dagger}$ is the operator that creates a π^+, etc. (as defined in the text).
   a. Show that this implies that $C|\pi^+\rangle = \eta|\pi^-\rangle$ and $C|\pi^-\rangle = \eta'|\pi^+\rangle$.
   b. Show that $|\eta|^2 = 1 = |\eta'|^2$ follows from the fact that C does not change the normalization of a state.
   c. Show that η' is the complex conjugate of η.
   d. Show that $\vec{\pi} \cdot \vec{\pi}$ is invariant under C conjugation for arbitrary η. Hint: The proof is simplified by writing $\vec{\pi} \cdot \vec{\pi}$ in terms of $\pi^{+\dagger}\pi^{-\dagger}$ rather than of $\pi_1^2 + \pi_2^2$.

8. Can a K meson decay into four π's? Why or why not?

9. a. Suppose the data for $\bar{K}^0$ intensity, as a function of time, in a beam of K^0's, fit the curve of Fig. 12.8. Verify that this curve corresponds to $\Delta m \, \tau_1 = 0.5$.
   b. Verify that, for $\tau_1 = .89 \times 10^{-10}$ s, this yields $\Delta m \approx 3.5 \times 10^{-6}$ eV.

10. a. Show that for a system of π^0 mesons, the relative ℓ value of any two π^0's cannot be odd. (Remember that the π^0's are bosons.)

---

40. It has been found, for example, that $|(m_{e^-} - m_{e^+})/m_e| < 4 \times 10^{-8}$.

    b. Consider a three-π^0 system which results from the decay of a neutral K. Show that, when we consider two π^0's as a subsystem, the third π^0 must have even ℓ with respect to that subsystem. (Remember that the K has spin 0.)

11. a. Show that the $\pi^0\pi^0$ state is even under P alone.

    b. Why can the K, which has odd parity, decay into that state?

12. Find the energy of a K^0 beam, which has the relative amplitude of its K_1 component equal to e^{-50}, 6 meters from its point of creation.

13. a. Justify the statement that a $\bar{K}^0$ interacts more strongly with matter than does a K^0 by writing down some of the reactions that the $\bar{K}^0$ can have with a proton or a neutron and explaining why a K^0 cannot have similar reactions.

    b. Explain how this leads to the inequality $|f_{\bar{0}}| < |f_0|$.

14. Compare the decay rate into two π's of the beam (K_f), emerging after a "pure" K_2 beam has gone through a regenerator, to the two-π decay rate of a fresh beam of K^0's. Express the answer in terms of $|f_0 - f_{\bar{0}}|^2$.

15. a. How much K_1 is left in a "pure" K_2 beam which has survived 100 K_1 lifetimes from the creation of the original K^0 beam?

    b. What two-π branching ratio would be expected if there were no CP violation?

16. Show that for $E = 100$ GeV, a (short-lived) K_1 travels 5.5 meters in one lifetime (τ_1), whereas a (long-lived) K_2 travels 3 km in one lifetime (τ_2).

17. Utilizing the requirement that

$$^T\langle\alpha|\beta\rangle^T = \langle\beta|\alpha\rangle = (\langle\alpha|\beta\rangle)^*, \text{ where } T|\alpha\rangle \equiv |\alpha\rangle^T,$$

show directly that applying T to the state $|\beta\rangle = b_1|\beta_1\rangle + b_2|\beta_2\rangle$ yields

$$T|\beta\rangle = b_1^* T|\beta_1\rangle + b_2^* T|\beta_2\rangle.$$

18. a. From the definitions: U is a unitary operator if and only if: for $|\alpha'\rangle = U|\alpha\rangle$ (all α), $\langle\beta'|\alpha'\rangle = \langle\beta|\alpha\rangle$ (all α,β); and A is an antiunitary operator if and only if: for $|\alpha'\rangle = A|\alpha\rangle$ (all α), $\langle\beta'|\alpha'\rangle = \langle\alpha|\beta\rangle$ (all α,β); prove that the product of any two antiunitary operators is unitary.

    b. From (a), prove that any antiunitary operator can be written as UK, where U is unitary and K is the (antiunitary) operator of taking the complex conjugate. Remember that $K^2 = 1$.

19. a. Show that $TJ_z = -J_z T$, $TJ_+ = -J_- T$ and $TJ_- = -J_+ T$.

    b. From the results in (a), show that $-T|\frac{1}{2},-\frac{1}{2}\rangle$ and $T|\frac{1}{2},+\frac{1}{2}\rangle$ form a spinor doublet, in the time-reversed system, with $J_z = \pm\frac{1}{2}$, respectively, so that we may write

$$T\begin{pmatrix}0\\1\end{pmatrix} = -\eta\begin{pmatrix}1\\0\end{pmatrix} \text{ and } T\begin{pmatrix}1\\0\end{pmatrix} = \eta\begin{pmatrix}0\\1\end{pmatrix}, \text{ with } |\eta| = 1, \text{ using } \begin{pmatrix}1\\0\end{pmatrix} \text{ and } \begin{pmatrix}0\\1\end{pmatrix}$$

    for spin up ($J_z = +\frac{1}{2}$) and spin down ($J_z = -\frac{1}{2}$), respectively.

    c. Using the results in (b), show that T^2 operating on a spin $\frac{1}{2}$ state yields -1.

    d. Show that the operator $(-i\sigma_y)\eta K$ may be used as the T operator for a spinor by checking that it anticommutes with $\vec{J}$. Check directly that it produces the spinors shown in (b).

    e. Using $T = (-i\sigma_y)\eta K$, verify that its square (T^2) is -1.

20. a. Using the result in Problem 19(c), show that the value of T^2 for a state of an odd number of spin $\frac{1}{2}$ particles is -1.

b. Show that **eigenstates with an odd number of spin $\frac{1}{2}$ particles of a T-invariant Hamiltonian must be doubly degenerate** by proving that: (i) $|\psi\rangle$ and $T|\psi\rangle$ have the same energy and (ii) if they were the same state, i.e., if $T|\psi\rangle = \eta|\psi\rangle$ (with $|\eta| = 1$), then $T^2|\psi\rangle = +|\psi\rangle$, so that $T^2 = -1$ implies that these are two *different* states. This is called **Kramers degeneracy**.

c. Show that a system of electrons keeps its Kramers degeneracy when placed in an external electric field, but that interactions with an external magnetic field can break the degeneracy. Hint: Look at the behavior (under T) of the new terms in the Hamiltonian, keeping the *external* fields unchanged.

21. Generalize the reasoning in Problem 19, which used the fact that T anticommutes with $\vec{J}$, to show that:

   a. The value of T^2 is $+1$ for a spin 0 state.

   b. The value of T^2 is $+1$ for a spin 1 state.

   c. The value of T^2 is $+1$ for bosons (j = integer) and -1 for fermions.

22. Show that $e^{-iH\delta t}T = Te^{+iH\delta t}$, Eq. (12.36), implies that $Ti = -iT$, Eq. (12.37), by considering small δt and expanding both sides of Eq. (12.36) to first order in δt.

23. Show that if two parts of the interaction matrix element for a transition behave oppositely under P (or T), then the observed rates will violate P (or T) invariance.

Part C

Experimental Concepts and Evidence

Chapter 13

Experimental Concepts

13.1 INTRODUCTION

The *beauty* and theoretical *decisiveness* of the gauge theories are indisputable; Yang-Mills theories have even become a field of interest to mathematicians who study *fiber bundles*. However, physics is a *science* which purports to characterize the *real* world. It is amazing that these gauge theories, constructed merely from (local) symmetries, really describe the world we live in. Be aware that these theories were not brought down to us from some analogue of Mt. Sinai nor were they the result of pure contemplation. The very success of these theories is due to their intimate relationship with experimental results. For several years, many parts of the standard model, including the idea of *quarks* itself, were thought by many to be a bunch of "crazy ideas." A kind of *selection by experiment* took place, which brought together the concepts which now make up the standard model from among the plethora of theoretical models and suggestions which had been proposed. It was the announcement (in November of 1974, referred to as "the November Revolution") of the discovery of the J/ψ (a $c\bar{c}$ resonance) that focused our attention on the standard model as a viable explanation of nature's structure.

Experiments have often been created and performed *before* knowledge of the relevant theoretical structures existed. Some of the ensuing discoveries were the result of great insight, most of great diligence and hard work, and some were just serendipity. (Serendipity without the knowledge to appreciate that something new has been discovered is, obviously, worthless.) Of course, there are often conceptual motivations for experiments, although the results may, in fact, have a completely different significance than intended. The interpretations of these data and the construction of theories, from the hints about nature which they provide, is a highly creative activity. Each significant discovery has its own story — some quite intriguing.

The results of high-energy experiments are usually tracks of sparks or bubbles, visible light, meter readings, etc. No neon sign appears saying "this is a strange particle," let alone proclaiming the supremacy of the gauge theories.

239

How experiments were designed and results interpreted to yield evidence for the standard model is a fascinating story of ingenuity and creativity. An appreciation of the development of our understanding of the world *requires* an understanding of how experiments are designed and interpreted. The culling of significant clues from experimental results is a tale of brilliance in its own right. I urge the reader not to ignore this second half of the story!

13.2   SCATTERING AND DECAY EXPERIMENTS

13.2.1   Collision Cross Sections

Cross sections are the numbers usually reported to describe the results of a scattering experiment. The basic idea of using cross sections to interpret the results of scattering experiments may be extracted from the following "proto-experiment." Suppose we have a target of *geometrical* cross-sectional area σ located somewhere on a wall of area A, as shown in Fig. 13.1. We are blindfolded before the target is put up, so that we do not know its location. We are given a number of darts, and we throw them randomly at the wall, trying to hit the target. What is the probability of hitting the target? Obviously, if the darts are distributed randomly over the wall, it is the fraction of the wall covered by the target, viz., σ/A. Now, let us count the number of hits per second $\Re$:

$$\Re = \frac{\text{No. hits}}{\text{sec}} = \frac{\text{No. reaching wall}}{\text{sec}} \times \frac{\sigma}{A} = \frac{\text{No. reaching wall}}{A\text{-sec}}\,\sigma = \Phi\sigma, \quad (13.1)$$

where $\Phi = $ (No. reaching wall/A-sec) and is called the flux of the "beam" of darts. Suppose that we throw the darts along directions perpendicular to the wall (neglect gravity) and that they all have the same speed v_i, as shown in Fig. 13.2. They are thrown at a steady rate so that they are uniformly distributed in space (before they hit). The number reaching the wall depends on how many are thrown, which we will express by looking at those in flight and counting the *number per unit volume n_a*. Let us ask how many reach the wall in a time Δt. A projectile, which is a distance $v_i\Delta t$ away to begin with, will just reach the wall in a time Δt (since it is moving towards the wall with speed v_i). Any projectile closer than that will also hit in the time interval Δt (before the aforemen-

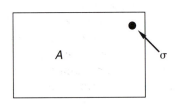

FIGURE 13.1
Target σ on wall of area A.

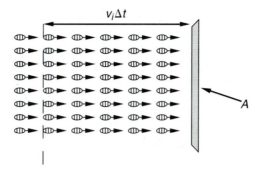

FIGURE 13.2
Sideview of wall of area A showing darts approaching. All those within a distance $v_i\Delta t$ reach the wall in a time Δt.

tioned one). So all the projectiles in the volume of lateral size A and length $v_i\Delta t$ will reach the wall in time Δt (and to find the number per second we need merely divide by Δt). The number hitting the wall in time interval Δt is, then, $n_a (Av_i\Delta t)$, so that

$$\Phi = \frac{n_a(Av_i\Delta t)}{A\Delta t} = n_a v_i. \tag{13.2}$$

In this experiment, we study the "beam" to determine the flux Φ, and then perform the "scattering" and count the number of hits of the target. Using Eq. (13.1), we see that we can determine the cross section from these results as

$$\sigma = \frac{\mathfrak{R}}{\Phi}, \tag{13.3}$$

which, of course **has units of (length)**^{2}. Obviously, the number of hits per second is proportional to the cross section, so that we can think of σ as a good measure of the effectiveness of the interaction — in this simple proto-experiment, the larger the target the more likely the occurrence of a hit.

Let us now consider a modification of this experiment in which the wall is replaced by a (three-dimensional) emulsion or solid in which targets are uniformly distributed, as shown in Fig. 13.3.

We will study the drop of flux as the beam traverses the material. There are many targets, so that the number of hits per second during the traversal of a thickness dx of material depends on the number of targets in that slice. Let W be the number of hits per second per unit area. Using our formula for the number of hits per second (Eq. 13.1) we obtain:

$$W = \left(\frac{\text{No. hits}}{\text{Area-sec}}\right) = \frac{\mathfrak{R}}{A} = \frac{\Phi\sigma}{A} \times (\text{No. of targets}). \tag{13.4}$$

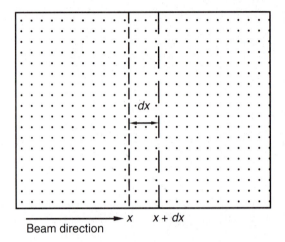

→ x   x + dx

Beam direction

FIGURE 13.3
Emulsion or solid with uniformly distributed targets.

Let us refer to the particles of the "beam" as a and the target particles as b. The number of targets present in the thickness dx (with lateral size A) is

$$\text{No. of targets in } dx = n_b A \, dx, \tag{13.5}$$

where n_b is the number of targets per unit volume. The flux of particles a drops because of collisions, and we shall assume that any particle which has a collision is removed from the beam. Therefore the change in flux between x and $x + dx$ is

$$d\Phi = -\frac{\text{No. hits}}{\text{Area-sec}} = -W = -\frac{\Phi\sigma}{A} \times (n_b A \, dx) = -\Phi(\sigma n_b) \, dx. \tag{13.6}$$

This has the same form as many other equations we have encountered for other physical phenomena; it describes exponential drop-off, which here corresponds to an attenuation of the beam. Its solution is

$$\Phi(x) = \Phi(0)e^{-\sigma n_b x}. \tag{13.7}$$

Thus, by studying the diminution of the beam through a material containing identical scatterers, we can obtain the cross section of an individual scatterer. Here, the larger the cross section, the greater the drop in the flux, so that again σ is a measure of the strength of the interaction.

Consider a more general situation in which particle a interacts with particle b to produce particles c and d, which may be different particles than a and b:

$$a + b \rightarrow c + d.$$

We can still ask how many of these events occur per second and use the same formula to calculate σ (Eq. 13.3), which is a measure of the effectiveness of the interaction; as we said above σ has the dimensions of (length)^2. It is no longer a geometrical cross section, but it is *more* relevant because it tells us the effectiveness of the interaction. It is referred to as the **cross section for $a + b \rightarrow c + d$**, and the letter σ is still used. Even for reactions in which the final particles are the same as the initial particles, the force generally has a complicated behavior with distance, so that σ is not the geometrical cross section. Beam-flux measurements through a material can also be made (as described above) to determine a cross section for removal of particles from the beam, which is the **total cross section σ_{total}**; this includes all reactions.

13.2.2   Types of Cross Sections and Their Ratios

Besides the total cross section, which we have just discussed, there are several partial cross sections and ratios that are of particular interest. We list those used most often in the determination of detailed information.

1. σ_{total} (the total cross section) measures all reactions which remove particles from the beam, as we have discussed already.
2. $\sigma_{elastic}$ (the elastic cross section) measures reactions where the particles only scatter, with no change of identity and no creation of other particles.
3. $\sigma_{inelastic}$ (the inelastic or reaction cross section) gives the combined effect of all reactions which are not elastic. It is related to the above cross sections by

$$\sigma_{inelastic} = \sigma_{total} - \sigma_{elastic}.$$

4. σ_f is the cross section for the particular elastic or inelastic channel f, such as $a + b \rightarrow c + d$ mentioned above.
5. The ratio of the sum of a particular subset of cross sections to the total cross section is referred to as the **branching ratio** for that subset of processes.
6. Some methods of collecting data lead to **inclusive** cross sections. These include all those final states which have a specified property, e.g., only one charged track or three charged tracks, etc. When *all* particles of the final state are specified, as in (4) above, then the results are sometimes referred to as **exclusive** cross sections to distinguish them from the inclusive cross sections under consideration. Of course, if we measure *all* the exclusive processes which contribute to particular inclusive data, their branching ratios must add to the inclusive branching ratio.
7. $d\sigma/d\Omega$ is called the differential cross section, and it exists for each possible channel (or sums of them). It arises when we look at scattering products appearing in a certain angular region. Consider an experiment

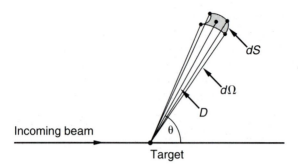

FIGURE 13.4
Small detector of area dS a distance D from the target at an angle θ from the beam direction.

of the type shown in Fig. 13.4. The detector subtends a small solid angle $d\Omega$ ($d\Omega = dS/D^2$) at an angle θ from the direction of the beam. The cross section measured by the detector is proportional to $d\Omega$ (for a small solid angle), so we write

$$d\sigma = \frac{d\sigma}{d\Omega}\, d\Omega = \frac{d\sigma}{d\Omega}\, d(\cos\theta)\, d\phi, \qquad (13.8)$$

since $|d\Omega| = |\sin\theta\, d\theta\, d\phi| = |d(\cos\theta)\, d\phi|$.

For an unpolarized beam and unpolarized targets, the experiment has symmetry under a rotation about the beam direction (cylindrical symmetry), so that the differential cross section depends only on θ (and is independent of ϕ).

In nuclear physics, where cross sections are often used, a very large cross section would be the geometrical cross section of the nucleus itself $\sim 10^{-24}$ cm^2. There is an old saying, used to describe a hunter who is a poor shot: "He can't hit the broad side of a barn!" So, a barn is a large target: **1 barn = 10^{-24} cm^2**.

13.2.3   Decay Rates and Lifetimes[1]

We can measure the lifetimes of unstable charged particles by measuring their tracks in detector chambers if they live long enough to produce tracks, but their lifetime is short enough so that they decay in the chamber. For an unstable, electrically neutral particle which decays into ionizing particles, we can instead measure the gap from its point of creation to the origin of its decay products. We can also measure the decay rates of particles by studying the drop of flux in a beam of them.

---

1. The fact that a fixed fraction (half) of a given number of unstable particles decay in a characteristic time (the half-life) immediately shows the probabilistic nature of *nature*. Remarkably, although Rutherford discovered this phenomenon experimentally near the turn of the century, he never remarked about its possible implications.

For particles which decay too rapidly to leave a track, we shall learn in Section 13.3 how an analysis of their decay products yields information about their lifetimes. These particles, in fact, appear only as *resonances* in reactions of other particles. Many of the elementary particles we have talked about do not leave tracks because they **decay strongly**, and we certainly cannot produce beams of them. Their **characteristic lifetimes are** $\tau \sim 10^{-23}$ **s**, which is the time it would take a particle moving at a speed near c to cross a nucleus (with a radius $\sim 3 \times 10^{-15}$ m). (Recall Problem 1 of Chapter 1.) We shall discuss resonances soon, and in that discussion we will show the connection between lifetimes and widths. We will see that **the corresponding resonance widths are $\Gamma \sim 100$ MeV**.

13.3   THEORETICAL PREDICTIONS

13.3.1   The Matrix Element

As we have learned in Chapter 2, the amplitude for a transition from one state (i) to another (f) depends upon the matrix element of the interaction term in the Lagrangian or Hamiltonian (H') between the two states:

$$M_{fi} = \int \psi_f^* H' \psi_i \, dV = \langle f | H' | i \rangle. \tag{13.9}$$

Just as was discovered in optics more than a century ago, **the rate of the reaction is proportional to the** *square* **of the amplitude**; therefore, for scattering $\sigma \sim |M_{fi}|^2$.

13.3.2   Phase Space and Fermi's Golden Rule

The transition rate to f also depends upon how many quantum states are included (or counted) in that final state.

Let us illustrate this concept with a *die analogy*. Suppose we paint the 1, 2, and 3 faces of a die red, the 4 face blue, and the 5 and 6 faces white. What are the relative probabilities of rolling red, white, or blue? For a true die (not loaded) they would be:

State	Relative Probability
red	3
white	2
blue	1

These relative probabilities have nothing to do with the "process" itself (it is not a loaded die) but just with the counting of the states. A similar counting situation occurs in scattering and decay experiments.

In order to investigate this equation, let us discuss a scattering experiment with two particles in both the initial and final states:

$$a + b \to c + d.$$

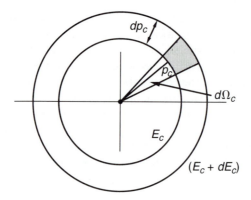

FIGURE 13.5
Momentum space of particle c showing the solid angle $d\Omega_c$ subtended by the detector.

Consider the simple situation where a detector determines the final energy E_c of particle c. A detector will always have a certain resolution, viz., it can determine the energy of the particles detected only within a certain range (dE_c here). We wish to determine how many states are in that range. In classical mechanics we can represent each state of a system as a point in phase space, the space containing axes for the positions and momenta of the particles of the system. For simplicity let us work in the center of momentum system, so that the momenta of the final two particles are equal and opposite to one another. Furthermore, the location in space of each final particle after the scattering is directly related to its momentum and actually cancels out of the calculation of the cross section. Thus, we can represent the final system by a point in a three-dimensional momentum space for particle c. Remember that after the scattering, the particles are free; thus, the surface representing a given energy E_c is a sphere in momentum space of radius $p_c = \sqrt{E_c^2 - m_c^2}$. (See Fig. 13.5.)

Let us now suppose that the detector subtends a certain solid angle $d\Omega_c$. Since each point corresponds to a state, the number of states being counted by the detector is proportional to the volume between the sphere for E_c and the sphere for $E_c + dE_c$, subtending the solid angle $d\Omega_c$, as shown in Fig. 13.5.

This is all rather straightforward, and it is easy to determine that volume, as we shall see. However, the world we live in is a quantum-mechanical one. We cannot determine precisely the position and momentum of a particle, as the uncertainty principle so succinctly states. This inability casts doubt on the utility of phase space, since we can never use a point in it to represent a system. Let us look at the restriction summarized by the uncertainty principle: $\Delta p_i \Delta x_i \geq h$. We will break each six-dimensional phase space (for each particle) into little volumes whose sides, $\delta p_i, \delta x_i$, are large enough so that

$$\delta p_i \delta x_i = kh, \text{ (for each value of } i) \qquad \text{where } k > 1. \tag{13.10}$$

We can then determine in which box that particle belongs by suitable measurement; in fact, quite precise simultaneous measurements of position and momentum of a particle are frequently made.[2] (The point is that h is a very small number in SI units, so that both the position and momentum of the particle may be determined very accurately on a macroscopic scale.) The number of states in a region of phase space is, thus, equal to the number of boxes enclosed, which equals the volume of that region divided by $(kh)^3$ (one factor for each Cartesian direction, corresponding to $\delta p_i \delta x_i$). Consequently, the number of states *is* proportional to the volume of phase space, just as in classical mechanics.

Let us now return to the scattering problem above. The momentum-space volume between the spheres is

$$dV_p = p_f^2 \, dp_f \, d\Omega_c,$$

using spherical coordinates (in momentum space). Since the number of states is proportional to this volume, we have

$$dn \sim p_f^2 \, dp_f \, d\Omega_c.$$

Because we are given an *energy* resolution for the detector, we convert by using

$$E_c^2 = p_f^2 + m_c^2 \Rightarrow 2E_c \, dE_c = 2p_f \, dp_f \Rightarrow dp_f = \frac{E_c}{p_f} \, dE_c = \frac{1}{v_f} \, dE_c \text{ to obtain}$$

$$\frac{dn}{dE_c} \sim \frac{p_f^2}{v_f} \, d\Omega_c.$$

We often perform experiments involving particles which have spin and usually do not distinguish different S_z values; these are referred to as **unpolarized experiments**. The multiplicity of the final states counted (g_f) would then be

$$g_f = (2s_c + 1)(2s_d + 1). \tag{13.11}$$

Combining these considerations, we find for the number of states per unit energy:

$$\frac{dn}{dE_c} \sim \frac{p_f^2}{v_f} \, g_f \, d\Omega_c.$$

The number of counts observed is ultimately expressed as a cross section, and so these factors will appear in the expression for the cross section. The number of counts also depends on the incident flux Φ, which is proportional to v_i (Eq. 13.2); in order to obtain a cross section, we must divide by Φ (Eq. 13.3) yielding a factor of v_i^{-1}. (This factor suggests the qualitative concept that the cross section is proportional to the time each particle of the incident beam is

---

2. The curved track in a bubble chamber measures simultaneously, with considerable accuracy, both position and momentum (from the curvature of the track) of the particle. In fact, almost any detector which measures the momentum of a particle also has made a determination of its position, which is confined to the small region of the detector during the measurement.

within the range of the interaction, since that time is inversely proportional to the incident speed.) Putting all these pieces together, we obtain

$$\frac{d\sigma}{d\Omega_c} \sim \frac{1}{v_i} |M_{fi}|^2 \left[\frac{p_f^2}{v_f} g_f \right]. \tag{13.12}$$

This relationship shows that **the rate** at which events occur is proportional to **[the absolute square of the matrix element] times [the phase-space factors]**; the generalized version of it is derived from quantum mechanics and is referred to as **Fermi's Golden Rule**.

Notice that for reactions which can just barely produce the final particles, i.e., with $p_f \approx 0$, the phase-space factors are very small ($p_f^2/v_f \sim p_f$); therefore, the number of such events will also be very small. Obviously, to produce these particles we should work at energies considerably above the threshold for their production.[3]

The same phase-space discussion applies to the rate of two-body decays, $a \rightarrow b + c$, and similar discussions arise for decays into many final products. It is left to the reader (Problem 10) to show that the rate of decay of a particle is proportional to the inverse of its lifetime (τ). In Eq. (13.39) we will show that the energy width (Γ) and the lifetime obey the uncertainty-type relationship $\Gamma\tau = \hbar$, so that we may write

$$\Gamma \sim \text{Rate of Decay} \sim |M_{fi}|^2 \times \text{[phase-space factors]}, \tag{13.13}$$

where the phase-space factors represent the size of phase space available to the final particles. (In Chapter 12, when discussing K^0 decay to pions, we expressed the phase-space dependence as a dependence on the amount of kinetic energy available to the final particles.)

For more complicated detection arrangements and other types of experiments, similar considerations show that phase-space factors are ubiquitous. In analyzing data, we must not jump to the conclusion that a peak corresponds to a resonance; one source of peaking could be the phase-space factors.[4]

13.3.3   Resonances and Their Interpretation as Elementary Particles

In atomic and nuclear physics a resonance seen in a scattering experiment is explained as the result of the excitation of a metastable state; it is a somewhat peripheral detail occurring in some experiments. As we shall see shortly, elementary particles which decay rapidly are detected *only* as resonances. This cate-

---

3. Of course, working at *very* high energies opens other channels which will compete with the channel under consideration. Thus, the branching ratio for the channel of interest drops at very high energies.

4. For spinless or spin-averaged three-body decays, the decay rate can be written $d\Gamma = (const.)|M|^2 d(m_{12}^2) d(m_{23}^2)$, where $m_{ij} \equiv (p_i - p_j)^2$. Consequently, using m_{12}^2 and m_{23}^2 as axes yields no phase-space factors, so that any bunching of points is due to $|M|^2$ itself. This transparent plot of three-body decays is called a *Dalitz plot*. See pp. III.48, 49 in the Particle Properties Data Booklet for details.

gory contains most hadrons and many of the fundamental particles we have been considering, including the gauge bosons ($W^\pm$ and Z^0) and the heaviest lepton (the τ). Many of the experimental clues, searches, and verifications of theoretical conjectures involve detailed analyses of resonances. Thus, the understanding and analysis of resonances is of prime importance.

Scattering, Partial Waves, and Phase Shifts   For simplicity we will consider the scattering of a spinless particle of momentum k from a spherically symmetric potential $V(r)$ of finite range R. The ideas we develop occur as well in the scattering of particles from one another, where the details are more elaborate but the essential features are the same.

We begin with an analysis which treats the incoming particle as a plane wave. (Recall that a wavepacket can always be Fourier analyzed.) Taking the z direction as the direction of motion of the incoming particle, the incident wave has the wavefunction:

$$\psi_{incident} = e^{i(kz - \omega t)}, \tag{13.14}$$

where k is the momentum and we have used 1 for the normalization constant. (Since the calculation of the scattering cross section involves a division by the incoming flux, any other choice would cancel out.) Far outside the range of the potential, we have a scattered *spherical* wave (moving in the r direction) with the same energy (this is elastic scattering), as shown in Fig. 13.6. We will write $\psi_{scatt.}$ for that part of the wavefunction.

The flux of the scattered wave is proportional to $|\psi_{scatt.}|^2$. As we follow a scattered wave out in the asymptotic region ($r > R$), the number of particles it represents must be independent of the distance from the scatterer, so that

$$|\psi_{scatt.}|^2 \times (\text{Area of large sphere}) = |\psi_{scatt.}|^2 \times (4\pi r^2) = \text{constant}.$$

Thus, we must have

$|\psi_{scatt.}|^2 \propto \dfrac{1}{r^2}$. Putting these requirements together, we may write

$$\psi_{scatt.} = f(\theta) \frac{e^{i(kr - \omega t)}}{r}, \tag{13.15}$$

which is, indeed, moving radially outward as can be seen by noting that its exponent may be written $k(r - vt)$. $f(\theta)$ is an unknown function representing the fact that $\psi_{scatt.}$ may not be the same in all directions. The undetermined factor of proportionality $f(\theta)$ is called **the scattering amplitude**. The scattering amplitude is only a function of θ here, since the cylindrical symmetry about the direction of the incoming particle implies that the scattered wave must be independent of the angle ϕ (on the cone determined by θ, as shown in Fig. 13.7). Thus, outside the scattering region, the wavefunction is:

$$\psi = e^{ikz} + \psi_{scatt.} = e^{ikz} + f(\theta) \frac{e^{ikr}}{r} \qquad r > R, \tag{13.16}$$

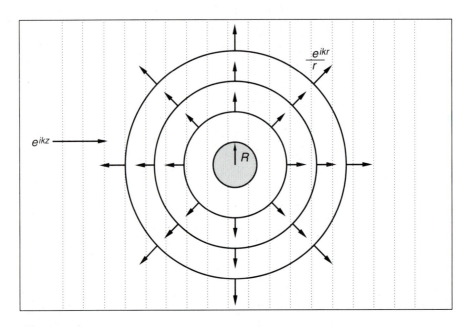

FIGURE 13.6
Scattering wavefronts. e^{ikz} is the incoming plane wave and e^{ikr}/r the outgoing spherical wave.

where we have dropped the common factor $e^{-i\omega t}$ which plays no role in our discussion. With the unit normalization of the incident wave, $|\psi_{scatt.}|^2$ gives the number of scattered particles per incident particle.

Consider a small detector located a distance r from the origin at an angle θ to the incoming direction, subtending a solid angle $d\Omega$; let us refer to the corresponding cross section as $d\sigma$. From Eq. (13.3) we see that

$$d\sigma = \frac{\mathfrak{R}}{\Phi},$$

where $\Phi = |\psi_{incident}|^2 = 1$ for the normalization chosen in Eq. (13.14).

(13.17)

The number of particles detected per second must equal **[the number scattered per second per unit area at angle θ]** × **[the area subtended by the detector]**:

$$\mathfrak{R} = [|\psi_{scatt.}|^2 \Phi]\,[r^2\,d\Omega]. \tag{13.18}$$

Inserting this into Eq. (13.17) yields

$$d\sigma = |\psi_{scatt.}|^2 r^2\,d\Omega = |f(\theta)|^2\,d\Omega$$

$$\text{or } \frac{d\sigma}{d\Omega} = |f(\theta)|^2. \tag{13.19}$$

(Notice that the incoming flux Φ cancels out as promised.)

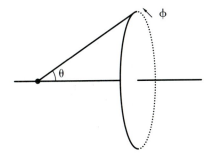

FIGURE 13.7
Scattering angles.

If we could solve the Schrödinger equation with the *actual* $V(r)$, we would obtain an *exact* solution for this scattering problem $\psi(r,\theta)$. The Legendre polynomials[5] $P_\ell(\cos\theta)$ form a complete set, so that we may expand the exact solution in a series:

$$\psi(r,\theta) = \sum_\ell R_\ell(r)P_\ell(\cos\theta), \tag{13.20}$$

where the coefficients $R_\ell(r)$ may be functions of r. This is called a **partial wave expansion**.

For $r < R$, $R_\ell(r)$ is the solution of the Schrödinger equation with $V(r) \neq 0$ and it depends on the detailed form of $V(r)$, whereas for $r > R$, $R_\ell(r)$ is the solution of the free Schrödinger equation (since $V = 0$ for $r > R$). At the detector, which is very far from the scattering region, that free solution takes the simple form:

$$R_\ell(r) \rightarrow A_\ell \frac{1}{kr} \sin(kr + \text{constant phase}) \text{ for } r \gg \ell k^{-1}. \tag{13.21}$$

(See Problem 3.) The constant phase and the normalization (A_ℓ) of R_ℓ would be determined by matching ψ and $\partial\psi/\partial r$, of the solution for $r > R$ with those of the solution for $r < R$, at $r = R$.

Suppose we solved the Schrödinger equation for $V = 0$ *everywhere*, i.e., *no scattering*. The solution obviously would be $\psi_{V=0} = e^{ikz}$, whose expansion in Legendre polynomials is well-known: [6]

$$\psi_{V=0} = e^{ikz} = e^{ikr\cos\theta} \rightarrow \sum_\ell [(2\ell + 1)e^{i(\ell\pi/2)}] \frac{1}{kr} \sin\left(kr - \frac{\ell\pi}{2}\right)P_\ell(\cos\theta)$$

$$\text{as } r \rightarrow \infty. \tag{13.22}$$

---

5. See Mathematics Table I for their definitions and orthogonality integrals.
6. See, e.g., Schiff, *Quantum Mechanics*, 3rd ed., p. 119.

As expected, this has exactly the *same form* at large r as we had in the presence of scattering (Eq. 13.21), since in the region outside R we had $V = 0$. The effect of the (elastic) scattering then will only show up in the phase and amplitude of $R_\ell(r)$ (Eq. 13.21). We shall see presently that, if the phase is $-\ell\pi/2$ as in Eq. (13.22), then the amplitude A_ℓ will coincide with the square bracket of that equation, so that ψ would be e^{ikz} and no scattering would have occurred: thus, *the effect of the scattering is to shift the phase* from $-\ell\pi/2$. It is convenient, therefore, to write the phase in Eq. (13.21) such that

$$R_\ell(r) \to A_\ell \frac{1}{kr} \sin\left(kr - \frac{\ell\pi}{2} + \delta_\ell\right), \qquad (13.23)$$

where δ_ℓ is a measure of the scattering and is called the **phase shift** (for that particular ℓ); we expect that the phase shifts depend upon the energy $-\delta_\ell(E)$. Inserting Eqs. (13.20), (13.22), and (13.23) into Eq. (13.16) yields

$$A_\ell = [(2\ell + 1)e^{i(\ell\pi/2)}]e^{i\delta_\ell}, \qquad (13.24)$$

which is determined by δ_ℓ[7] and also yields (see Problem 4) an expression for $f(\theta)$ in terms of the phase shifts:

$$f(\theta) = \frac{1}{k} \sum_\ell (2\ell + 1)f_\ell P_\ell(\cos\theta),$$

$$\text{where } f_\ell = \frac{(e^{2i\delta_\ell} - 1)}{2i}. \qquad (13.25)$$

Note that if a particular phase shift δ_ℓ is 0, then the corresponding f_ℓ is 0 as expected, since there should not be scattering for that partial wave.

You may be wondering what has been accomplished; we have expressed an unknown function $f(\theta)$ in terms of an infinite number of numbers δ_ℓ. In order to explore the utility of this expansion and to gain some insight concerning phase shifts, let us consider a semiclassical picture of the scattering.

In Fig. 13.8 we see a particle, with momentum $\vec{k}$ outside the scattering region, "aimed" along a line which misses the center of the region by an amount b. That distance b is called the **impact parameter**, and the result of the scattering should definitely depend on its value; for $b > R$ there is no scattering at all. The angular momentum of the incoming particle (about the scattering center) is

$$|\vec{L}| = |\vec{r} \times \vec{p}| = r_\perp k = bk. \qquad (13.26)$$

The largest angular momentum for which there can be scattering occurs when $b = R$, which yields $L_{max} = Rk$.

Now for the magic implied by the prefix "semi-." Recall that quantum mechanics has quantized angular momenta with

$$L = \sqrt{\ell(\ell + 1)} \approx \ell, \text{ within the validity of a semiclassical argument.}$$

---

7. When δ_ℓ is 0, A_ℓ coincides with the amplitude for no scattering, as we have asserted above.

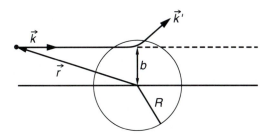

FIGURE 13.8
Semiclassical trajectory; b is the impact parameter.

Therefore, we expect very little scattering for ℓ above Rk, which in SI units reads

$$\ell_{max} \approx \frac{Rp}{\hbar}. \tag{13.27}$$

Suppose we take the range of the scattering to be of the size of a nucleus $R \approx 1.4$ Fermis $= 1.4 \times 10^{-15}$ m; this yields

$$\ell_{max} \approx 1.4 \times 10^{19}\, p\left(\text{in } \frac{\text{kg}\cdot\text{m}}{\text{s}}\right) \approx 7 \times 10^{-3}\, p\left(\text{in } \frac{\text{MeV}}{\text{c}}\right).^8$$

This tells us that

$$\text{for } p < 140\, \frac{\text{MeV}}{\text{c}}, \text{ we have } \ell_{max} < 1.$$

Therefore, *below 140 MeV* we would have only one term, with $\boldsymbol{\ell = 0}$, differing appreciably from 0 in the partial wave expansion of the scattering amplitude. This is usually referred to as **S-wave scattering**. Substituting the expression for f_0 into Eq. (13.25) (recalling Eq. 13.19) we obtain

$$\frac{d\sigma}{d\Omega} = \frac{1}{4k^2}\, |e^{2i\delta_0} - 1|^2, \tag{13.28}$$

which is *isotropic* (independent of θ).

As we increase the energy (consequently, p), higher partial waves contribute and produce their characteristic angular dependence (by each contributing a term $\sim P_\ell(\cos\theta)$ to $f(\theta)$). From an analysis of the data for the angular distribution ($d\sigma/d\Omega$), the phase shifts can be determined.[9]

---

8. 1 MeV/c $\approx 5 \times 10^{-22}$ kg·m/s, so that a 1 on the right-hand side gives the same result as 5×10^{-22} on the left.

9. Due to the inherent experimental uncertainties, there is sometimes ambiguity in the choice of sets of phase shifts which fit the data; reasonableness criteria, like the semiclassical argument above, are utilized to resolve such ambiguities.

Let us calculate the differential and total cross sections in terms of the phase shifts:

$$\frac{d\sigma}{d\Omega} = |f(\theta)|^2 = \frac{1}{k^2} \sum_{\ell\ell'} (2\ell + 1)(2\ell' + 1) f_\ell f_{\ell'}^* P_\ell(\cos\theta) P_{\ell'}(\cos\theta)$$

$$= \frac{1}{4k^2} \sum_{\ell\ell'} (2\ell + 1)(2\ell' + 1)(e^{2i\delta_\ell} - 1)(e^{-2i\delta_{\ell'}} - 1) P_\ell(\cos\theta) P_{\ell'}(\cos\theta);$$

$$\sigma = \int \frac{d\sigma}{d\Omega} d\Omega = \frac{4\pi}{k^2} \sum_\ell (2\ell + 1)|f_\ell|^2 = \frac{4\pi}{k^2} \sum_\ell (2\ell + 1)\sin^2\delta_\ell, \qquad (13.29)$$

where the orthogonality of the Legendre polynomials has produced a single sum in σ, and we have substituted $|f_\ell|^2 = \sin^2\delta_\ell$. Recall that the phase shifts are functions of energy $-\delta_\ell(E)$. For a particular ℓ, $|f_\ell|^2$ goes through a maximum at energy E_R if $\delta_\ell(E_R) = \pi/2$. (Note that: $|f_\ell|^2_{max} = 1$.) The contribution of that partial wave to σ will then be a maximum; this is considered the signal of a **resonance** at E_R.

For particle-particle scattering, the magnitude of the scattered (exponential) term in the elastic amplitude f_ℓ can be less than 1 due to the loss of particles to inelastic channels:

$$f_\ell = \frac{(\eta_\ell e^{2i\delta_\ell} - 1)}{2i}, \quad \text{with } 0 \le \eta_\ell \le 1. \qquad (13.30)$$

The limitation on the magnitude of f_ℓ would still be

$$|f_\ell|^2 \le 1, \qquad (13.31)$$

which is called **the partial wave unitarity limit** because it can be shown to be a direct result of the unitarity of the S-matrix. This limit places severe constraints on theoretical conjectures.

The Optical Theorem   For potential scattering (where $\eta_\ell = 1$ for all ℓ), let us consider the imaginary part of the scattering amplitude f in the forward direction; from Eq. (13.25) we obtain

$$\text{Im} f(0) = \frac{1}{k} \sum_\ell (2\ell + 1)\text{Im} f_\ell P_\ell(1)$$

$$= \frac{1}{k} \sum_\ell (2\ell + 1)\left(\frac{-\cos 2\delta_\ell + 1}{2}\right) = \frac{1}{k} \sum_\ell (2\ell + 1)\sin^2\delta_\ell, \text{[10]} \quad (13.32)$$

which is proportional to the total cross section (Eq. 13.29). The same result is obtained in particle-particle scattering for the forward *elastic* amplitude (inde-

---

10. Note that $P_\ell(1) = 1$.

pendent of the values of the η_ℓ). This relationship is usually written:

$$\operatorname{Im} f_{elastic}(0) = \frac{k\sigma_{total}}{4\pi}, \tag{13.33}$$

where $\sigma_{total} = \sigma_{elastic} + \sigma_{inelastic}$. (See Problem 7.) This is often called the **optical theorem** because the relationship between forward (elastic) scattering and the total scattering was first realized in the study of optics.[11] In fact, even when an obstacle blocks a beam of light, a bright spot is found in the center of the shadow. This is explained qualitatively by saying that the light is diffracted around the obstacle. The peak in the forward direction is, therefore, often referred to as the **diffraction peak**. In particle-particle scattering we expect the forward peak in the elastic channel to become larger at energies corresponding to the opening of new channels for the final state.

The optical theorem has a simple intuitive explanation. *The total cross section represents processes which remove particles from the incident beam.* Thus, we expect the wavefunction to be *depleted* in the forward direction. So why do we talk of a forward *peak*? This peak results from the way we have chosen to partition the wavefunction; we refer to one part as an incident (plane) wave and the rest as a scattered wave. In other words, wherever the wave differs from the incoming plane wave, that difference is considered a scattered wave. In the forward direction, although that difference is actually due to a depletion of the number of particles in the beam, we still use its absolute square to calculate the differential scattering cross section. In fact, the optical theorem can be derived from the relationship of the total cross section to the depletion of the forward beam (i.e., from particle current conservation) *without* performing a partial wave expansion.[12] (See Problem 8.)

Although the break-up of the wavefunction into an incident and scattered wave seems arbitrary, in actual experiments $\psi_{scatt.}$ *alone* is detected for $\theta \neq 0$. This occurs because the beam is not exactly a plane wave in transverse directions but is confined laterally, i.e., it is a wavepacket (with a continuous range of Fourier components). The detectors are located outside the beam, so that they detect only the scattered portion of the wavefunction. As we approach the beam we detect a peaking, which extends beyond the forward direction because of the wavepacket nature of the actual beam. (In any actual situation, observations never have infinitely sharp jumps, so that we expect to see the scattering amplitude continuously approaching the value predicted in the forward direction by the optical theorem.)

Resonances and Decaying States — The Breit-Wigner Form   Suppose that the phase shift of the ℓ_1th partial wave goes through $\pi/2$ at energy E_R. Assume further that the phase shifts of all other partial waves ($\ell \neq \ell_1$) are small; we

---

11. In optics the wavelength λ appears (in Eq. 13.33), which is related to k by: $\lambda = 2\pi/k$.

12. The left-hand side of the expression in Eq. (13.33) comes directly from the cross-term between the incident wave and the scattered wave, so that the optical theorem is a result of the wave-particle duality of the beam "particles."

shall neglect their contribution to the scattering amplitude. Let us expand the partial wave amplitude around E_R. First, we rewrite it from Eq. (13.25) as

$$f_{\ell_1} = \frac{(e^{2i\delta_{\ell_1}} - 1)}{2i} = \frac{e^{i\delta_{\ell_1}}(e^{i\delta_{\ell_1}} - e^{-i\delta_{\ell_1}})}{2i} = e^{i\delta_{\ell_1}} \sin \delta_{\ell_1}$$

$$= \frac{\sin \delta_{\ell_1}}{e^{-i\delta_{\ell_1}}} = \frac{\sin \delta_{\ell_1}}{\cos \delta_{\ell_1} - i \sin \delta_{\ell_1}} = \frac{1}{\cot \delta_{\ell_1} - i}. \qquad (13.34)$$

Let us do a Taylor series expansion of $\cot \delta$ about E_R, dropping the subscript ℓ_1 for convenience, since it is the only partial wave being considered. We obtain[13]

$$\cot \delta(E) \approx \cot \delta(E_R) + (E - E_R)\left[\frac{d}{dE}\cot \delta\right]_{E_R} = 0 + (E - E_R)\left(\frac{-1}{\Delta}\right),$$

$$\text{since } \delta(E_R) = \frac{\pi}{2}, \qquad (13.35)$$

where we have introduced the unknown Δ which has units of energy, since $[d/dE \cot \delta]_{E_R}$ is an inverse energy (from the derivative with respect to energy). Substituting in Eq. (13.34), we obtain

$$f_{\ell_1} = \frac{1}{\cot \delta_{\ell_1} - i} \approx \frac{1}{(E - E_R)\left(\frac{-1}{\Delta}\right) - i} = \frac{\Delta}{(E_R - E) - i\Delta}. \qquad (13.36)$$

Using Eqs. (13.19), (13.25), and (13.36), we find

$$\sigma = \int |f|^2 \, d\Omega \approx \left[\frac{4\pi(2\ell_1 + 1)}{k^2}\right] \frac{\Delta^2}{(E_R - E)^2 + \Delta^2}. \qquad (13.37)$$

For $\Delta \ll E_R$, in a region $O(\Delta)$ around E_R, the expression in the square bracket is very nearly constant and the dependence of σ on energy is determined by the last factor in Eq. (13.37), which is called the **Breit-Wigner form**. In Fig. 13.9 we show the shape of the Breit-Wigner resonance cross section. Notice that the full width at half maximum is 2Δ.

Now let us consider a decaying state of energy E_R with lifetime τ. Since the decay rate is proportional to $|\psi|^2$, we have to modify the time behavior of ψ:

$$\psi(t) = \psi(0)e^{-(i/\hbar)E_R t - (t/2\tau)},$$

$$\text{whose form does yield } |\psi|^2 \sim e^{-(t/\tau)}. \qquad (13.38)$$

This wavefunction is identical to

$$\psi(t) = \psi(0)e^{-(i/\hbar)(E_R - i(\Gamma/2))t},$$

$$\text{for } \Gamma \equiv \frac{\hbar}{\tau}. \qquad (13.39)$$

---

13. The minus sign in the definition of Δ is a detail of no importance or consequence to our discussion.

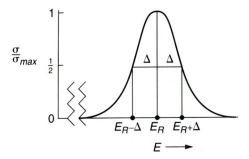

FIGURE 13.9
Breit-Wigner resonance.

The relationship between Γ and τ looks like the usual minimum uncertainty relation between ΔE and Δt. Recall that the wavefunction in frequency space (call it $\chi(E)$, since $E = hf$) is related to $\psi(t)$ by a Fourier transform. Therefore, we may write

$$\chi(E) \propto \int_0^\infty \psi(t)e^{(i/\hbar)Et}\, dt = \int_0^\infty \psi(0)e^{-(i/\hbar)(E_R - i(\Gamma/2))t}e^{(i/\hbar)Et}\, dt$$

$$= \frac{\psi(0)e^{-(i/\hbar)((E_R - E) - i(\Gamma/2))t}}{\dfrac{-i}{\hbar}\left[(E_R - E) - i\,\dfrac{\Gamma}{2}\right]}\Bigg|_0^\infty$$

$$= \left[\frac{\psi(0)}{\dfrac{i}{\hbar}}\right]\frac{1}{\left[(E_R - E) - i\,\dfrac{\Gamma}{2}\right]}. \qquad (13.40)$$

Noting that the cross section is proportional to the absolute square of the wavefunction, we have

$$\sigma \sim \frac{1}{(E_R - E)^2 + \dfrac{\Gamma^2}{4}},$$

which yields $\sigma(E) = \sigma_{max.}\left[\dfrac{\dfrac{\Gamma^2}{4}}{(E_R - E)^2 + \dfrac{\Gamma^2}{4}}\right].$[14] $\qquad (13.41)$

This is a **Breit-Wigner resonance** with **full width at half maximum** equal to Γ, such that the lifetime (τ) is $\tau = \boldsymbol{\hbar/\Gamma}$. (Notice that the width $\Gamma = \tau^{-1}$ in natural

---

14. We have pulled a factor of $\Gamma^2/4$ from the constant of proportionality, so that the remaining constant is the maximum value that σ reaches (at $E = E_R$).

units, so that it is equal to the decay constant we introduced with the same letter in Section 12.4.2.) We already have seen in Eq. (13.37) that

$$\sigma_{max.} = \frac{4\pi}{k_R^2}(2\ell_1 + 1).$$
(13.42)

For the *elastic scattering* $a + b \rightarrow c^* \rightarrow a + b$, where c^* is a resonance, the multiplicity $(2J + 1)$ of the resonance should appear, where $(2\ell + 1)$ appears for the spinless case. For each m state of the c^*, only one particular linear combination of the a-b system emerges; therefore, we should remove their individual spin multiplicities. This yields a division by $(2S_a + 1)(2S_b + 1)$. Thus, we obtain

$$\sigma_{max.} = \frac{4\pi}{k^2} \frac{(2J + 1)}{(2S_a + 1)(2S_b + 1)}.$$
(13.43)

Example of a Baryon Resonance — The Δ(1232)   The $\Delta(1232)$, which is a member of the spin $\frac{3}{2}$ baryon decuplet (discussed in Chapter 6) was first observed as a resonance in πN (pion-nucleon) scattering.

Consider first the scattering of a π^+ from a proton. The cross-section data is shown in Fig. 13.10. There we see a large peak, whose shape resembles the Breit-Wigner form (slightly distorted because background events are not separated out and there is some decay to inelastic channels) with its maximum at $E_R = 1232$ MeV. Its width is $\Gamma \approx 115$ MeV, which corresponds to a lifetime $\tau \approx 5 \times 10^{-24}$ s. It is indeed a very short-lived particle. The size of the cross section at resonance is $\sigma_{max.} \approx 8\pi/k_R^2$. From Eq. (13.43), recalling that $S_\pi = 0$ and $S_p = \frac{1}{2}$, we learn that this is a $J = \frac{3}{2}$ resonance. An analysis of the *angular distribution of the scattered pion* is more complicated than it is for the spinless situation discussed above; it is discussed in Appendix F where we see that the data agree with that predicted for $J = \frac{3}{2}$. This π^+p resonance is doubly charged; it is the Δ^{++}.

Now, let us determine the isotopic spin of the Δ. The π^+ is the highest member of the pion isotopic triplet $(I = 1)$ with $I_z = +1$, and the p is the highest member of the nucleon isotopic doublet $(I = \frac{1}{2})$ with $I_z = +\frac{1}{2}$. They combine to form a state with $I = \frac{3}{2}$ and $I_z = \frac{3}{2}$, as we have studied in Section 6.4 and in Chapter 3, so these must be the isotopic spin values of the Δ^{++}. We expect, then, that **the Δ is an $I = \frac{3}{2}$ resonance** and infer the existence of the other members of the multiplet: a Δ^+ with $I_z = +\frac{1}{2}$; a Δ^0 with $I_z = -\frac{1}{2}$; and a Δ^- with $I_z = -\frac{3}{2}$.

Recall that a state composed of a $\pi^-(I = 1, I_z = -1)$ and a proton $(I = \frac{1}{2}, I_z = +\frac{1}{2})$ has a part which is $I = \frac{3}{2}$ (with $I_z = -\frac{1}{2}$), so that the Δ resonance should show up in π^-p scattering as well (as the Δ^0). Indeed, Fig. 13.10 shows such a peak for π^-p scattering at $E_R = 1232$ MeV. (It also shows resonances in the π^-p experiments at energies for which there are none in the π^+p; they are the N^* $I = \frac{1}{2}$ resonances.)

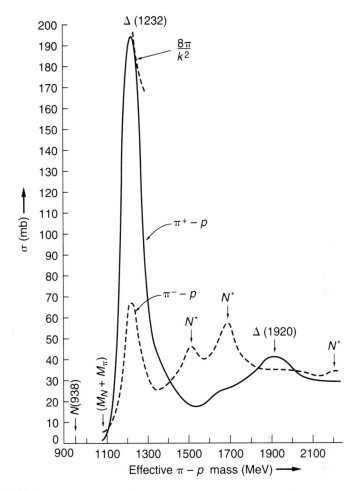

FIGURE 13.10
Total cross sections for $\pi^{\pm}p$ scattering. The N^*'s are $I = \frac{1}{2}$ states and so do not appear in $\pi^{+}p$ scattering.

We have learned (in Chapter 3) that the Clebsch-Gordan coefficients relate the $|1,-1\rangle|\frac{1}{2},+\frac{1}{2}\rangle$ (the $\pi^{-}p$ state) and the $|1,0\rangle|\frac{1}{2},-\frac{1}{2}\rangle$ (the $\pi^{0}n$ state) to the $I = \frac{3}{2}$ and the $I = \frac{1}{2}$ states:

$$|\tfrac{3}{2},-\tfrac{1}{2}\rangle = C^{\frac{3}{2}\ \ 1\ \ \frac{1}{2}}_{-\frac{1}{2}\ -1\ +\frac{1}{2}}|1,-1\rangle|\tfrac{1}{2},+\tfrac{1}{2}\rangle + C^{\frac{3}{2}\ \ 1\ \ \frac{1}{2}}_{-\frac{1}{2}\ \ 0\ -\frac{1}{2}}|1,0\rangle|\tfrac{1}{2},-\tfrac{1}{2}\rangle$$

and

$$|\tfrac{1}{2},-\tfrac{1}{2}\rangle = C^{\frac{1}{2}\ \ 1\ \ \frac{1}{2}}_{-\frac{1}{2}\ -1\ +\frac{1}{2}}|1,-1\rangle|\tfrac{1}{2},+\tfrac{1}{2}\rangle + C^{\frac{1}{2}\ \ 1\ \ \frac{1}{2}}_{-\frac{1}{2}\ \ 0\ -\frac{1}{2}}|1,0\rangle|\tfrac{1}{2},-\tfrac{1}{2}\rangle, \qquad (13.44)$$

where $C^{I\ \ I_\pi\ \ I_N}_{M\ M_\pi\ M_N}$ are Clebsch-Gordan coefficients. The upper state $(I = \frac{3}{2})$ should have a resonance at $E = E_R = 1232$ MeV — the Δ^0. The Clebsch-Gordan coefficients are numbers which have been calculated and tabulated, as discussed in Chapter 3. The values of those appearing in Eq. (13.44) may be found in Mathematical Table II (at the end of this book) to be

$$C^{\frac{3}{2}\ \ 1\ \ \frac{1}{2}}_{-\frac{1}{2}\ -1\ +\frac{1}{2}} = \sqrt{\tfrac{1}{3}} \qquad C^{\frac{3}{2}\ \ 1\ \ \frac{1}{2}}_{-\frac{1}{2}\ 0\ -\frac{1}{2}} = \sqrt{\tfrac{2}{3}}$$

$$C^{\frac{1}{2}\ \ 1\ \ \frac{1}{2}}_{-\frac{1}{2}\ -1\ +\frac{1}{2}} = -\sqrt{\tfrac{2}{3}} \qquad C^{\frac{1}{2}\ \ 1\ \ \frac{1}{2}}_{-\frac{1}{2}\ 0\ -\frac{1}{2}} = \sqrt{\tfrac{1}{3}}. \tag{13.45}$$

The equations in Eq. (13.44) can be solved to express the product states as a linear combination of the states on the left-hand side. This yields

$$|\pi^-\rangle|p\rangle = |1,-1\rangle|\tfrac{1}{2},+\tfrac{1}{2}\rangle = \sqrt{\tfrac{1}{3}}|\tfrac{3}{2},-\tfrac{1}{2}\rangle - \sqrt{\tfrac{2}{3}}|\tfrac{1}{2},-\tfrac{1}{2}\rangle$$

$$|\pi^0\rangle|n\rangle = |1,0\rangle|\tfrac{1}{2},-\tfrac{1}{2}\rangle = \sqrt{\tfrac{2}{3}}|\tfrac{3}{2},-\tfrac{1}{2}\rangle + \sqrt{\tfrac{1}{3}}|\tfrac{1}{2},-\tfrac{1}{2}\rangle. \tag{13.46}$$

Let us now examine the prediction for $\pi^- + p \to \pi^- + p$, from this decomposition into $I = \frac{1}{2}$ and $I = \frac{3}{2}$ states. Utilizing the expressions in Eq. (13.46), we find

$$_{out}(\langle p|\langle \pi^-|)(|\pi^-\rangle|p\rangle)_{in} = \tfrac{1}{3}M_{\frac{3}{2}} + \tfrac{2}{3}M_{\frac{1}{2}},$$

where M_I represents the amplitude for the Ith isotopic spin part: [15]

$$M_I = {_{out}}\langle I_Z, I|I, I_Z\rangle_{in}. \tag{13.47}$$

The amplitude for $\pi^- + p \to \pi^0 + n$ may be obtained in the same way:

$$_{out}(\langle n|\langle \pi^0|)(|\pi^-\rangle|p\rangle)_{in} = \tfrac{\sqrt{2}}{3}M_{\frac{3}{2}} - \tfrac{\sqrt{2}}{3}M_{\frac{1}{2}}. \tag{13.48}$$

In this notation, the amplitude for $\pi^+ + p \to \pi^+ + p$ is

$$_{out}(\langle p|\langle \pi^+|)(|\pi^+\rangle|p\rangle)_{in} = 1M_{\frac{3}{2}}, \tag{13.49}$$

since $\pi^+ p$ is a pure $I = \frac{3}{2}$ state.

We have learned that

$$\sigma_{fi} = K|M_{fi}|^2,$$

where K is determined by the energy and the phase space available,

$$\text{and } M_{fi} = S_{fi} = {_{out}}\langle f|i\rangle_{in}. \tag{13.50}$$

---

15. States with different I values are orthogonal, and the M_I can be shown to be independent of I_z.

(See Chapter 2.) K has the same value in all these reactions, because the final (and initial) particles in each are one pion and one nucleon. For a resonance in the $I = \frac{3}{2}$ state, $M_{\frac{3}{2}} \gg M_{\frac{1}{2}}$. Neglecting $M_{\frac{1}{2}}$, we obtain

$$\sigma_{\pi^+ p \to \pi^+ p} : \sigma_{\pi^- p \to \pi^- p} : \sigma_{\pi^- p \to \pi^0 n} = (1)^2 : \left(\tfrac{1}{3}\right)^2 : \left(\tfrac{\sqrt{2}}{3}\right)^2 = 1 : \tfrac{1}{9} : \tfrac{2}{9}. \tag{13.51}$$

At energies near $E_R = 1232$ MeV, other channels may be neglected, so we find that the total cross sections have the ratio

$$\sigma_{\pi^+ p} : \sigma_{\pi^- p} = 1 : \left(\tfrac{1}{9} + \tfrac{2}{9}\right) = 3 : 1, \tag{13.52}$$

where $\sigma_{\pi^- p}$ includes both final channels ($\pi^- p$ and $\pi^0 n$). Looking at the data in Fig. 13.10, we see this factor of 3 in the comparison of the total cross sections at $E = E_R = 1232$ MeV. (It is not exact since some nonresonant scattering is included in the data.)

By studying the angular distribution of the final particles, i.e., from the differential cross section $d\sigma/d\Omega$, we can determine the phase shifts. By performing the scattering over a range of energies, the energy dependence of the phase shifts is determined. It has been found that $\delta_{\frac{3}{2}}$ goes through $\pi/2$ at $E = E_R = 1232$ MeV, verifying that the Δ is a true resonance with $J = \frac{3}{2}$.

We have learned here, by example, that careful analysis of a peak in a cross section is required to determine the existence of a resonance or unstable particle. The Δ was among the first strongly decaying, short-lived particles to be discovered (in 1949). Today we understand **the Δ as an $I = \frac{3}{2}$, $J = \frac{3}{2}$ bound state of three quarks**, just as **the nucleon is an $I = \frac{1}{2}$, $J = \frac{1}{2}$ bound state of three quarks**.

Resonances in a Subset of the Final Particles — The Discovery of the ρ (Vector Meson)   In high-energy scattering experiments, the final state is often quite complicated because of the production of many particles. When the products of the decay of a short-lived particle are only a *subset* of the final particles, the analysis of resonances described above is not appropriate. An analysis of subsets of the final particles, though, can reveal that a very short-lived particle *was* present which decayed rapidly into a subset. In order to search for short-lived particles, we must know how to determine that a subset of the observed particles has resulted from the decay of such a particle. As an example of this kind of analysis, we will study the discovery of the ρ, which was the first boson resonance discovered (1961). At the time, the nucleon and pion were believed to be fundamental particles and the strong interactions were thought to be generated by the basic vertex of a nucleon current interacting with the pion field, as discussed in Chapter 6. The discovery of this $\pi\pi$ resonance was the first indication that the strong interactions were more complicated. The mass of the ρ is $M_\rho \approx 770$ MeV and its width is $\Gamma \approx 150$ MeV, so that it is also one of the very short-lived particles ($\tau \approx 10^{-24}$ s) which decays via the strong interaction.

The reactions studied were inelastic channels in $\pi^- p$ scattering:

$$\pi^- + p \to \pi^+ + \pi^- + n$$

$$\pi^- + p \to \pi^- + \pi^0 + p. \tag{13.53}$$

The pairs of pions in the final state were studied. It was supposed that they had been produced by the decay of a short-lived particle, and the consequences of such a decay were compared with the actual (experimental) results.

Let us examine the decay of an unstable particle of mass M^* into two particles of masses m_1 and m_2. The rest frame of the unstable particle is the center-of-momentum (CM) frame of the two emerging particles, by conservation of momentum:

$$\vec{p}_1^{\,CM} + \vec{p}_2^{\,CM} = 0. \tag{13.54}$$

Conservation of energy yields

$$M^* = \sqrt{(p^{CM})^2 + m_1^2} + \sqrt{(p^{CM})^2 + m_2^2}, \tag{13.55}$$

where we have dropped the subscript on the momenta, since they are equal in size. An unstable particle has a width associated with its mass distribution, as we have learned in Section 13.3, and so we expect the sum of the energies in the CM frame of its decay products to form a peaked distribution around M^*. (Actually, the total CM energy is Lorentz invariant and can be expressed in a covariant form; its square is called s.[16] Therefore, we can plot the number of pairs vs. $\sqrt{s}$, calculated in *any* Lorentz frame, and look for peaks.) We also should be aware that the main reaction being observed ($\pi^- p$ scattering in the above) can produce the two pions directly, so we expect to see that contribution as well.

In Fig. 13.11 we show the data collected on the effective mass of $\pi^+ \pi^-$ pairs in $\pi^+ p$ scattering to a particular final state. (The resolution is $dE = 0.05$ GeV.) In our discussion of phase space, we emphasized the fact that we must consider the shape of the phase-space factors when searching for resonances. Notice that a phase-space curve is drawn on the data; it tells us what to expect for the nonresonant production of the two π's. Two clear peaks appear above the prediction for nonresonant production of the pair of pions; the first of them is the ρ^0. (The prediction marked phase space lies above the observed number of counts, for masses larger than 1.5 GeV, which does raise the question of the efficiency of the detection.[17] However, since the quantum theory is probabilistic, we can attribute the discrepancy to statistical fluctuations. The ρ^0 stands

---

16. In any frame of reference: $\vec{p}^* = \vec{p}_1 + \vec{p}_2$, $E^* = E_1 + E_2$, and $M^* = \sqrt{E^{*2} - (\vec{p}^*)^2}$. As discussed in Appendix E, we can write this in Lorentz-covariant form using the four-momenta p_μ^*, $p_{1\mu}$, and $p_{2\mu}$, where p_μ^* is the four-momentum of the resonance. Conservation of energy and momentum is contained in the equation $p_\mu^* = p_{1\mu} + p_{2\mu}$. Squaring both sides yields $M^{*2} = (p_{1\mu} + p_{2\mu})^2 \equiv s$, which is obviously a Lorentz-invariant quantity.

17. This question would not arise if $|M|^2$ had a sudden drop in the energy region above 1.5 GeV for some dynamic or conservation reason.

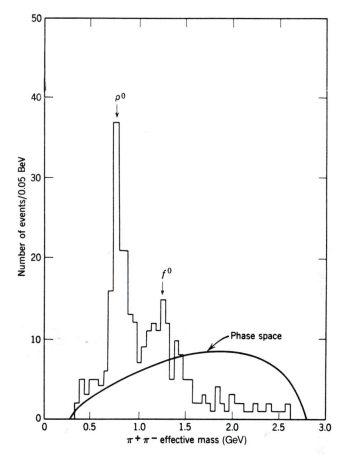

FIGURE 13.11

The mass distribution M^* of the $\pi^+\pi^-$ subset of final particles of the reaction $\pi^+ + p \rightarrow \pi^+ + \pi^- + \Delta^{++}$.

well above that supposed fluctuation.) The ρ (and the f^0 appearing in the figure as well) have been found in many other pion subsets (see Fig. 13.12) and so are well-established resonances.

The discoveries of the Δ and the ρ, as well as many other resonances, were not predicted by any theoretical structure. In fact, their discoveries and the determinations of their properties provided crucial clues for the eventual development of the standard model.

Some of the *fundamental* particles, such as the W's, Z, and τ, are also found in this indirect manner of resonance analysis. They decay via the weak interaction, but because of their large masses, they have many channels of decay; consequently, they decay rapidly. We shall discuss their respective discoveries in Chapter 15.

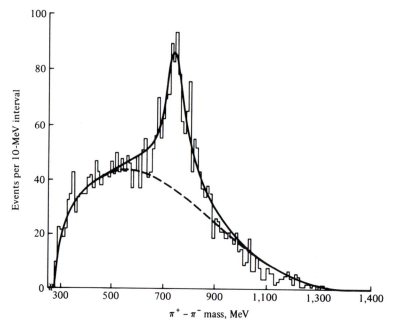

FIGURE 13.12

Effective mass (M^*) distributions of final $\pi^+\pi^-$ pairs in the reaction $\pi^+ + p \rightarrow \pi^+ + \pi^+ + \pi^- + p$. The peak is at $M^* \approx 765$ MeV. The dashed curve is due to the phase-space factors alone.

PROBLEMS

1. Suppose that, while blindfolded, we flash a laser gun (which produces a very narrow beam) at a wall of area 15 m^2 and try to hit a target somewhere on the wall. We flash the laser 2 times per second randomly at the wall. After 1 hour, 18 hits have been recorded. Find the flux (Φ) of flashes, the number of hits per second ($\mathfrak{R}$), and the size of the target?

2. A beam of particles (call them type x) enters a slab of atoms (call them type Y) in a cubic array with a lattice spacing of 2A°. The thickness of the slab is 10 cm. The incident flux is 10^{10} particles/m$^2 \cdot$s and the emerging flux is 5×10^9 particles/m$^2 \cdot$s. What is the total cross section for the interaction of the x particles with the Y atoms?

3. The Schrödinger equation for a free particle ($V = 0$) may be written, using spherical coordinates (r, θ, ϕ):

$$\frac{-1}{2m}\left[\frac{1}{r^2}\frac{\partial}{\partial r}r^2\frac{\partial}{\partial r}\psi - \frac{L_{op}^2}{r^2}\psi\right] = \frac{k^2}{2m}\psi,$$

where we have expressed the (kinetic) energy E as $\dfrac{k^2}{2m}$.

For a separated solution of the form $\psi \sim R_\ell(r)Y_\ell^m(\theta,\phi)$, show that, as $r \to \infty$,

$$R_\ell(r) \to N \frac{\sin(kr + a)}{kr} \text{ , where } a \text{ is an arbitrary phase.}$$

(The normalization N is of no significance to us here.)

It may be simplest to let $R_\ell(r) \to u_\ell(r)/r$ and then to show that $u_\ell'' \cong -k^2 u_\ell$.

4. Verify that $f(\theta) = 1/k \sum_\ell (2\ell + 1)f_\ell P_\ell(\cos\theta)$ (Eq. 13.25), by performing the substitutions outlined in the text.

5. Find the value of δ_ℓ, which makes $|f_\ell|^2$ a maximum, and verify that $|f_\ell|^2 \leq 1$.

6. Verify, from Eq. (13.27), that $\ell_{max} = 7 \times 10^{-3}p$ (in MeV/c).

7. a. Substitute Eq. (13.24) for A_ℓ into R_ℓ (Eq. 13.23) and show that the incoming wave part $(\sim e^{-ikr}/r)$ is unaffected by the scattering, i.e., is independent of δ_ℓ. Call it ψ_{in}. Write R_ℓ for the case when there are inelastic channels by multiplying the amplitude of the outgoing wave part $(\psi_{out} \sim e^{+ikr}/r)$ by η_ℓ.

   b. The inelastic cross section may be defined as $\sigma_{inelastic} = \int(|\psi_{in}|^2 - |\psi_{out}|^2)r^2 \, d\Omega$. Show that $\sigma_{inelastic} = \pi/k^2 \sum_\ell (2\ell + 1)(1 - \eta_\ell^2)$.

   c. Show that $\sigma_{elastic} = \pi/k^2 \sum_\ell (2\ell + 1)(\eta_\ell^2 - 2\eta_\ell \cos 2\delta_\ell + 1)$.

   d. Show that $\sigma_{total} \equiv \sigma_{elastic} + \sigma_{inelastic} = 2\pi/k^2 \sum_\ell (2\ell + 1)(1 - \eta_\ell \cos 2\delta_\ell)$ and verify the optical theorem by showing that $\text{Im } f_{elastic}(0) = k\sigma_{total}/4\pi$.

8. For the scattering of a spinless particle from a spherical potential, we have found that we may write $\psi = e^{i\vec{k}\cdot\vec{r}} + f(\theta)(e^{ikr}/r)$ for large r. One way to express the fact that all the particles which enter the scattering region emerge from it, is to demand that **the net flux of particles crossing a large sphere surrounding the scattering region be 0**. Since the particles follow the probability predictions of quantum mechanics, this may be expressed as

$$\int\int_{large\ sphere} \vec{J}_p \cdot \hat{r} \, r^2 \, d\Omega = 0,$$

where $\vec{J}_p = (1/2im)(\psi^*\vec{\nabla}\psi - (\vec{\nabla}\psi^*)\psi)$ is the probability current defined in Eq. (2.12). Show that when the radius of the sphere is taken to ∞, the **optical theorem** may be proven, viz., show that this integral yields

$$\text{Im } f(0) = \frac{k}{4\pi} \int\int_{large\ sphere} |f(\theta)|^2 \, d\Omega = \frac{k}{4\pi} \sigma.$$

It may be easiest to write $d\Omega$ as $d\mu d\phi$, with $\mu = \cos\theta$ running from -1 to $+1$. Note also that only the radial component of $\vec{J}$ is required, so that only radial derivatives appear. You may also need:

$$\lim_{r\to\infty} \int_{-1}^{1} d\mu \, e^{ikr(\mu-1)}g(\mu) = \int_{-1}^{1} d\mu \, e^{ikr(\mu-1)}g(1),$$

which follows from the fact that in (infinitesimal) regions where $\mu \neq 1$, the exponential oscillates infinitely rapidly when $r = \infty$, giving 0 contribution to this integral. (Note that $f(\theta)$ evaluated at $\mu = 1$ is $f(0)$.)

9. The optical theorem is also true in quantum field theory. Show that the optical theorem implies that an amplitude calculated from a tree-level diagram alone has no imaginary part. Hint: Assume that the corresponding amplitude is proportional to g^n.

10. Show that the rate of decay of a state with time behavior $|\psi|^2 \sim e^{-t/\tau}$ is proportional to $1/\tau$.

11. Find τ for $\Gamma = 100$ MeV; for $\Gamma = 0.1$ MeV.

12. Explain why the phase-space curves in Figs. 13.11 and 13.12 for the $\pi^+\pi^-$ effective mass are:

    a. 0 below a certain value of the effective mass, and explain what that value of the effective mass should be and

    b. 0 above a certain value of the effective mass.

13. a. Should we expect to see the ρ resonance in the study of $\pi^+\pi^+$ pairs? $\pi^-\pi^-$? Explain.

    b. If we could perform $\pi\pi$ scattering experiments, what would the following ratios be?

$$\sigma_{\pi^+\pi^-\to\rho} : \sigma_{\pi^0\pi^0\to\rho} : \sigma_{\pi^+\pi^0\to\rho} : \sigma_{\pi^-\pi^0\to\rho} : \sigma_{\pi^+\pi^+\to\rho} : \sigma_{\pi^-\pi^-\to\rho} = ?$$

    Hint: Look up the relevant Clebsch-Gordan coefficients in Mathematical Table II (at the end of the book).

Chapter 14

Experimental Devices and Facilities

14.1 INTRODUCTION

From the modest sizes and scale of the apparati and experiments of the 1930s, experimental devices and facilities have grown so in scale and complexity that they require major involvement of the federal and state governments. Detectors as large as office buildings and accelerators the size of major highways are required. Enterprises of this size are often referred to as "Big Science." There are a number of other areas of science which also require huge sums of money and vast resources, as we shall mention before the end of this chapter. Let us begin by outlining the major types of equipment used and sketching their general features.

14.2 DETECTORS

Many of the detectors used today are very large, composite structures. Here we will survey the basic features only of simple detectors. Those who wish to know more about modern-day detectors will find a more thorough treatment in Appendix H.

14.2.1 Particle Interactions Employed in Detectors

To be detected, a particle must interact with the detector material. This interaction necessarily removes some energy from the particle, but a good detection system is designed so that it will not distort the measurement. Let us list the interactions which occur most often:

1. *Ionization.* Charged particles and photons ionize atoms in the detector material. The distance a charged particle covers before it stops will tell us its energy, as will dE/dx (the energy deposited per unit length). Trails

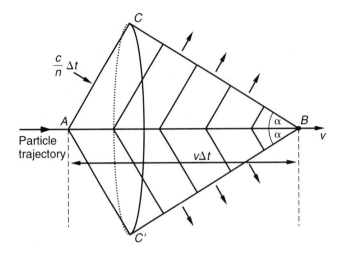

FIGURE 14.1
Čerenkov radiation from a charged particle which has moved from A to B. The wave-front is a cone of opening angle α.

of ions can be studied to determine the paths and, from them, the momenta of the particles which produced them, as explained below. The value of dE/dx can often be used, in conjunction with the momentum, to determine particle type by employing standardized dE/dx vs. p curves.

2. *Čerenkov radiation.* When, in a dielectric medium, a charged particle moves faster than light does in that medium, $v > c/n$, a wake of light is emitted by the particle. The geometry involved is shown in Fig. 14.1. In a time Δt the particle has moved from point A to point B, which is the distance $v\Delta t$. In that same time interval the light emitted at A travels a distance $(c/n)\Delta t$. The particle has been continuously emitting light during its journey to B. The wavefronts of the emitted light are perpendicular to the rays, as shown in Fig. 14.1. Simple trigonometry determines the wake angle α to be: $\alpha = \sin^{-1}[(c\Delta t/n)/v\Delta t] = \sin^{-1}(1/\beta n)$. The angle of the wake then determines the speed of the particle. (Notice that for $v > c/n$, $1/\beta n < 1$ so that the expression for α does make sense.) Since the index of refraction (n) depends on frequency,[1] so does the angle α. A commonly used dielectric medium is water, which has $n \approx \frac{4}{3}$ for visible light. Thus, any particle moving through water with a speed $v > \frac{3}{4}c$ will produce a wavefront of light in the visible region.

3. *Coulomb scattering.* When a charged particle moves through a detector, it is scattered by the nuclei and electrons of the detector material. This virtually random Coulomb scattering can spoil the measurement,

1. Recall dispersion.

especially the directional measurement. Momentum determination is often made by measuring the curvature of a track in a magnetic field (as discussed below); thus, Coulomb scattering can interfere with momentum resolution as well.

4. *Absorption of γ-rays.* There are many different processes that can occur when a photon interacts with matter:
   (i) For $E_\gamma < 1$ MeV, there is the photoelectric effect in which the photon is absorbed and the absorber is consequently excited.
   (ii) For 1 MeV $< E_\gamma < 10$ MeV, Compton scattering dominates over the photoelectric effect. The photon scatters with a resulting shift in its energy (or frequency).
   (iii) For $E_\gamma > 10$ MeV, pair production is the dominant process. The photon produces an electron-positron pair.

5. *Bremsstrahlung.* Charged particles will emit photons when accelerated by Coulomb interactions, primarily with the nuclei of the detector material. This radiation is copiously emitted by high-energy ($E > 1$ MeV) electrons.

   High-energy e^-'s and γ's produce *cascade showers* consisting of bremsstrahlung and pair-production events, as shown in Appendix H.

6. *Scintillation.* The electromagnetic field of a fast-moving, charged particle can eject electrons from inner shells of the atoms of some materials. Electrons in higher-energy states will subsequently make transitions to these empty shells with the concomitant emission of characteristic photons. These photons can then be detected in various ways. Scintillation material is used in many high-energy detectors, such as calorimeters and scintillation counters, which will soon be described. (Also see Appendix H.)

14.2.2 Elementary Detectors

1. Bubble Chamber Bubble chambers played a major role in many of the discoveries we have discussed. Even though they have been superceded by other devices and are no longer used in high-energy experiments, we shall take a brief look at them. Charged particles and photons leave tracks of bubbles in a large, liquid-filled chamber. The tracks are photographed by several cameras, so that a three-dimensional reconstruction of the tracks can be made. The chamber contains a pure liquid which is superheated[2] by a sudden expansion. If an ionizing particle moves through the superheated liquid, it leaves a trail of ions which serve as centers for bubble formation. Thus, the path of the particle is marked by a trail of bubbles. The bubbles are ~ 10 μm in size and last for between 10 and 50 ms, so that they are easy to photograph with a suitable flash. The recom-

---

2. "Superheated" means that it is at a temperature above its boiling point. A sudden expansion accomplishes this by lowering the pressure so rapidly that it is still at its original temperature, which is higher than the boiling point of the liquid at that reduced pressure.

pression of the liquid and subsequent sudden expansion can be achieved in ~1 s so that the chamber can be used to study a large number of events.[3] Liquid hydrogen is frequently used in the chambers, the protons serving as target particles. Light liquids like hydrogen have small Coulomb-scattering errors. Heavy liquids (like freon) are used for special purposes, since many more particles will stop in them. (They are good for the detection of e^-'s and γ's.)

A magnetic field is set up in most bubble chambers. Charged particles move in a helical path around the field lines, and a three-dimensional (computer) reconstruction of the track from multiple photographs (from different directions) allows a determination of the momentum of the particle.

Computer programs[4] are employed to select events of interest and to decide what kinds of particles produced the various tracks as well as what non-ionizing (neutral) particles were emitted. These programs usually have built-in hypotheses concerning the reactions that are based upon theoretical considerations. Of course, these programs require conservation of energy, momentum, and angular momentum.

In Fig. 14.2 we show a picture of a typical large bubble chamber, to acquaint the reader with its size and complexity.

2. Scintillation Counters   Certain materials such as ZnS emit visible light when hit by a charged particle or photon. In his classic experiments, Rutherford used a ZnS screen and a couple of people, who were watching it in a dark room and counting flashes. Each observer was good for less than an hour, the story goes, so that frequent changes of the person part of the apparatus had to take place.[5] Needless to say, this business has been automated by now, with phototubes (and counters) replacing the human detectors. In present-day scintillation counters, ZnS has been replaced with other materials, both inorganic and organic.

3. Čerenkov Counters   These counters employ the Čerenkov effect to determine the velocities of particles. Rarely will the particle be moving directly at the photomultipliers, so that the cone of light emitted will be received at an angle. The actual reconstruction of the particle's track has turned out to be quite complicated. (In Ring Imaging Čerenkov Counters (RICH), sophisticated imaging techniques have been developed to aid in this reconstruction.) More typi-

---

3. This repetition rate is considered low by modern standards. Also, there is no ability to do on-line analysis (while the experiment is proceeding) of events. The photographs must be analyzed after the fact to search for events of interest; usually, most pictures are discarded (as uninteresting).

4. An amusing anecdote from the early days (1950s) concerns the fact that each picture contains numerous events. Those were the days before computers had been perfected, and the scanning of pictures for events of interest was done by hand (by people!). It was quickly found that it was much more efficient to hire people who knew nothing about physics and to teach them to search for certain patterns of tracks, than to employ physicists at the same task. This was because the physicists knew what many of the other events were and wasted time *gazing* at them, severely slowing down the scanning. So I was told, by someone who was there.

5. When the technician came into the room to change the source, the observers would go into a closet to keep their eyes dark adapted.

FIGURE 14.2
Argonne National Laboratory bubble chamber (no longer in existence). The volume of
the chamber was 25 m³. It contained about 20,000 liters of liquid hydrogen and used a
superconducting magnet to produce a magnetic field of about 1.8 Tesla.

cally, Čerenkov counters are used as threshold indicators of particle type—at
a given momentum, a pion will be moving fast enough to produce light, whereas
a (heavier) kaon will not.

4. Spark Chambers and Streamer Chambers   A series of plates, with large
potential differences between neighbors, are employed with a gas such as neon
or argon. When a charged particle goes through the detector, the ions produced
cause sparks between the plates. These sparks cause the gas to emit visible light,
so that a visible track is produced. When very short (~10 ns), high-voltage pulses
are used, short sparks are produced and the device is called a streamer chamber. These devices, are rarely used any more because it takes time to clear the
chambers of charge, thus, limiting the rate of data acquisition.

5. Proportional Counters and Drift Chambers   A proportional counter consists of a gas-filled metallic tube at a negative potential with respect to a fine
wire in the center. When a charged particle or photon enters the tube, ions are
produced in the gas. The liberated electrons rush towards the anode wire while

the ions move towards the outer tube. For a suitably large potential difference (but below the ionization potential of the gas) between the anode and cathode, the secondary electrons and the ions reach kinetic energies high enough to cause further ionization of the gas. This leads to an avalanche of electrons and ions, which produces a measurable current. The potential may be chosen so that the current is proportional to the initial ionization caused by the incoming particle, hence, the name "proportional counter."[6] The ionization information is used to determine the energy of the ionizing particle. Multiple wire proportional counters, drift chambers (in which the electrons drift in a low field before approaching the anode wire), and vertex time projection chambers are elaborate, ingenious variations of the basic venerable proportional counter. See Appendix H for a discussion of these detectors.

6. Calorimeter   This is a generalization of the word originally used for devices which measure heat. Here it is a device which measures the total energy deposited by a particle or jet. These devices have very good resolution at high energies and are very fast, so that on-line computers can use their readings to decide quickly whether to trigger and read out the event. There are electromagnetic calorimeters for photons or charged leptons and hadronic calorimeters for individual hadrons or jets. A typical calorimeter might be made of passive absorber plates (Pb or Fe) with active scintillator material sandwiched between them to read the energy deposited. Monte Carlo simulations (to be discussed) are often used to interpret the data. A detailed discussion of electromagnetic calorimeters and hadron calorimeters may be found in Appendix H.

This is only a quick survey of the types of devices most frequently used. A more complete discussion may be found in Appendix H. Many devices with specialized applications exist, and new devices are continually being created as needed.

14.2.3   Large Hybrid Detectors

Large hybrid detectors have been constructed that incorporate many of the above-mentioned detectors in creative, cooperative ways. They are expensive, complex structures which take years to build. There exist about two dozen such large hybrid detectors in the world today, and they themselves are major facilities. Indeed, they dwarf the first accelerators and their detectors combined. These detectors require large numbers of physicists to design, construct, and operate them as well as to analyze their results, explaining why journal articles often have long lists of authors from several institutions (sometimes ~100

---

6. Geiger counters have this same basic structure; however, larger potentials are used, for which the avalanche is independent of the initial ionization (called the plateau region of the potential). A simple resistor in the outer circuit is often used to cause a drop in the potential, thus, quenching the avalanche. The result is a current pulse for each particle which enters the tube. These pulses are then counted to measure the radioactivity of a sample.

FIGURE 14.3
The CDF (Collider Detector at Fermilab). The beam tube of the accelerator goes through the center when this detector is in place. The calorimeter arches have been pushed to the sides for maintenance.

authors). Each of the four detectors at LEP/CERN[7] (ALEPH, DELPHI, OPAL, and L3) involve hundreds of physicists from dozens of institutions. The names of the large hybrid detectors are usually acronyms, and in Table 14.1 we include a list of those most frequently mentioned.

In Fig. 14.3 we show a picture of the Collider Detector at Fermilab (CDF), which has been characterized as "a 5,000-ton Swiss watch,"[8] because of the precision of its construction and the precise measurements it makes (10,000 of them per second). Note the size and complexity of this device. For a detailed "tour" of this huge detector, see Appendix H.

---

7. We will soon describe the accelerator facilities.

8. This characterization was made by Leon L. Lederman in "The Tevatron," Scientific American, March 1991, p. 48.

TABLE 14.1
Major Detectors

Detector	Accelerator/Facility†
UA1 (Underground Area 1)	$\bar{p}p$ collider (S$\bar{p}p$S)/CERN
UA2 (Underground Area 2)	$\bar{p}p$ collider (S$\bar{p}p$S)/CERN
ALEPH (Apparatus for LEP PHysics)	LEP/CERN
DELPHI (DEtector with Lepton, Photon and Hadron Identification)	LEP/CERN
OPAL (Omni-Purpose Apparatus for LEP)	LEP/CERN
L3	LEP/CERN
ARGUS	DORIS/DESY
ZEUS	HERA/DESY
H1	HERA/DESY
CDF (Collider Detector at Fermilab)	Tevatron Collider/Fermilab
D0	Tevatron Collider/Fermilab
MARK III	SPEAR/SLAC
TPC	PEP/SLAC
MARK II	SLC/SLAC
SLD	SLC/SLAC
CUSB	CESR/Cornell
CLEO	CESR/Cornell
TOPAZ	TRISTAN/KEK
VENUS	TRISTAN/KEK

†See Tables 14.2 and 14.3.

14.2.4   Monte Carlo Programs

Elaborate computer programs are written that contain portions representing each piece of equipment in these enormously complicated experiments. The programs generally contain some sort of random number generator to simulate the distribution of particles in the initial beam. The particle distributions through each part of each detector can be studied to aid in the design of the experimental setup and to check on the performance of actual pieces of equipment in an experiment already running. For example, it was a Monte Carlo study which determined that the width of the upsilon was probably much smaller than the raw data indicated, as we shall see in Chapter 15. Yes, these programs *are* named after that famous European gambling mecca.

14.2.5   Particle Identification

There are many methods of particle identification; we now list those most often used.

1. Measurements of both p and v yield the mass of the particle. The momentum p can be measured from the particle's trajectory in a mag-

netic field or by use of conservation of energy and momentum at the vertex of its production or decay. The speed may be determined by use of Čerenkov counters or time-of-flight measurements (for lower speeds). Usually, the value of its mass is enough to identify the particle.

2. A measurement of the energy of the particle and either p or v can likewise be used to determine its mass.

3. As we have already mentioned, the dE/dx vs. p curves can often be used to determine particle type. See Appendix H for details.

4. We can determine the lifetime of an unstable, charged particle from (i) the length of its track from production to decay, if it lives long enough to leave a track, or (ii) the widths of the resonances occurring in its decay products (as discussed in Chapter 13).

5. For unstable, neutral particles which are produced and decay in the detector, we can use the gap length as its path length or (just as for charged particles) we can study its decay products.

6. The penetration, or lack of penetration, of dense absorbers (e.g., steel) can sometimes be used for particle identification. The most common example is muon (vs. electron or pion) identification; the muon is a very penetrating particle that will go through thick layers of steel, which stop electrons, pions, and other strongly interacting or charged particles.

7. We can study the interaction of a neutral particle with matter that leads to particle creation (sometimes referred to as the creation of secondaries).

14.2.6 Problems

The usual problems of making equipment work properly and eliminating spurious noise is ever present. Once this is accomplished, there remain the following concerns.

1. Triggering An enormous number of events occur in the detectors, and it is physically impossible to measure and record them all. Most must be ignored (typically less than 1 in 10^5 are of interest) or not measured at all, and this is often accomplished by using *triggers*. Quick-response elements of the detector are employed to decide whether to turn on the entire detector, which will then measure and record the details of the event. The parts of the device that make these rapid preliminary readings along with the computer programs that make these decisions are usually referred to as triggers. An example is the use of scintillators with fast response times, to determine whether a charged particle has gone through a chamber (from a particular region), and computer software, which then decides whether to apply the voltages across the plates in the chamber and record the track.

The decisions, which veto the recording of events, must be built into the computer programs. These programs are integral parts of the triggers. The physicists who write these programs use our understanding of the interactions under investigation and our expectations of what is a reasonable result, to throw out uninteresting events. That is the problem! Perhaps there is something wrong with our understanding in the realm of energies (or other properties) under study, and we are throwing away evidence of completely unknown, unexpected properties of nature. This possibility is always kept in mind (we trust), so that the overwhelming number of uninteresting events *can* be removed but not the new or "bizarre." In the design of triggers and the study of candidate events, the "chemistry of intuition"[9] of the experimenters plays a significant, though elusive role.

2. Backgrounds   Any other events, which will produce tracks just like those of the reaction under study, are called background events. Knowledge of detailed probability distributions (for certain variables) in the background events is used to cut out large amounts of data, so that the data remaining is rich in the desired reaction. A good understanding of the background events is important, since the final results are interesting only after the expected number of background events has been subtracted. Sometimes a combination of tracks form a distinctive **signature** for the events sought (which implies a low background), e.g., a μ and an e in the final state was the signature used to discover the τ, as we shall discuss in Chapter 15.

3. Counting Rates   For some experiments the expected counting rates may be so low that a graduate student, let alone his or her thesis advisor, may not have the fortitude to wait for the data to be recorded. Counting rates of the order of one per year(s) are *usually* not acceptable (unless someone is willing to devote his or her life to that experiment), and great effort is made to revise the experiment to increase the number of events. Low counting rates also run into the problem of competition with and probable "swamping" by background events.

4. Decrease of σ with Energy   Recall from Eq. (13.43) that the maximum cross section at a resonance is proportional to $1/k^2$. This came from general considerations and so is a predominant feature of most exclusive cross sections of interest. At high energies $E \approx k$, therefore, we expect

$$\sigma_{exclusive} \sim \frac{1}{k^2} \sim \frac{1}{s} \text{ at high energy.} \tag{14.1}$$

This means that when a new realm of (higher) energy is to be explored, the cross sections to particular channels will be significantly smaller. The next machines and experiments must be designed to give acceptable counting rates in spite of this inevitable precipitous decrease in these cross sections. (Further discussion of this crucial question must await our discussion of colliders in Section 14.3.3.)

---

9. This expression was used by L. Lederman, while director of Fermilab, in the NOVA program "Race for the Top." See Chapter 15.

5. *Rumors* Experimenters must avoid letting their experimental decisions or analyses be influenced by unsubstantiated rumors about what other groups have found. Is this really a problem? Not seriously, although considerable anxiety is sometimes aroused!

14.3 ACCELERATORS

14.3.1 Why High Energies?

In order to study high-energy scatterings, larger and larger accelerators have been built. Let us first ask: Why high energies? The most important answers to this question are:

1. Since $\lambda_{De\ Broglie} = h/p$, high p effectively probes short distances.
2. High energy is needed to create and study the gauge particles, W's and Z, of the standard model.
3. There are still missing pieces of the standard model: The top quark and the Higgs particle are yet to be found. Higher-energy searches are necessary.
4. With high energies, experimenters can *try* to produce free quarks and gluons. See the discussion of the quark-gluon plasma in Chapter 16.
5. Copious beams of π, K, ν, etc., can be made to perform detailed lower-energy studies.
6. Perhaps beams of short-lived particles, e.g., hyperons, heavy mesons, etc., can be produced.
7. Completely new interactions and particles can be sought in new realms of energy, viz., *it is an exploration of the unknown!*

14.3.2 Fixed-Target Machines

Cyclic Accelerators Charged particles repeatedly traverse a circular or oval path. They are periodically given a small "kick" and after many cycles of the path have gained a lot of energy. A magnetic field is used to keep the particles on the "circular" path (there often are straight sections in which no field is needed). The cyclic, high-energy machines in use today are synchrotrons. Both the magnetic field keeping the particles in the correct path and the rf field providing the acceleration must be kept in synchronization with the particles as they speed up. The name *synchrotron* results from the fact that these accelerators are designed to automatically synchronize the motion of the particles with the increasing frequency of the rf pulses and the rising magnetic field.

Early Cyclic Accelerators. In order to understand the nature of these machines, let us examine the structure of the first cyclic machines built for particle research.

(i) *The Cyclotron.* In Fig. 14.4 we see a photograph of E. O. Lawrence, the inventor of the cyclotron, with one of his larger cyclotrons. Notice the size of

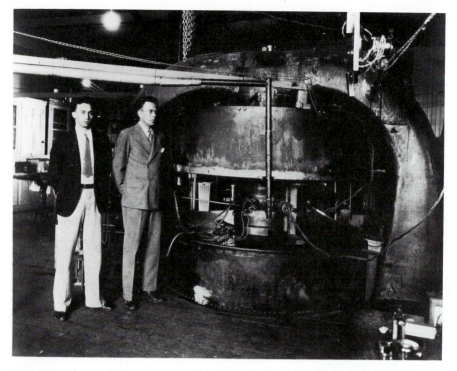

FIGURE 14.4
The $27\frac{1}{2}$ inch "huge" (in 1933) cyclotron, shown with its inventor E. O. Lawrence (right) and M. S. Livingston. It reached an energy of 4 MeV.

that accelerator; keep the image in your mind for comparison with the accelerators of today and the supercollider scheduled to be built in Texas during the 1990s. (The latter, the SSC, will soon be discussed.)

In Fig. 14.5 we see the basic design of the cyclotron—two Dees, back to back, with a gap separating them. A magnetic field is applied orthogonal to the plane, as shown. Particles from a source in the center start at low speed, and their paths in the magnetic field are arcs of circles (ignoring any motion along the B-field direction). Let us suppose for simplicity that the particles are bunched together, so that they cross the gap between the Dees at the same time. An electric voltage is applied across the gap with the correct polarity, so that there is a force on the particles crossing at that time in the direction of their motion. They are then moving faster as they next enter a Dee, and the radius of their motion in the magnetic field increases. In the Dee, they move in a circle and after traveling 180° they come to the gap again, about to cross in the opposite direction. If the polarity of the potential were the same as that during the previous crossing, the particles would be slowed down. Of course, it is arranged that the polarity has reversed, and the particles again receive a kick, so that they

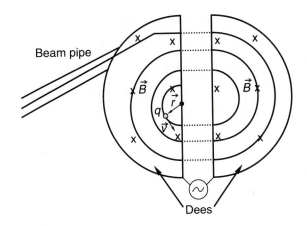

FIGURE 14.5

Sketch of cyclotron showing the path of an ion (q). The X's indicate that the direction of the magnetic field $(\vec{B})$ is into the plane.

move even faster. Thus, the particles go around and around, receiving kick after kick, until their energy is quite high. The magnetic field can then be suddenly changed so as to eject the particles from the accelerator into a beam line. The timing of the polarity reversals and the bunching and focusing of the particles are the crucial aspects which determine the practicality of this device as an accelerator.

The timing of the alternation of the voltage is easily understood. A particle, moving at speed v in a circle of radius r, must be experiencing a centripetal (center-pointing) force of size mv^2/r. The only force on the particles, when they are in the Dees, is the magnetic force qvB. That force must be the centripetal force keeping them moving in a circle, so that

$$qvB = \frac{mv^2}{r}.\tag{14.2}$$

Solving Eq. (14.2) we obtain

$$r = \frac{mv}{qB},\tag{14.3}$$

which shows that the radius increases (linearly) with v. However, the problem of timing the reversals of the electric polarity across the gap turns out to be very simple, for acceleration below relativistic speeds, since

$$\text{the period of repetition} \equiv T = \frac{2\pi r}{v} = \frac{2\pi m}{qB} = \text{a constant.}\tag{14.4}$$

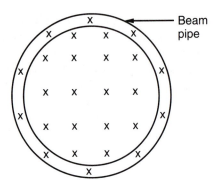

FIGURE 14.6
Betatron.

Therefore, a simple AC voltage with frequency $f = T^{-1} = qB/2\pi m$ is all that is needed. (Equation 14.2 must be modified at relativistic speeds, so that for $v \to c$ some synchronization of the frequency is required.)

(ii) *The Betatron.* In Fig. 14.6 we show a simplified sketch of the betatron. The particles are kept in the beam pipe by a magnetic field which is increased as the speed of the particles increases. This field need only extend over the small region of the doughnut-shaped beam pipe. The acceleration is achieved by applying a changing magnetic field in the region encircled by the beam pipe. An induced ℰℳℱ along the circular beam path accelerates the particles:

$$\mathcal{EMF} = \frac{d\Phi}{dt} = \int \frac{dB}{dt}\, dA. \qquad (14.5)$$

Synchrotrons. As experiments were done at higher and higher energies, larger and larger accelerators were employed. It became not only impractical but impossible to use either the cyclotron or betatron designs, since they require large magnetic fields over large regions of space. Today's accelerators, called **synchrotrons**, employ features of both of these early devices to avoid this problem.

Synchrotrons employ the narrow, doughnut-shaped evacuated beam pipe of the betatron, which requires a magnetic field (B) only in the region of the pipe; the radius of the motion is kept constant by increasing B appropriately. This magnetic field is set up by a large number of dipole magnets positioned around the ring. For very high energy machines, such as the Tevatron at Fermilab, superconducting bending magnets are used to provide the large magnetic field ($\sim$5 Tesla)[10] needed. Because of the narrowness of the beam pipe, focus-

---

10. A Tesla is a very large unit of magnetic field: 1 Tesla = 10^4 Gauss. The Earth's magnetic field is $\sim$0.5 Gauss.

ing is a crucial requirement to avoid loss of the beam by collision with the walls; the intense beams could even damage the walls of the beam pipe. For the larger synchrotrons a method of strong focusing has been devised to supplant the weak focusing that the dipole magnets provide. This is accomplished by using quadrupole magnets which produce stronger focusing than do the dipole magnets. Their fields have large gradients transverse to the beam direction, and the magnets are arranged so that the gradients alternate from one to the next. The quadrupole magnets alternately focus and defocus the beam both vertically and horizontally; a magnet which focuses vertically and defocuses horizontally is followed by one which defocuses vertically and focuses horizontally and vice versa. Consequently, the particles oscillate around the desired orbit, but the amplitude of that oscillation is kept small by the alternation of the direction of the gradient of the field. This oscillation around the desired orbit is referred to as the **betatron oscillation**.

Acceleration is achieved by pumping energy directly into the beam pipe via rf fields; the particle path goes through rf cavities and the charged particles absorb energy from the fields. The particles circulate in bunches, as in the cyclotron, in order to be in phase with the "kicks" as they go through the rf cavities. Typically, a particle arriving late gets a smaller kick, so that it goes into a smaller orbit and arrives earlier at the next rf cavity. Eventually, that particle will be in synchronization with the pulse. Likewise, a particle arriving early gets a greater kick that eventually brings it into synchronization. The particles typically will circle through the accelerator 100,000 times from injection to ejection. Actually, the particles oscillate around the ideal (synchronous) bunching but stay reasonably well bunched. This oscillation around the desired synchronous bunching is referred to as the **synchrotron oscillation**.

The largest accelerators receive particles which have already been accelerated by a succession of smaller ones, and, each can, therefore, be designed to deal with particles in a restricted energy range from injection to ejection. (See Fig. 14.7 showing the succession of accelerators leading to the main ring of the Tevatron at Fermilab.) This is an important advantage for design and operating purposes.

One problem that exists for circular machines is due to what is called **synchrotron radiation**. Charged particles radiate electromagnetic energy when they are accelerated. A particle moving at a constant speed in a circle *is* being accelerated (centripetal acceleration), so that it *does* radiate; this is synchrotron radiation. For a given particle energy, the energy radiated is inversely proportional to the fourth power of the mass. Thus, the energy radiated in an electron synchrotron exceeds that in a proton synchrotron of the same energy by a factor of $(m_p/m_e)^4 = (1836)^4 \approx 10^{13}$. For energies greater than 10 GeV, the synchrotron radiation in an electron synchrotron becomes significant; this was a prime motivation for the construction of a large linear accelerator for electrons near Stanford University in California. It was named SLAC, an acronym standing for $\underline{S}$tanford $\underline{L}$inear $\underline{A}$ccelerator $\underline{C}$enter.

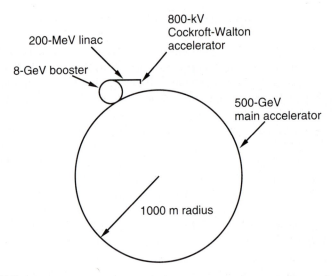

200-MeV linac

800-kV
Cockroft-Walton
accelerator

8-GeV booster

500-GeV
main accelerator

1000 m radius

FIGURE 14.7
Succession of accelerators at the Fermilab synchrotron. The Cockcroft-Walton accelerator is an electrostatic device.

Linear Accelerators (Linacs)   These are simply straight-line accelerators. The most primitive ones employed an electrostatic voltage through which the charged particles accelerated. These electrostatic accelerators could reach several million volts without breakdown (sparking), thus, imparting a few MeV of energy to singly charged particles. Larger linacs use rf cavities to provide the acceleration. Successive cavities are suitably spaced to efficiently pump more energy into the beam particles. Again, bunching and focusing play an important role in design considerations. One of the most celebrated linear accelerators in this last part of the twentieth century is SLAC. It accelerates electrons to about 25 GeV in a two-mile-long path, using 82,560 rf cavities (separated by many drift tubes). An aerial view of this accelerator is shown in Fig. 14.8.

Major Fixed-Target Accelerators   There are a large number of fixed-target accelerators, including many proton synchrotrons and some electron synchro-trons, as listed in Table 14.2. Although they do not reach the highest energies attainable in colliders (to be discussed in Section 14.3.3), fixed-target experiments are indispensable because (i) beams of unstable particles, such as pions and kaons, can be used; and (ii) for many types of experiments, the number of tar-get particles exceeds, by several orders of magnitude, the corresponding num-ber of target particles in a colliding-beam experiment. Consequently, many experiments requiring a greater number of events, either to reduce experimen-tal error or to study rare (low cross section) events, can be done only at fixed-target machines. Notice, from Table 14.2, that the CM energy is much less than

FIGURE 14.8
Aerial view of the Stanford Linear Accelerator (SLAC). The two-mile long linear accelerator tube is the straight structure running from the center of the picture upward (actually passing under an expressway).

the beam energy and grows more slowly than does the latter. This will be explained in detail in Section 14.3.3, where we discuss colliders.

14.3.3   Colliding Beams

Instead of bombarding a stationary target with an accelerated beam, these machines, called **colliders**, accelerate two beams in opposite directions and allow them to collide. The great advantage of colliders over fixed-target machines is the energy available for the production of new particles (or resonances). How-

TABLE 14.2
Major Fixed-Target Accelerators

Accelerator	Date of First Operation	Particles Accelerated	Beam Energy (GeV)	Center-of-Mass Energy (GeV)
PS (Proton Synchrotron) (CERN [Conseil Européen pour la Recherche Nucléaire], Geneva, Switzerland)	1959	protons	28	7.4
AGS (Alternating Gradient Synchrotron) (Brookhaven National Laboratory, New York)	1961	protons	33	8
DESY (Deutches Elektronen-Synchrotron) (Hamburg, Germany)	1964	electrons	7	3.8
SLAC (Stanford Linear Accelerator Center) (Stanford University, Stanford, Calif.)	1966	electrons	22	6.5
Cornell Electron Synchrotron (Cornell University, Ithaca, N.Y.)	1967	electrons	12	4.9
Serpukhov Proton Synchrotron (Serpukhov, Russia)	1967	protons	76	12
Fermilab (Fermi National Accelerator Lab) (Batavia, Ill.)	1972	protons	500	30.7
SPS (Super Proton Synchrotron) (CERN, Geneva)	1976	protons	500	30.7
KEK (Oho-machi, Tsukuba-gun, Japan)	1977	protons	12	5
Tevatron (Fermilab, Batavia, Ill.)	1985	protons	1,000	43
UNK (Serpukhov, Russia)	1990s†	protons	3,000	75

†Proposed

ever, beams are much less dense (by factors $\sim 10^6$) than fixed targets, so that the counting rate and hence the statistics of the data suffer.

Let us first examine the energy comparison between a collider and a fixed-target machine. In Fig. 14.9 we show the fixed target arrangement, where a beam particle of momentum p^L and mass m is about to interact with a target particle of mass M (at rest). To create a single particle or resonance of mass M^* requires more energy than just M^*; the particle created must be moving with momentum p^L because momentum must be conserved. It is simpler, therefore, to look at the reaction in the CM frame, where the created particle would be at rest (since $\vec{p}_{TOTAL}^{CM} = 0$), so that all the energy would contribute to M^*. Let

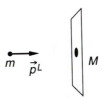

FIGURE 14.9
Scattering in a fixed-target machine.

us calculate the Lorentz-invariant quantity $s = (p_1 + p_2)^2$ in both the CM frame and the Lab frame:

$$E^{\mathrm{CM}2} = s = E^{L2}_{TOTAL} - (\vec{p}^{L}_{TOTAL})^2, \tag{14.6}$$

where the first expression is s evaluated in the CM frame. To just produce (at rest) a particle of mass M^*, E^{CM} must be equal to M^*. Thus, setting $E^{\mathrm{CM}} = M^*$, we obtain

$$M^* = \sqrt{E^{L2}_{TOTAL} - (\vec{p}^{L}_{TOTAL})^2} = \sqrt{(E^{L}_{BEAM} + M)^2 - (\vec{p}^{L})^2}$$

$$= \sqrt{E^{L2}_{BEAM} + 2ME^{L}_{BEAM} + M^2 - (\vec{p}^{L}_{BEAM})^2} = \sqrt{2ME^{L}_{BEAM} + M^2 + m^2},$$

$$(\text{since } E^{L2}_{BEAM} - (\vec{p}^{L}_{BEAM})^2 = m^2). \tag{14.7}$$

At high beam energies, the squared masses are negligible on the right-hand side, so that we may write

$$M^* \approx \sqrt{2ME^{L}_{BEAM}} \qquad \textit{high-energy, fixed-target machine.} \tag{14.8}$$

Alternatively, we may write (from Eq. 14.7):

$$E_{threshold} = \frac{M^{*2} - m^2 - M^2}{2M} \approx \frac{M^{*2}}{2M} \qquad \textit{high-energy, fixed target,} \tag{14.9}$$

where we have written $E_{threshold}$ for the lowest beam energy which can produce the resonance. Notice that the largest[11] mass which can be produced grows as $\sqrt{E^{L}_{BEAM}}$, and $E_{threshold}$ grows as M^{*2}. To produce a resonance of twice the mass of that which can be produced by a given beam energy, we need four times as much energy, and an increase in resonance energy by a factor of 10 requires 100 times as much beam energy. At the highest energies being probed, the latter is impractical!

For colliding beams, assuming the particles in both beams have mass m, the Lab frame *is* the CM frame so that

$$M^* = 2E_{BEAM} \text{ or } E_{threshold} = \frac{M^*}{2} \qquad \textit{collider.} \tag{14.10}$$

---

11. Lower mass resonances can be produced if other particles are produced as well.

Here the threshold for production of a resonance grows linearly with its mass, so that it is practical to consider the production of resonances, whose masses are many times those already produced.[12]

As an example, consider the collision of the p beam and the $\bar{p}$ beam of the Fermilab Tevatron collider, the particles of each having energies of *1 TeV*. This can produce a resonance with $M^* = 2$ TeV. The threshold beam energy of a fixed-target machine able to produce such a resonance is obtained from Eq. (14.9):

$$E_{threshold} \approx \frac{M^{*2}}{2M} = \frac{(2 \times 10^3 \text{ GeV})^2}{2(.94 \text{ GeV})} \approx 2 \times 10^6 \text{ GeV} = 2{,}000 \text{ TeV}.$$

This factor of 2,000 puts this mass range of resonances out of the realm of possibility for fixed-target machines.

Cyclic Colliders   Cyclic colliders use geometries similar to those described above for cyclic accelerators to produce a large number of beam crossings. Many crossings per second are required in order to obtain a reasonable number of interactions because of the diffuseness of the beams compared to the density of targets in a solid. Many bunches are injected into the ring(s). They then circulate for many hours to increase the intensity of the beam as more and more bunches are fed in. Consequently, cyclic colliders are also called **storage rings**.

Although they reach very high energies, there are several limitations on the use of these colliders:

(i)   The particles must be stable because the beams circle for hours and, in fact, there are many millions of intersections of the beams. Thus, only protons (or heavy ions), electrons, or their antiparticles can be used.

(ii)   The energy loss to synchrotron radiation grows as E^4; therefore, large amounts of energy must be supplied to the beam. This requirement is expected to be the crucial limiting factor on the use of cyclic accelerators at high energies.

(iii)   It is difficult to produce $\bar{p}$ beams. Antiprotons are produced by bombarding nuclei with high-energy protons. The resulting antiprotons have momenta which are nearly random in both size (around the average value) and direction, much like a hot gas. Those employing $\bar{p}$ beams (like the CERN collider and the Tevatron) have developed methods to cool the $\bar{p}$'s and produce a beam. The antiprotons enter an **antiproton accumulator**, where they circulate in a magnetic field; special sensors detect their average deviation from the desired orbit and initiate magnetic kicks to reduce that deviation. These statistically operative methods, known as **stochastic cooling**, work quite well.

---

12. The linear dependence is also true in accelerators designed for beam particles of unequal mass.

Let us consider the rate (R) for a particular reaction.[13] It will be proportional to the cross section (σ) for that reaction and it is written:

$$R = L\sigma,\qquad(14.11)$$

where the coefficient of proportionality (L) incorporates all the relevant properties of the accelerator or collider and is called the **luminosity**. L has dimensions $A^{-1}t^{-1}$, which is usually given in cgs units as **$\text{cm}^{-2}\text{sec}^{-1}$**, and σ **is given in cm^2**, so that the units of R are inverse seconds as expected. For colliders, assuming that the two beams have the particles uniformly distributed over an area A and are circling at the same frequency (f), the definition of L (the luminosity) leads to

$$L = n\,\frac{fN_1 N_2}{A}\qquad\text{for a collider,}\qquad(14.12)$$

where n is the number of intersection points and N_i is number of particles in beam i. Typical values for the best $p\bar{p}$ colliders are $L \sim 10^{30}\,\text{cm}^{-2}\text{sec}^{-1}$, for e^+e^- colliders are $L \sim 10^{31}\,\text{cm}^{-2}\text{sec}^{-1}$, and for pp colliders are $L \sim 10^{32}\,\text{cm}^{-2}\text{sec}^{-1}$. For a typical fixed-target machine, the primary beam has a particle flux $\sim 2 \times 10^{12}$ particles per second which, for a one-interaction length target, would have an effective luminosity $L \sim 10^{37}\,\text{cm}^{-2}\text{sec}^{-1}$. For experiments using secondary beams, the particle flux can be $\sim 10^7$–10^{10} particles per second with luminosities of $10^{32}\,\text{cm}^{-2}\text{sec}^{-1}$–$10^{35}\,\text{cm}^{-2}\text{sec}^{-1}$. Thus, in many types of experiments, the luminosities of fixed-target machines are several orders of magnitude greater than those of colliding beams.

To find out how many events a particular experiment will detect (assuming 100 percent efficiency), we must multiply the rate R by the running time. Thus, we obtain

$$\text{number of events} = (Lt)\sigma.\qquad(14.13)$$

The quantity in the parentheses (Lt) has dimensions of inverse area and is often reported in inverse picobarns (10^{-12} barns $= 10^{-36}\,\text{cm}^2$). For $L = 10^{33}\,\text{cm}^{-2}\text{sec}^{-1}$ (a value projected for the SSC), an experiment which runs for 1 year, with 10^7 seconds of actual running time, will have an Lt value of 10^4 inverse picobarns. Thus, a reaction with a cross section of 1 picobarn will produce 10^4 events, but a rarer reaction with a cross section of 1 femtobarn (10^{-15} barns) will produce only 10 events. These numbers must be reduced by the efficiency of the experiment, and the background must also be taken into account. If a year of running would produce only a few events, the experiment becomes unfeasible (at that machine).

Recall that $\sigma \sim 1/s$ for most of the events of interest. Thus, to create resonances (particles) of large mass, which requires large s, we must increase the luminosity accordingly. Obviously, the required luminosities scale as s, which

---

13. Our considerations will apply to all types of cyclic accelerators.

equals the square of the mass of the resonance sought. In Chapter 15, we shall learn that Z^0 production has a cross section

$$\sigma_Z^0 \approx 10^{-33} \text{ cm}^2 \ (M_{Z^0} \approx 91 \text{ GeV}).$$

Therefore, we expect the cross section for the production of a 9-TeV particle (up by a factor of 10^2) to be of the order of $10^{-37} \text{ cm}^2 = 0.1$ picobarns. This will yield 10^3 events per year at the SSC (with the luminosity given in Table 14.3).

Linear Colliders   A linear collider is a machine that accelerates beams in a straight path to high energies and then causes them to collide with one another. It has a distinct advantage over cyclic machines because no centripetal acceleration is required; the only acceleration is that which increases the kinetic energy of the beam particles. Therefore, there is no synchrotron-radiation energy loss. In fact, it is expected that the next large colliders to be built, after the SSC, will be linear colliders.

The linear accelerator at Stanford (SLAC) has been modified to accelerate two beams, one e^- and one e^+, down the two-mile path. Their paths are then turned by bending magnets to cause a head-on collision, as shown in Fig. 14.10. This is called the Stanford Linear Collider (SLC). It is designed to achieve 50-GeV energies in each beam and a luminosity of $6 \times 10^{30} \text{ cm}^{-2}\text{sec}^{-1}$.

FIGURE 14.10
The Stanford Linear Collider.

TABLE 14.3
Colliders

Collider (Location)	Particles	Energy (GeV)	Luminosity $(cm^{-2}sec^{-1})$
BEPC (Beijing, China)	e^+e^-	2.8 + 2.8	1.7×10^{31}
SPEAR (Stanford Linear Accelerator Center, Stanford, Calif.)	e^+e^-	4.2 + 4.2	10^{32}
DORIS (DESY, Hamburg, Germany)	e^+e^-	5.6 + 5.6	10^{32}
VEPP−4M (Novosibirsk, Russia)	e^+e^-	6 + 6	5×10^{31}
CESR (Cornell University, Ithaca, N.Y.)	e^+e^-	6.5 + 6.5	9×10^{31}
PEP (Stanford University, Stanford, Calif.)	e^+e^-	15 + 15	$5 \to 8 \times 10^{31}$
PETRA (DESY, Hamburg, Germany)	e^+e^-	23 + 23	10^{31}
TRISTAN (Tsukuba, KEK, Japan)	e^+e^-	30 + 30	10^{31}
HERA (DESY, Hamburg)	ep	30 *(e)* + 820 *(p)*	1.5×10^{31}
SLC (Stanford University)	e^+e^-	50 + 50	6×10^{30} (10^{28} reached)
LEP (CERN [The European Laboratory for Particle Physics], Geneva, Switzerland)	e^+e^-	60 + 60	1.7×10^{31}
Spp̄S (CERN, Geneva)	$p\bar{p}$	315 + 315	3×10^{30}
VLEPP, INP† (Serpukhov, Russia)	e^+e^-	500 (1000) + 500 (1000)	10^{32} (10^{33})
Tevatron (Fermilab, Batavia, Ill.)	$p\bar{p}$	1,000 + 1,000	10^{30}
UNK† (Serpukhov, Russia)	pp	3,000 *(p)* + 500 *(p̄)*	400×10^{30}
LHC† (CERN, Geneva)	pp, ep	pp: 8,000 + e: 50; p: 8,000	pp: 1.4×10^{33} ep: 2×10^{32}
SSC† (Waxahachie, Texas)	pp	20,000 + 20,000	10^{33}

†Not yet completed

FIGURE 14.11

Aerial view of Fermilab. The main accelerator ring, which is easily seen from planes taking off or landing at O'Hare International Airport, contains the Tevatron.

Major Colliders   A large number of colliders either exist or are in the planning stages at this time. The Tevatron (at Fermilab), LEP (at CERN), the SLC (at Stanford), HERA (at DESY), the $Sp\bar{p}S$ (at CERN), etc., are some of the larger ones already in operation. In Table 14.3 we show the energies and luminosities for the colliders now in operation and for those which are beyond the planning stage at this time (1992).

At Fermilab, a second ring was added in the main tunnel to convert it into a collider. In Fig. 14.11 we show an aerial view of the Fermilab Tevatron, and in Fig. 14.12 we show a picture taken inside the tunnel; these photographs give some idea of the scale of the colliders now in existence.

At CERN (The European Laboratory for Particle Physics) the SPS proton synchrotron is used as a collider (the $Sp\bar{p}S$) and a large electron-positron collider (LEP) is also in operation. The 8 TeV + 8 TeV LHC will be built in the LEP tunnel. (See Fig. 14.13.)

14.3.4   The Superconducting SuperCollider (SSC)

We shall see in Chapter 15 and in PART D — BEYOND THE STANDARD MODEL that the determination of the validity of the standard model and pro-

FIGURE 14.12
This photograph, taken inside the main ring at Fermilab, shows the bending magnets of the original synchrotron above (inside of which is the beam pipe) and those of the Tevatron below.

posed extensions of it require an investigation of the TeV energy region. In order to investigate matter at such high energies, an enormous **pp collider** has been designed and is scheduled to be built in Texas (about 30 miles southwest of Dallas). This is the **SSC**, which will have **40 TeV** CM energy and a luminosity per interaction region of $L \approx 10^{33}$ cm^{-2}sec^{-1}.

There were, in fact, a number of sites proposed; detailed studies of geological features, land acquisition strategies, construction problems and costs, radiation emissions and environmental impacts, etc., had to be undertaken before site selection was made. None of these are trivial matters for such a large facility. The main ring will have a **circumference of about 54 miles**, located approximately 120 feet below the surface, and the ring cross section will be about 10 feet in diameter. (See Figs. 14.14 and 14.15.) In the actual construction of this gigantic ring, plumb lines (which are used to determine the vertical direction) will not be adequate because the Earth's *curvature* will have a significant effect.

The SSC will contain two separate beam pipes, with a ring of about 10,000 superconducting bending magnets and 2,000 quadrupole magnets to guide and focus the beam, respectively. The bending magnets will create magnetic fields

FIGURE 14.13
Aerial view of CERN. The smaller ring in the foreground shows the location of the
7-km-circumference underground SPS tunnel. The larger ring shows the location of the
27-km-circumference LEP tunnel. The LEP tunnel is flanked on the left by the Jura
Mountains in France and on the right by Lac Léman in Switzerland.

≥6 Tesla at the highest energy of the beam.[14] The acceleration will be accomplished by rf cavities in the rings. Each ring will contain about 15,000 bunches
of protons, with 10^{10} protons in each bunch. It is estimated that it will take
40 minutes to fill the rings and 20 minutes to reach the energy of 20 TeV in each.
The counter-rotating beams will then be made to intersect at several interaction
regions around the ring. Although only one or two collisions are expected at
each crossing, we expect 10^8 collisions per second due to the high speed ($\sim c$)
of the protons. The entire cycle should be repeatable once every 24 hours.

The particles injected into the rings will come from a succession of four
accelerators located in an injector complex. It will contain a 600-MeV linac, fol-

---

14. The design and construction of these large and powerful superconducting magnets has
proven to be an extraordinarily difficult undertaking, involving many scientists at several different laboratories. There are extremely large forces on the current-carrying wires in these elec-
tromagnets. During the acceleration of particles, the currents rapidly rise to about 4,000 amps.
Concomitant twisting and shaking of the wires may produce enough frictional heat to destroy their
superconductivity.

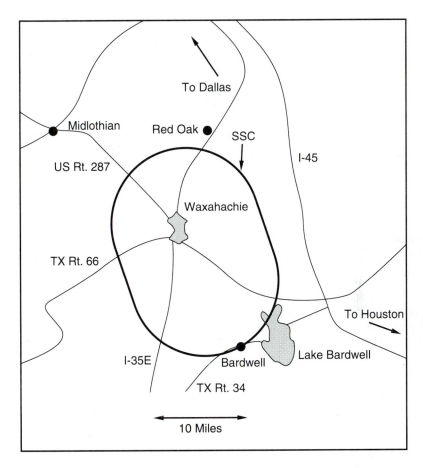

FIGURE 14.14
Map showing the proposed location of the SSC.

lowed by an 8-GeV circular accelerator, followed by a 100-GeV ring and then a 1-TeV ring. (The latter is the size of the Fermilab Tevatron; the 100-GeV ring is three times the energy of the large accelerator at Brookhaven, N.Y.) See Fig. 14.16. The facility will also require the building of new, complex, and costly detectors. This is indeed a colossal project!

The estimated cost of this project is about $8 billion.[15] This is small compared to our GNP but it is among today's costlier Big Science projects. Using information from the National Aeronautics and Space Administration (NASA), the Department of Energy (DOE), and the National Science Foundation (NSF),

---

15. This is the estimate at this writing (1990). Remarkably, at the writing of the last draft, in 1992, it is still the same! In proof (1993): The effects of extreme political pressure on the viability and cost of this and other Big Science projects are uncertain.

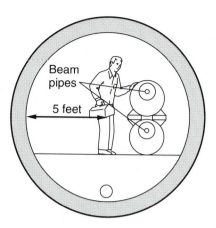

FIGURE 14.15
Cross section of the main ring of the SSC.

the *New York Times* (May 27, 1990) compiled a list of the largest and most expensive Big Science projects. They had a total projected cost of $64.8 billion, so that the SSC was ≈13% of the total. This general (newspaper) source is cited to emphasize the fact that such projects, of the size and expense of these Big

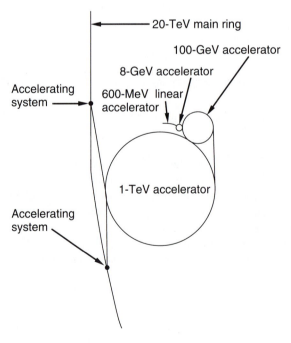

FIGURE 14.16
Injector complex of the SSC.

Science facilities, have become the subject of public scrutiny; their priorities will be decided by the general public. There is even controversy within the scientific community as to whether these large projects will result in the denial of funds for smaller and potentially more productive (all together) projects. In defense of this facility, it should be pointed out that construction of the SSC appears to be the only feasible way to investigate the fundamental questions facing particle physics. It is definitely a leap into the unknown as important as the other projects being proposed. Completely new ways of understanding the world might very well result from this exploration beyond the frontier of our knowledge, of the basic forces and constituents of nature, whose ramifications for all the sciences cannot be imagined.

PROBLEMS

1. a. Find the CM energy from the given beam energies at (i) the AGS (Brookhaven), (ii) the fixed-target accelerator at Fermilab and the SPS at CERN, (iii) the fixed-target Tevatron, (iv) the Tevatron collider, (v) the SLC, (vi) LEP, (vii) the LHC (pp), and (viii) the SSC.
   b. Check that, for fixed-target machines, the ratios of the CM energies grow as the square roots of the ratios of the beam energies, by comparing the PS and the SPS at CERN.
2. Find $\lambda_{DeBroglie}$ for a momentum equal to the CM value (treating the particles as if they are massless, so that $p = E/c$) at: (a) the AGS (Brookhaven), (b) the fixed target accelerator at Fermilab and the SPS, (c) the fixed target Tevatron, (d) the Tevatron collider, (e) the SLC, (f) LEP, (g) the LHC (pp), and (h) the SSC. This gives a sense of the corresponding distances probed.
3. Suppose we wanted to design a cyclotron for particles moving at *relativistic* speeds.
   a. Calculate the frequency required for the AC voltage across the Dees as a function of particle speed.
   b. How would you modify the design of the cyclotron so that the frequency modulation of the AC voltage across the Dees is synchronized properly with the particles' speeds?
4. For the simple betatron shown in Fig. 14.6, show that the magnetic field in the beam pipe (B_b) must change with time at half the rate that a uniform magnetic field inside the circle (B_c) does, when the particles are nonrelativistic.
5. Calculate the luminosity per interaction region for the SSC, assuming that the diameter of each beam will be 1 mm. Assume also that the protons in the beam are moving at speed c in the calculation of the frequency of revolution around the ring.
6. Find the energy that a beam in a fixed-target proton accelerator would need, so that it would have as much CM energy as the projected SSC value.

Chapter 15

Evidence for the Standard Model with Three Generations

There is a large body of detailed experimental evidence verifying the predictions of the standard model. We have already introduced some of it in our development of the ideas and structure of the theory. Here, we shall discuss some of the major experimental evidence for the standard model. The material will be representative rather than exhaustive, as is appropriate for an introductory book. Again, we remind the reader of the crucial role played by experiment in determining which model or theory has anything to do with the *real* world.

15.1 EVIDENCE FOR THE THREE GENERATIONS

15.1.1 Leptons

First Generation The first lepton discovered was the **electron**, which was first shown to be a particle by J. J. Thomson at the end of the nineteenth century. Experimental studies moved more slowly in that century, and this work was the culmination of 40 years of studying cathode rays. This achievement was the direct result of tremendous increases in the vacua attained and in the energy of the rays. (The pace of discovery quickened in the 1890s.[1]) Studies of beta decay in the early part of the twentieth century indicated missing energy,[2] momentum, and angular momentum; in 1930 Pauli proposed an explanation for all of this by postulating the existence of a neutral, weakly interacting, spin $\frac{1}{2}$ parti-

---

1. α-rays and β-rays (i.e., radioactivity) as well as X-rays were also discovered at that time. It is interesting to note that Maxwell felt that the atoms of nature could not be broken or worn down, so that the ideas concerning the discreteness of charge (i.e., the existence of electrons) and concerning the source of the energy of radioactivity were merely passing fads.

2. There were many, including Becquerel (the discoverer of uranic radioactivity), M. Curie, Rutherford, and Heisenberg, who speculated that the new phenomena being found in the early part of the twentieth century might be evidence that energy was not conserved.

FIGURE 15.1
The FMMF neutrino detector at Fermilab. The eight crates lined up on the floor (which look like loudspeakers) are each 7 ft. high. They contain digitizing electronics.

cle. It was named "neutrino" (little neutral one) and incorporated in a quantum field theory for the weak interactions by Fermi; today it is referred to as the **electron-neutrino**. Our belief in the conservation of energy, momentum, and angular momentum, together with the fact that a particle which only interacted weakly would not have been observed, was convincing enough for physicists to believe that the neutrino did exist. This belief, without direct evidence, was firmly held for almost 25 years until a large enough detector was built.[3] (In Fig. 15.1 we see the massive size of an electron-neutrino detector.) We refer to the electron and its neutrino as the **first generation** of leptons.

Second Generation In 1936 the **muon** was found in cosmic rays. It was natural to think of this particle as the meson predicted by Yukawa to be the carrier

---

3. It took almost a decade to design and build the detector (near a nuclear reactor) in which the reaction $\bar{\nu} + p \rightarrow n + e^+$ occurred; the $\bar{\nu}$ came from the reactor and the p was in the (tons of) material contained in the detector. The cross section is $\sim 10^{-43}$ cm^2.

of the nuclear force, since its mass (106 MeV) was very close to the value predicted. In fact it was called the "mu meson" well into the 1950s. However, it was a highly penetrating particle, which was even detected in cosmic rays penetrating deep mines. This was clear evidence that it did not interact strongly, as would be expected of Yukawa's particle. It wasn't until 1947 that the Yukawa-like particle, the pion, was discovered in cosmic rays and subsequently produced in collisions at cyclotrons. It was well known in the 1950s that the muon was a weakly interacting particle, which seemed to be a heavy copy of the electron ($m_\mu \approx 207 m_e$). I. I. Rabi, a renowned physicist[4] of that time, is said to have remarked concerning the muon: "Who ordered *that*?" The decays of the charged pions are overwhelmingly to muons and mu-neutrinos (branching ratio = 0.99988):

$$\pi^+ \to \mu^+ + \nu_\mu \text{ and } \pi^- \to \mu^- + \bar{\nu}_\mu. \tag{15.1}$$

The evidence for the creation of the ν's in these decays is the same as that found in ordinary beta decay. We have inserted the sub-μ and the bar over the ν produced with the μ^-, using information and conventions not known at the time; the conservation of lepton number appears in a simple form when the bar (anti-neutrino) designation is employed as shown.

The muon is not a truly stable particle; it decays with a lifetime of 2×10^{-6} seconds. However, when traveling near the speed of light, it travels many meters, so that it is easily identified from the tracks it makes as well as by its ability to go through dense materials. Particles (including the muon) which **do not decay via the strong interaction** are considered **stable particles,** as long as they travel measurable distances before decay. The characteristic decay time for **strong decays of 10^{-23} seconds** is always in the backs of our minds when we consider the "stability" of particles.

For several years, it was not known whether the neutrinos, created along with muons, were the same as those found in ordinary beta decay. Experiments studying the reactions

$$n + \nu \to p + \textbf{(lepton)}^- \text{ and } p + \bar{\nu} \to n + \textbf{(lepton)}^+$$

showed that they were different, and that one type (now called ν_e) occurs only when electrons are involved, the other (ν_μ) only when muons are involved, and never vice versa. Using ν_e's (from ordinary beta decay) produces only e's, whereas ν_μ's produce only μ's. Also, **the *absence* of the reaction** $\mu \to e + \gamma$ is evidence that the neutrinos are different, since otherwise that reaction would result from a slight modification of the diagram which produces $\mu \to e + \nu + \bar{\nu}$. (In the modified diagram the virtual ν and $\bar{\nu}$ annihilate, and a photon (γ) emerges from one of the charged lines. See Problem 1.)

---

4. He pioneered the use of molecular beams for detailed experiments and was also among those responsible for the introduction, earlier in the century, of (the then modern) quantum mechanics to American physics.

Thus, a **second generation** of leptons, the **muon** and the **muon-neutrino**, was found that looked like a copy of the first but with a more massive charged member.

Third Generation   Once two generations of leptons were known to exist (as well as more than one generation of quarks), it became a natural and exciting prospect to search for more. As the search probed higher and higher energies, the expected lifetime of the lepton sought dropped rapidly because of the larger number of decay channels available. (A more massive particle can decay into many more final states.) It became clear that the researchers would have to search for resonances. Resonances are generally hard to find and are usually attributed to the strong interaction, so that a decay channel with a distinctive experimental signature had to be chosen. It is also important to be able to produce that channel with sufficient counting rate, which depends directly on the luminosity of the collider.

The experiments which discovered the tau were done at the SPEAR e^+e^- storage ring at SLAC a couple of years after it began operating.[5] The events sought were

$$e^+ + e^- \rightarrow \tau^+ + \tau^- \text{ followed by } \tau^+ \rightarrow \mu^+ + \nu_\mu + \bar{\nu}_\tau \text{ and } \tau^- \rightarrow e^- + \bar{\nu}_e + \nu_\tau,$$

as expected for a heavy lepton decaying via the weak interaction. As explained in Chapter 13, high-energy exclusive cross sections vary as $1/s$, so that we expect the cross section for production of the τ's to be of the order of magnitude:

$$\sigma \sim \frac{\alpha^2}{s} \sim \frac{(137)^{-2}}{(2m_\tau)^2} \approx \frac{(137)^{-2}}{16 \text{ GeV}^2} \sim 10^{-33} \text{ cm}^2$$

(using: 1 GeV $\leftrightarrow$ 5.06 $\times$ 10^{15} (meters)$^{-1}$, as discussed in Chapter 1). Using the SPEAR luminosity (see Table 14.3) of 10^{32} cm^{-2}sec^{-1}, we obtain a rate of 10^{-1} per second, which yields thousands of events in one day. In Problem 2, you may calculate the branching ratio to the μ^+e^- channel sought, using the details given there.

The tau pair production is via a virtual photon, as shown in Fig. 15.2. At a collider the detector is in the CM frame, consequently, the τ's will be produced back-to-back, as shown in Fig. 15.3. However, the subsequently produced four neutrinos will be undetected, so that these events will look like

$$e^+ + e^- \rightarrow \mu^+ + e^- + \text{ much missing energy and momentum.}$$

You can see that this has a very distinctive signature, since it appears to violate conservation of muon number and electron number as well as conservation of

---

5. M. Perl, the leader of the group that discovered the τ, attributes the discovery to the opening of the new technology of high-luminosity, high-energy colliders. He points out, in seminars and colloquia, that many new discoveries are the results of new technologies. For example, Thomson's discovery of the electron was made possible by the higher vacua that he was able to produce.

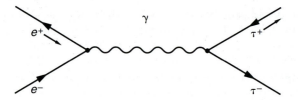

FIGURE 15.2
Tree-level Feynman diagram for tau pair production at e^+e^- colliders.

energy and momentum. The first 24 such events found (out of 10,000 events of all types) were treated with skepticism for several reasons. First, a misidentification of a hadron as an electron or muon was possible; remember that no device is ideal—experimentalists must have an intimate knowledge of the quirks of their equipment! It was determined that, at most, six of the events could be attributed to such a misidentification. Second, the resonance (which is being identified as a τ) might be a hadron, as so many are. In fact, this possibility was more serious because the $c\bar{c}$ resonances (ψ's) are produced in the same energy region and their details were not fully determined at that time. Furthermore, some of the ψ decays[6] produce a D meson (with a c quark) at approxi-

---

6. Of special note is $\psi(3770)$ which is above the $D\bar{D}$ threshold, so that it decays rapidly into D's. This will be discussed in Section 15.3.2; see Fig. 15.18.

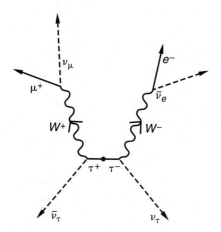

FIGURE 15.3
Sketch of the decays of the taus, produced at the point in the center of the diagram, showing the virtual W's involved (although not a Feynman diagram). The tau lifetime is $\sim 10^{-13}$ seconds, so that they do not produce tracks. Note that four (the ν's) of the six emerging particles are undetected.

mately 1800 MeV, right near the mass of this resonance: $m_\tau = \mathbf{1784\ MeV}$. However, the decay of the D would be a two-body decay, whereas the decay of a lepton to lighter leptons is a three-body decay. The momenta of the emerging μ's and e's would follow a completely different distribution in the two cases. To distinguish between these two alternatives, many more events were needed. After the detection of a few hundred events, it was determined that it was a three-body decay. Other detailed evidence, including a plot of the rate of production vs. $\sqrt{s}$ (CM energy), showed a behavior expected for lepton production rather than for hadron production. Much work was done analyzing the details of these studies. It was more than a year later that confirming evidence for the τ was found at the DORIS storage ring in Hamburg. Without that verification, it would have been necessary for those claiming the discovery to explain why it was not showing up at DORIS, which was very similar to SPEAR. Indeed, the discovery of a new particle at a large facility requires hard and painstaking work by hundreds of physicists as well as corroboration by other experimental searches around the world. In confirmation of these results, the same resonance (the τ) has been discovered in all the other channels expected. In fact, dozens of modes of τ decay have been measured.

The existence of the tau-neutrino has been inferred by the same kind of indirect means that was used for 25 years for the electron-neutrino, viz., energy, momentum, and angular momentum conservation. (No experiments have as yet been set up to directly detect the emerging ν_τ's.) However, if the τ were not a part of a third-generation doublet but instead a singlet, it would not couple to the SU(2) gauge particles (W's). Then the observed decays would require its mixing directly into the electron and muon generations. Models with such mixing would predict modes of decay which have not been found, despite several searches. Also, the agreement of the decay rate of the $\tau^\pm$ with the prediction of the standard model (with the assignment $T_Z = \pm\frac{1}{2}$) is sometimes cited in this regard. The τ and the τ-neutrino make up the **third generation** of leptons.

It should be mentioned that there is, at the present time, a mismatch between the measured branching ratios for the exclusive 1-charged particle decays of the τ and the corresponding inclusive branching ratio. Recall that the inclusive branching ratio should be the sum of the exclusive (to a particular final state) branching ratios. However, the inclusive cross section is usually measured by different methods and often in different experiments. The discrepancy appears to be a few percent and may have to do with experimental accuracy. With the thoroughness and inquisitiveness characteristic of good experimentalists, researchers are studying this discrepancy to see if it is a signal of the existence of higher generations or a signal of new physics.

Higher Generations The existence of higher generations is considered doubtful because of the following:

1. **The ρ parameter**, $\rho = M_W/(M_Z \cos\theta_W)$, discussed in Section 11.2.1, has been experimentally determined to be 1.0. Higher generations (with

light neutrinos) would shift the value of ρ by contributing to the masses in ρ through radiative corrections (as virtual particles).

2. At LEP, the large e^+e^- collider at CERN, and the SLC (Stanford Linear Collider), enough Z^0's (millions) have been made to determine **the shape of the Z^0 resonance**. In Eq. (10.26) we saw that the Z^0 couples to a pair of neutrinos. If there are higher generations (with neutrinos of mass less than half the mass of the Z^0), then the Z^0 could decay into a lepton pair of that type. Thus, its lifetime would be shortened, hence, enlarging its width since $\Gamma = 1/\tau$. Present data on **the Z^0 width**, measured at LEP's four different detector complexes and at the SLC, give **number of neutrino families $\approx 2.99 \pm 0.04$**, ruling out more than three generations.

3. **Cosmological considerations of primordial nucleosynthesis** (the production of nuclei before star formation) show that four (or more) light neutrino species would be in conflict with the observed **abundance of ^{4}He**. This is a consequence of the fact that ^{4}He production in stars is estimated to account for $<10\%$ of the observed abundance; therefore, most of what we see should have originated during that epoch ($t \approx 2$ min after the Big Bang) when primordial nucleosynthesis occurred. Virtually all of the neutrons in existence at that time are in that primordial ^{4}He.

The relevant parts of the Big Bang scenario are:

1. After their formation, neutrons and protons were in equilibrium via the reversible processes:

$$n + \nu_e \leftrightarrow p + e^-, n + e^+ \leftrightarrow p + \bar{\nu}_e \text{ and } n \leftrightarrow p + e^- + \bar{\nu}_e.$$

Because of their mass difference, the neutron to proton ratio (n/p) was

$$n/p = e^{-\Delta m/kT} \text{ in equilibrium}, \tag{15.2}$$

where $\Delta m \equiv m_n - m_p = 1.3$ MeV. At large kT ($\gg 1$ MeV), this ratio is essentially 1. At about **10 seconds after the Big Bang, when kT dropped to about 1 MeV**, the rate of occurrence of the $p \to n$ processes (above) dropped significantly, because most of the electrons and antineutrinos no longer had enough energy to change a proton back to a neutron. **The processes which create neutrons "freeze out" as the reaction rate falls below the expansion rate of the Universe**, presumably because the particles are too far apart for the reactions to occur. Thus, the effective time and temperature at which the freeze-out occurs depend on how fast the Universe is expanding. Furthermore, we see from Eq. (15.2) that n/p at freeze-out depends exponentially on the inverse of the effective freeze-out temperature.

The expansion rate of the Universe varies as the square root of the energy density (of relativistic particles), which in turn varies linearly with the number of degrees of freedom of relativistic particles. Thus, **for a larger number (N_ν) of light neutrino species ($m_\nu < 100$ MeV), the**

expansion rate is greater and the freeze-out occurs earlier at a higher effective T. This means that there would be **a larger n/p ratio at freeze-out, the greater the N_ν.** With three light neutrino generations ($N_\nu = 3$), the n/p ratio at freeze-out is about $\frac{1}{6}$.

However, nucleosynthesis does not begin at this time because the first nuclei to be produced, deuterons (D), readily photodisintegrate with a rate $\approx n_\gamma/n_b \, e^{-E_B/kT}$, where n_γ/n_b is the photon-to-baryon ratio[7] and E_B is the binding energy. Deuterons are very loosely bound ($E_B \sim 2.2$ MeV), and the large density of energetic photons ($n_\gamma/n_b \approx 10^{10}$) causes photodisintegration of the deuterons before they can combine with others (p, n, or D) to produce heavier nuclei. We expect the time from freeze-out to nucleosynthesis to be larger, the larger the relative number of photons present; this would provide more time for the neutrons to decay. These decays cause the n/p ratio as nucleosynthesis begins to be smaller than the n/p freeze-out ratio. (The amount of time available for neutron decay also varies slightly with the exact freeze-out time, which must be taken into account in a detailed calculation.)

2. **At $t \approx 2$ minutes after the Big Bang, when $kT \approx 0.1$ MeV**, the D's are stable enough (in the enormous photon background) for nucleosynthesis to begin. (By this time, some of the free neutrons have decayed ($\tau_{mean} = 888.6 \pm 3.5$ s ≈ 15 min.) and the n/p ratio has dropped to about $\frac{1}{7}$.) The nature of the nucleosynthesis is such that **virtually all the neutrons** in existence during that epoch **wind up in ^{4}He.** Hence, **the greater the n/p ratio, the more ^{4}He produced.** Thus, the helium mass fraction (^{4}He$/(n + p)$) is a sensitive measure of n/p at nucleosynthesis; that n/p ratio is greater, the greater is **n/p at freeze-out** which, we have seen, **is greater for greater N_ν.** Consequently, **the greater the number of light neutrino generations (N_ν), the more ^{4}He produced.**

Small amounts of D, ^{3}He, and ^{7}Li, which were created during this primordial nucleosynthesis period, still remain today. Since they are essentially not produced in stellar interactions, their present abundances are sensitive indicators of conditions during that epoch. In fact, the presence of free protons and neutrons causes the destruction of some of the D, ^{3}He, and ^{7}Li (via processes like $p + {}^7$Li $\to 2 \, {}^4$He, etc.), so that their present-day abundances provide a reliable estimate of the relative number of baryons (n_b/n_γ) in the Universe.

In Fig. 15.4 the vertical lines show the implications of recent determinations of the abundances of D, ^{3}He, and ^{7}Li on the relative number of baryons (n_b/n_γ). The horizontal line is the upper limit on the ^{4}He mass fraction. We can see that if $N_\nu = 4$, the primordial ^{4}He abundance would be about 1% more than for $N_\nu = 3$, and that value lies well outside the experimental uncertainties. (We also see the increase of the ^{4}He mass fraction with n_b/n_γ when N_ν

---

7. Cosmologists usually use the baryon-to-photon ratio rather than the photon-to-baryon ratio.

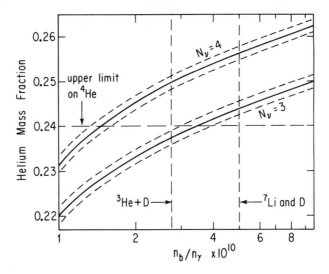

FIGURE 15.4
Determination of N_ν, the number of light neutrino species, from the (primordial) abundances of ^{4}He, ^{3}He, D, and ^{7}Li.

is fixed.) From Fig. 15.4, we may conclude that **$N_\nu = 3$ is compatible with these observations, whereas $N_\nu = 4$ is not**. This is one instance of the considerable overlap of the interests of cosmologists and particle physicists. Roughly speaking, the energies attained in high-energy experiments reproduce the energy densities which existed shortly after the Big Bang. (We will have more to say about this when we discuss quark-gluon plasmas in Chapter 16.) The overlap of these research areas is so great that in one recent cosmology paper,[8] after a discussion of the determination of N_ν, the following appears: "The new SLC results seem to experimentally support these cosmological results. The particle accelerators are now verifying cosmological predictions." This is *not* the way a particle physicist would characterize the relationship of the experimental high-energy results to cosmological predictions.

15.1.2   Quarks

First Generation — Evidence for the Existence of Quarks   The discoveries of the proton and neutron were made early in the twentieth century. Charged pions were found in cosmic rays in 1947; charged pions leave tracks and so were easier to find than the π^0, found in 1950. Pions are easily produced and (for the

---

8. "Cosmology and the Weak Interaction" by D. N. Schramm, from the Proceedings of the Wein 89 International Symposium on Weak and Electromagnetic Interactions in Nuclei, Montreal, Canada. (May, 1989)

next fifteen years) were thought to be the Yukawa particles. As we have already seen, these particles (p, n, and π) are bound states of the u and d quarks.

The u and d form the first generation of quarks.

Let us trace the experimental evidence for this substructure of these familiar particles, which make up ordinary matter. If the proton and neutron were elementary particles, then the magnitudes of their magnetic moments should be

$$\mu = +\frac{e\hbar}{2m_p c} \equiv \mu_{Nuclear} \text{ for a "proton"}$$

and $\mu = 0$ for a "neutron," as elementary particles.

The observed values are

$\mu_p = +2.79 \, \mu_{Nuclear}$ (for the proton), and $\mu_n = -1.91 \, \mu_{Nuclear}$ (for the neutron),

so that $\dfrac{\mu_n}{\mu_p} = -.685$ (measured to many more places).

This is a strong indication that these are not elementary particles. When we add (vectorially) the magnetic moments of the quarks, of which these spin $\frac{1}{2}$ particles are composed, assuming $m_u = m_d$, we obtain $\mu_n/\mu_p = -\frac{2}{3}$ (as you may verify in Problem 6), which is very close to the observed value. (We do not expect exact agreement, since there may be some orbital contribution as well, and the mass assumption is not strictly correct.) This is strong evidence for the quark structure of the neutron and proton.

In Chapter 6, we saw that the growing list of "elementary" particles seemed to belong to representations of flavor SU(3). They were mostly **8**s, and a monograph at the time was even titled "The Eightfold Way" (as discussed in Chapter 6). It was conjectured[9] that the role of SU(3) could be understood if there existed three types of more fundamental particles, quarks (of similar mass), of which the observed particles were made. We have already discussed (in Chapter 9) the fruitless searches for *free* quarks. Here, we are exploring the evidence for their existence as *bound* particles.

By the end of the 1960s it was clear that quarks did not exist as ordinary particles. Most people thought that the failure to find any, coupled with poor predictions for differential scattering cross sections, precluded their existence. Nevertheless, scattering experiments were performed to delve into the structure of the nucleons and to seek their constituents. The hypothetical constituents were called **partons**, so as to allow the physicists involved to remain uncommitted in the controversy concerning the existence of *quarks*. Electrons, photons, and neutrinos (which are considered to have no structure of their own) were used as probes in various experiments, somewhat analogous to those from which Ruth-

---

9. This hypothesis was suggested independently by M. Gell-Mann and G. Zweig. Gell-Mann actually conjectured that these *particles* may be infinitely tightly bound.

erford deduced the existence of the nucleus. In order to find the tightly bound constituents of the nucleon, it seems reasonable to try to knock one of them out with a high-energy probe. In fact, the higher the energy, the more likely that the scattering would be from only one parton. A semiclassical understanding of this comes from recalling the De Broglie connection between momentum and position; in order to probe small sizes within the nucleon, we need to use high-momentum probes. (At low energies the entire nucleon would interact with the probe in a coherent fashion.) These experiments are referred to as **deep inelastic scattering**.[10]

For simplicity, let us consider the scattering of a spinless charged point particle from a region of distributed charge. We recall that the differential cross section found by Rutherford for the scattering of a charged (ze) particle from a nucleus (of charge Ze) is

$$\frac{d\sigma}{d\Omega} = K\alpha^2 \left(\frac{zZ}{E}\right)^2 \frac{1}{\sin^4\left(\frac{\theta}{2}\right)}, \quad \text{from a nucleus (treated as a point),} \quad (15.3)$$

where K is a dimensionless constant, E is the energy of the point particle, and θ is the angle of scattering. When the target has a distribution of charge, the effective charge seen by the probe depends on the four-momentum (q) it transfers in the collision, i.e., on how close in it probes. Thus, the cross section must have Z replaced by a factor $F(q^2)$, which depends upon the four-scalar q^2 and is called the form factor. Let us generalize Rutherford's formula to the case of a charge ($z =$) 1 projectile and a distributed charge in the target:

$$\frac{d\sigma}{d\Omega} = K\alpha^2 \left(\frac{1}{E}\right)^2 \frac{|F(q^2)|^2}{\sin^4\left(\frac{\theta}{2}\right)}. \quad (15.4)$$

(In quantum mechanics F could be complex, so we use the absolute square in Eq. 15.4.) **The q^2 dependence of the form factor is an indication of the fact that the target is not a (fundamental) point particle.**

This generalized Rutherford equation is for elastic scattering, whereas we are considering inelastic scattering to a hadronic system of equivalent mass M_f, as shown in Fig. 15.5. Thus, we expect the deep inelastic form factor to depend upon M_f as well:

$$F(q^2) \rightarrow F(q^2, M_f).$$

From conservation of energy and momentum we obtain

$$M_f^2 \equiv p_f^2 = (p_1 + q)^2 = p_1^2 + 2q\cdot p_1 + q^2 = M_N^2 + 2q\cdot p_1 + q^2,$$

$$\text{since } p_1^2 = M_N^2. \quad (15.5)$$

---

10. In this context "deep" refers to the high momentum transferred to the parton.

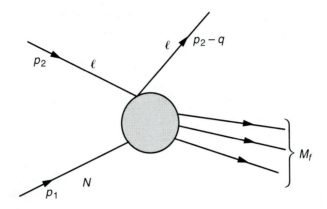

FIGURE 15.5
Inelastic scattering of a lepton and a nucleon.

It is convenient to consider the two variables in the form factor to be q^2 and $x \equiv -q^2/2q \cdot p_1$, since we can write Eq. (15.5) as

$$M_f^2 = M_N^2 - \frac{q^2}{x} + q^2. \qquad (15.6)$$

Consequently, **the form factor is usually written as $F(q^2, x)$.**

At very high energies, we expect that the probe particle will effectively scatter from a point constituent (parton) of the nucleon, with the other partons being merely **spectators** of that simple elastic scattering, as shown in Fig. 15.6. (This is like the sudden or impulse approximation in quantum mechanics.) Since the effective target particle is a point constituent, the q^2 dependence of the form

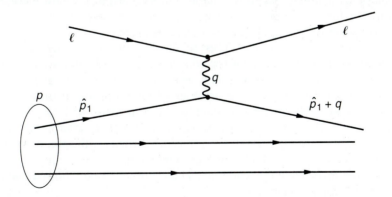

FIGURE 15.6
Elastic scattering of a lepton and one of the quarks in the proton. $\hat{p}_1$ is the initial four-momentum of the scattered quark; the other quarks are spectators in the collision. The higher the energy, the more likely that this will occur.

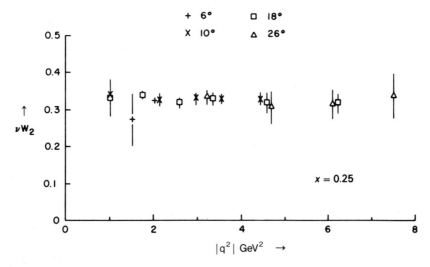

FIGURE 15.7
One of the (effective) structure functions (νW_2) in electron-proton scattering, which is expected to show scaling, as a function of $|q^2|$, for x held fixed at the value 0.25. (SLAC data)

factor disappears. This high-energy behavior is referred to as **Bjorken scaling**, where the word "scaling" refers to the fact that the form factor depends only on the dimensionless quantity x and so is independent of (mass or length) scale; it remains constant (independent of q^2) at very high q^2, for fixed x. At these high energies the form factor essentially depends on the probability that a parton (in the proton of momentum p_1) has momentum $\hat{p}_1 = xp_1$, because x becomes the fraction of the proton's momentum carried by the hard-scattering parton. The form factor is usually written $F(x)$, and $x \equiv \hat{p}_1/p_1$ is referred to as **the Feynman x**. This scaling has been observed in many deep inelastic, high-energy experiments. See Fig. 15.7. These results show that indeed **there are point-like constituents in the nucleon** (proton or neutron) which scatter at high energies, as if they were free particles. Actually, QCD corrections give a q^2 dependence to the form factors. The observed insignificance of those effects show that **the effective coupling constant for this high-energy scattering is small**. This is further evidence that the perturbation (Feynman diagram) description of high-energy scattering is valid. For projectile particles with spin, more than one form factor arises, and relationships between them depend upon the spin of the parton involved in the scattering. The scattering results showed that the charged constituents have spin $\frac{1}{2}$.

Careful studies of deep inelastic scatterings have even found the fraction of the proton's momentum carried by each type of constituent. Recall that from quantum field theory we expect virtual gluons to be exchanged by the quarks that are present, and these virtual gluons can produce pairs of virtual quarks.

These virtual particles sometimes interact with the probe and have been detected and identified. The proton, as an example, has been found to have two u quarks and one d quark as we expect, but gluons and pairs of virtual quarks ($u\bar{u}$, $d\bar{d}$, $s\bar{s}$, $c\bar{c}$, $b\bar{b}$, and possibly $t\bar{t}$) have been found as well.[11] It is usual to refer to the predominant u, u, and d quarks as "valence quarks," in analogy to the terminology of atomic physics, and the pairs as "sea quarks."

Second Generation In Chapter 6 we discussed the evidence for the existence of strangeness as well as flavor SU(3). There, the famous bubble chamber picture (Fig. 6.4) showing the discovery of the (quasi-stable) Ω^- was displayed, and a discussion of the vector mesons was presented. We noted that the ϕ meson was nearly pure $s\bar{s}$ due to the fact that the s was considerably more massive than the u or d. As we might guess, with hindsight, heavier flavors of quarks might most easily be sought by looking at the spectrum of vector meson resonances at higher and higher energies. In fact, in the 1970s, S. Ting et al., looked for heavier vector mesons by making precise measurements of effective-mass spectra of electron and muon pairs. His group at the AGS (at Brookhaven) found an extremely narrow, enormous resonance in the inclusive reactions:

$$p + \text{Be} \rightarrow e^+ + e^- + X,$$

where X means that *any* other particles may be in the final state, i.e., all events which produced an e^+e^- pair were counted. (Protons, accelerated by the AGS, bombarded a beryllium target.) The effective mass of the electron-positron pairs was found to have a large, extremely narrow peak at 3.1 GeV, which was attributed to a short-lived meson. The meson was named J by this group. (This group did not report its findings immediately, but proceeded to study the entire experiment in order to be sure of the results.) The narrowness of the peak (which implies a long lifetime) is what would be expected due to Zweig suppression, if the meson were made of quarks of a new flavor, as we have discussed in Chapters 6 and 9.

Meanwhile, at the SPEAR e^+e^- collider (at SLAC), the resonance was independently found in the study of

$$R = \frac{\sigma(e^+ + e^- \rightarrow hadrons)}{\sigma(e^+ + e^- \rightarrow \mu^+ + \mu^-)}, \tag{15.7}$$

where (i) the numerator includes all hadronic channels and is essentially the inclusive cross section for all final states with more than two particles and (ii) the denominator requires just a simply detected, exclusive final state of two back-to-back muons. The value of R at the resonance was found to be about 100 times its value outside the resonance with a width of $\Gamma \leq 1.3$ MeV. (The widths of the particles made of u and d quarks, discovered earlier, were ~ hundreds of MeV.) The observed upper limit, 1.3 MeV, is due to experimental lim-

---

11. The gluons have the largest fraction of the momentum of the proton.

itations, and the formula for cross section in terms of width has been used (considering the area under the curve) to determine the width to be $\Gamma \approx .06$ MeV. (This extremely narrow, pronounced peak was found unexpectedly; we shall presently discuss the original reason for studying R of Eq. 15.7 and the evidence it provides for the standard model.) This tiny width implies that this resonance lives more than 1,000 times as long as it should if it were made solely of the u, d, and s quarks. A follow-up study of the channels

$$e^+ + e^- \rightarrow hadrons$$

$$\rightarrow e^+ + e^-$$

$$\rightarrow \mu^+ + \mu^-,$$

was performed, and the same resonance was found in all of them. It was called the ψ at SPEAR. They also found a smaller resonance in R, very close to the ψ, and called it ψ' (at 3.7 GeV).

The joint announcement of the discovery by these groups, on the weekend of November 10, 1974, is referred to as "the November Revolution" because it presented such convincing evidence of the validity of the quark hypothesis, that the attention of the entire high-energy community focused on quark theories. The papers announcing this discovery appeared consecutively in the same issue of *Physical Review Letters*.[12] The resonance (particle) is called J/ψ: it is a bound state of $c\bar{c}$, and the ψ' is an excited state of the $c\bar{c}$ system. These particles have hidden charm, and the verification of their interpretation as bound states of a new flavor of quark came with the discovery of the D mesons, analogous to the K mesons for strangeness, which have one charmed quark and one lighter quark. (The ability to produce charmonium states above the $D\bar{D}$ threshold (e.g., the $\psi(3770)$) has led to the operation of SPEAR as a D "factory.") In Table 6.6 we have listed these charmed mesons and the masses at which they have been found. Notice that the mass of the $c\bar{c}$ vector meson, the ψ, is about three times that of the $s\bar{s}$, viz., the ϕ. This leads us to believe that the mass of the c quark is about three times that of the s quark, so that a flavor SU(4) symmetry, i.e., an invariance to the change of the flavors of the quarks making up observed particles, is badly broken in nature. We have already seen that flavor SU(3) is only a broken symmetry and that the masses of the particles in the SU(3) multiplets are not very close to one another. Nevertheless, the existence of particles which fill those multiplets make flavor SU(3) a useful broken symmetry. In fact, group theoretical arguments, concerning the possible form of the symmetry breaking, successfully predicted the mass splittings in the multiplets of SU(3). Although this is not the case for SU(4), we did show the lowest-lying SU(4) multiplets of mesons and baryons in Fig. 6.8. The status of the experimental data is: All 16 of the 0^- mesons and of the 1^- mesons, which

---

12. Shortly after hearing about the discovery at Brookhaven, three experimental groups at the ADONE collider (at Frascati) pushed the energy above the nominal design limit of 2×1.5 GeV and verified the existence of this very narrow resonance.

each form a **15 + 1** of SU(4), have been found. This is not true of the charmed baryons. The *Review of Particle Properties*, compiled in 1992, lists only: $\Lambda_c^+ = cud$ (2285 MeV), $\Sigma_c^{++} = cuu$ (2453 MeV), $\Sigma_c^+ = cud$ (2453 MeV), $\Sigma_c^0 = cdd$ (2453 MeV), $\Xi_c^+ = csu$ (2466 MeV), and $\Xi_c^0 = csd$ (2473 MeV). The doubly and triply charmed baryons have not been found in this energy region. It is possible that the SU(4) symmetry is so badly broken that the multiplets are severely mixed and we cannot find physical states which correspond to all the members of the SU(4) multiplets. Mass formulas to calculate the splittings have not been sought because the symmetry breaking is so large that they would be useless.

The *s* and *c* **quarks** comprise the **second generation**.

Third Generation   A group of researchers at Brookhaven had studied muon pairs as early as 1967 and had found a "shoulder" in that data for the mass of the "virtual photon," which they assumed had produced the muon pair, at about the J/ψ mass (long before its definite discovery). However, it was poorly resolved because of the more than 10 feet of iron used to block all particles but muons from reaching the (muon) detectors. Their work was continued at higher energies at Fermilab, using beryllium absorbers instead of iron. (Since Be has only four protons and four electrons, it hardly deflected the muon pairs, and the J/ψ peak was easily resolved. Unfortunately for them, that was *after* it had already been discovered.) The J/ψ resonance (at 3.1 GeV) was used to check the performance of the equipment. They then proceeded to explore a new (higher) energy region, and at about 10 GeV they found an "intriguing bump." The experimenters made many detailed checks to convince themselves that the results were not spurious. Each square centimeter of the muon detectors was examined to see that the results were smoothly distributed and had not been generated by the detectors themselves. They even mixed μ^+ data from one day with μ^- data of another to test this smoothness. A Monte Carlo calculation of the entire experiment was performed, tracing tens of thousands of muon pairs through the apparatus. (Each piece of equipment of the apparatus had been programmed into the calculation.) The results showed that the actual apparatus had spread the mass distribution, so that the resonance was considerably narrower than indicated by the data. Again, evidence for a new quark flavor (bottom) had been deduced from the extreme narrowness of a resonance. The newly discovered particle was called the Υ (Upsilon); its mass is 9.4 GeV (ten times the mass of a proton), and its width $\Gamma \approx .05$ MeV. In fact, two more Upsilon levels were discovered along with the Υ, and were named Υ' and Υ''. The new quark was found to behave like a charge $-\frac{1}{3}$ particle (as we shall soon discuss). To be consistent with the structure of the electroweak theory, it would have to be the bottom of a new doublet generation of quarks. It was given the name "bottom" by most, although some call it "beauty." However, we might have echoed Rabi's question concerning the muon. The charmed quark was anticipated on theoretical grounds long before its discovery, as we have discussed in Section 11.3 (recall the GIM mechanism), but "Who ordered *this*?" Well, it is recognized today that renormalizability requires as many generations of quark isodoublets as there are leptons (as we have already discussed in Section 11.4). Thus, not only should we have expected the b quark, but also that a "top" quark should exist as well.

You might think that we should now investigate the multiplets of SU(5), since five flavors of quarks have been discovered. However, we have already seen that the SU(4) structures were not of much use beyond the designation of mesons and baryons by quark content, which can be made without reference to a broken flavor symmetry. The high mass of the b quark certainly has precluded the investigation of flavor SU(5).

Of course, experimental searches have been made for (mesons and baryons with) "bare bottom" or "naked beauty."[13] The bottom mesons found so far are the $B^+(u\bar{b})$ and the $B^0(d\bar{b})$, both at 5279 MeV, and their antiparticles.

Where is the top quark? The top quark mass appears to be so large that its decay should be rapid enough to preclude hadronization or formation of toponium; thus, searches for $t\bar{t}$ mesons would be futile. However, searches for the top quark *are* in progress. As we learned in Chapter 11, the $\binom{t}{b}$ doublet couples to the W (in the weak interaction); consequently, we expect the decays $t \to b + W^+$ and $\bar{t} \to \bar{b} + W^-$. In fact, each of these decay modes is expected to have a branching ratio of 100%. Among the subsequent decays, there should be events where the W from the t produces an e^+, and that from the $\bar{t}$ produces a μ^- (as well as events where the e and μ are from the $\bar{t}$ and t, respectively). The b and $\bar{b}$ should each decay into a high-energy hadron jet, whose beginning (vertex) is slightly displaced from the origin of the muon and electron lines, since the b and $\bar{b}$ travel a short distance before decay. Events with a high-energy μ, a high-energy e, plus two displaced high-energy hadron jets have a very distinctive signature, and those are the events being sought. (See Fig. H.15 for the first candidate $t\bar{t}$ event, which was found at the CDF detector.)

Towards the end of the 1980s, two groups of researchers, one at Fermilab using the CDF detector and one at CERN using the UA2, searched for the top (t) quark. The collection and analysis of the data were very laborious, involving the selection of about 1 of the 50,000 collisions per second as possible candidate events, the storage of the data on computer tapes which filled large rooms, the sending of this data to universities and laboratories (some thousands of miles away) for analysis, and the winnowing out of those events which merely *mimic* a top quark event (background). These simultaneously running experiments produced a friendly rivalry, which was documented in a sociologically as well as scientifically fascinating *NOVA* TV program.[14] In that program it was clearly shown how the challenge, to claim as much as possible from the data obtained, was tempered by the likelihood that any excess claim might be shown invalid by the data of the other group. At stake, of course, was the discovery of a new quark (flavor), and the excitement of exploration radiates from this documentary. Neither group found any evidence of the t quark's existence. The experi-

---

13. Who could resist the urge to include these commonly used designations in his or her book?
14. This program was called "Race for the Top."

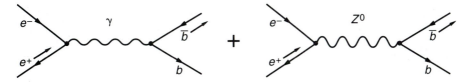

FIGURE 15.8
Tree-level diagrams in the amplitude for the production of a $b\bar{b}$ pair.

ments were not wastes of time, however, as it was determined that the top quark, if it exists (with the expected standard model decay channel $t \to b + W^+$), must have a mass $m_t > 91$ GeV (determined by CDF).[15] One physicist on the CDF team expressed her motivation for taking part in the experiment as the desire to "get to that [high] place and look around. . . . It is totally *cool!*"[16] Two further searches are underway at Fermilab (one using the CDF and the other the D0 detector).

Detailed studies of the ρ parameter, etc., indicate that **$m_t \approx 160 \pm 30$ GeV** (the precise value depends on the mass of the Higgs), and electroweak calculations for precision experiments indicate that **$m_t < 200$ GeV**. There is further indirect evidence of the existence of the t quark, obtained from detailed studies of the b quark. Study of the angular distribution of $b\bar{b}$ production has shown an interference term between the term for the creation from a virtual photon and that from a virtual Z; these terms are added in the amplitude, as shown in Fig. 15.8. The coupling to the photon is proportional to the **charge of the b**, which has been found in several experiments to be $(-\frac{1}{3})e$. As we saw in Eq. 10.36, the coupling to the Z is $\sim (T_Z - Q \sin^2 \theta_w)$, from which the data yield **$T_Z = -\frac{1}{2}$ for the b quark**. We learned (in Section 3.3) that *all* members of a multiplet of a symmetry group must exist, so that we expect a $T_Z = +\frac{1}{2}$, $q = +\frac{2}{3}e$ partner, viz., **the top quark**. Also, similar to what was said in our discussion of the τ, the decay pattern of the b is that expected of a standard model doublet member; the extra decays, expected if the b were an SU(2) singlet, do not occur.

We have mentioned that the $c\bar{c}$ resonance (J/ψ) was found at SPEAR during a study of

$$R = \frac{\sigma(e^+ + e^- \to hadrons)}{\sigma(e^+ + e^- \to \mu^+ + \mu^-)}.$$

In Fig. 15.9 we show the most recent plot of R vs. energy. **The value of R, away from the resonant peaks, provides evidence of the existence of the first five flavors of quarks as well as strong evidence for the number of colors for each flavor.** The constant value, for energies above all the resonances, is indirect

---

15. If the top quark has decay modes not yet searched for, there is a chance that 55 GeV $\leq$ $m_t \leq$ 91 GeV.

16. The statement was made by Melissa Franklin.

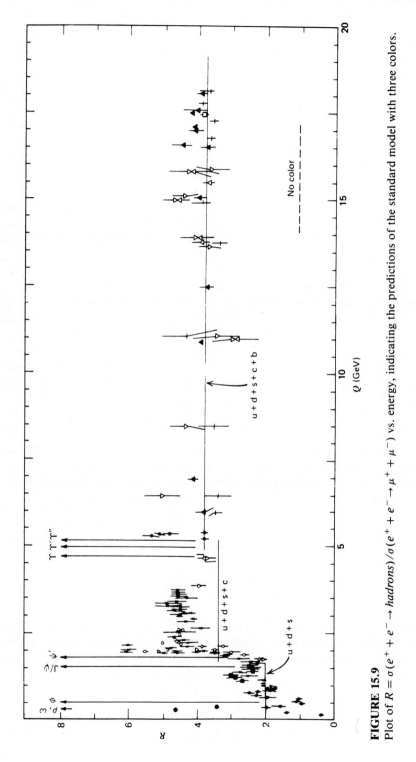

FIGURE 15.9

Plot of $R = \sigma(e^+ + e^- \rightarrow hadrons)/\sigma(e^+ + e^- \rightarrow \mu^+ + \mu^-)$ vs. energy, indicating the predictions of the standard model with three colors.

FIGURE 15.10
The production of a fermion pair $f\bar{f}$ (f could be a quark or a lepton) by the virtual photon in an e^+e^- collision. Recall Fig. 15.2.

evidence of the pointlike nature of the quarks themselves. Let us examine the basic process, which is shown in Fig. 15.10 (in which f can be a quark or a lepton). The quarks produced by the virtual photon "hadronize," i.e., emerge as jets of colorless hadrons (because of confinement, as we have already discussed in Chapter 9). This hadronization is not understood in detail, but if we sum over all hadronic channels, all quarks are included. We see from Fig. 15.10 that the Feynman diagram for the production of a quark pair is identical to that for a muon pair except that the charges of the quarks ($|Q| = \frac{1}{3}e$ or $\frac{2}{3}e$) are different from that of the muon ($|Q| = 1e$). Remember that each cross section is proportional to the absolute square of the amplitude obtained from the corresponding Feynman diagram, so that the ratio of the e^+e^- inclusive hadronic cross section to the cross section for μ pair production is

$$R = \frac{\displaystyle\sum_{quarks} Q^2_{quark}}{e^2}. \tag{15.8}$$

(The mass of the fermions produced by the virtual photon is different for different quark flavors and for muon production, but this has a negligible effect on the phase-space factors for E well above the threshold of their production.) Below the J/ψ, no c's or b's are made so that we expect

$$R = \left(\frac{2}{3}\right)^2 + \left(\frac{-1}{3}\right)^2 + \left(\frac{-1}{3}\right)^2 = \frac{2}{3},$$

for one u,   one d,   and one s quark.   (15.9)

The data is 3 times this value, as you can see in Fig. 15.9, which is the value we would obtain if each flavor comes in 3 colors. With three colors, we expect the data above the J/ψ to settle down to a value close to

$$R = 3\left[\left(\frac{2}{3}\right)^2 + \left(\frac{-1}{3}\right)^2 + \left(\frac{-1}{3}\right)^2 + \left(\frac{2}{3}\right)^2\right] = \frac{10}{3},$$

due to u, d, s,   and c quarks in three colors.   (15.10)

R does appear to be approaching that value when the Υ peak appears at about 9.5 GeV. Beyond the Υ, i.e., for $E > 10$ GeV, we see R approaching the value predicted for five flavors of point quarks, with three colors:

$$R = 3\left[\left(\frac{2}{3}\right)^2 + \left(\frac{-1}{3}\right)^2 + \left(\frac{-1}{3}\right)^2 + \left(\frac{2}{3}\right)^2 + \left(\frac{-1}{3}\right)^2\right] = \frac{11}{3},$$

due to u, d, s, c, and b quarks in three colors.

(15.11)

The jump in R, by the amount expected as each new set of resonances is found, is convincing evidence that the resonances are indeed due to the production of a new flavor. The fact that the values are all three times those expected in the absence of color, provides strong evidence that there are indeed **three colors of each flavor**.

  The detection of two-jet events also provides evidence for the existence of quarks, and we shall discuss both two-jet and three-jet events in our discussion of the evidence for QCD. Indeed, all of the evidence for the various parts of the standard model are implicit evidence for the existence of quarks.

Higher Generations No evidence of higher generations of quarks has been found in any experiment. Furthermore, the requirement that the number of quark generations must equal the number of lepton generations (for anomaly cancellation) brings the evidence that $N_\nu = 3$ to bear on this question.

15.2 EVIDENCE FOR THE ELECTROWEAK THEORY

15.2.1 Quantum Electrodynamics

In 1905, following the discovery of the quantization of the action by M. Planck (in units of h), Einstein found,[17] within the resulting blackbody radiation law, strong reason to believe that light came in energy quanta (of size $h\nu$). It wasn't until 1916 that Einstein realized that they were momentum quanta as well, so the existence of the gauge particle of electrodynamics was tortuously established over a long period. The photoelectric effect and Compton scattering confirmed the existence of the photon; furthermore, an enormous amount of detailed evidence of the quantization of electrodynamics was accumulated during the first half of the twentieth century. The first theoretical formulation of quantum electrodynamics was due to Dirac, even before his discovery of the equation for a spin $\frac{1}{2}$ particle. However, the modern form, including the understanding of higher-order terms using renormalization, was created in the 1940s by

---

17. He noticed a Boltzmann factor $e^{-h\nu/kT}$ in Wien's blackbody radiation law and used classical (Boltzmann) statistics in his argument; Bose-Einstein statistics hadn't yet been invented. Luckily, the two statistics give the same results in the circumstances considered — serendipity!

Tomanaga, Schwinger, and Feynman (independently of one another). The predictions, which have been verified in many detailed experiments, are too numerous to mention. Here, we shall discuss the two results which are reputed to have convinced the entire physics community.

1. In 1947 Lamb and Retherford precisely measured an energy difference between the $2S_{\frac{1}{2}}$ level and the $2P_{\frac{1}{2}}$ level of hydrogen, which the Dirac equation (with no quantization of the electromagnetic field) predicts to be degenerate. They found the energy difference to be $\sim 10\%$ of the fine-structure splitting (between levels of different j), which agreed with the prediction of the new quantum electrodynamics (produced by the renormalization of radiative corrections). Agreement between theory and experiment has been verified to many decimal places. This is referred to as **the Lamb shift.**

2. In an atom, the magnetic moment of a spinless particle of charge q and mass m is proportional to its orbital angular momentum: $\vec{\mu}_\ell = g_\ell(q/2m)\vec{L}$ and $g_\ell = 1$. If the particle has spin, its spin angular momentum contributes to the magnetic moment but with a different g-factor:

$$\vec{\mu}_S = g_S \frac{q}{2m}\,\vec{S} \text{ and } g_S = 2 \text{ is predicted from the Dirac equation.}$$

Detailed QED calculations, including radiative corrections, predicted that $g_S \approx 2.0023$; the deviation from 2 has been calculated to many more (nine) decimal places than shown here and is slightly different for electrons and muons.[18]

One class of experiments utilizes polarized electrons (or muons) in a magnetic field. Let us consider a spin $\frac{1}{2}$ particle of charge q, mass m, and momentum $\vec{p}$ in a magnetic field ($\vec{B}$). It is left to the reader (Problem 11) to show that, nonrelativistically:

$$\frac{d(\vec{S}\cdot\vec{p})}{dt} = (\vec{\mu}_s \times \vec{B})\cdot\vec{p} + \vec{S}\cdot\left(q\,\frac{\vec{p}}{m} \times \vec{B}\right) = \frac{g_s - 2}{2}\,\frac{q}{m}\,\vec{p}\cdot(\vec{S} \times \vec{B}),$$

(15.12)

so that the extent to which the particle's helicity changes is a direct measure of the amount that g_s differs from 2.

Extremely ingenious **(g − 2) experiments** for both electron and muon have been done over the years, whose results agree with the QED prediction (to nine decimal places).

These experimental verifications of QED have checked that theory to a much higher accuracy than any other theory has ever been tested.

---

18. This difference is due to their disparate masses and a very small, but significant, contribution from virtual hadrons.

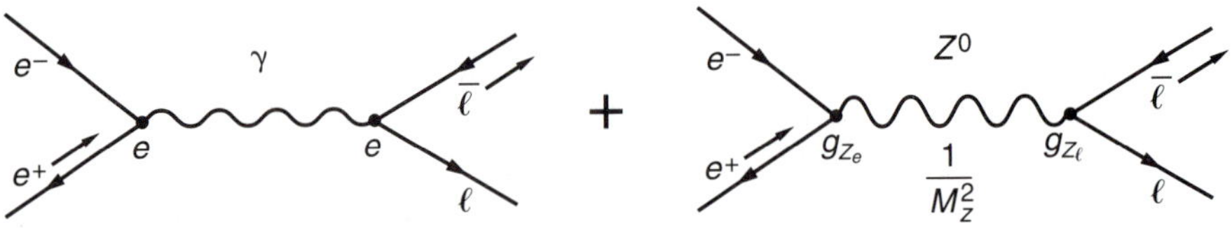

FIGURE 15.11

Tree-level Feynman diagrams contributing to the amplitude for charged-lepton (ℓ) pair production in an e^+e^- collision, showing coupling constants ($g_{Z_e} \sim e$) and the effective propagator ($\sim 1/M_Z^2$) of the virtual Z^0.

15.2.2   Weak Interactions

The weak interactions had been studied experimentally since the beginning of the twentieth century and very intensively since the 1930s. We have already talked about the charged-current decays and the handedness discoveries. Before the electroweak theory was proposed, it was not known how to construct a logically consistent (renormalizable) theory. The prediction and experimental discovery of the violation of parity (P) and the apparent CP conservation followed by the discovery of a small CP violation has been discussed in Chapter 12 (Discrete Symmetries). Here, we will look at some of the evidence which convinced people of the validity of the electroweak theory.

The electroweak theory was constructed (in the mid-1960s) to be consistent with all the experimental observations made to that time. In addition, the theory contains the Z^0 which leads to the prediction of the existence of neutral-current interactions (due to Z^0 exchange). The search for the latter was the first major test of the electroweak theory. In 1973, at CERN, a large number of $\nu_\mu + N \to \nu_\mu +$ hadrons events (see Fig. 15.5) were seen, verifying the existence of the neutral-current interaction. Furthermore, the sine squared of the undetermined mixing angle (θ_W) in the electroweak theory was determined from the data to be $\sin^2 \theta_W \approx .22$.

The contribution of Z^0 exchange to the scattering of *charged* leptons is considerably smaller than that of photon exchange (pure QED) because of the large mass of the Z^0, as we have seen in Chapter 11. However, we must add the two amplitudes before squaring. Actually, $e^+e^- \to \mu^+\mu^-$ was studied instead of e-μ scattering because $e^+e^- \to \mu^+\mu^-$ has a signature that makes it more amenable to experiments. (From our discussion in Section 4.4.2 we infer that the same Feynman diagram describes both of these processes.) The diagrams contributing to the amplitude for the production of charged-lepton pairs is shown in Fig. 15.11. For the contributions of these diagrams, we may symbolically write:

$$\frac{d\sigma}{d\Omega} = K|\alpha M(\gamma) + GM(Z)|^2 = \left(\frac{d\sigma}{d\Omega}\right)_{QED} + \alpha G \left(\frac{d\sigma}{d\Omega}\right)_{crossterm} + O(G^2),$$

$$\text{with} \left(\frac{d\sigma}{d\Omega}\right)_{QED} \sim \alpha^2, \tag{15.13}$$

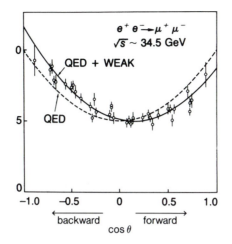

FIGURE 15.12
Angular distribution for muon pair production at an e^+e^- collider (PETRA). θ is the angle between the incoming and the outgoing directions in the CM frame.

where α is the fine-structure constant ($\sim e^2$) and $G(\sim e^2(m_\ell^2/M_Z^2))$ represents the very small effective coupling constant of the weak interaction (Z^0 exchange diagram). The cross-term $O(\alpha G)$ is large enough to have a small effect on the results, as shown in the data in Fig. 15.12. The asymmetry around $\cos\theta = 0$ (forward-backward asymmetry) due to the interference term is clearly evident in the data.

These data also give a value $\sin^2\theta_W \approx .22$, as do the data for deep inelastic scattering discussed in Section 15.1.2. The fact that all the data are explained by the same value of this free parameter gives us confidence in the validity of the theory. The **currently (1992) accepted world value** (with some hidden phenomenological assumptions) is $\sin^2\theta_W = .2325 \pm .0008$.

15.2.3   The Gauge Bosons — $W^\pm$ and Z^0

A *gauge* theory is not verified, in the minds of some, until the gauge particles are found. The predictions for the masses of the $W^\pm$ and Z^0 were discussed in Chapter 11, where we saw that

$$M_W = \frac{g_2 v}{2} \text{ and } M_Z = \frac{v}{2}\sqrt{g_1^2 + g_2^2}. \tag{15.14}$$

We also saw, at the end of Chapter 10, that the determination of $\sin^2\theta_W$ yields

$$g_1 \approx 1.14e \text{ and } g_2 \approx 2.07e. \tag{15.15}$$

The one number remaining to be determined, before we can predict the masses, is the vacuum-expectation-value v. This is easily determined using the weak interaction data, which have been accumulated for most of the twentieth century. The data for leptonic decay, e.g., $\mu^- \to \nu_\mu + e^- + \bar{\nu}_e$, are most easily inter-

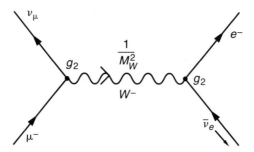

FIGURE 15.13
Tree-level diagram for μ^- decay: $\mu^- \rightarrow e^- + \nu_\mu + \bar{\nu}_e$, showing the couplings (g_2) and the effective propagator of the virtual W (since $q^2 \sim m_\mu^2 \ll M_W^2$).

preted, since leptons are fundamental particles. The electroweak theory predicts these charged-current decays via tree-level Feynman diagrams, like that shown for μ decay in Fig. 15.13.

Since these decays occur at energies far below the expected W mass, the denominator of the propagator may be approximated by $(M_W)^2$. This gives an *effective* coupling G_F (the F is for Fermi) of the four leptons: $G_F \propto g_2^2/M_W^2 = 4/v^2$. The Fermi constant ($G_F$) has been very well determined by decades of measurements of the rate of μ decay. From the value of G_F, that of v has been determined to be $ve \approx 75.7$ GeV, yielding the predictions (from Eq. 15.14) $M_W \approx 78$ GeV $\rightarrow 80$ GeV (with radiative corrections) and $M_Z \approx 89$ GeV $\rightarrow 91$ GeV (with radiative corrections).[19] In order to produce these very massive particles, an accelerator must have at least that much energy in the CM frame; as we have seen, this is only practical for colliders. In 1983 two different detector facilities, UA1 and UA2 at the $S\bar{p}pS$ CERN collider, found the $W^\pm$ and Z^0 right at the masses predicted.

These discoveries were, again, a tribute to the ingenuity of the experimental groups involved. From the enormous flood of events occurring, they selected those with a unique signature to leave no doubt that they had discovered these "relatives" of the photon. These particles are expected to decay with a width ~ 2.7 GeV, which corresponds to a lifetime $\tau_{W,Z} \sim 10^{-24}$ s, so that they can be found only by studying resonances of their decay products. The events searched for were

$$W^\pm \rightarrow \ell^\pm + \nu \text{ and } Z^0 \rightarrow \ell^+ + \ell^-,$$

whose signatures are completely different from just about all other events, as we shall soon discuss. The process of production and decay is shown in Fig. 15.14. They occur at a rate of 1 in 10^9 scatterings, but their unique signatures make them stand out from the crowd! (In technical language: They have very small branching ratios, but their signatures have very low backgrounds.)

---

19. The contribution of radiative corrections to the masses depends somewhat upon the mass of the top quark, since it must be included as a virtual particle.

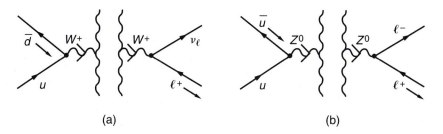

FIGURE 15.14
Sketches (not Feynman diagrams) of (a) a typical production mechanism for the W^+ in a high-energy $p\bar{p}$ collision (the u is a valence quark in the p and the $\bar{d}$ in the $\bar{p}$) followed by a leptonic decay and (b) a similar production mechanism for the Z^0 followed by a leptonic decay. **A tear is placed in the middle of each, to indicate the production of real W's and Z's, which subsequently decay.**

The charged lepton(s) produced in these two-body decays are easily distinguished from those produced in other ways, since they have very high energies, e.g., for a W decay, $p \approx M_W/2 \approx 40$ GeV,[20] producing a single isolated high momentum track, and similarly for Z decay, producing two 45-GeV tracks. The identities of the leptons are easily determined: *electrons* from the showers they produce in the electromagnetic calorimeter, with no accompanying large energy deposition in nearby hadronic detectors and no penetration of material to muon detectors; and *muons* by their penetration to reach the muon detectors. In Fig. 15.15 we show an angular plot of one of these events, which are often referred to as "lego" plots due to their resemblance to those popular toy blocks. It took only a handful[21] of such events (at both UA1 and UA2) to convince the world.

A great bonus was obtained from *details* of the accumulated data. Studies of the asymmetry in decay (to leptons) direction verified the left-handed $(V - A)$ coupling and the spin 1 property required of gauge bosons. As an example, let us examine

$$p + \bar{p} \to W^+(+X) \to \ell^+ + \nu_\ell(+X) \text{ or } p + \bar{p} \to W^-(+X) \to \ell^- + \bar{\nu}_\ell(+X).$$

Our analysis is shown in Fig. 15.16, where the momenta are shown as single arrows and the spins as double arrows. We have taken the direction of the p to

---

20. Actually, those quarks (in the proton and antiproton) which do not produce the W hadronize to particles, some of which continue down the beam pipe. This leads to a loss of some energy and longitudinal momentum, so that the identification of the neutrinos (produced in W decay) as lost energy and momentum is accomplished by measuring only transverse momenta. Also for Z decay, transverse quantities are measured. The analysis of such data for resonances requires detailed considerations beyond the scope of this book.

21. At UA1, six events with the signature of the W were found, out of 10^9 recorded events, as expected from the standard model. (Their first publication had 135 authors from 14 institutions.) Five Z events at UA1 and eight at UA2 were reported shortly thereafter.

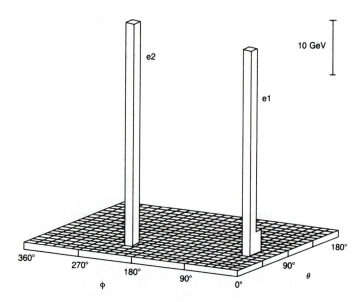

FIGURE 15.15

$Z^0 \rightarrow e^+ + e^-$ event found at UA1. Transverse energy is plotted as a function of angles. The two isolated peaks at very high energies are characteristics of an electron-positron pair.

be the z direction, so that the $\bar{p}$ is moving in the $-z$ direction. The constituents, although they may have some small transverse motion, are essentially moving in the same direction as the composite particles at these high energies. The theory says that a W has vertices only to left-handed particles or right-handed antiparticles. Consequently, the quark (u for W^+ production and d for W^- production) must be left-handed (spin antiparallel to momentum) to interact at a vertex with the W, so that $S_z^q = -\frac{1}{2}$. Likewise, the antiquark ($\bar{d}$ for W^+ production and $\bar{u}$ for W^- production) must be right handed (spin parallel to momentum). Since the antiquark is moving in the $-z$ direction, $S_z^{\bar{q}} = -\frac{1}{2}$. Thus, the W produced has $S_z^W = -1$. This decays to a left-handed lepton and a right-handed antilepton, whose S_z values must add to -1. Since they are both $S = \frac{1}{2}$ particles they must each have $S_z = -\frac{1}{2}$.[22] Thus, the lepton (ν_ℓ for W^+ or ℓ^- for W^-) cannot be moving in the $-z$ direction, since then it would be right-handed.[23] As you can see from Fig. 15.16, we expect **the ℓ^- to be mov-**

---

22. You might be wondering whether we have suddenly invented some sort of spin conservation here, but this is not the case. For particles moving in the $\pm z$ direction, $L_z = 0$, so that this is the usual conservation of angular momentum.

23. We have studied the fact that naive handedness arguments are strictly true only for massless particles, but the leptons produced here are at such high energy that these are good qualitative statements. In fact, as we have explained in Chapter 10, the form of the interaction is exact, and the left-handed projection operator gives a definite result for massive particles.

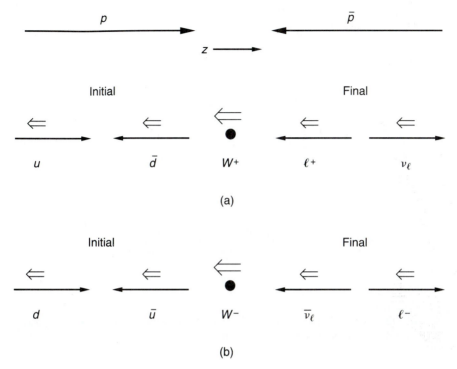

FIGURE 15.16
$W^\pm$ production and (forward) leptonic decay at an e^+e^- collider. The single arrows are the momenta and the double arrows are the spin directions.

ing predominantly in the same direction as the q, which is the direction of the proton, and similarly the ℓ^+ should be **moving predominantly in the $\bar{q}$ or antiproton direction.** The detailed electroweak calculation predicts that the decay distribution of the produced leptons varies as $(1 + \cos\theta)^2$, where θ is the angle between the lepton and the proton beam or the antilepton and the antiproton beam. (This *is* 0 for $\theta = \pi$, so that indeed no lepton will be produced moving in the $\bar{p}$ direction nor antilepton in the p direction, as we have just asserted.) In Fig. 15.17 we see that the data confirm this prediction, which was based upon the handedness of the leptons and the spin 1 designation of the gauge particles.

As of 1992, a combination of the detailed experimental studies of reactions at LEP, SLC, etc., has verified the standard model predictions of the couplings of the gauge particles to quarks to remarkable precision: W-u-d + W-c-s to 1%; Z-e-e, Z-μ-μ, and Z-τ-τ, each to $\frac{1}{2}$%; Z-ν-ν (all three generations) to 1%; Z-q-q (all three generations) to $\frac{1}{2}$%, which even includes Z-t-t to about 20% (using $m_t = 130 \pm 30$ GeV).

The discoveries and analyses of the W's and Z, along with these detailed results about the vertices and the evidence for three generations discussed in Section 15.1, have convinced even the most cynical skeptics that the electroweak

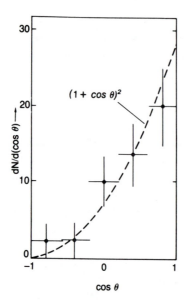

FIGURE 15.17

Angular distribution of electrons produced in W decay: $W \to e + \nu$. θ is the angle between e^- and p directions or between e^+ and $\bar{p}$ directions.

theory is valid for energies below about 300 GeV. Any attempt at a broader, more expansive theory must reduce to the electroweak theory in this range of its validity, just as the mechanics based upon Einstein's special theory of relativity had to reduce to Newtonian mechanics at speeds well below c.

15.3 EVIDENCE FOR QUANTUM CHROMODYNAMICS

15.3.1 The Number of Colors

We have already discussed (in Section 15.1.2) the observation, at e^+e^- colliders, that an extra factor of 3 is required to account for the ratio (R) of the inclusive hadronic production cross section to the muon pair-production cross section. In Chapter 6, we saw that the dilemma of the wrong[24] spin-statistics connection for quarks is resolved by this new degree of freedom with three colors.

Further evidence for color was found in the decay $\pi^0 \to 2\gamma$. Its amplitude had originally been calculated[25] without considering color, and it yielded a result which was a factor of 3 too small. When we include the three colors, we

---

24. This spin-statistics connection is the opposite of that expected from very general quantum field theory considerations.

25. This calculation used properties of the gauge theories which are beyond the level of this book.

obtain that missing factor of 3. This yields **an extra factor of 9 in the decay rate** (since observations always involve the absolute squares of amplitudes), so that **the observed rate of decay of the π^0 is strong evidence for the three-color prediction**.

All of these observations involve *large* factors, which jump out at us from the data and certainly will not disappear if improved experimental precision yields accuracy to more decimal places. Furthermore, these factors are not easily accounted for in any other way. They certainly point to the existence of another degree of freedom with three possible values.

15.3.2   The Gauge Theory

The verification that the new degree of freedom is associated with a gauge theory, requires just the other kind of data, viz., *detailed* experimental studies of QCD predictions.

Very precise and beautiful work has been done studying the resonance states of those $q\bar{q}$ systems in which the quark and antiquark have the same flavor, namely $b\bar{b}$, $c\bar{c}$, and $s\bar{s}$. These systems are collectively referred to as **quarkonium systems** (named in analogy to the name used for the $e^- e^+$ bound system — positronium). For each type of quarkonium system, many different energy levels have been observed, and transitions between those levels have been measured as well. For example, the $c\bar{c}$ levels ψ and ψ', whose discoveries we have already mentioned, are part of a large charmonium spectrum. In Fig. 15.18 we see the lowest-lying levels of charmonium and bottomonium (the primes in the names have been dispensed with). The $^3S_1$ states are the ones produced directly at $e^+ e^-$ machines (via quark pair production, as shown in Fig. 15.10) since the virtual photon has $J^{PC} = 1^{--}$. The other states result from transitions from the $^3S_1$ states. For charmonium, those with energies (masses) below $2m_D$ cannot decay into charmed mesons, and so decay much more slowly due to Zweig suppression (the OZI rule).[26] That is why they are produced as very narrow peaks. Particles with masses above the threshold for decay into charmed mesons have widths which are ~100 to 1,000 times as large (so that the lifetimes are correspondingly shorter). Thus, the detailed values of the masses are responsible for the qualitatively dramatic phenomena that led to our discovery of this new flavor. Similar remarks apply to the $b\bar{b}$ system and the B meson mass. As we have already discussed in Section 15.1.2, the top quark decays so rapidly (because of its large mass) that toponium would not form (as a measurable state).

The calculations for quarkonium systems require considerably more skill and insight than is required for positronium for two major reasons:

1. As we learned in Chapter 9, the gluons themselves carry color, so that we have gluon-gluon vertices (both three-gluon and four-gluon vertices

---

26. That rule tells us that processes in which no initial quark lines appear in the final state, i.e., which have the initial and final sets of quarks connected only by virtual gluons, are suppressed, as we have already discussed in Chapters 6 and 9.

The Charmonium System

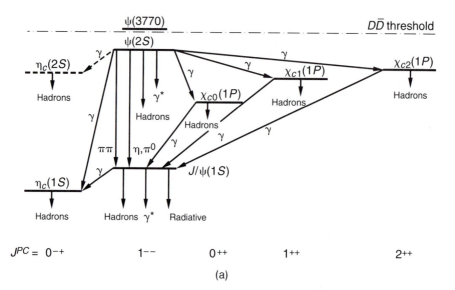

FIGURE 15.18

Energy levels of (a) the charmonium ($c\bar{c}$) system and (b) the bottomonium system ($b\bar{b}$). Numbers in parentheses are the masses in MeV. Dashed levels have been predicted theoretically, but not yet found. Transition modes are shown, with γ^* meaning via a virtual photon, which includes decays to lepton pairs. Levels above the $D\bar{D}$ or $B\bar{B}$ thresholds, respectively, can decay into those pairs (via non-Zweig suppressed decays). [Source: *Review of Particle Properties*, 1992, Phys. Rev. D **45, Part II**]

occur). Thus, the types of Feynman diagrams in QCD are much more numerous, and some require more subtlety to calculate (including renormalization) than those of QED (which are not trivial to calculate either).

2. The QCD potential should have a linear tail for large r, corresponding to confinement, so that we expect it to have the form $V = (\ )/r + kr$. The $1/r$ part is calculated (perturbatively as in QED) by the use of Feynman diagrams, but the linear part cannot be determined that way.

Using part of the data, the linear coefficient has been determined ($k \approx 1$

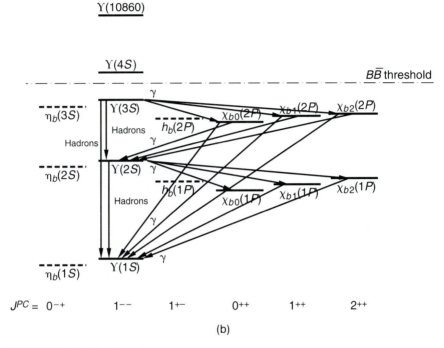

FIGURE 15.18 *Continued*

GeV/fm), and the short-range Feynman diagram calculations have been done to such accuracy as to match the observed energy levels and widths, within experimental error.

Another verification of the gauge nature of the strong interaction is the observation of bremsstrahlung-like events. In QED, for every process of elastic scattering of *charged* particles from one another, there is a bremsstrahlung process (with the emission of one extra photon). These bremsstrahlung events occur at a rate and with an angular distribution which are calculable in QED, and the predictions have been experimentally verified. Likewise, in QCD, for each type of elastic quark-quark scattering event, there is scattering with the emission of a gluon (therefore inelastic). Predictions have been made from QCD for the rate of these bremsstrahlung events relative to the elastic-scattering rate; angular distributions have also been calculated. The reader might think that such calculations are moot, since color confinement tells us that the quarks and gluons will hadronize. Thus, all we will see is a large number of hadrons in (possibly overlapping) jet distributions, each jet being the result of the hadronization of one emerging colored particle. Well, that's all right! Studies have been made

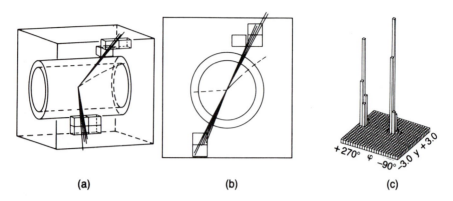

(a)   (b)   (c)

FIGURE 15.19
A two-jet event in the UA1 detector at CERN. (a) A perspective drawing showing only
the high-energy hadron jets. The blocks represent activated calorimeter modules. (b) View
down the beam pipe, in which the event looks like elastic (back-to-back) scattering.
(c) Lego plot of the jets. (y is the rapidity, a measure of the longitudinal speed, as discussed in Appendix E.)

of jets, and criteria have been developed to decide which particles belong to
which jet. Although there sometimes is ambiguity in these assignments, they have
been very successful. In Fig. 15.19 we show one of the many two-jet events,
whose analyses were part of the evidence for the quark (parton) structure of the
nucleon. In Fig. 15.20 we can see a three-jet event and the plot of the angular
distribution which verifies that the gluon has spin 1.

15.3.3   Glueballs (Gluonium) and Non-$q\bar{q}$ States

Well, what about the gluon-gluon states (discussed in Section 9.5) or $q\bar{q}g$ states?
There have been a number of reported non-$q\bar{q}$ candidates in the experimental
literature. In order to be considered gluonium states, such states should have
flavor-singlet couplings (i.e., they should have the same couplings to all flavors of quarks); enhanced production in gluon-rich decays, such as the OZI-
suppressed J/ψ decay; and reduced couplings to the two-photon state (since
gluons do not couple to photons). Experimentally, we might expect the mixing
of states, so that the identification of an observed state as a non-$q\bar{q}$ state may
not be certain. The mixing of such a state with a conventional ($q\bar{q}$) state may
not even be observable, so that the only clear signal might be an enhancement
of the *number* of states. There are several dozen such states, listed in the 1992
Review of Particle Properties, whose common property is that there has been
some difficulty in their identification as ordinary $q\bar{q}$ states, but no firm theoretical predictions have been embraced by a consensus of researchers so far.
There has been *some* advancement made, as was already seen in the 1990 tables,
by the creation of a more conventional name: **gluonium states** for what had been
originally referred to as "glueballs."

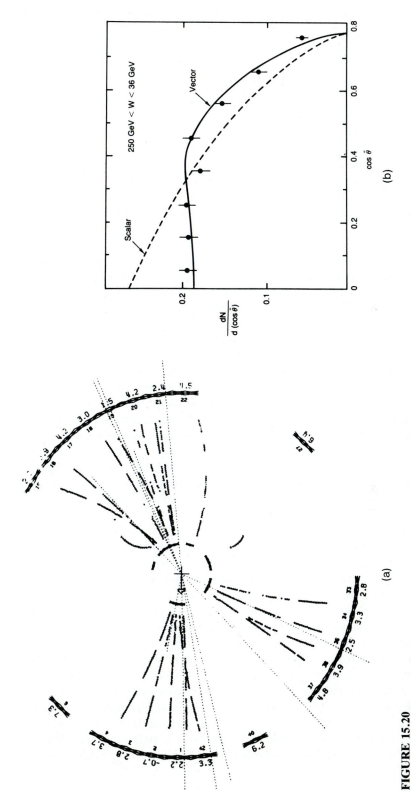

FIGURE 15.20

(a) Three-jet event observed in the JADE detector at the PETRA e^+e^- collider (DESY). (b) Distribution of the highest-energy jet with respect to the line of motion of the other two in their CM frame. The solid curve is the prediction for vector gluons, the dashed curve for scalar gluons. (Data are from three-jet events in the TASSO detector at PETRA.)

15.4   WHERE IS THE HIGGS?

There is another missing piece of crucial evidence for the standard model besides the missing top quark (which we have already discussed) — **there is no experimental evidence for the Higgs particle!**

As we have seen in Chapter 11, the Higgs coupling to fermions is expected to be proportional to the fermions' masses. Therefore, we can justify the absence of evidence of the Higgs particle, in the detailed experiments done so far (on relatively low-mass fermions). It is suspected, from a combination of calculations and experimental results, that the mass of the Higgs particle of the standard model exceeds 56 GeV. We have seen in Eq. (11.20) that the mass of the Higgs depends upon the Higgs self-coupling coefficient λ, whose value is completely unknown. Thus, it is not even known what range of masses to investigate. We expect that either (i) detailed experiments with the *heaviest* fundamental fermions will be required in order to find the Higgs or (ii) since the coupling of the Higgs to the transverse (W and Z) gauge bosons is proportional to their masses and its coupling to the longitudinal gauge bosons $\sim M_H/M_G$ (which would be large for large M_H), detailed studies of high-energy scatterings of gauge particles from one another may yield the evidence we seek.

In models that are extensions or alternatives to the standard model further spontaneous symmetry breaking is included, so that more Higgs particles are predicted, as we shall learn in Chapter 17. (We shall see that one class of such theories, GUTs with supersymmetry, has the attractive feature that the Higgs mechanism (μ^2 becoming negative) arises in it automatically.) So, a lot rides on the experimental discovery of this supposed cause of spontaneous symmetry breaking.

Of course, the existence of a field operator which has a non-zero vacuum-expectation-value does not require that operator to correspond to a fundamental (Higgs) field. The operator might be a *product* of operators of fundamental fields (such a combination is sometimes referred to as a "condensate"). In fact, theories have been constructed which do not contain a fundamental Higgs, most notably the technicolor theories we mentioned in Chapter 11 (footnote 12) and will discuss further in Chapter 19. There have been technical difficulties within these theories, in their present forms, that make them inconsistent with experiment. However, we learn from them that theories with no fundamental Higgs would most likely have a rich spectrum of particles in the TeV range. (So would a supersymmetric world, which *does* have fundamental Higgs particles.) Perhaps, data from the SSC will resolve this question.

If the Higgs is very heavy (several TeV), it will have many decay channels and, consequently, will show up only as a wide resonance, which will be extremely difficult to find. Furthermore, the existence of a very heavy Higgs would imply that the quartic coupling constant λ in the standard model is very large (since $m_H \sim \sqrt{\lambda}$, as we found in Eq. 11.20). Consequently, Feynman diagrams with virtual Higgs particles would make very large contributions for some processes, rendering our perturbation calculations invalid because of apparent violations of the unitarity limit. We would thus be unable to calculate amplitudes for processes such as W-W scattering. So, **the nonexistence of the Higgs or a very heavy Higgs would tell us that there is something very wrong with the standard model**.

These considerations, along with experience gained in the development of our understanding of the weak interactions, lead some researchers to believe that we will certainly find new, enlightening physics at the SSC — the Higgs and/or a spectrum of new particles.

PROBLEMS

1. Show how the existence of the $\mu \rightarrow e + \nu + \bar{\nu}$ process implies the existence of the $\mu \rightarrow e + \gamma$ process, if the neutrino and antineutrino appearing were antiparticles of one another, by constructing the corresponding Feynman diagrams. (Remember that neutrinos have no charge.)

2. Consider the SPEAR e^+e^- storage ring. The basic processes are the production of fundamental particle pairs, as shown in Fig. 15.10.
   a. Can a $b\bar{b}$ pair be produced at SPEAR? Explain.
   b. Consider all the pairs that the virtual photon can produce at SPEAR to find the branching ratio for τ pair production.
   c. About 17.8% of τ decays are to a μ and about 17.7% of τ decays are to an e. Find the effective branching ratio for the process used to discover the τ's:

   $$e^+ + e^- \rightarrow \mu^+ + e^- + \text{much missing energy and momentum.}$$

   d. Find the number of $e^+ + e^- \rightarrow \mu^+ + e^- +$ much missing energy and momentum events produced per hour at SPEAR.

3. a. For a particle with total energy E and mass m, show that in one lifetime (t_{life}) it travels a distance: $d = \sqrt{\gamma^2 - 1} \, (ct_{life})$, where $\gamma = E/mc^2$. Note that the lifetime is the time measured in the rest frame of the particle and E includes the rest energy (mc^2).
   b. How far will a muon with 100 MeV kinetic energy travel in one lifetime?
   c. The lifetime of the τ is about 3×10^{-13} s. How far will a τ produced at SPEAR (whose beams have energy: 4.2 GeV + 4.2 GeV) travel in one lifetime? Do you think that the researchers should have looked for tracks? Note that the lifetime is the time measured in the rest frame of the τ.

4. Deuterons formed in the early Universe are unstable in the large photon background until $(n_\gamma/n_b)e^{-E_B/kT} < 1$, where $E_B = 2.2$ MeV. Find the value of kT in MeV for which $(n_\gamma/n_b)e^{-E_B/kT} = 1$. What is T in °K?

5. The mean lifetime of the neutron is about 889 s. Find the time (in the early Universe) for the n/p ratio to drop from $\frac{1}{6}$ to $\frac{1}{7}$. Remember that each n which decays produces a p.

6. It has been shown (see the miscellaneous references in the Bibliography) that the state of a proton with spin up may be written: $N\varepsilon_{\alpha\beta\gamma}(d_\downarrow^\alpha u_\uparrow^\beta - d_\uparrow^\alpha u_\downarrow^\beta)u_\uparrow^\gamma$ and the state of a neutron with spin up may be written: $N\varepsilon_{\alpha\beta\gamma}(d_\uparrow^\alpha u_\downarrow^\beta - d_\downarrow^\alpha u_\uparrow^\beta)d_\uparrow^\gamma$. Here, each quark symbol stands for the quark that is in the state, the spin is shown with arrows, and the α, β, and γ are color labels. Show that this produces the prediction $\mu_n/\mu_p = -\frac{2}{3}$, assuming $m_u = m_d$, when we merely add vectorially the spin magnetic moments

of the three quarks in each term and then *add* vectorially those results. Be aware that the relative minus signs *between* terms are irrelevant for this calculation, since $(|a_i\rangle - |a_j\rangle)^\dagger A(|a_i\rangle - |a_j\rangle) = a_i + a_j$, where $|a_i\rangle$ and $|a_j\rangle$ are orthonormal eigenstates of A (with corresponding eigenvalues). Recall that $\vec{\mu} = (g_s\hbar/2mc)q\vec{S}$ for each quark in each product, where q is the charge of the quark, and note that the constants in the parenthesis of this expression cancel out of the ratio.

7. a. Find the sign of q^2 in Fig. 15.5. Hint: It may be useful to call the four-momentum of the outgoing lepton p_4. Then, we have $q = p_2 - p_4$, where p_2 and p_4 are the four-momenta of physical particles.

    b. Show that for deep inelastic scattering $M_f^2 > M_N^2$. Hint: Recall baryon number conservation.

    c. Use the results of parts a and b to show that the range of $x \equiv -q^2/2q \cdot p_1$ is $0 \le x \le 1$.

8. Find the lifetime (τ) for the J/ψ ($\Gamma = .06$ MeV) and for the Υ ($\Gamma = .05$ MeV).

9. a. Find the expected value of $R = \sigma(e^+ + e^- \to hadrons)/\sigma(e^+ + e^- \to \mu^+ + \mu^-)$, above the top quark mass.

    b. Why is a constant value of R away from resonances an indication of the point-like nature of the quarks?

10. Calculate g_Z for the b quark and for the t quark.

11. Verify Eq. (15.12), which shows that $d(\vec{S} \cdot \vec{p})/dt$ is proportional to $(g - 2)$.

12. a. For polarized charged particles in a magnetic field, show nonrelativistically that $d^2(\vec{S} \cdot \vec{p})/dt^2 \sim (g_s - 2)^2 F$, where F is independent of g_s.

    b. Show that $d^n(\vec{S} \cdot \vec{p})/dt^n \sim (g_s - 2)^n f$, where f is independent of g_s.

    This shows that the change of helicity of the particle as it moves through the magnetic field is a direct measure of $(g_s - 2)$; it is 0 for g_s exactly equal to 2. The result is valid relativistically as well. (The difference of the precession angular frequency of a particle's spin and that of its momentum, in the simultaneous presence of magnetic and electric fields, has also been shown to depend on g_s only in the combination $(g_s - 2)$.)

13. In a classic $(g - 2)$ experiment, created by H. R. Crane, longitudinally polarized electrons ($\vec{S}$ parallel to $\vec{p}$) are injected into a uniform magnetic field orthogonal to $\vec{p}$. Their momenta and spins precess in the magnetic field, and the difference of their ($\vec{p}$ and $\vec{S}$) angular rates is a direct measure of $(g_s - 2)$. Let us analyze the nonrelativistic situation. We will take the z axis in the direction of the magnetic field, $\vec{B} = B\hat{z}$, and the x axis along the initial direction of motion. Do this problem in natural units.

    a. Verify that $\vec{p} = p\cos\omega_c t\,\hat{x} + p\sin\omega_c t\,\hat{y}$, $\vec{S} = \frac{1}{2}\cos\omega_s t\,\hat{x} + \frac{1}{2}\sin\omega_s t\,\hat{y}$ satisfy the classical equations for $d\vec{p}/dt$ and $d\vec{S}/dt$ by finding ω_c and ω_s.

    b. Find the helicity, as a function of time, and show that it depends on g_s solely in the combination $(g_s - 2)$. Verify that it remains $+1$ for $g_s = 2$.

    c. For electrons, $(g_s - 2)/2 \approx 10^{-3}$. Approximately how many oscillations of helicity will be observed during a time interval of $t = 10^6\, 2\pi/\omega_c$?

14. Estimate G/α of Eq. (15.13) for muons.

15. Find the electroweak v, from $ve = 75.7$ GeV, where e is in Heaviside-Lorentz units, in which the fine-structure constant is defined as $\alpha = e^2/4\pi\hbar c \approx \frac{1}{137}$.

16. Find the τ that corresponds to $\Gamma = 2.7$ GeV.

Part D

Beyond the
Standard Model

Introduction

Every book of this type should have a discussion of the ideas that go beyond the **standard model**, even though there is no experimental data in conflict with it at this time. After all, this model, **or theory**, has **a large number of undetermined constants** (as we have pointed out), and its very structure seems quite arbitrary: **SU(3) × SU(2) × U(1) is not one simple group, but a product of** *three independently acting* **gauge symmetries**. Furthermore, there are the **missing Higgs** and **missing top quark** to contend with, although all the concepts we shall discuss assume that the top quark exists. Furthermore, we may ask:

1. **Why are there three generations?**
2. **Why is there a handedness property** (left-handed particles couple to W and Z, but right-handed particles do not)?
3. **What is the nature of baryon-number and lepton-number conservation?**
4. **The Y of the U(1) can be arbitrary, so why do the observed charges appear to be quantized?**
5. **Are neutrinos really massless, and, if they are not, where are the right-handed neutrinos?**

Other questions will arise as we proceed.

When I remarked to a colleague that this was the most difficult part of the book to write because of the constant invention of new ideas, he responded that one could easily write a whole book on the developments "beyond the standard model." Whatever would appear in such a book and, indeed, the material that appears here could not possibly remain up to date. However current the material may be at the writing, just after the last key is struck, there will be new ideas proposed and there may be new discoveries as well! Our aim here is to survey the ideas which have been proposed to date, looking at the rationale behind them

335

and their likelihood of being correct. We may presume that many of the ideas which will be proposed in the future will fall into the categories outlined here, although, of course, a completely new insight or discovery may be "just around the corner."

Chapter 16

QCD Revisited

Here we present (1) a brief qualitative discussion of nonperturbative developments in QCD, which may or may not be considered part of the standard model, and (2) the rationale for and a description of a planned experimental search for quark-gluon plasma.

16.1 NONPERTURBATIVE QCD – INSTANTONS, AXIONS, AND THE STRONG CP PROBLEM

The quantum field theory structure described in this book, by its very nature, produces perturbative solutions, e.g., via Feynman-diagram techniques. There is another approach to quantum mechanics and thence to quantum field theory, which can find states unobtainable from the perturbative treatment. It is referred to as the *path integral formulation* and is based on the consideration of classical paths, with the action $\left(\int \mathcal{L} \, d^4x\right)$ playing a major role. It is in this formulation of classical field theories that instanton states are understood (and were first found). Their important role in physical considerations shows up in the resulting quantum field theory. Although the path integral formulation of quantum field theory is beyond the level of this book, we shall describe the basic features of these nonperturbative solutions and show their connection to degenerate vacua, the strong CP problem, and axions.

Among the most important results obtained from the path integral formulation is that there exist degenerate vacua (soon to be described) and fields which produce transitions from one to another. Those fields can be found by first considering solutions to a modified classical field theory, with t(ime) replaced by $i\tau$ in the action, which changes the metric from Minkowskian to Euclidean $(g_{\mu\nu} = \delta_{\mu\nu})$, because solutions with **finite Euclidean action**[1] dominate the transitions from one of these vacua to another. In order for the Euclidean action

---

1. This is an analytic continuation of the action in the complex plane for time. The importance of the Euclidean action in path integral considerations is beyond the level of this book.

(which is an integral over all space-(τ)time) to be finite, $\mathcal{L}$ must vanish as we go to infinity in any direction in the Euclidean four-space. Therefore, the fields must satisfy

$$\lim_{r \to \infty} F_{\mu\nu} \to 0, \tag{16.1}$$

where $r(\to \infty)$ is the radius of a large three-sphere (at ∞) in the four-dimensional Euclidean space. Many solutions with these properties have been found, and they are called **instantons** because they are "localized" in time as well as in space.[2] They are important because transitions between the degenerate vacua of the theory are mediated by instantons.

Because of the gauge structure of QCD, the gauge field (G_μ) need not be 0 at Euclidean infinity to satisfy Eq. (16.1) but may instead be related to $G_\mu = 0$ by a gauge transformation; such a gauge field is referred to as a *pure gauge*, and the resulting $F_{\mu\nu}$ is still 0. In fact, from our discussion in Chapter 8, it is easily seen that, for $G_\mu = 0$, a gauge transformation produces a $\partial_\mu \alpha$ term (to order α) such that $F_{\mu\nu}$ remains 0; the exact result for the local gauge transformation $U(x)$ is (see Problem 4 in Chapter 8)

$$\vec{T} \cdot \vec{G}'_\mu = \frac{i}{g} [\partial_\mu U(x)] U^{-1}(x) \qquad \text{for a pure gauge,}[3] \tag{16.2}$$

$$\text{for which } F_{\mu\nu} = 0,$$

where, as x changes, U can run through all the elements of the gauge group. The elements of the group may be thought of as forming a group space,[4] so that U runs through group space as x changes. Looking along the surface at Euclidean infinity, we would not be surprised to find different values for G_μ in different directions as long as they were all pure gauges. As we move continuously around that surface at infinity, we expect that U would move continuously through the group space. However, as we proceed around and back to the same point on the infinite space-time surface, the U reached must match the operator determined there already, so that we would have gone in a closed loop in group space. There are closed loops in group space which are continuously deformable to a point, but those that cycle completely around group space may not be (continuously) shrunk down to a point, while remaining in group space.

As a simple example, consider the group U(1). Recall that one parameter (call it α) is needed to define the elements of the group ($e^{i\alpha}$). Thus, group space is one-dimensional; only one axis is needed to plot α. Furthermore, the

---

2. Finite *energy* solutions to field equations, which are fairly localized in space and remain localized as they move and interact, are sometimes referred to as solitons. Instantons are analogous to them, having finite Euclidean action instead of energy.

3. Here, x is a shorthand for all four space-time components.

4. In order to visualize the group space of SU(3), we can think of eight coordinate axes in an eight-dimensional group space, representing the eight parameters (α^a) that define the group elements ($\exp(i\lambda^a \alpha^a)$).

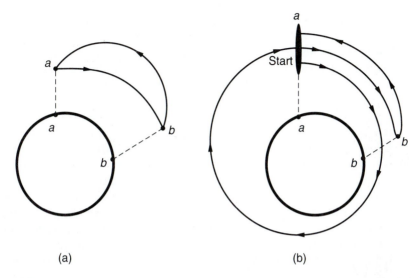

(a)   (b)

FIGURE 16.1

Inequivalent closed paths on the circumference of a circle, with the paths lifted and spread out for clarity. (a) A path $a \to b \to a$, which can be continuously shrunk to a point (as b approaches a), hence having winding number $= 0$. (b) A path $a \to a$ (once around) $\to b \to a$, which can be continuously shrunk to a single cycle (as b approaches a), but not to a point, hence having winding number $= 1$.

group elements repeat as α goes through 2π, so that we can plot the values required on a line segment which has been bent so that the 2π end coincides with the 0 end. More simply for this case, you can think of plotting $e^{i\alpha}$ in the complex plane, where all its values lie on a circle. Consequently, the group space is the circumference of a circle. A closed loop in this group space may be of the form: a to b and then follow the same path backwards to a, as shown in Fig. 16.1(a). We can continuously move the point b towards a, with each corresponding path entirely in the group space, until the paths have shrunk to a point. However, we may have a closed loop which goes completely around past a and then to b, as shown in Fig. 16.1(b). In this case, moving b to a continuously shrinks the path to a single cycle a to a.

In fact, as we move along the space-time surface at infinity, we may cycle through group space many times. The number of such cycles is called the **winding number**. Topological considerations show that, for the SU(3) gauge group, **a given path can be continuously changed into any other path with the same winding number**, but paths with different winding numbers cannot be connected in this way. This may be thought of as a ratchet effect, whereby once we have ratcheted up n complete turns, we cannot continuously go below that number. Mathematicians refer to these sets of paths, each set corresponding to a different winding number, as **homotopy classes**.

When we seek the vacuum state(s) (or lowest-energy state(s)) of the system, which must have $F_{\mu\nu} = 0$ everywhere, we find that there are an **infinite number of degenerate vacua**, one for each winding number ($n = 0,1,2,3,\dots$). This may be understood as follows: *For a physical state*, consider a gauge transformation ($g(x)$, where g are elements of the gauge group), which changes $U(x) \to g(x)U(x)$ in Eq. (16.2). The gauge transformations $g(x)$ at space-time infinity must be obtainable continuously (in group space) from the gauge transformation at the origin, as we move continuously (in space-time) out to space-time infinity. Therefore, $g(x)$ on the sphere at Euclidean infinity is in the same homotopy class as the gauge transformation on a "collapsed" (degenerate) sphere at the origin. On the "sphere" at the origin, the gauge transformation must be independent of angle, because that sphere is a single point; hence, it is a single (constant) transformation on that degenerate sphere. Moreover, a constant transformation is a *point* in group space and so has winding number 0. (It is in the same homotopy class as the identity element ($g = I$) of the gauge group.) Therefore, the physical gauge transformation $g(x)$ has winding number 0 and cannot change the winding number of the state under consideration. Consequently, **states with different winding numbers are not gauge equivalent to one another,** since they cannot be connected by a gauge transformation with winding number 0. In particular, vacua with different winding numbers are not gauge equivalent. Now, let us consider the fact that a vacuum state must be invariant to continuous coordinate transformations. Continuous coordinate transformations change **the mapping of the points in space** (in particular, on the surface at Euclidean infinity) **to the elements of the gauge group**, which is equivalent to keeping space unchanged but moving the paths in group space continuously. Thus, in a given homotopy class (i.e., for a given winding number), the vacuum must be composed of that linear combination (there is only one) of all possible paths (in that homotopy class) in group space which is invariant to a continuous sweep through those paths. We conclude that there is precisely one vacuum state for each homotopy class. However, path integral considerations show that these inequivalent vacua are not stable; instead, transitions occur from one to another, mediated by the instantons.

In order to understand the significance of the classical solutions in a Euclidean metric to vacuum-vacuum transitions, let us consider an analogy in quantum mechanics. Suppose the potential is

$$V = 1 - \cos x, \tag{16.3}$$

as shown in Fig. 16.2. Here, we have an infinite number of potential wells. If we had only one such well instead, the ground state (analogous to the vacuum state in a quantum field theory) would have a wavefunction fairly well localized within the well, as shown in Fig. 16.3. In the case of this periodic potential we expect that a state of lowest energy should have its wavefunction localized in one of the wells, with only a small part outside that well. (Imagine beginning with one well and the system in its ground state. Introducing further wells in the regions outside the original well should not perturb the state much, since

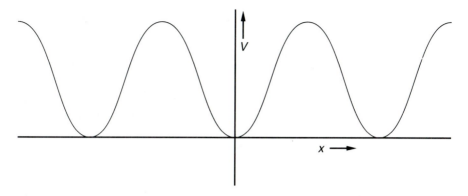

FIGURE 16.2
The periodic potential $V = 1 - \cos x$.

the wavefunction is small in those regions.) However, since there are an infinite number of wells, the localized wavefunction can be chosen to reside in any one of them. Each such choice produces a different ground state, but they all have the same energy. Thus, there are an infinite number of degenerate (same energy) ground states. If the potential V also had a term which would cause transitions among these degenerate ground states, our analogy would be complete. In that circumstance, the degeneracy of the levels would disappear, with the energies shifted to form a band. Consequently, there would be only *one* lowest-energy state.

Classically, a particle in one such well *could not* move to another, so that semiclassical solutions are not relevant to the transition process. However, if

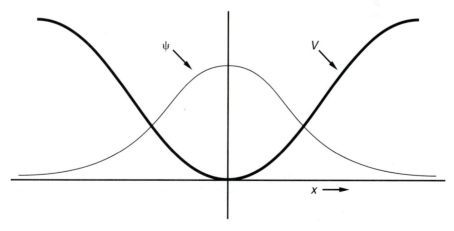

FIGURE 16.3
A single-well potential, showing the ground-state wavefunction (ψ).

we analytically continue to a Euclidean metric, the potential (in the new problem) is turned over so that the vacuum states are at maxima (as you may show in Problem 1) and *classical* fields (analogous to instantons) can connect one solution to another. After discussing the vacuum state in this way, we can then analytically continue back to Minkowski space.

In (the quantum field theory) QCD there exists a local *gauge transformation* (U_1) which represents the instability of the original, degenerate vacua (via instantons), as follows: Let $|n\rangle$ represent a vacuum state with winding number n. The operator U_1 raises the winding number by one, i.e., it has winding number 1 itself, hence,

$$U_1|n\rangle = |n + 1\rangle, \tag{16.4}$$

and, since H is gauge invariant,

$$[U_1, H] = 0, \tag{16.5}$$

so that after the degeneracy is removed, the vacuum would be an eigenstate of U_1. (See Problem 2.) A linear combination of the original, degenerate vacua which is stable (under U_1) is

$$|\theta\rangle \equiv \sum_n e^{in\theta}|n\rangle, [5] \tag{16.6}$$

which is indeed an eigenstate of U_1:

$$U_1|\theta\rangle = e^{-i\theta}|\theta\rangle, \tag{16.7}$$

where θ is arbitrary.

Instead of the explicit use of the state $|\theta\rangle$ as the vacuum state, it can be shown, in the path integral formulation, that this complicated vacuum state can be subsumed into an effective Lagrangian density with a θ-dependent term of the form:

$$\theta \, \text{tr}(F_{\mu\nu}\tilde{F}^{\mu\nu}) = \theta \, \text{tr}(F_{ij}\tilde{F}^{ij} + F_{0i}\tilde{F}^{0i} + F_{i0}\tilde{F}^{i0}),$$

$$\text{where } F_{\mu\nu} = \sum_{a=1}^{8} F_{\mu\nu}^a \lambda_a$$

$$\text{and } \tilde{F}_{\mu\nu} = \sum_{a=1}^{8} \varepsilon_{\mu\nu\rho\lambda} F^{a\rho\lambda} \lambda_a; [6,7] \tag{16.8}$$

---

5. To a student of condensed matter physics, this should be reminiscent of a Bloch wave in a periodic potential. There, the translation operator plays the role that U_1 plays here.

6. $\varepsilon_{\mu\nu\rho\lambda}$ is totally antisymmetric to any interchange of two indices, with $\varepsilon_{0123} = +1$. Thus, all the indices must be different for $\varepsilon_{\mu\nu\rho\lambda}$ to be non-zero, and its values are ± 1, depending on whether their order is an even or odd permutation of 0123.

7. λ_a are the 3×3 matrices representing the generators of SU(3) defined in Chapter 5.

"tr" means the trace of the matrix,[8] we have used the summation convention for repeated space-time indices, 0 is the time direction, and ordinary (Roman) letters i and j stand for space (1,2,3) directions.[9]

Looking solely at the space properties of this θ-dependent term (Eq. 16.8), it is not hard to see that **this term violates P** (inversion symmetry), since the field **F_{ij} has even parity and $\tilde{F}_{ij}$ has odd parity**:

$$F_{ij} \sim \partial_i G_j \rightarrow (-\partial_i)(-G_j) \rightarrow +\partial_i G_j - \text{even under } P - \text{and, similarly, } F^{0k} \text{ is odd;}$$

$$\text{whereas, } \tilde{F}_{ij} \sim \varepsilon_{ij\rho\lambda} F^{\rho\lambda} = \varepsilon_{ijk0} F^{k0} + \varepsilon_{ij0k} F^{0k} \rightarrow \varepsilon_{ijk0}(-F^{k0}) + \varepsilon_{ij0k}(-F^{0k})$$

$$= -(\varepsilon_{ijk0} F^{k0} + \varepsilon_{ij0k} F^{0k}) - \text{odd under } P,$$

so that $F_{ij}\tilde{F}_{ij}$ is odd under P. The remaining terms (which have one 0 subscript) in Eq. (16.8) are also odd under parity transformations, as you may show in Problem 3. All these terms are invariant under charge conjugation (C) because they are colorless (invariant to QCD transformations (see Section 12.3)), so that **the θ-dependent term in the effective QCD Lagrangian (Eq. 16.8) violates CP**. In fact, even for small values of θ, this term predicts a *large* electric dipole moment for the neutron, whereas no experiment has found evidence of its existence. The **upper limit of the neutron electric dipole moment**,[10] determined to date, **implies that $\theta < 10^{-9}$**, which leads us to believe that θ is probably 0. Again, a feasible parameter has been found to be 0, and again we seek a fundamental reason.

There is another source for a term in the effective Lagrangian of the same form as that in Eq. (16.8), and we are led to the hope that these two terms cancel, so that $\theta_{effective} = 0$. It arises from considerations, beyond the level of this book, of chiral symmetries[11] and the anomalies therein. We are still left with the question of why two seemingly independent parameters have identical values (of opposite sign). The question of the smallness of $\theta_{effective}$ is referred to as **the strong CP problem**.

A theoretical solution to this problem has been created; it introduces interactions, such that $\theta_{effective}$ becomes 0 because of a chiral type of global symmetry.[12] However, this symmetry is spontaneously broken, leading to the existence of a nearly massless[13] Goldstone boson called the **axion**, which should have been observed by now. Consequently, a modification of the original formulation has been constructed in which the interaction of the axions with ordinary matter is very small, so that the failure to detect them to date does not conflict

---

8. $\text{tr}(\lambda_a \lambda_b) = 2\delta_{ab}$; consequently, this addition to the Lagrangian is **colorless**.

9. Recall that, since $F_{\mu\nu}$ is antisymmetric, there is no F_{00}.

10. As of 1992, experiments have determined that the electric dipole moment of the neutron $<1.2 \times 10^{-25}$ e-cm. See Section 12.5.

11. Chiral symmetries will be discussed in Chapter 17.

12. This is known as Peccei-Quinn symmetry, named after the creators of the idea.

13. The axion is expected to have a very small mass arising from instanton effects.

with this theoretical structure. Axions are candidates for the dark matter, which is believed to permeate the Universe, as we shall discuss in Chapter 18.

16.2   QUARK-GLUON PLASMA

We have learned that quarks and gluons are confined inside hadrons because unconfined colored particles would yield infinite energies. However, it has been conjectured that at about 10^{-5} seconds after the Big Bang, when the Universe was very hot and very small (so that there was a very large energy density), quarks and gluons may have existed as free particles—a quark-gluon plasma. It is also conjectured that suitable conditions exist at the centers of neutron stars for the existence of a quark-gluon plasma.

In Chapter 9 we learned that QCD calculations, using Feynman diagrams, are a perturbative expansion in powers of α_3. The calculations assume that the quarks and gluons are free, so that they are valid only for high-energy scatterings[14] or in a quark-gluon plasma. In fact, we have seen that the effective α_3 becomes very large at low energies, so that the power series generated by Feynman calculations are no longer meaningful as we approach the phase transition from free quarks and gluons to hadrons. However, nonperturbative studies have shown definite indications of the existence of a quark-gluon plasma:

1. Lattice gauge calculations of QCD, for models with only gluons present as well as models with infinitely massive quarks, show a first-order phase transition from hadronic matter to a quark-gluon plasma as the temperature is raised (to $kT \approx 200$ MeV). (In more realistic lattice calculations, it is unclear whether a phase transition occurs; however, we know that, for other phases of matter, it is possible to go smoothly from one phase to another by taking a path in the PV diagram which passes above the corresponding critical point.)
2. In order to study the hadronic phase of matter, effective low-energy theories have been constructed which possess the expected features. In these models it has been shown that, at *finite T* (higher energies), a transition occurs from the hadronic phase to the quark-gluon plasma and the transition is first order under certain conditions. (Furthermore, it is found that the breaking (in the hadronic phase) of separate handedness (chiral) symmetry disappears at this transition (because all masses vanish). This chiral symmetry is discussed in Chapter 17.)

Thus, several different QCD-based considerations lead us to believe that there should be a quark-gluon phase of matter at high temperatures.

Not only are cosmologists, astrophysicists, and elementary particle physicists interested in this intriguing question, but nuclear physicists are as well. For

---

14. An impulse approximation is employed for high-energy scattering, which assumes that individual quarks hard-scatter. Hadronization is assumed to occur after the hard scattering.

many years nuclear physicists have been asking whether, at very high nuclear densities and temperatures, a new state of matter might exist. In particular, when the internucleon distance is less than the size of a nucleon (r_N), we might expect new physics to occur; since $r_N \approx 0.8$ fm and ordinary nuclear matter has internucleon distances of 1.8 fm, we are talking about mass densities (ρ) which are at least 10 times that of ordinary nuclear matter $(\rho_0) - \rho/\rho_0 = (1.8/0.8)^3 \approx$ 10. With the advent of QCD, it became apparent that this new state of matter (if it exists) would be a quark-gluon plasma. In their quest for this exotic state of matter, nuclear physicists have been carrying out high-energy, heavy-ion collisions — the high energies to produce high effective temperatures and the heavy ions to produce large mass densities. We expect that, if a quark-gluon plasma were to be produced in such a collision, it would exist only for a very short time, since the particles fly apart explosively. The time interval, during which the quark-gluon plasma could exist, would be $\sim 10^{-23}$ seconds. The transparency of nuclear matter to very high energy incident ions is also a consideration in the design of such experiments, since the likelihood of the formation of a baryon-rich, quark-gluon plasma may depend on the extent to which the beam is stopped; consequently, for a given ion, increasing the beam energy may, in fact, decrease the probability of producing a baryon-rich, quark-gluon plasma. However, the energy density is expected to increase in the central region, even without significant stopping, and this is expected to lead to the production of a baryon-free, quark-gluon plasma.

Several questions arise about the ability of such experiments to produce meaningful evidence of the production of a quark-gluon plasma. First, we could ask, **what experimental results would be convincing evidence** that a quark-gluon plasma had existed during the short interval that the matter was compressed? Second, **how could we extract this evidence from the wild melee** (including copiously produced secondaries, mainly pions) resulting from the scattering of the hundreds of nucleons (in the ions) from one another? Third, **how could we show that the system had time to reach equilibrium** before the particles flew apart? (If it had not, then the concept of a temperature would not be valid nor could we talk about a new (plasma) phase of matter.)

16.2.1 Possible Evidence for the Quark-Gluon Plasma

Many ramifications of the existence of a quark-gluon plasma have been investigated, including detailed considerations of how to distinguish the effects from those of very high energy multihadron interactions. Here, we list the most prominently discussed effects, as of this writing, although we will not elaborate on the detailed experimental features necessary to distinguish them from hadronic events. Any one, or even a few, such observations may not be convincing evidence of quark-gluon plasma production; however, should a large number of diverse results fit the scenario, we would conclude that it was highly likely that a quark-gluon plasma had been produced. The most prominent effects expected are:

1. A larger number of particles will emerge from the plasma in transverse directions because the plasma at equilibrium will have very little "memory" of the direction of motion of the incident ions. Hence, we expect to see an unusually large amount of transverse energy compared to what would be obtained by superposing nucleon-nucleon scattering data for the scatterings of the nucleons in the ions. However, multiple scattering by the many nucleons present may have a similar effect.

2. In a quark-gluon plasma we expect to see an anomalously high production rate of strange quarks, since there would be so many u and d quarks (from the original nucleons) in a very small spatial region that the effect of the exclusion principle would be to raise the energy level, to which u and d states are filled, above the mass of the s quark (120 MeV). This is sometimes expressed by saying that the chemical potential of the u and d quarks (or their Fermi momentum) is very high. Furthermore, the large number of high-energy gluons in the plasma would lead to copious production of $s\bar{s}$ pairs from gluon-gluon fusion $-g + g \to s + \bar{s}$. Because the many quarks in the quark-gluon plasma would effectively keep the positive- and negative-strangeness quarks apart, there would be many less annihilations of strangeness. Thus, an anomalously large number of strange hadrons would be a signal of the existence of a quark-gluon plasma. (K^+ and Λ production rates, achieved in experiments already performed, appear to be about two times the value for simple nucleon-nucleon (hadronic) collisions of the nucleons in the colliding ions.) We also expect to see enhanced production of higher-strangeness hyperons, so that Ξ/Λ and Ω/Λ production ratios should be anomalously large. Because of the large production of $\bar{s}$ (from gluon-gluon fusion), without a similar production of $\bar{u}$ and $\bar{d}$, we would expect to see many more K^+'s ($u\bar{s}$) than K^-'s ($s\bar{u}$) produced. Likewise, the production of $\bar{\Lambda}$, $\bar{\Sigma}$, $\bar{\Xi}$, and $\bar{\Omega}$ (strange antihyperons) would be enhanced.

3. We expect that many of the large number of s quarks produced would not readily find $\bar{u}$ and $\bar{d}$ quarks with which to form mesons. If a region with a large number of strange quarks remained stable for a significant time interval, a new, very strange (pun intended) form of matter would have existed as a metastable entity called a *strangelet*. The most probable ratios of the numbers of each light-flavor quark in a strangelet have been shown to be $u : d : s = 1 : 1 : 0.8$. Consequently, the charge-to-mass ratio of this hadronic matter would be only about 10% of the charge-to-mass ratio of the proton (since the s is very much heavier than the u and d quarks and its charge is the same as that of the d). In fact, the strangelet with a baryon number of 2 ($ssuudd$) has already been named (H) and searched for. If its mass is less than twice the mass of the Λ, we would expect the H to be stable (under the strong interaction).

4. We would expect a suppression of J/ψ production because the c's and $\bar{c}$'s (produced mainly from gluon-gluon fusion) would be separated by many quarks of other flavors, leading instead to the production of

charmed mesons (D's and D_s's). This suppression should occur for those c and $\bar{c}$'s which remain in the central region of the plasma for a non-negligible time, which in the CM system would have low transverse momenta (p_T); those with high p_T are swept aside to less populated regions of space where they can more readily form a bound $c\bar{c}$ system (the J/ψ, etc.). In preliminary experiments evidence for this detailed behavior of J/ψ suppression (by a factor of 2) has been observed. However, recent results from nucleon-nucleus scattering show such a suppression as well, suggesting that interactions with a large number of nucleons can lead to the destruction of J/ψ's.

5. Experiments based on interference effects, analogous to those used to determine the sizes of stars (Hanberry-Brown/Twiss), can be used to determine the effective size of the region from which the interfering particles were emitted; they employ two-particle correlations. Studies of pion two-particle correlations in heavy-ion collisions have found an effective size, which is about the size of the colliding ions. This result is thought to be a consequence of the fact that pions interact strongly with nuclear matter and so have very short mean-free-paths. Consequently, very few may emerge from a quark-gluon plasma, and the *observed* pions would have been produced *after* the very brief (highly compressed matter) plasma epoch, when the nuclear matter had already spread out. Similar observations for K^+'s are planned as well. K^+'s are expected to be produced copiously in the center of the quark-gluon plasma, as explained above, but those $\bar{s}$ quarks which are in the outer regions are less likely to be separated from s quarks, with which they can combine to produce nonstrange products. Therefore, the effective size of the region in which K^+'s are produced may, in fact, be *smaller* than the size of the quark-gluon plasma region. Since the K^+ does not interact readily with ordinary matter (i.e., it has a long mean-free-path),[15] K^+'s are virtually unaltered by subsequent interactions and, consequently, are expected to provide information about the quark-gluon plasma in which they were formed.

6. In order to obtain detailed information, unmodified by the subsequent numerous high-energy hadronic interactions, about the quark-gluon plasma, we should study those emerging particles which have long mean-free-paths. The K^+ has already been mentioned in this regard, and we expect that the leptons produced in the plasma ("prompt" leptons) and prompt photons would also be good messengers, since they do not take part in the strong interactions. Examples of the kinds of information these particles can bring us about the quark-gluon plasma are: (a) as the quarks slow down and thermalize in the plasma, we expect the emission

---

15. The K^+, like the K^0 (as discussed in Section 12.4.3), possesses an $\bar{s}$ and so will not readily produce a strange baryon (which contains an s) when it collides with (nonstrange) nuclear matter.

of photons characteristic of the temperature of the plasma; and (b) studies of prompt leptons will provide information about the number of degrees of freedom of the medium in which they were produced. Theoretical studies of the features which would distinguish the prompt particles from those subsequently produced have determined the kinds of experimental information (including cuts of the data in CM rapidity, p_T, etc.) needed and the most promising signatures to search for.

16.2.2   The Relativistic Heavy Ion Collider (RHIC)

Many experiments at accessible nuclear densities and temperatures have already shown indications that something other than hadronic interactions is occurring, as we have mentioned above. (Several factors of 2 appear in the experimental results.) In order to pursue the matter (unintended pun) further, nuclear physicists are building a high-energy, heavy-ion collider (known as RHIC) in a tunnel originally built for a particle physics accelerator that was cancelled (the Isabelle tunnel) at Brookhaven, Long Island. It will accelerate uranium nuclei to 100 GeV per nucleon and is expected to be operational in the late 1990s. The heavy nuclei and high CM energies (characteristic of a *collider*) will produce the high density and high temperature required for a more definitive attempt to produce a quark-gluon plasma.

PROBLEMS

1. For a single particle with unit mass in a one-dimensional potential well, $L = \frac{1}{2}(dx/dt)^2 - V$. Show that, when we replace t by $i\tau$, the Lagrangian (in terms of τ) looks like that for a particle in a potential that is turned over compared to the original (V).

2. Show that $[U_1, H] = 0$ implies that, for a nondegenerate vacuum $|0\rangle$, the latter is an eigenstate of U_1. Hint: You can prove, more generally, that if $H|i\rangle = E_i|i\rangle$ and E_i is a nondegenerate level, then $U_1|i\rangle = (\text{constant})|i\rangle$. In fact, since U_1 is unitary, the constant must be of the form $e^{i\alpha}$.

3. Show that $F_{0i}\tilde{F}^{0i}$ is odd under parity transformations.

Chapter 17

Larger Symmetries

In this chapter we shall qualitatively survey the most prominent extensions of the standard model to larger symmetries. We begin with the smallest extension, chiral symmetries, which provide a rationale for the small mass of the pions and also serve as a jumping off point for the technicolor theories (which we will meet in Chapter 19). Then we discuss the grand unified theories and the supersymmetric theories which extend the GUTs.

17.1 CHIRAL SYMMETRIES, PSEUDO-GOLDSTONE BOSONS, AND DYNAMICAL SYMMETRY BREAKING

Chiral symmetries are symmetries which mix fields of the same handedness among themselves, such that **the amount of mixing of the right-handed fields among themselves is independent of the amount of mixing of the left-handed fields among themselves**. If the up, down, and strange quarks were all massless, then presumably $SU_{flavor}(3)$[1] would be an unbroken symmetry. In fact, with no mass terms present in the Lagrangian, there would be no terms with both left-handed and right-handed particle operators.[2] Consequently, the Lagrangian would be invariant to independent rotations of the left-handed quarks among themselves and right-handed quarks among themselves, i.e., invariant to $SU_L(3) \times SU_R(3)$ (we omit the subscript flavor for this group from here on). This symmetry would require the existence of massless left-handed and right-handed quarks with identical weak interactions, which are *not seen in nature*. However, suppose that there is **spontaneous breaking of this global (not gauge) symmetry** to a single $SU_{flavor}(3)$, which contains only those transformations which mix the left-handed and right-handed particles by the *same* amount. Generalizing what we have learned in Section 11.1.2, we can see that **the Goldstone theorem predicts the existence of a massless particle for each broken sym-**

---

1. Recall that this is the global symmetry which mixes u, d, and s quarks.
2. See Section 11.2.2 for a discussion of the left-right structure of mass terms.

metry generator. The Goldstone theorem would then predict the existence of a massless SU(3)-octet of spin 0 particles (actually pseudoscalars), since 8 of the 16 generators in $SU_L(3) \times SU_R(3)$ are broken when the symmetry is reduced to $SU_{flavor}(3)$. Even though the original $SU_L(3) \times SU_R(3)$ was not an exact symmetry (presumably due to the masses of the quarks), there might very well exist an octet of light massive particles, which are called **pseudo-Goldstone bosons**[3] — *and they have been found*, viz., the pseudoscalar meson octet of pions, eta, and kaons.

Since the up and down quarks are very light, the $SU_L(2) \times SU_R(2)$ corresponding to them is nearly unbroken, which explains in a simple way the low mass (140 MeV) of the corresponding (nonstrange) pseudo-Goldstone bosons, the pions. Thus, the pion may be thought of as the light pseudo-Goldstone boson resulting from chiral symmetry breaking. Here the spontaneous breaking of the chiral symmetry occurs because

$$\langle 0 | u\bar{u} + d\bar{d} | 0 \rangle \neq 0. \tag{17.1}$$

Notice that the field operator, whose *vev* is non-zero, is a composite of the fundamental fermion field operators (sometimes referred to as a **condensate**), rather than a fundamental-particle operator. This is described as a **dynamical symmetry breaking**, since it results from the attraction between the quarks in the condensate, due to the (strong) QCD color force.

17.2   GRAND UNIFIED THEORIES (GUTs)

17.2.1   Motivation

It has long been the quest of science to explain the world around us in terms of a few simple, underlying principles. In the early part of the twentieth century, attempts were made to combine electromagnetism and gravitation into a more fundamental, unified field theory. The standard model, especially the electroweak theory, appears to be a step in that direction (although gravitation is not involved). However, it certainly does not reach that ultimate goal, even for the electromagnetic and weak interactions, since the electroweak theory still has two independent coupling constants.

Recall that the cross-product structure of SU(3) × SU(2) × U(1) yields three independent coupling constants (g_1, g_2, and g_3). A simple[4] GUT gauge group (like SU(n) or SO(n)) would have only one independent coupling constant and so would be more "fundamental," the story goes.[5] Well, how can all the evi-

---

3. Pseudo because the original $SU_L(3) \times SU_R(3)$ is not an exact symmetry.

4. Technically, a simple Lie group is a group with no subgroup (aside from the identity element itself) which is invariant to operations of the entire group, i.e., the resulting similarity transformations of the elements of any of its subgroups yield elements outside that subgroup.

5. Gauge groups are the only field theories with vector particles which are sensible, renormalizable theories.

dence in favor of the standard model be explained by such a group? As has been done successfully many times in physics, we invoke a correspondence principle, viz., the simple GUT group must reduce to the standard model at the energies we have already investigated. The script followed is analogous to that of the electroweak theory, where a large symmetry (SU(2) × U(1)) is spontaneously broken at low energies to a residual symmetry ($U_{EM}(1)$). The aim here is to start with a large symmetry, **described by a simple group**, at high energies, which is then spontaneously broken to SU(3) × SU(2) × U(1) as the energy decreases. The three coupling constants which result will not be independent, but instead will be directly determinable from the one coupling constant of the simple group. Of course, this scenario requires the introduction of more Higgs fields in order to bring about the spontaneous symmetry breaking of that simple group. (We haven't found the *first* Higgs particle yet!) Furthermore, as we shall learn, new gauge particles, with their corresponding vertices, are an inevitable result of these theories; all GUTs but one have new (matter) particles as well.

17.2.2   Preliminary Considerations

How can one group break into another group? It should not be surprising to the reader that the original group must *contain* the resulting group; the latter is technically referred to as a subgroup of the former. Let us take, as a simple example of this structure, a group which we have already discussed, namely, SU(3). We have discussed two physically different SU(3) groups, but the mathematical properties we found are properties of any SU(3). We have seen that the gauge group $SU_c(3)$ requires **only one coupling constant** and have discussed some of the subgroups of SU(3) in our treatment of flavor SU(3). We saw that flavor SU(3) contains the SU(2) of isotopic spin; but besides the three generators $\vec{I}$ of that SU(2), it also contains, among its generators, the hypercharge Y, which is the generator of a U(1) of "internal" rotations. Consequently, we conclude that the mathematics of the group is such that:

$$SU(3) \supset SU(2) \times U(1).$$

Let us explore the essential differences between SU(3) and SU(2) × U(1).

1. We have just said that as a gauge theory, SU(3) requires only **one independent coupling constant**, whereas SU(2) × U(1) requires two.
2. In our discussion of flavor SU(3) in Chapter 6, we saw that several states, each with a definite isotopic spin (an SU(2) multiplet) and Y value, occurred together in a single representation of SU(3); remember the octets. Thus, **the larger group has larger multiplets, within which** *several* **multiplets of the smaller (sub)group SU(2) × U(1) reside.**
3. Furthermore, SU(3) has **many more generators** than just the three of SU(2) $\vec{I}$ ($\sim\lambda_1$, λ_2, and λ_3) and the $Y(\sim\lambda_8)$ of U(1) (see Section 5.3); in fact, we have seen in Chapter 5 that there are eight generators. So **the large group has many more elements (symmetry operations)**

$(e^{i \sum_{j=1}^{8} \alpha_j \lambda_j})$ than its subgroup (SU(2) × U(1)) and, therefore, represents a larger symmetry.

4. Also, **the larger group must possess more gauge particles** than its subgroup, since the number of gauge particles is the same as the number of generators.

17.2.3   Coupling Constants

Before seeking a symmetry group larger than SU(3) × SU(2) × U(1), we must address a crucial question: Can a single independent coupling constant (of a larger *simple* group) account for the three that appear in the standard model? The question is not just one of the detailed mathematics of group theory, because the electroweak couplings (α_1 and α_2) and that of QCD (α_3) differ from one another by large factors:

$$\alpha_1 = \frac{g_1^2}{4\pi} \approx \frac{1}{100}, \; \alpha_2 = \frac{g_2^2}{4\pi} \approx \frac{1}{30} , \text{ and } \alpha_3 = \frac{g_3^2}{4\pi} \approx \frac{1}{10} \text{ at about } 10^2 \text{ GeV.}$$

(17.2)

The factors of 3 and 10 that appear here cannot be obtained by spontaneous symmetry breaking of the kinds of groups we are seeking, which instead predict ratios not far from 1. However, as we have learned in Chapter 9, the effective coupling constants depend on the value of the "energy" possessed by the gauge particles, so that the coupling constants "run" with energy. In fact, it has been shown that

$$\alpha_i(p^2) = \frac{\alpha_i(\mu^2)}{1 + b_i \alpha_i(\mu^2) \ln \frac{|p^2|}{\mu^2}}.$$

(17.3)

We have seen in Chapter 9 that the coefficients of the logarithm term in the denominator (b_i) are different for the three groups, being considerably smaller than 1 in magnitude for all as well as negative for U(1), positive for SU(2), and larger positive for SU(3). Thus, α_1 *increases* very slowly with energy (governed by a slowly changing logarithm), α_2 *decreases* very slowly, and α_3 *decreases somewhat more rapidly* with energy, as shown for g_i in Fig. 17.1. These changes are of the right nature for the **three couplings to be approximately equal at some extremely high energy.** Starting with the values known (in the 1970s) at low energies, the values of the three α_i fortuitously appeared to converge to ~ the *same* value (with $\alpha_2 = \alpha_3$) at an energy $\sim 10^{15}$ GeV.[6] For energies $\sim 10^{15}$ GeV and above, we may have a simple-group GUT, which is spontaneously broken by appropriate Higgs fields, with *vevs* $\sim 10^{15}$ GeV, to SU(3) × SU(2) × U(1) (the standard model).

---

6. Actually, by 1991, it was determined that they miss meeting by enough to suggest that some modification of GUTs is necessary. In fact, they meet, within experimental error, for a supersymmetric GUT-SU(5), as we shall soon see.

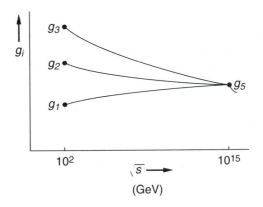

FIGURE 17.1
Extrapolation of the running coupling constants to extremely high energies. By 1991, the precision of the measurements of α_i indicated that they do not exactly meet without some modification of GUTs. (See footnote 6.) g_5 shows how the coupling constant for an SU(5) would run.

All of the physics we know is at energies $\sim 10^2$ GeV or below, so how can we expect an extrapolation in energy by 13 orders of magnitude to be valid? How, indeed! If we read the history of the advances that have been made in physics, we find that bold new guesses, sometimes based on wrong or confused ideas, sometimes containing stark contradictions, have turned out to be valid. (Bohr's model of the atom, a semiclassical model, comes immediately to mind.) Of course, most bold guesses are *wrong*. Actually, we have no obvious contradiction here, merely an assumption that above the (low[7]) energies we have explored lies a virtual **desert** as far as new physics is concerned, until we reach $\sim 10^{15}$ GeV. Why not go along for the ride and see what has been developed?

17.2.4   General Properties

In a simple Lie group the diagonal generators are quantized[8] (as opposed to the generator of a U(1), which we have seen can have any value), so that in any GUT, Y must be quantized. Therefore, in a GUT, Q $(= I_z + \frac{1}{2} Y)$ is a quantized (diagonal) generator,[9] and we conclude that **electric charge is quantized**. As we learned in Chapter 2 when we studied J_z, in any irreducible representation (of a simple unimodular group) the eigenvalues of the diagonal generators must add to 0:[10]

$$\sum_i Q_i = 0 \text{ in any irreducible representation.} \tag{17.4}$$

---

7. extremely low compared to the unification energy which is $\sim 10^{15}$ GeV.
8. Recall SU(2), where the $J^\pm$ operators raise or lower the eigenvalues of the diagonal generator J_z by 1.
9. In a simple group any linear combination of generators is also a generator.
10. In fact, all the generators must be traceless.

Within a given generation, the sum of the charges of the *leptons* is $0 + (-1) = -1$. In theories which do not contain their antiparticles in the same irreducible representation,[11] Eq. (17.4), implies that quarks must be in the same irreducible representation as the leptons. Furthermore, for the sum of the charges of the irrep to be 0, **the size of the fundamental unit of charge (e) would have to be equal for baryons and leptons.** Thus, most GUTs would *automatically* contain this equality, which arises in the standard model solely from the requirement of anomaly cancellation (Eq. 11.63).[12] This has the consequence that **the charge of the proton is exactly the same size as that of the electron.** We need not include all the particles of a given generation in one irreducible representation, and, in fact, the simplest attempt at building a grand unified theory (with SU(5), as we shall discuss in the next section) uses two irreducible representations. The sum of the charges of the particles within each is indeed 0, and each contains quarks and/or antiquarks as well as leptons.

We learned in Chapter 5 that the nature of irreducible representations is that the operations of the symmetry group mix all the members together. If there are quarks and leptons in the same *irreducible* representation of the GUT gauge group, then there are operators in the group which can change quarks into leptons and vice versa. Consequently, there are generators of the GUT gauge group which, along with their corresponding gauge particles, can change quarks into leptons and vice versa (just as the I^- accompanying the corresponding $W^{+\dagger}$ changes the $I_z = +\frac{1}{2}$ quarks or leptons into the $I_z = -\frac{1}{2}$ members of the same isodoublet). The **gauge particles**, which **change quarks into leptons** and vice versa, obviously change the baryon number and lepton number of a state. The covariant derivative of a simple GUT gauge theory is

$$D_\mu = \partial_\mu - ig G^{(\alpha)} \mathbf{G}^{(\alpha)}, \tag{17.5}$$

where $G^{(\alpha)}$ are the generators of the GUT gauge theory, $\mathbf{G}^{(\alpha)}$ are the corresponding gauge particles, g is the coupling constant, and α is summed over all the generators. We know that the terms in the sum produce the (interaction) vertices of the theory; consequently, there will be vertices which change baryon number and lepton number. Thus, we expect that **baryon number and lepton number are not conserved in most grand unified theories.** This violates no deep principle for, as we have said in the introduction to this book (Chapter 1), the conservation of these numbers is not connected to any gauge theory. Their "conservation" at low energies may be due merely to the particle content of the multiplets of the standard model and may be true only in the *low-energy limit* of the full gauge symmetry.

As we have already learned, the larger group must possess more gauge particles than we have found in the energy region $E \leq 10^2$ GeV, i.e., more than

---

11. We already have such a situation in the standard model because of its handedness structure: ν is in a separate multiplet from $\bar{\nu}$ (similarly for e^- and e^+). These two multiplets could be in one larger irrep of a GUT, but it is not required.

12. For groups other than SU(2) × U(1), anomaly cancellation may not lead to Eq. (17.4).

those of the standard model. Some of these, in fact, correspond to the generators of the lepton-quark mixings we have just talked about. These gauge particles have not been seen at low energies, so that they must have masses which are much larger than those of the W's and Z. That requirement is not a problem, since the symmetry is to be broken by Higgs particles at energies $\sim 10^{15}$ GeV. In fact, usual field theory results are at the opposite extreme; when the Higgs vacuum-expectation-values are inserted, *all* the gauge particles will acquire masses of that order of magnitude unless there are very detailed cancellations of contributions from the various Feynman diagrams (often due to unbroken symmetries). This is referred to as **the hierarchy problem**, since there must be enormously different (mass) hierarchies of gauge particles in the theory, viz., the gluons, γ, $W^{\pm}$, and Z must be "massless" compared to the grand unified mass (GUM) $\sim 10^{15}$ GeV.

Recall that although the masses of the W's and Z are $\sim 10^2$ GeV, their effects (as virtual particles) were seen in weak decays at energies $\sim$ MeV. Do the newly predicted, extremely heavy ($\sim 10^{15}$ GeV) gauge particles similarly have ramifications, as virtual particles, which we can detect at the "low" energies of our experiments? Consider the fact that all baryons appear to decay to the lightest one, namely the proton. Since baryon number is no longer conserved, we may ask whether the GUT predicts the proton's eventual decay into (less massive) leptons and photons. We shall see that in particular examples the answer is "yes," and, in fact, quite a few large experiments were set up in the 1980s to search for proton decay.

17.2.5   A Simple Model: SU(5)

The standard model $SU_c(3) \times SU_L(2) \times U_Y(1)$ has a total of four diagonal generators: two in $SU_c(3)$ ($\sim \lambda_3$ and λ_8), the I_3 of $SU_L(2)$, and the Y of U(1). The smallest *simple* group (with no cross-products between simple subgroups) containing this group would have four diagonal generators also.[13] There are only a finite number of such groups, and only one of those contains SU(3) $\times$ SU(2) $\times$ U(1) and representations compatible with the particle content of the standard model as well. This is SU(5). A unitary group is usually defined by looking at its defining representation, and for SU(5) that is a **5**. It seems natural to take two of the five "directions" as pure $SU_L(2)$ and the other three as pure $SU_c(3)$. Thus, two entries in the **5** should be an *SU(2) doublet of (colorless)* **left-handed leptons or right-handed antileptons**, and the other three should be an *SU(3)* **3**, *which is unchanged by SU(2), i.e., an SU(2) singlet*, which are **right-handed quarks**.[14]

Let us consider the first generation. Recall that **the sum of the charges in the multiplet must be 0** (Eq. 17.4). The sum of the charges of the e^- doublet

---

13. They are called **rank 4 groups**.

14. The **5** of SU(5) contains a **3** and not a $\bar{\mathbf{3}}$ of SU(3), hence, left-handed antiquarks cannot be in the **5**.

is -1, that of the e^+ doublet is $+1$, that of the d_{RH} in each of three colors is -1, and that of the u_{RH} in each of three colors is $+2$. The particle content which satisfies the criteria of the previous paragraph is, therefore,

$$\mathbf{5} = \begin{pmatrix} \bar{v}_e \\ e^+ \\ d_r \\ d_g \\ d_b \end{pmatrix}_{RH}, \text{ for the first generation,} \tag{17.6}$$

where RH means that these are *all* right-handed particles and r, g, and b stand for red, green, and blue, respectively. A corresponding left-handed antimultiplet, a $\bar{\mathbf{5}}$, also exists. The other ten right-handed particles belong to a $\overline{\mathbf{10}}$ of SU(5), and their electric charges also add to 0.[15,16] This theory has only particles of the same handedness in a given multiplet, as expected, since internal symmetry operations should not change the handedness (as they mix together the components of an irreducible representation).

The number of generators in an SU(n) theory is $n^2 - 1$, which can be shown in the same way as was done for SU(3) (see Problem 19 in Chapter 5), so that SU(5) has 24 generators. Consequently, the gauge theory has 24 gauge particles. These include the 12 of its subgroup SU(3) $\times$ SU(2) $\times$ U(1), which are the eight gluons, the three W's, and the B. There are 12 *new* gauge particles here; they form a doublet under SU(2), which is also a triplet under SU(3), so that they can simultaneously change the flavor and color of a particle. These gauge particles are named

$$\begin{pmatrix} Y^\alpha \\ X^\alpha \end{pmatrix} \quad \begin{array}{l} q_Y = -\tfrac{1}{3} \\ q_X = -\tfrac{4}{3} \end{array} \quad \text{and their antiparticles,} \tag{17.7}$$

where the SU(2) doublet structure is shown explicitly and α is the color index which runs from 1 to 3. It has been shown that X can cause a rise and $\bar{X}$ can cause a drop of one unit each of B and L via processes like

$$u + u \to \bar{X} \to \bar{d} + e^+. \tag{17.8}$$

Similarly, we have for $\bar{Y}$:

$$u + d \to \bar{Y} \to \bar{u} + e^{+}.[17] \tag{17.9}$$

---

15. As we have mentioned in Chapter 11, renormalizability of these field theories depends upon cancellation of certain types of Feynman diagrams called anomalies. The anomalies of these two irreps are not 0, but they *do* cancel to yield a renormalizable theory.

16. Many other texts talk about the left-handed particles which are antiparticles to those shown here and belong to a $\bar{\mathbf{5}} + \mathbf{10}$.

17. We display here those processes which have an initial state of two quarks, since they are relevant to proton decay, soon to be discussed.

They are referred to as **diquarks** (due to the left-hand sides of Eqs. 17.8 and 17.9) as well as **leptoquarks** (due to the right-hand sides of those equations).[18]

As we have already mentioned, in GUTs with simple groups there is only one independent coupling constant g_{GUT}. (In SU(5) it is called g_5.) Those which were used in the standard model (g_1, g_2, and g_3) should be proportional to it. Recall (from Chapter 10) that the weak mixing angle θ_W is expressed by the *ratio*:

$$\sin^2 \theta_W = \frac{g_1^2}{g_1^2 + g_2^2}, \tag{17.10}$$

so that, for any simple GUT, g_{GUT} cancels out and the prediction for $\sin^2 \theta_W$ comes solely from the group structure. For SU(5), it has been found that $g_1 = \sqrt{\frac{3}{5}} g_5$ and $g_2 = g_3 = g_5$, so that $\sin^2 \theta_W = \frac{3}{8} = 0.375$. This is quite far from the measured value of 0.23, but the prediction is the value at the unification mass $\sim 10^{15}$ GeV. Using the measured values of g_1^2 and g_2^2 at $E \sim 10^2$ GeV and "running" them up to 10^{15} GeV (using Eq. 17.3) yields $\sin^2 \theta_W \approx 0.375$, in agreement with the prediction of the theory. Furthermore, the SU(5) prediction that $g_2 = g_3$ ($= g_5$) at 10^{15} GeV agrees with the experimental values, measured at $\sim 10^2$ GeV and then "run" up in energy.

17.2.6 The Baryon Number of the Universe

Everything around us is made of "particles" and not their antiparticles. But you may say that that is no surprise, since we have *chosen* to call what we see around us "particles." However, all evidence indicates that we can interpret the word "around" to apply to the galaxy and, due to the absence of certain types of astronomical occurrences,[19] to surrounding galaxies as well. Cosmological considerations, including the discovery of the 2.7 °K background radiation, have led us to believe that our universe began as a space-time point at which there was a singular explosive event about 15 billion years ago, which is referred to as the Big Bang. The enormously high energy densities and temperatures during the first instant ($\sim 10^{-43}$ s to 10^{-35} s after the Big Bang), were just right for the creation of the fundamental elementary particles we have been studying. Thus, cosmology and elementary particle theory overlap, as we have already seen in Chapters 15 and 16 (where we discussed the conjectured phase of these primordial particles, viz., a quark-gluon plasma). Astronomical studies tell us that the density of antibaryons in the Universe (at least in our "corner" of the Universe) is negligible; however, although the density of baryons is many orders of magnitude less than that of photons, it is *not* 0. Thus, **the Universe has a net nonzero baryon number**. Cosmological theories generally assume that the Universe evolved from a completely symmetric initial state, which had baryon number

---

18. In SU(5) we find, in fact, that there is a conserved number remaining, which is $(B - L)$.

19. One such occurrence would be the complete annihilation of a star of matter with one of antimatter, which has never been seen.

0. Such a scenario could be consistent with the non-zero baryon number we see today if:

1. there were interactions which do not conserve baryon number. This is an obvious requirement.
2. there were a violation of CP. For if CP were conserved, no interaction could change the baryon number of the Universe (since CP changes a particle to its antiparticle, thus, reversing the sign of its baryon number).
3. the Universe were not in thermodynamic equilibrium while the baryon-number-violating (CP-violating) events were occurring. This is necessary, since if it were at equilibrium there would be the same number of reactions going each way; for each event raising the baryon number there would be one lowering it, so that the net change would be 0.

We can see that those who prefer a symmetrical Big Bang were delighted with the ramifications of GUTs, which provide a rationale for both (1) and (2).[20]

These considerations are part of the growing interrelationship of cosmology and elementary particle physics, which can be expressed simply by stating that the early Universe was the first high-energy physics laboratory. In fact, the energy of the largest accelerator which we envision building, the SSC, was astoundingly exceeded during the first 10^{-13} seconds after the Big Bang.

17.2.7   Proton Decay

As we have already mentioned in Section 17.2.4, in a GUT the new heavy gauge particles can mediate baryon-number and lepton-number violations. In SU(5) we have the X^{α}, Y^{α} colored SU(2) doublet whose exchange mediates the decay:

$$p \to \pi^0 + e^+$$
$$\downarrow$$
$$\gamma + \gamma$$

via the diagrams shown in Fig. 17.2, as well as

$$p \to \rho^0 + e^+$$
$$\to \omega^0 + e^+$$
$$\to \pi^+ + \bar{\nu}_e,$$

followed by $(\pi^0, \rho^0, \omega^0,$ and $\pi^+) \to \gamma$s, ℓs, and ν_ℓs.

The ultimate final states have no hadrons; they contain only leptons, some with accompanying photons. This predicts the demise of the Universe as we know it. But we are made of nucleons and we *do* exist, so that you might think that this prediction is a fatal flaw of the theory. However, the lifetime of the Universe is $\sim 10^{10}$ years, so that we cannot be sure that the proton is completely

---

20. Requirement (3) can be understood in terms of the expansion rate of the Universe.

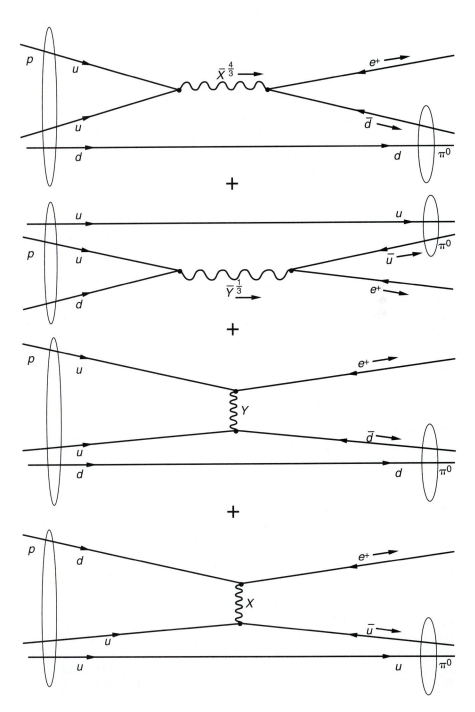

FIGURE 17.2
$p \rightarrow \pi^0 + e^+$ in SU(5).

stable. Actually, early experiments had not found proton decay, with a lower limit for its lifetime of $\tau_p > 10^{25}$ years, before this SU(5) theory was proposed. However, the prediction of the theory is $\tau_p \approx 10^{31}$ years, which is consistent with that experimental lower limit.

How do we measure lifetimes that are many orders of magnitude larger than the age of the Universe? *In principle* it is very simple. Remember that the protons will not all last as long as their lifetime, but instead have a probability of decay at an exponential rate governed by that lifetime. If the probability that a proton will decay during one year of observation were $\sim 10^{-31}$ ($\sim 1/\tau_P$), then we should see many decays in a year if we watch $\sim 10^{33}$ protons. That is not so hard to arrange because there are 6×10^{23} molecules in each mole of a substance, and the nuclei in these molecules contain many protons. Obviously, we need about 10^9 moles; however, a mole of molecules of atomic weight A is in fact only A grams of that substance. Therefore, several thousand tons[21] would do the trick. This is not a small amount, but it is certainly within the realm of possibility.

In the 1980s, approximately a dozen such experiments were set up around the world to search for proton decay. The preparation and monitoring of large quantities of matter was accomplished with great skill. The main difficulty of these experiments was the task of obtaining a signal in the presence of a large background. Incoming cosmic ray particles can easily cause interactions in the detectors, with signatures identical to proton decay events. In order to screen out cosmic rays, these experiments were set up in mines or in tunnels under mountains. Nevertheless, if the proton lifetime were to exceed 10^{32} years, the expected number of decays would be no more than the false events of background.

You can rest easy! The experimental results were indistinguishable from pure background, so that we can say only that

$$\tau_p > 9 \times 10^{32} \text{ years} -$$

hence, **the demise of SU(5)** as a viable grand unified theory. There do exist "grander" grand unified theories which are still in the running because they predict a longer lifetime for the proton, and it is their creators' good (or bad, according to your point of view) fortune that no viable experiment can be done at this time to test such a prediction. It was instructive for us to study the features of SU(5), since it is the simplest GUT which can be constructed and more elaborate ones possess most of the features we have emphasized. SU(5) is an explicit demonstration that consistent grand unified theories can be constructed. It has shown us how the Higgs "sector" of the theory is necessarily more complicated than that of the standard model; however, it has the inelegance to contain the matter fields in two irreducible representations $(\mathbf{5} + \overline{\mathbf{10}})$. Most of the larger GUTs have all the matter fields in one irreducible representation, but pos-

---

21. A metric ton is 10^3 kg.

sess more matter particles in each generation than have actually been seen, so that special arguments are required to account for this (e.g., by arranging for the masses of the extra particles to be very large). There are interesting arguments in favor of SO(10) as the grand unified theory, and so we now turn to a brief overview of that GUT.

17.2.8 SO(10)

If we seek a gauge group that can accommodate the 15 fundamental right-handed fermions ($e_R^+, e_R^-, \bar{\nu}_R, u_R^\alpha, \bar{u}_R^\alpha, d_R^\alpha, \bar{d}_R^\alpha$) in *one* irreducible representation,[22] we are led to groups with five diagonal generators (i.e., rank 5). The group of "rotations" in 10 dimensions, SO(10), is one of those which has been carefully studied. It contains SU(5), so that it certainly has SU(3) $\times$ SU(2) $\times$ U(1) as a subgroup. In fact,

$$\text{SO}(10) \supset \text{SU}_c(3) \times \text{SU}_L(2) \times \text{SU}_R(2) \times \text{U}(1),$$

where the first two subscripts are those of the standard model (*c* for color and *L* for left-handed) and the $\text{SU}_R(2)$ has gauge particles analogous to those in $\text{SU}_L(2)$ (the *W*'s), which couple instead to right-handed particles (and left-handed antiparticles). At this stage, the group structure appears **left-right symmetric**. The spontaneous symmetry breaking would then take place in stages (in a manner analogous to that of the standard model), with each stage requiring its own set of Higgs scalars:[23]

1. SO(10) $\rightarrow$ $\text{SU}_c(3) \times \text{SU}_L(2) \times \text{SU}_R(2) \times \text{U}(1)$ at extremely high energies ($E > 10^{15}$ GeV) and
2. $\text{SU}_R(2) \times \text{U}(1) \rightarrow \text{U}_Y(1)$ at very high energies ($\sim 10^{15}$ GeV), where sub-*Y* indicates that this is the U(1) of the standard model (in a manner analogous to the breaking to $\text{U}_{EM}(1)$ of the standard model), so that we reach the standard model $\text{SU}_c(3) \times \text{SU}_L(2) \times \text{U}_Y(1)$.

Stage (2) breaks left-right symmetry and, consequently, one set of these gauge particles is *expected* to be heavier than the other because they obtain their masses via different spontaneous symmetry breakings. In fact, it would require explanation if they both broke at the same energy. We might question, however, why there are any very *light* particles in the world, i.e., the hierarchy problem arises here also.

For the fundamental fermion fields, SO(10) has a right-handed **16**, which contains the **5**, $\overline{\mathbf{10}}$, and a **1** of its SU(5) subgroup. The $\mathbf{5} + \overline{\mathbf{10}}$ contain the 15 fundamental fermions of the standard model, as in SU(5). The **1** is a right-handed ν, so that the **16** multiplet has a right-handed ν as well as the right-handed $\bar{\nu}$, which was in the **5** of SU(5). (Corresponding antiparticles are in a

---

22. Their left-handed antiparticles would be in the conjugate irrep.
23. Sequences of symmetry breaking, other than the one listed here, have been studied as well.

left-handed $\overline{16}$.) Thus, *all* the Fermi particles and antiparticles come in both handednesses. So, this theory is **left-right symmetric**. The handedness we see at low energies results from spontaneous symmetry breaking, in which the SU(2) gauge particles that couple to one (the right) handedness are much heavier than the other set. Furthermore, **the fundamental matter fields are all in one irreducible representation**, a feature which *appears* simpler and, perhaps, more fundamental. Of course, a larger group contains more generators, hence, SO(10) has many more gauge particles than does SU(5); it also has one more diagonal generator, so there is another conserved quantum number. The presence of these new generators and gauge particles leads to many more requirements and complexities.

17.2.9   GUT Magnetic Monopoles

It has been shown that the grand unified theories built from simple groups must contain magnetic monopoles among their predictions. Let us ask first, what is a monopole? In classical electrodynamics we learn about multipoles, and in previous chapters we have already discussed electric and magnetic dipoles. You may not be familiar with the word "monopole" because, for a distribution of charges, we generally refer to the electric monopole as the "total charge." Let us consider Maxwell's equations, which (in a vacuum) are

$$\vec{\nabla}\cdot\vec{E} = 4\pi\rho,\ \vec{\nabla}\times\vec{E} = -\frac{1}{c}\frac{\partial\vec{B}}{\partial t},\ \vec{\nabla}\cdot\vec{B} = 0,\ \text{and}\ \vec{\nabla}\times\vec{B} = \frac{4\pi}{c}\vec{J} + \frac{1}{c}\frac{\partial\vec{E}}{\partial t},$$

in Gaussian units (they have the same form, even though the constants are different, in any other units). Notice that the equations for the divergence of $\vec{E}$ and curl of $\vec{B}$ are of a different nature than those for the divergence of $\vec{B}$, and curl of $\vec{E}$ because of the existence of electric charges (hence ρ) which can move (hence $\vec{J}$). During the 1930s, Dirac questioned whether there might exist magnetic-type charges, which would then lead to the modification of these equations by the existence of appropriate terms with ρ_m and $\vec{J}_m$:

$$\vec{\nabla}\cdot\vec{E} = 4\pi\rho,\ -\vec{\nabla}\times\vec{E} = \frac{4\pi}{c}\vec{J}_m + \frac{1}{c}\frac{\partial\vec{B}}{\partial t},$$

$$\vec{\nabla}\cdot\vec{B} = 4\pi\rho_m,\ \text{and}\ \vec{\nabla}\times\vec{B} = \frac{4\pi}{c}\vec{J} + \frac{1}{c}\frac{\partial\vec{E}}{\partial t}. \tag{17.11}$$

He was able to prove, within the context of a general quantum field theory, that the allowed values of the magnetic charge (g) were related to e by

$$\frac{eg}{\hbar c} = \frac{n}{2}, \tag{17.12}$$

where n is an integer.

That proof is beyond the level of this book, but we can sketch a semi-classical reasonableness argument for this quantization condition. Consider a system composed of an *electric charge q and a magnetic charge g at rest*. The electric field (set up by the electric charge) and the magnetic field (set up by the magnetic charge) together have a non-zero angular momentum density ($\vec{r} \times (\vec{E} \times \vec{B})/4\pi c$) such that the total angular momentum in the field is gq/c. In our quantum theories, angular momentum is quantized in half-integral units of $\hbar$, so that we expect

$$\frac{gq}{c} = \frac{n}{2}\,\hbar,$$

where n is an integer, which is identical to Eq. (17.12) for $q = e$.

From Eq. (17.12) we learn that the value of g is quantized, because q is quantized in units of e. We also learn that the **smallest** non-zero value of g is very large, as can be seen easily in natural units: $ge = \frac{1}{2}$, with $e = \sqrt{\alpha} = \sqrt{\frac{1}{137}}$, yields $g = \frac{\sqrt{137}}{2}$, so that the effective coupling between two magnetic monopoles has strength $\alpha_g = g^2 = \frac{137}{4} \approx \mathbf{34}$, which yields **an enormously strong attraction for opposite sign monopoles**. Thus, it would be understandable that pairs produced just after the Big Bang would have recombined by now, and it would be extremely difficult to produce pairs in experiments.

It has been proven (using topology and group theory) that whenever a simple GUT group breaks into a group of the form (any group) $\times$ U(1), there must exist magnetic monopoles in the theory. They must have the very large magnetic charges we have just discussed and (it turns out)[24] a mass $M_m \geq M_x g_3^{-1} \sim 10^{16}$ GeV (where M_x is the unification mass). Direct searches of bulk matter have not uncovered even one magnetic monopole. There are various ways to search for moving magnetic monopoles, such as by recording the effects of the magnetic flux change as one of them moves through a loop, looking for the catalysis of proton decay, observing the ionization effects from the motion of their large magnetic charge, etc. Only one candidate event has ever been reported—a current jump was recorded at Stanford for an unattended loop in 1982. The flux size, suggested by the occurrence of this event, would have produced a number of such events in the much larger and more isotropic loops built since that time; but no further events have been found, casting doubt that the jump was caused by a monopole. However, cosmological models do not require that the monopole flux be large enough to be measurable (when "inflation" is included in the Big Bang model), and the required interaction strength is so large that it is extremely difficult to produce magnetic monopole pairs. Therefore, their existence as a type of particle is still an open question.

---

24. We expect all masses to be of the order of M_x except for those which have exact cancellations in the calculation of their masses, as we have stated in our discussion of the hierarchy problem.

17.2.10   Limitations of GUTs

Although grand unified theories appear to be more elegant than the standard model, a successful GUT will only be a small step towards unification and a full physical understanding of the fundamental particles and their interactions. The major limitations of GUTs are

1. The grand unified theories do not include gravitation, so that they do not achieve full unification.
2. Although they reduce the number of coupling constants, GUTs introduce new parameters and new (Higgs) particles (with associated complications) into the Higgs sector of the theory.
3. They also introduce new gauge particles and their associated interactions as well as new matter particles[25] (in each generation), none of which have been seen; of course we can observe only at energies which are much smaller than the unification mass.
4. They do not address the question of multiple generations of particles, which seem to be merely heavier copies of the first generation.
5. They do not explain the CKM mixing angles or the possible CP violation associated with the phase angle in the CKM matrix.
6. As we have already discussed, they assume a "desert" stretching from 10^2 GeV all the way up to 10^{15} GeV, which may be experimentally tested at the 10^3 GeV level by the huge accelerators being planned, like the SSC and the LHC.

17.3   SUPERSYMMETRY (SUSY) AND SUPERGRAVITY

Let us set the stage for the arrival of supersymmetry and supergravity theories as viable, although conjectural, theoretical structures:

1. All experimental evidence indicates that the fundamental matter particles are spin $\frac{1}{2}$ fermions and the quantum particles of the force fields are integral spin—spin 1 or spin 2 (gravity)—bosons. The standard model, which is a theoretical structure consistent with all experimental data obtained to date, also contains a spin 0 particle (thus a boson) crucial to its structure, viz., the Higgs particle. Grand unified theories have more Higgs particles as well.

    Fermions obey the Pauli exclusion principle, which (in a field theory) results from the anticommutation relations of the corresponding field operators, whereas boson field operators obey commutation relations instead. For this reason, it did not appear that there could exist

---

25. except for SU(5).

internal symmetry operations which could mix them, i.e., we did not expect fermions and bosons in the same multiplet.[26] Thus, the quest for full unification was limited.

2. We have seen that the extensions of the standard model known as grand unified theories also contain Higgs particles, which produce spontaneous symmetry breaking at much higher scales. In Chapter 11 we saw that the breaking in the electroweak theory has $v \approx 250$ GeV, whereas the breaking in grand unified theories has $vevs \approx 10^{15}$ GeV — the disparity of these numbers is part of the so-called **hierarchy problem**. The structure of GUTs is based on a perturbative treatment of the standard model, which assumes that the self-coupling (λ) of the electroweak Higgs field is not very large. From Eq. (11.20) we see that this implies that the electroweak Higgs mass cannot be very large either. However, as we have mentioned earlier, **radiative corrections tend to raise small masses to values characteristic of the masses of the heavy particles and of the scale of the theory**. There is an elegant way (using chiral symmetry) to have massless fermions in our theories, in spite of radiative corrections, but no such mechanism exists for bosons.[27] At the scale of the GUTs, 10^{15} GeV, all the matter fermions of the standard model are essentially massless, so that **chiral symmetry is assumed to be present**. (Their small masses can then be accounted for by low-energy terms, as shown in Chapter 11.) However, in grand unified theories, where the breaking of the symmetry and the mass scales are of the order of 10^{15} GeV, it was hard to understand how the mass of the electroweak Higgs particle could be light enough (so that the self-coupling parameter λ would be small) for the perturbative treatment of the electroweak theory to have any validity. A detailed "fine tuning" of the free parameters of the theory could, in principle, produce a cancellation of the large radiative corrections to the electroweak Higgs mass, but **fine tunings are distasteful**. Perhaps there is a deeper reason for the cancellations required.

3. It had been long known that internal symmetries and space-time symmetries could not be consistently combined in a nontrivial way.

4. The unification of gravity with the other three basic interactions had not been achieved. Furthermore, all attempts to quantize Einstein's theory of gravity, known as general relativity, produced formulations which were unrenormalizable, with intractable infinities arising in non-tree-level calculations.

---

26. Recall that particles in the same multiplet may be thought of as different states of the same particle, e.g., the proton and the neutron are thought of as two possible states of the nucleon in discussions of isospin.

27. In fact, for bosons, the infinities which arise in diagrams with loops of virtual particles are more virulent than those in the corresponding fermion calculations.

Supersymmetry and supergravity theories (a) contain transformations which **successfully mix fermions and bosons**, leading us to hope for more unification than is accomplished by the grand unified theories; (b) **provide, in a natural way,** for the necessary **cancellations of the large radiative corrections to Higgs masses,** preserving the validity of our present understanding of the standard model; (c) contain a **fundamental connection between their internal symmetry operations and space-time transformations** (space-time translations, rotations, and Lorentz transformations), such that a quantized version of **general relativity**[28] **is automatically unified with the gauge theories of the other basic interactions.** Furthermore, theories with several supersymmetric operators (like Q, which is discussed below) have been found to produce finite results, so that those theories need not be renormalizable.

Let us look at a simplified model in which there is one kind of boson with $b^\dagger$ as its particle creation operator, and one kind of fermion with $f^\dagger$ as its particle creation operator; b and f are the corresponding annihilation operators, respectively. We will take our Hamiltonian operator (whose eigenstates are states of definite energy) as

$$H = E_1(b^\dagger b + f^\dagger f). \tag{17.13}$$

$b^\dagger b$ is the number operator for bosons and, similarly, $f^\dagger f$ is the number operator for fermions (see Problem 5), so that states with definite numbers of fermions (n_f) and bosons (n_b) have definite energies:

$$H|n_b n_f\rangle = E_1(n_b + n_f)|n_b n_f\rangle. \tag{17.14}$$

Obviously, this is a simple model in which there is no interaction. Now, let us recall that the Pauli exclusion principle should, in fact, restrict n_f to the values 0 or 1 only. In fact, quantum field theory tells us that the fermion operators must obey anticommutation relations, viz.,

$$\{f^\dagger, f^\dagger\} = \{f, f\} = 0 \text{ and } \{f, f^\dagger\} = 1. \tag{17.15}$$

The first of these shows that there cannot be a state with more than one fermion since

$$0 = \{f^\dagger, f^\dagger\} = f^\dagger f^\dagger + f^\dagger f^\dagger = 2f^\dagger f^\dagger, \tag{17.16}$$

so that attempting to create two fermions (using $f^\dagger f^\dagger$) or more yields 0, i.e., no such state can exist. Contrastingly, bosons obey commutation relations:

$$[b^\dagger, b^\dagger] = [b, b] = 0 \text{ and } [b, b^\dagger] = 1, \tag{17.17}$$

and the b's and f's commute with one another, since they refer to independent fields. Now let us consider a new operator Q (not the charge operator), defined as

$$Q = b^\dagger f + f^\dagger b, \tag{17.18}$$

---

28. although still not renormalizable.

which is a spinor operator (because of the f and $f^\dagger$ in its definition) and, hence, can change a boson state into a fermion state (and vice versa). As you may verify in Problem 6, Q commutes with the Hamiltonian:

$$[Q,H] = 0, \qquad (17.19)$$

so that Q may be thought of as the generator of a symmetry of the Hamiltonian. The associated transformations $e^{iQ\alpha}$ mix fermions and bosons, and this type of symmetry is called a **supersymmetry**.

Because it is a spinor operator, we might expect Q to obey anticommutation relations, and you may verify in Problem 7 that

$$\{Q,Q^\dagger\} = \frac{2}{E_1} H. \qquad (17.20)$$

In our discussions of Lie groups (starting in Chapter 5), we saw that the commutator of any two generators produced a linear combination of generators. This is usually expressed by saying that the generators form a closed Lie algebra (with commutation as the multiplication operation). Since Q obeys an anticommutation relation instead of a commutation relation, it does not appear that it could be incorporated into the Lie algebras associated with the generators of the symmetry groups we have studied so far. However, mathematicians have already studied systems of operators which contain some anticommutation "products"; they have even given them a name, "graded Lie algebras." Thus, although this involves more complicated mathematics, it is tractable.

Since Q is a spinor operator, the α's in $e^{iQ\alpha}$ must be spinors as well (just as we learned for rotations, where $e^{i\vec{J}\cdot\vec{\theta}}$ is the rotation operator and $\vec{\theta}$ must be a vector). Furthermore, because the operations $e^{iQ\alpha}$ with different values for α should obey commutation relations[29] (even though Q obeys anticommutation relations), the *entries* in α are not ordinary numbers, but instead must anticommute.[30] Thus, these are not the kinds of symmetry groups considered in the past, which explains why it was believed that no symmetry could mix fermions and bosons.

Notice that the "product" in Eq. (17.20) shows that the closed (graded) Lie algebra includes the generator H (in a nontrivial way), so that this simple example has an intertwining of the internal symmetry (generated by Q) and the time translations (generated by H, which is the 0th component of the operator P^μ). In a supersymmetric *quantum field theory*, the Lie algebra would contain the generators of all the space-time transformations. When these global supersymmetry theories are extended to local theories (in the above: $\alpha \to \alpha(x^\mu)$), the gauge particle must absorb $\partial_\mu\alpha$, which behaves like a $J = \frac{3}{2}$ object (since α is now a spinor), so that the gauge particle must have spin $\frac{3}{2}$. Furthermore, to maintain the supersymmetry, that (fermion) gauge particle must have a boson

---

29. meaning that $e^{-iQ\alpha_1}e^{-iQ\alpha}e^{+iQ\alpha_1}$, which for small α's gives commutators, is also an element of the group.

30. These are called Grassman numbers.

partner which has spin 2. Remarkably, due to the structure of the theory, this spin 2 particle interacts just as the graviton (the quantum of the gravitational field) should, so that these **local theories are referred to as supergravity theories**. They contain a **quantized theory of gravity**, whose classical limit is indeed Einstein's general theory of relativity — gravity has arisen here in a natural way. Furthermore, a unification of gravity with the other three basic interactions has been automatically achieved. The **spin 2** particle is referred to as the **graviton**, its **spin $\frac{3}{2}$ partner** as the **gravitino**. In larger theories, with several supersymmetric generators like Q above, it appears as if all radiative corrections yield finite results, so that the gauge-theory description of the world may contain a finite quantum field theory of gravity.

Supersymmetric versions of various grand unified theories have been constructed and studied; they are often referred to as **SUSY-GUTs**. In fact, for SUSY-SU(5), the running coupling constants meet more exactly (within experimental limits) than in ordinary GUTs and at a somewhat higher value — 10^{16} GeV.[31] That higher value (plus new channels of decay) lead to a prediction for the lifetime of the proton which is several orders of magnitude longer than the SU(5) prediction and, consequently, is not in conflict with the experimental lower limit ($\sim 10^{33}$ years). Furthermore, the existence of new channels of decay with large branching ratios suggests unique signatures to search for in future proton-decay experiments. SUSY-SU(5) also contains the remarkable property that when the (real) masses of the Higgs particles (at the GUT scale of 10^{16} GeV) are "run" down in energy, one μ^2 becomes negative, producing the Higgs mechanism in a natural way.

An important property of our primitive model above is the presence of one type of fermion and one type of boson of the same energy (E_1). In *any* supersymmetric quantum field, bosons and fermions[32] appear as equal-mass pairs, so that we would expect a boson "partner" for each lepton and quark, and a fermion partner for each gauge particle and Higgs. **No evidence of the existence of such supersymmetric partners of the particles of the standard model has ever been detected (as of 1992).** Nevertheless, there is considerable interest in these theories because of the beautiful theoretical properties we have outlined above. Furthermore, with the usual ingenuity of model builders, it has been possible to create *broken* supersymmetric theories, in which the (so far) undetected supersymmetric partners are too heavy to have been detected in any of the experiments performed to date. However, in most of these models the masses of the supersymmetric partners are of the order of a TeV, so that we expect to observe them when the SSC is operational.[33]

31. Gluinos, fermion partners of the gluons, produce shielding (as the other fermions did) which moderates the effective antishielding and consequent asymptotic freedom. This slows down the drop of the QCD running coupling constant. (Look at Fig. 17.1.)

32. Actually, the number of boson *degrees of freedom* and the number of fermion *degrees of freedom* must be the same.

33. Recall that the SSC will have 40-TeV CM energy.

The partners of the fundamental fermions are spin 0 particles and have been named: **squarks, sleptons, selectron, or scalar-quark**, etc. **The spin $\frac{1}{2}$ fermion partners of the gauge particles** are named, using "ino" as a suffix (pronounced to rhyme with "neutrino"), viz., **photino, gluino, wino, bino, zino, etc.**; they are collectively referred to as **gauginos**. Likewise, **the spin $\frac{1}{2}$ partner of the (scalar) Higgs particle** is called **the higgsino** and the **gravitino is the spin $\frac{3}{2}$ partner of the spin 2 graviton**, as discussed above. The charged gauginos and higgsinos are referred to as **charginos**, the neutral ones as **neutralinos**. For some of these, it is expected that the *observed* particles will be linear combinations, hence the names **wiggsino, ziggsino**, etc.

Since, for each (scalar) Higgs boson, there is a corresponding fermion, we might expect the chiral mechanism which keeps fermions "massless" to have a similar effect on the appropriate Higgs bosons' masses and *vev*s as well (e.g., the Higgs particle which breaks the electroweak symmetry should have a *vev* ≈ 0 on the GUT scale). Detailed investigations of the radiative corrections to their masses and *vev*s, indeed, show that loops involving virtual bosons and similar loops involving virtual fermions give the same size contributions but with opposite signs. Consequently, there is a cancellation of the undesirable renormalization effects and the hierarchy of different scales arising in GUTs is preserved in a natural way.[34]

It is interesting to note that most supersymmetric theories have implications for cosmology. This is because in most SUSYs *exactly two* supersymmetric partners (of ordinary matter) appear in the vertices which contain them. Thus, the lightest supersymmetric particle (LSP) cannot decay because it does not have enough energy to produce any of the heavier ones, and there is no vertex which can connect it solely to ordinary matter (i.e., connecting *one* supersymmetric partner with ordinary matter). Since they have not been detected at accelerators so far, we expect these particles to be quite massive (e.g., $M_{photino} > 5$ GeV). Many LSPs should have been produced in the early Universe and, because their interactions are expected to be only electroweak, they would not form radiant matter, but would be part of a dark matter background. Halos of these particles should exist around galaxies (due to gravitational attraction), forming the massive dark matter that astronomers believe would account for the behavior of star systems and that cosmologists "need" to close the Universe.[35]

PROBLEMS

1. Consider the simple group SU(2). Show that the subgroup $U_{I_3}(1)$, whose elements are $e^{iI_3\theta}$, is not an invariant subgroup, i.e., find a group operation (in SU(2)) which changes an element in the subgroup to one not in that subgroup. Remember that $\vec{I} = \frac{1}{2}\vec{\sigma}$.

---

34. without fine tuning.
35. Dark matter in the Universe will be discussed in Chapter 18.

2. How many more gauge particles does SO(10) have than SU(5)? Hint: See Problem 19 in Chapter 5.

3. For the first generation, what particles are in the right-handed $\overline{10}$ of SU(5)? Remember that the right-handed fundamental fermions are contained in $5 + \overline{10}$.

4. Find the angular momentum contained in the fields set up by an electric charge q sitting at the origin and a magnetic charge g at $(0,0,d)$. Remember that the magnetic field due to g has the same relation to g as the electric field has to q.

5. a. Recall that the state $b^{\dagger}|0\rangle$ is a state with one boson, and the corresponding bra is $\langle 0|b$. Check its normalization, using the commutation properties of b and $b^{\dagger}$ to pull the b across the $b^{\dagger}$ and annihilate the vacuum ket (viz., $b|0\rangle = 0$). Show that this state is an eigenstate of the operator $b^{\dagger}b$ with eigenvalue 1.

   b. A state with 2 bosons is $N_2 b^{\dagger}b^{\dagger}|0\rangle$. Show that $N_2 = 1/\sqrt{2!}$ and $(b^{\dagger}b)$ operating on the state yields 2. (Again, pull the b's across the $b^{\dagger}$'s and annihilate the vacuum ket.)

   c. A state with n bosons is $N_n(b^{\dagger})^n|0\rangle$. Show that $N_n = 1/\sqrt{n!}$ and $(b^{\dagger}b)$ operating on the state yields n, so that $(b^{\dagger}b)$ is the number operator.

   d. For fermions, since $f^{\dagger}f^{\dagger}$ is 0, you need only show that $f^{\dagger}f$ yields 1 acting on a state with one fermion and 0 acting on the vacuum, to verify that $(f^{\dagger}f)$ is the number operator. Do it.

6. Show that $[Q,H] = 0$, where $Q = b^{\dagger}f + f^{\dagger}b$ and $H = E_1(b^{\dagger}b + f^{\dagger}f)$.

7. Verify that $\{Q,Q^{\dagger}\} = (2/E_1)H$.

Chapter 18

Neutrino Mass, Mixing, and Oscillations

18.1   MASS

Are the neutrinos truly massless? Only upper limits can be determined experimentally (because of experimental error). To date, the experimental results are

$m_{\nu_e} < 9$ eV, $m_{\nu_\mu} < 0.27$ MeV, and $m_{\nu_\tau} < 35$ MeV for the left-handed neutrinos;

no right-handed neutrinos have ever been observed.

As we have seen in Chapter 17, the question of neutrino mass arises in the context of grand unified theories. In fact, many of those theories, such as SO(10), possess neutrinos of both handedness. The right-handed neutrinos, in such theories, must have very large masses, since they have not been detected during more than half a century of weak interaction experiments. However, the subject of neutrino mass, along with the related ideas of possible generation mixing and oscillation, extends beyond GUTs; the neutrinos may possess mass even in the absence of a GUT.

The question of neutrino mass leads us to wonder about the absence of right-handed neutrinos at low energies, which was related to masslessness in our discussion in Chapter 11. In fact, we saw there that the usual mass term in the Lagrangian is

$$m\bar\psi\psi = (m\bar\psi_R\psi_L + m\bar\psi_L\psi_R), \text{ called a \textbf{Dirac mass term},} \qquad (18.1)$$

so that the two handednesses are mixed in each term. The left-hand side of Eq. (18.1) shows us that the state (or operator) $\psi = \psi_R + \psi_L$ has a definite mass; since this has equal parts of the two handednesses, both left-handed and right-handed particles will have the same mass. However, we have seen that there is an approximately massless left-handed neutrino in each generation, but there is no light right-handed one. This leads us to suspect that there is no Dirac mass term at all for the neutrino in $\mathcal{L}$, since such a term would *require* the existence of a right-handed neutrino; consequently, the mass of the left-handed neutrino

371

is *identically* 0. We have seen this to be the case in the standard model and in SU(5) as well.[1]

However, as we now show, right- and left-handed neutrinos *can* have different masses because neutrinos have the value 0 for all the truly conserved internal numbers (electric charge and color λ_3, λ_8 of $SU_c(3)$). They are, in fact, the only fundamental fermions which are electrically neutral as well as colorless. For the masses of a left-handed particle and the corresponding right-handed particle to be different, the Lagrangian must have a mass term of the form $\bar{\psi}_L \chi_L$ (a left-left mass term) and/or $\bar{\psi}_R \chi_R$ (a right-right mass term). However, such terms, using the usual fields (e.g., $\bar{\psi}_L \psi_L$), would be identically 0 (as shown in Problem 11 of Chapter 11). It is instructive to re-examine this result:

$$\bar{\psi}_L \psi_L = [(P_L \psi)^\dagger \gamma^0] P_L \psi = \psi^\dagger P_L \gamma^0 P_L \psi = \psi^\dagger \gamma^0 P_R P_L \psi = 0, \qquad (18.2)$$

since, as we have already learned (in Chapter 11), $P_L \gamma^0 = \gamma^0 P_R$, and $P_R P_L = 0$ (as P_R and P_L are orthogonal projection operators). Similarly, we can see that $\bar{\psi}_R \psi_R = 0$. We seem to need *another* field that refers to the same particle.

Let us recall the structure of the field operator:

$$\psi \sim \text{(creation of an antiparticle)} + \text{(annihilation of a particle)}. \qquad (18.3)$$

If we apply the charge conjugation operator C to this we obtain

$$C\psi \sim \text{(creation of a particle)} + \text{(annihilation of an antiparticle)}. \qquad (18.4)$$

$$\text{Let us define } \psi^C \equiv \gamma^0 C\psi^*, \qquad (18.5)$$

where the star means complex conjugation. The complex conjugation is necessary so that the signs of the frequencies are correctly correlated with the particle and antiparticle, and the γ^0 is needed so that we can have a **left-left Lorentz invariant mass term**[2] in the Lagrangian, which is of the form:

$$a\bar{\chi}_L \chi_L = a(\bar{\psi}_L \psi_L^C + \bar{\psi_L^C} \psi_L), \qquad (18.6)$$

with $\chi_L \equiv \psi_L + \psi_L^C$, a self-conjugate field ($\chi_L^C = \chi_L$),

$$\text{and } \psi_L^C \equiv (P_L \psi)^C = \gamma^0 C(P_L \psi)^* = P_R \gamma^0 C\psi^*.[3]$$

We have not written the terms $\bar{\psi}_L \psi_L$ or $\bar{\psi_L^C} \psi_L^C$ on the right-hand side of Eq. (18.6) because they are each identically 0. (Check that the latter is 0.) In order to check that Eq. (18.6) is *not* identically 0, we may write

$$a\bar{\psi}_L \psi_L^C = a[(P_L \psi)^\dagger \gamma^0] \psi_L^C = a[(P_L \psi)^\dagger \gamma^0] P_R \gamma^0 C\psi^*$$

$$= a\psi^\dagger P_L \gamma^0 P_R \gamma^0 C\psi^* = a\psi^\dagger P_L C\psi^* \neq 0,$$

$$\text{since } \gamma^0 P_R = P_L \gamma^0, \ P_L^2 = P_L, \text{ and } \gamma^{02} = 1.$$

---

1. Recall that SU(5) was a highly successful, consistent model. (It failed because it predicted too short a lifetime for the proton.)

2. A proof of the Lorentz invariance of the term is beyond the level of this book.

3. P_L commutes with C but changes to P_R when brought across γ^0.

(It is the presence of the γ^0 in the definition of ψ^C that accounts for the fact that this left-left term is non-zero.) Likewise, the second term in Eq. (18.6) is not 0 (nor does it cancel the first term).

Similarly, a term can be added to the Lagrangian of the form:

$$b\bar{\chi}_R\chi_R = b(\bar{\psi}_R\psi_R^C + \bar{\psi}_R^C\psi_R), \tag{18.7}$$

with $\chi_R \equiv \psi_R + \psi_R^C$, also a self-conjugate field ($\chi_R^C = \chi_R$).

These expressions (Eqs. 18.6 and 18.7) are called **Majorana mass terms**.[4] We shall include these Majorana mass terms in our Lagrangian along with the Dirac mass term for χ:

$$\mathcal{L}_{Dirac\,mass} = M_D(\bar{\chi}_R\chi_L + \bar{\chi}_L\chi_R),$$

yielding

$$\mathcal{L}_{\nu\,mass} = M_D(\bar{\chi}_R\chi_L + \bar{\chi}_L\chi_R) + a\bar{\chi}_L\chi_L + b\bar{\chi}_R\chi_R$$

$$= (\overline{\chi_L},\overline{\chi_R})\begin{pmatrix} a & M_D \\ M_D & b \end{pmatrix}\begin{pmatrix} \chi_L \\ \chi_R \end{pmatrix}.^5 \tag{18.8}$$

Suppose that $a = 0$, $b = \boldsymbol{M_R} \gg \boldsymbol{M_D}$. Eq. (18.8) then becomes

$$\mathcal{L}_{\nu\,mass} = M_D(\bar{\chi}_R\chi_L + \bar{\chi}_L\chi_R) + M_R\bar{\chi}_R\chi_R. \tag{18.9}$$

This can be recast in the form:

$$\mathcal{L}_{\nu\,mass} = M_1\bar{\chi}_1\chi_1 + M_2\bar{\chi}_2\chi_2, \text{ with}^{6,7} \tag{18.10}$$

$$M_1 \approx -M_D\frac{M_D}{M_R} \approx 0; \quad \chi_1 \approx \chi_L - \frac{M_D}{M_R}\chi_R \text{ (essentially left-handed)},$$

and

$$M_2 \approx M_R \textbf{ very large}; \quad \chi_2 \approx \chi_R + \frac{M_D}{M_R}\chi_L \text{ (essentially right-handed)}.$$

(Check that the $\mathcal{L}$'s of Eqs. (18.9) and (18.10) are the same to $O(M_D/M_R)$.) Thus, this **"see-saw"** mechanism allows us to have large masses for the right-handed neutrinos and very small masses for the left-handed neutrinos (which have already been observed). It also provides an uncontrived explanation of the smallness of the left-handed neutrino masses relative to those of the charged leptons.

---

4. The "j" is pronounced like a "y" in Majorana.

5. This Lagrangian can be written in the form:

$$\mathcal{L}_{mass} = M_1\bar{\chi}_1\chi_1 + M_2\bar{\chi}_2\chi_2, \text{ with } M_{1,2} = \tfrac{1}{2}[(a + b) \pm ((a - b)^2 + 4M_D^2)^{1/2}],$$

by diagonalizing the matrix in Eq. (18.8).

6. M_1 is negative, but a transformation beyond the level of this book changes that sign.

7. This result can be obtained from the exact solution, which is obtained by diagonalization, displayed in footnote 5.

However, let us examine the structure of these new terms from a different point of view. Consider the term $b\bar{\psi}_R \psi_R^C$ of Eq. (18.7). Remember that

$$\bar{\psi} \sim \text{(creation of a particle)} + \text{(annihilation of an antiparticle)}. \quad (18.11)$$

Equations (18.4) and (18.11) show us that $\bar{\psi}_R \psi_R^C$ has the form:

[(creation of a particle) + (annihilation of an antiparticle)]

$$\times \text{[(creation of a particle)} + \text{(annihilation of an antiparticle)],} \quad (18.12)$$

where it is the usual "one from square bracket 2 and one from square bracket 1." This innocent-looking term is, in fact, quite different from any of the terms we have previously considered; for instance, it can annihilate an antiparticle and create a particle in its place or create two particles, etc. Consequently, **such a term would violate the conservation of any additive number for which the particle has a non-zero value.**

First let us consider electric charge. As far as we know, $U_{EM}(1)$ is not broken at all (so that electric charge is conserved). But neutrinos are electrically neutral, so that a Majorana mass term for neutrinos does not violate charge conservation (or $U_{EM}(1)$ conservation). Next, we consider color charge. $SU_c(3)$ is also an unbroken symmetry of nature, but the neutrino is colorless (a singlet), so that such a Majorana mass term is invariant to color transformations as well. $U_{EM}(1)$ and $SU_c(3)$ are the *only* unbroken particle symmetries in nature; consequently, a Majorana mass term for neutrinos *can* be present in the Lagrangian. However, **this term definitely violates lepton number (L) conservation,** since replacement of an antineutrino by the corresponding neutrino or the creation of two neutrinos raises L by 2.

Since our analysis (above) of the first term of Eq. (18.7) has shown that an antiparticle can change into a particle and a similar analysis of the second term of that equation would have a particle changing into an antiparticle, we may say that the particle and antiparticle are equivalent. Alternatively, recall that the right-right mass term of Eq. (18.7) is written in terms of the self-conjugate field χ_R ($\chi_R^C = \chi_R$), which corresponds to **a spin $\frac{1}{2}$ particle that is its own antiparticle.** Majorana studied such particles shortly after Dirac created the idea of antiparticles; such particles (and fields) are, therefore, referred to as Majorana particles (and fields). (That is why this type of term is referred to as a Majorana mass.)

In the standard model the neutrino is a member of an isodoublet with $I_z = +\frac{1}{2}$ and the antineutrino has $I_z = -\frac{1}{2}$, so that a Majorana mass term for the neutrino would break isospin; it also would break Y invariance (since $Y \neq 0$ for the neutrino) and, in a GUT, a Majorana mass term for the neutrino would, therefore, break the larger symmetry. However, we cannot have terms in the Lagrangian which directly break our gauge symmetries, since they destroy renormalizability; we have painstakingly studied *spontaneous symmetry breaking* for that reason. But, just as masses for the gauge particles arise as the result of Higgs particles acquiring non-zero vacuum-expectation-values, we can envision that

Majorana mass terms of the type displayed in Eqs. (18.6) and (18.7) may arise as well. (In fact, the global breaking of lepton number produces massless Goldstone particles known as Majorons.) The sizes of the coefficients a and b will depend upon the vacuum-expectation-values which spontaneously break the various symmetries, hence, upon the mass scales at which those breakings occur. The inequality used to obtain Eq. (18.9), $M_R \gg M_D$ (and $M_L = 0$), has been shown to be very plausible and is directly related to the mass hierarchies that arise in GUTs. Thus, the existence of both Dirac and Majorana mass terms for the neutrinos can lead, in a natural way, to the *nearly* massless left-handed neutrinos we know and love along with the very heavy right-handed neutrinos required in most GUTs.

And, so, as we learn from Eq. (18.10), it is possible that the masses of the (observed) very light left-handed neutrinos we see are not identically 0. If the neutrino masses are not identically 0 and the neutrinos of different generations have masses different from one another, it is possible to have generation mixing and oscillations, as we shall now discuss.

18.2   MIXING AND OSCILLATIONS

We have already seen the mixing phenomenon in our discussion of K^0 decay (in Chapter 12). In that chapter we learned that the specie-content of the physical state may vary with time, if it is a linear combination of mass eigenstates which have different masses.

Let us consider the neutrinos of the first two generations. We will call the mass eigenstates $|\nu_1\rangle$ and $|\nu_2\rangle$. If an electron neutrino, or alternatively a muon neutrino, is produced at $t = 0$, although we do not know the initial mixture, we can write:

$$|\nu_e(0)\rangle = \cos\theta_\nu|\nu_1\rangle + \sin\theta_\nu|\nu_2\rangle \equiv |\boldsymbol{\nu_e}\rangle \text{ and}$$

$$|\nu_\mu(0)\rangle = -\sin\theta_\nu|\nu_1\rangle + \cos\theta_\nu|\nu_2\rangle \equiv |\boldsymbol{\nu_\mu}\rangle, \tag{18.13}$$

where θ_ν is some mixing angle (and the states are constructed to be orthogonal to one another). Then at a later time t, we have

$$|\nu_e(t)\rangle = \cos\theta_\nu e^{-iE_1 t}|\nu_1\rangle + \sin\theta_\nu e^{-iE_2 t}|\nu_2\rangle$$

$$= (e^{-iE_1 t}\cos^2\theta_\nu + e^{-iE_2 t}\sin^2\theta_\nu)|\boldsymbol{\nu_e}\rangle + \sin\theta_\nu\cos\theta_\nu(e^{-iE_2 t} - e^{-iE_1 t})|\boldsymbol{\nu_\mu}\rangle$$

and

$$|\nu_\mu(t)\rangle = -\sin\theta_\nu e^{-iE_1 t}|\nu_1\rangle + \cos\theta_\nu e^{-iE_2 t}|\nu_2\rangle$$

$$= \sin\theta_\nu\cos\theta_\nu(e^{-iE_2 t} - e^{-iE_1 t})|\boldsymbol{\nu_e}\rangle$$

$$+ (e^{-iE_2 t}\cos^2\theta_\nu + e^{-iE_1 t}\sin^2\theta_\nu)|\boldsymbol{\nu_\mu}\rangle, \tag{18.14}$$

where the second form of each equation results from inverting Eq. (18.13) to express $|\nu_1\rangle$ and $|\nu_2\rangle$ in terms of $|\boldsymbol{\nu_e}\rangle$ and $|\boldsymbol{\nu_\mu}\rangle$.

Thus, an e-neutrino, produced at $t = 0$, has a non-zero probability of being a μ-neutrino ($|\nu_\mu\rangle$) at a later time (and vice versa). From Eq. (18.14) we obtain

$$P_{\nu_e \to \nu_\mu} = |\langle \nu_\mu | \nu_e(t)\rangle|^2 = |\sin\theta_\nu \cos\theta_\nu (e^{-iE_2 t} - e^{-iE_1 t})|^2$$

$$= \sin^2\theta_\nu \cos^2\theta_\nu (1 - e^{i(E_1 - E_2)t} - e^{-i(E_1 - E_2)t} + 1)$$

$$= 2\sin^2\theta_\nu \cos^2\theta_\nu [1 - \cos(E_1 - E_2)t]. \qquad (18.15)$$

In our experiments, the neutrinos have very well-defined momenta (p) and are moving at high speeds, so that

$$E_1 - E_2 = \sqrt{p^2 + m_1^2} - \sqrt{p^2 + m_2^2} \approx \left(p + \frac{m_1^2}{2p}\right) - \left(p + \frac{m_2^2}{2p}\right) = \frac{m_1^2 - m_2^2}{2p}.$$

$$(18.16)$$

Inserting Eq. (18.16) into Eq. (18.15), we obtain

$$P_{\nu_e \to \nu_\mu} \approx 2\sin^2\theta_\nu \cos^2\theta_\nu \left[1 - \cos\left(\frac{m_1^2 - m_2^2}{2p}\right)t\right] \qquad \text{in natural units,}$$

yielding

$$P_{\nu_e \to \nu_\mu} \approx 2\sin^2\theta_\nu \cos^2\theta_\nu \left[1 - \cos\left(\frac{m_1^2 - m_2^2}{2p}\right)\frac{c^3}{\hbar}t\right], \qquad (18.17)$$

since the argument of the cosine must be dimensionless. We can rewrite this as a function of the distance (x) of the detector from the point of creation of the neutrino by noting that the time traveled is

$$t = \frac{x}{v} \approx \frac{x}{c},$$

yielding

$$P_{\nu_e \to \nu_\mu} \approx 2\sin^2\theta_\nu \cos^2\theta_\nu \left[1 - \cos\left(\frac{m_1^2 - m_2^2}{2p}\right)\frac{c^2}{\hbar}x\right]$$

$$= 2\sin^2\theta_\nu \cos^2\theta_\nu \left[1 - \cos 2\pi \frac{x}{\ell_{12}}\right], \qquad (18.18)$$

where ℓ_{12} is the distance through which the electron neutrino undergoes a complete oscillation into a μ neutrino and back:

$$\ell_{12} = \frac{4\pi p\hbar}{(m_1^2 - m_2^2)c^2} = 2.5 \text{ meters} \left[\frac{p \text{ (in MeV/}c)}{(m_1^2 - m_2^2)(\text{in}(\text{eV}/c^2)^2)}\right]. \qquad (18.19)$$

For a momentum (p) of a few MeV/c and $(m_1^2 - m_2^2) \approx (10^{-2} \text{ eV}/c^2)^2$, ℓ_{12} is of the order of meters, so that this oscillation of the number of electron neutrinos in the beam should be detectable near reactors, where large numbers of electron antineutrinos[8] are created. To date, no experiments have detected any such oscillation for terrestrially created neutrinos.

On the other hand, the study of solar neutrinos has revealed a strong possibility that, before they reach the earth, the neutrinos undergo a significant oscillation of identity, which may have been dramatically enhanced by the dense plasmas within the sun. This enhancement presumably results from coherent forward scattering of the neutrinos from electrons and is referred to as the *MSW effect*.[9] The masses and mixing angles required by the MSW mechanism to account for the solar neutrino data are so small that they are consistent with the absence of detectable mixing in terrestrial experiments.

Measurements of the flux of solar neutrinos have not yet led to any definitive conclusions; most facilities have been operational for only a short time and the statistical uncertainties are still large. One long-running experiment, using 600 tons of cleaning fluid as the detector material, has been operating in the Homestake gold mine of South Dakota for the past 20 years. The process occurring is

$$\nu_e + Cl^{37} \rightarrow e^- + Ar^{37}.$$

The argon is periodically chemically separated, and its amount is precisely determined in order to deduce the flux of electron neutrinos reaching the earth from the sun. There appears to be a paucity (ranging from 25% to 40%) of such solar neutrinos compared to the predictions of stellar model calculations. In 1987 the proton decay detector, in the Kamiokande (Japan) zinc mine, was converted into a neutrino detector. It is a three-kiloton water-Čerenkov detector, which is being used to detect electrons scattered elastically by the solar neutrinos. This study is also finding many less ($\sim \frac{1}{2}$) solar neutrinos than predicted for their detector. These two detectors are looking at neutrinos in different energy regions, which might explain their mutual discrepancy, but both are detecting neutrinos produced by the same process in the sun, viz., the beta decay of B^8. These are high-energy neutrinos, which are not produced abundantly and whose production rate depends very sensitively upon the temperature in the core of the sun ($\sim T^{18}$). We can question how well that temperature is known.

The most abundant source of neutrinos in the sun is the fusion reaction taking place:

$$p + p \rightarrow H^2 + e^+ + \nu_e.$$

The flux of these neutrinos is a very definite prediction of our stellar models, and a number far below that value cannot be obtained theoretically without wreaking havoc on our basic understanding of main sequence stars like our

8. The same phenomenon must occur for the antineutrinos, as well.
9. after its creators: Mikheyev and Smirnov (together) and Wolfenstein.

sun.[10] However, these neutrinos have energies below the thresholds of the Homestake and Kamiokande detectors. Two new experiments are now in operation with low enough thresholds to detect these fusion-produced neutrinos. One has recently begun running under the Caucasus mountains, which is a collaborative project called Soviet American Gallium Experiment (SAGE). Many tons of gallium are used to capture neutrinos and produce Ge^{71}, which can be chemically separated with the same superb accuracy as is obtained at the Homestake mine. This experiment has detected a smaller number ($\sim 45\%$) of electron neutrinos than expected. Another new gallium experiment, GALLEX in the Gran Sasso tunnel in Italy, has just reported a paucity of solar neutrinos as well. Only about $\frac{2}{3}$ of the number expected for the detector has been found.[11] (Statistical and systematic uncertainties are still large in the gallium experiments, since they have been running for a relatively short time.)

The evidence is beginning to indicate that the flux of electron neutrinos reaching the earth is far below the prediction of the standard solar model. Does this mean that our understanding of main sequence stars is severely flawed? Perhaps not. We could account for the lower detection rate by noting that the detection processes are only sensitive to electron neutrinos. If the identities of some of the neutrinos have changed during their journey through the sun and then to the earth, the discrepancy can be accounted for without questioning our basic understanding of stars. The *word* is not in yet, so we cannot say whether or not there are neutrino oscillations.

Other explanations of the paucity of solar neutrinos have been proposed. One with several versions is based upon the fact that, although the neutrino has 0 charge, it can have an effective magnetic moment ($\sim 10^{-11}\mu_B$) in one loop (see Problem 2) and, thereby, would interact with the strong magnetic fields in the sun. The precession of the magnetic moment of ν_e, during its journey from the center of the sun, would yield some right-handed (SU(2) singlet) ν_e's which would not have weak interactions. These "sterile" neutrinos would not be detected, since our detectors are designed to search for weak interactions initiated by the incoming particle.

18.3   DARK MATTER

Astronomical observations indicate that the behavior of stars and galaxies, as well as the abundances of light elements, cannot be understood without postu-

---

10. Some nonstandard solar models are being studied, including the possibility that heretofore unknown, weakly interacting massive particles (WIMPs) are being emitted from the center of the sun. This would lower the central temperature, thus changing the neutrino production rate, and the WIMPs might also be the "missing" dark matter thought to permeate the Universe (to be discussed in Section 18.3).

11. There is no obvious discrepancy due to the different results of different detectors because their sensitivities differ across the neutrino energy spectrum.

lating the existence of a lot of dark, unseen matter. The existence of dark matter would also have profound implications for the expansion of the Universe; without the existence of considerable dark matter, the Universe's expansion would not be the asymptotic type (with $v \to 0$ as $t \to \infty$) preferred by many cosmologists. The matter density for this asymptotic type of expansion is called **the critical matter density**. Cosmological arguments show that **the ratio (Ω) of the actual matter density to the critical matter density** changes with time. The evolution of Ω is such that the present-day value ($0.01 \le \Omega \le 4$) implies that Ω must have been extremely close to 1 at early times. ($|\Omega - 1| \le 10^{-60}$ at $t = 10^{-43}$ s, $|\Omega - 1| \le 10^{-16}$ at $t = 1$ s, etc.) If Ω were within 10^{-60} of 1 (at any time), we might guess that it probably was exactly 1. Furthermore, it has been shown that, for $\Omega < 1$ at early times we expect that $\Omega \to 0$ rapidly with time, whereas for $\Omega > 1$ at early times we would have $\Omega \to \infty$ rapidly. However, if Ω had been exactly 1, it would not have changed with time, so that we would expect that $\Omega = 1$ now too; this would imply that there must be a lot of dark matter (unidentified, so far) in the Universe, since $\Omega_{luminous}$ is only about 0.01.

In fact, the abundances of the light elements produced during primordial nucleosynthesis indicate that there is a much larger amount of baryonic matter (mostly dark) with $\Omega_{baryon} \approx 0.05$. This value of Ω is consistent with the motion of stars and galaxies on relatively small scales. However, the velocities[12] of clusters of galaxies on the very large scale indicate that $\Omega \approx 1$. Hence, astronomers and astrophysicists are very interested in the possibility that the neutrinos may have mass, since we expect these neutral leptons to pervade all of space. It has also been conjectured that there exist other weakly interacting massive particles (WIMPs), which have not yet been observed, but which play important roles in several theories (as mentioned in a footnote to the previous section). As we have already mentioned, axions (Chapter 16) and the lightest supersymmetric particle (LSP discussed in Chapter 17) are also candidates for nonbaryonic dark matter.

PROBLEMS

1. Show that $\overline{\psi_L^C}\psi_L^C$ is identically 0.
2. Show how a neutrino can have an effective magnetic moment in one loop by showing that a photon can effectively couple to a neutrino in a Feynman diagram with one loop.
3. Verify Eq. (18.19) for ℓ_{12}.

---

12. measured with respect to the corresponding Hubble velocities; they are called "peculiar" velocities by cosmologists.

Chapter 19

Further Developments

Here we present only brief discussions of further developments because a quantitative understanding of them is beyond the level of this book, and some are only in rudimentary form at this time. These discussions will complete our introduction to the field of elementary particles.

19.1   TECHNICOLOR

We have seen in Chapter 17 that the pion's existence can be understood to result from the spontaneous symmetry breaking of chiral symmetry by a *composite of quark field operators*. It is natural to wonder whether, analogously, the spontaneous symmetry breaking in the standard model may be accomplished by a composite field operator constructed of heretofore undiscovered fermions, with no Higgs boson in the theory at all. This would explain the failure to discover the Higgs and might render the hierarchy problem moot. The technicolor theories have taken this approach to spontaneous symmetry breaking.

The main reason for the creation of technicolor theories was the lack of experimental evidence of the existence of fundamental Higgs scalars. In these theories their roles are performed by condensates, composed of heavy fermions called **techniquarks**. These techniquarks experience a heretofore unknown (technicolor) interaction considerably stronger than the color forces. In the technicolor scenario the non-zero *vev* of a condensate of the very heavy techniquarks breaks the SU(2) × U(1) gauge theory of the electroweak theory. This **dynamical symmetry breaking** is due to the very strong technicolor force between the techniquarks in the condensate. However, we would expect low-mass pseudo-Goldstone bosons **(technipions)** to exist, analogous to the pseudo-Goldstone bosons (pions) which were a consequence of chiral symmetry breaking. **None has ever been experimentally detected!** Technicolor theories *have* been constructed in which no light pseudo-Goldstone boson (technipion) is present. These theories utilize the fact that, when a *gauge* theory is spontaneously broken, "the gauge field eats the Goldstone boson and gains weight," as we learned in Section 11.1.3; consequently, **there is no technipion** in the spectrum. However, **the technicolor theories predict the existence of technihadrons in the TeV range.** These predictions will certainly be experimentally testable in SSC experiments.

Technicolor theories can be incorporated into grand unified theories, with strong forces occurring in certain fermion-fermion channels. The **most attractive channel** (mac) can then serve as the condensate which breaks the large symmetry to a smaller one. This is followed, as we go down in energy, by a succession of such mac-tumblings (or breakings) to smaller and smaller symmetries. These **tumbling gauge symmetries** eliminate the hierarchy problem in a natural way.

Thus, if a technicolor theory is valid, the lack of experimental evidence of the Higgs particle and the theoretical concerns related to its bosonic nature[1] need not bother us. Furthermore, grand unification is no longer flawed by the hierarchy problem.

There is a serious problem which arises when the basic technicolor theories are extended to explain the widely different masses of the matter fermions (light and heavy quarks — u, d, s, c, b, and t; light and heavy leptons — e, μ, and τ). These **extended technicolor theories** must contain new particles which couple to both color and technicolor in order to provide mass for the quarks. They invariably contain vector (gauge) particles playing that role, which leads to a large **extended technicolor symmetry** (containing the technicolor group). Techniquark multiplets of the extended technicolor group must *also* contain color as well as technicolor. Together with technileptons, a generation of technifermions would have the structure:

$$\text{Techniquarks: } \begin{pmatrix} U_L^{a,\alpha} \\ D_L^{a,\alpha} \end{pmatrix}, U_R^{a,\alpha}, D_R^{a,\alpha}$$

$$\text{Technileptons: } \begin{pmatrix} N_L^{\alpha} \\ E_L^{\alpha} \end{pmatrix}, N_R^{\alpha}, E_R^{\alpha}, \tag{19.1}$$

where $a = 1,2,3$ is the color index and α is the technicolor index; the left-handed particles form a doublet and the right-handed particles are singlets under the SU(2) of the electroweak theory.

For the simplest technicolor structure, where α need not be used since there is only one technicolor, we see (from Eq. 19.1) that each handedness has eight technifermions. The theory is (approximately) invariant to the global chiral group $SU_L(8) \times SU_R(8)$, which has $63 + 63$ generators.[2] The breaking of this chiral symmetry to SU(8) then produces 63 Goldstone bosons. There are **several light pseudo-Goldstone bosons in the spectrum** of the fully constructed theory (after suitable symmetry breaking), **which are incompatible with experiment.** These theories also contain flavor-changing vector bosons, which invariably lead to **(unobserved) flavor-changing neutral currents.** (Of course, with larger tech-

---

1. Recall that, because the Higgs is a boson, we would expect very large radiative corrections to the Higgs mass, as explained in our discussion of supersymmetry in Chapter 17.

2. SU(N) has ($N^2 - 1$) generators.

nicolor groups (α having several possible values), the number of pseudo-Goldstone bosons grows, exacerbating the problem.) Attempted modifications to alleviate these problems have led to a loss of asymptotic freedom for QCD or to other undesirable effects (as of this writing).

Another approach being considered is the introduction of scalars with both color and technicolor, instead of vector bosons and their concomitant higher symmetries. These theories incorporate supersymmetry to "protect" the scalars from the destructive effects of radiative corrections (as discussed in Chapter 17). Their structures are, thus, quite involved.

When a beautiful, simple idea needs very elaborate modifications to be compatible with experiment, we are reminded of "epicycles," which made better predictions than Copernicus' original model, but comprised a fruitless dead-end. Many physicists have given up on technicolor because of the failure of the extended technicolor theories. However, see Section 19.4.

19.2 HORIZONTAL SYMMETRIES

Attempts have been made to account for the existence of three generations of matter fermions, the second and third generations appearing to be heavier copies of the first (u, d, e, ν). In essence, they attempt to answer Rabi's question, "Who ordered *that?*" Most models with horizontal symmetry contain a family (or generation) symmetry, which has an irreducible representation of dimension three, so that each of the four fermions u, d, e, ν would have a generation index with three possible values. Discrete groups, global continuous groups, and gauge groups have been employed in this quest, the latter containing gauge particles called **familons**. These groups are called *horizontal symmetries* to distinguish them from the GUTs, SUSY, and technicolor extensions of the standard model — the name makes visual sense if we write the three generations *across* the page. Because of the ad hoc nature of these theories, with no experimental evidence to "latch on to," the spotlight is not on this area of research at present, and the multiple-family question remains a mystery in most physicists' minds. However, see Section 19.4.

19.3 STRING THEORIES

String theories grew out of models based on arcane properties of amplitudes (their analytic properties and crossing symmetry when they are written in terms of the relativistic invariants discussed in Appendix E). It was shown that equivalent results could be obtained by considering the basic structures of matter to be one-dimensional strings rather than point particles. These strings sweep out two-dimensional world sheets, in contrast to the world lines followed by point particles. Following the procedures, analogous to those that lead from classical point particles to quantum field theory, produces theories of quantized strings and string field theories. The most successful of these structures, viz., superstring theory, possesses the promise of leading to the *theory of everything* (**TOE**), its proponents tell us, whereas skeptics believe that these structures are indeed beautiful mathematics but have no connection to the physical world.

The string nature of the underlying structures in these string theories would be discernible only at energies $\sim 10^{19}$ GeV (the Planck mass) or greater. Amazingly, below these energies, the string theories look like those SUSY-GUTs which we have believed most likely to be valid. Furthermore, there are indications that self-consistency considerations within the string theories will force this "lower" energy behavior to match *exactly one* of those SUSY-GUTs (perhaps SO(32)). Thus, the form of the (lower-energy) GUT and the resulting (very low energy) SU(3) × SU(2) × U(1) would be understood to be a consequence of basic principles. These string theories also contain within them a theory of gravity which is perturbatively renormalizable, since (as we have said) the string structures become relevant at energies where gravity becomes strong. There are, however, anomalies within the string theories, such that they are consistent only in ten (nine space) dimensions.[3] In order to make such theories applicable to our four-dimensional space-time, it is assumed that the extra six space dimensions are compact (closed) and very small ($\sim$ Planck length $\approx 10^{-35}$ m), so that they are not detected in ordinary experience.[4] The compactification of six dimensions is not a trivial mathematical operation, and we are led into very advanced, even some yet undiscovered, topological considerations.

Unhappily, **no string theory has ever produced a prediction that can be experimentally tested**, its difference from GUTs being manifested at energies (10^{19} GeV) which are 17 orders of magnitude above those (10^2 GeV) of present-day experiments. Even GUTs themselves have produced very few testable predictions. (At the present time, the most dramatic one is proton decay, for which a null result has been obtained so far. Others are the successful predictions: (1) of $\sin^2 \theta_W$ and (2) that $m_b = m_\tau$ at the GUT scale.)

19.4 COMPOSITE MODELS—PREONS

We have already made an excursion into compositeness above, inasmuch as the technicolor theories have the effective Higgs boson as a composite of fundamental techniquarks. The next questions which naturally arise are whether the quarks, leptons, heavy gauge bosons, and/or even the massless gauge bosons (photons, gluons, and gravitons) are composites. Note that if the quarks and

---

3. Recently, more elaborate superstring theories have been formulated in four dimensions, but they introduce comparable difficulties.

4. In the early part of the century a theory of gravity in five dimensions ($g_{\rho\sigma}$ with $\rho, \sigma = 0, \ldots, 4$) was developed by Kaluza and Klein. In that theory electrodynamics (A_μ) arose in a natural way with $g_{4\mu}$ becoming A_μ ($\mu = 0, \ldots, 3$), after the extra dimension was compactified. These types of theories are called **Kaluza-Klein** theories.

leptons are composites, the second and third generations may merely be excited states of these bound systems. This would extend the conjectures about horizontal symmetries and attach them to a more solid foundation. Thus, technicolor (Section 19.1) and horizontal symmetry (Section 19.2) may consequently be dealing with the ramifications of the compositeness of the fundamental particles of the standard model. Composite theories could also contain GUTs within them in a natural way, with naturally occurring mass hierarchies (therefore, no hierarchy problem or fine tuning). The quarks and leptons would be connected in a natural way because they would all be bound states of more fundamental constituents.

The fundamental particles in these theories[5] are generally called **preons** and, in one particular elaborate theory, **rishons**. They interact via very strong forces usually referred to as **hypercolor** or **metacolor**. It is natural to expect that these theories have **the standard model as their low-energy limits**. They require a very strong force (even compared to that required in technicolor theories), yet must have bound state fermions corresponding to the particles which are fundamental in the standard model, hence, which are "massless" on that energy scale. Chiral symmetry or the possibility that those fermions are goldstinos (fermion supersymmetric partners of Goldstone bosons) have been suggested as mechanisms that could accomplish this. **To date, no satisfactory theory has been constructed that meets all the theoretical requirements.**[6]

There are also severe constraints on these theories from the data which have explored *beyond* the standard model predictions, such as those of proton decay and the baryon asymmetry of the Universe. Furthermore, **there is absolutely no experimental evidence to suggest the compositeness of the fundamental particles of the standard model**. Consequently, it is not at all apparent how to proceed to construct a theory. One prominent contributor to considerations of compositeness has written, "It is difficult to believe that the correct theory can be found without some experimental hints."[7] Furthermore, it is not clear what the signatures for compositeness of quarks would be because they never appear as *free* particles.

PROBLEMS

1. Use dimensional analysis to construct a mass from the gravitational constant (G), $\hbar$ and c. Show that its value $\sim 10^{19}$ GeV.

---

5. A different kind of underlying theory with nonlinearly self-interacting fundamental fermions has also been proposed.

6. including renormalizability and the avoidance of anomalies.

7. H. Harari, in *Composite Models For Quarks and Leptons* preprint WIS-82/60 Dec-Ph.

Postscript

As the final draft of this book was prepared for mailing, an article appeared in the *Science Times* section of the *New York Times* (Jan. 5, 1993) with the head-line **"315 Physicists Report Failure In Search For Supersymmetry"** and with the sub-headline "The negative result illustrates the risks of Big Science, and its often sparse pickings." The original paper reporting the experiment was: F. Abe *et al.* (CDF collaboration) "Search for Squarks and Gluinos from $\bar{p}p$ Collisions at $\sqrt{s} = 1.8$ TeV," Phys. Rev. Lett. **69**, 3439 (1992). The experiment analyzed events with jets and large missing transverse energy, at the $p\bar{p}$ CM energy of 1.8 TeV, and found an event rate consistent with the standard model. It excluded, in several SUSY models, the existence of squarks and gluinos in certain mass ranges.

This was an exploration into the *unknown*, which found new information about the nature of the Universe. It was not an attempt to "find gold in them thar hills" or an attempt to build a new widget. (Of course, those involved in such experiments *hope* to find something startling; this is part of the fun and excitement of exploration.) The search for the fundamental particles and their interactions involves the exploration of new regions of energy, etc., and the headlines in the *New York Times* display a complete misunderstanding of basic research. The standard model and conjectures beyond the standard model must be tested experimentally before we can advance our understanding of the under-lying structure of nature. As was enunciated long ago by Sir Francis Bacon (in his *Novum Organum*), "If, however, the common judgement of the logicians has been so laborious, and has exercised such great wits, **how much more must we labour in this which is drawn not only from the recesses of the mind, but the very entrails of nature**." The new knowledge, obtained in experiments of the sort reported as a "failure" by the *New York Times*, illuminates realms of real-ity, unseen heretofore. *In order to continue the age-old search for the funda-mental structure of matter, such investigations must be undertaken, despite the difficulty.*

In this book we have described the standard model and have indicated its experimental validation in myriad experiments. We have also examined its aesthetic shortcomings and have explored the conjectures which take us beyond the standard model, even to the question of compositeness of quarks and leptons. Those considerations bring us to the same turn of mind with which this book opened. We seem to be descending a spiral staircase, arriving at the analogous position again and again. Is this an endless quest — does the staircase go on and on or will we reach the bottom (the truly fundamental particles)? Instead of pursuing *imaginary* musings, we shall let *real time* answer further questions.

Bill Rolnick
January 8, 1993

Appendix A

$\vec{L}$ Matrices for $\ell = 1$

The functions $Y_1^{+1}(\theta,\phi)$, $Y_1^0(\theta,\phi)$, and $Y_1^{-1}(\theta,\phi)$ (spherical harmonics) are used as basis functions to construct these 3×3 matrices. We may calculate

$$(L_i)_{m_1 m_2} \equiv \int Y_1^{m_1^*}(\theta,\phi) L_i Y_1^{m_2}(\theta,\phi) \, d\Omega, \qquad (\text{A.1})$$

where L_i are the differential operators defined in Chapter 2, although we shall not do this calculation here. The results obtained may be displayed in matrix form as

$$L_x = \begin{pmatrix} 0 & \frac{1}{\sqrt{2}} & 0 \\ \frac{1}{\sqrt{2}} & 0 & \frac{1}{\sqrt{2}} \\ 0 & \frac{1}{\sqrt{2}} & 0 \end{pmatrix}, \; L_y = \begin{pmatrix} 0 & \frac{-i}{\sqrt{2}} & 0 \\ \frac{i}{\sqrt{2}} & 0 & \frac{-i}{\sqrt{2}} \\ 0 & \frac{i}{\sqrt{2}} & 0 \end{pmatrix}, \; \text{and}$$

$$L_z = \begin{pmatrix} +1 & 0 & 0 \\ 0 & 0 & 0 \\ 0 & 0 & -1 \end{pmatrix}. \qquad (\text{A.2})$$

Check the commutation relations: $[L_x, L_y] = iL_z$, $[L_z, L_x] = iL_y$, and $[L_y, L_z] = iL_x$ for these matrices. Also *verify* that the spherical harmonics may be represented here by the columns:

$$Y_1^{+1} = \begin{pmatrix} 1 \\ 0 \\ 0 \end{pmatrix}, \; Y_1^0 = \begin{pmatrix} 0 \\ 1 \\ 0 \end{pmatrix}, \; \text{and} \; Y_1^{-1} = \begin{pmatrix} 0 \\ 0 \\ 1 \end{pmatrix}, \qquad (\text{A.3})$$

by applying L_z to each.

In Chapter 2 we stated that

$$L^2 Y_\ell^m = \ell(\ell + 1) Y_\ell^m, \qquad (\text{A.4})$$

387

for any m. Therefore, $L^2 \ (= L_x^2 + L_y^2 + L_z^2)$ should be diagonal with the same number, $[\ell(\ell + 1)] = 2$ for $\ell = 1$, down the diagonal. We leave it to the reader to *check this* for the matrices in Eq. (A.2).

Notice that L_x and L_y have off-diagonal elements, which tells us that operating on one Y_l^m they produce the others. This is most easily understood by constructing the following linear combinations of them:

$$L_+ \equiv L_x + iL_y \text{ and } L_- \equiv L_x - iL_y. \tag{A.5}$$

Calculate L_+ and L_- from the 3 × 3 matrices (Eq. A.2) for $\ell = 1$. Find out what $L_\pm$ do to the states displayed in Eq. (A.3), i.e., beginning with one of these states, find which states are generated when $L_\pm$ are applied. Can you see why L_+ is referred to as a raising operator and L_- as a lowering operator?

Appendix B

Four-Vectors in Electrodynamics

B.1 VERIFICATION THAT J^μ IS A FOUR-VECTOR

Suppose that in frame S we have a static charge density $\rho(\vec{r})$ with no current flowing ($\vec{J} = 0$). Let us concentrate on an infinitesimal chunk of charge Δq at rest in S at the point P, occupying a region $\Delta x \Delta y \Delta z$. We shall consider two other frames as well: S' moving with speed v' and S'' moving with speed v'' in the $+x$ direction, as shown in Fig. B.1(a). Let us first consider the view in frame S', as shown in Fig. B.1(b). The amount of charge (e.g., number of electrons) Δq in the region under consideration is a physical quantity, which does not depend upon who (or if anyone at all) is looking at it. Thus, the charge inside the region is the same (Δq) in S':

$$\Delta q = \rho \Delta x \Delta y \Delta z = \rho'(\Delta x)'(\Delta y)'(\Delta z)'. \qquad (B.1)$$

But in S', the size of the volume is reduced by the usual Lorentz-contraction factor:

$$(\Delta x)' = \frac{\Delta x}{\gamma'}, \ (\Delta y)' = \Delta y \text{ and } (\Delta z)' = \Delta z, \text{ where } \gamma' = \frac{1}{\sqrt{1 - \dfrac{v'^2}{c^2}}}, \qquad (B.2)$$

so that Eq. (B.1) becomes

$$\Delta q = \rho \Delta x \Delta y \Delta z = \rho'\left(\frac{1}{\gamma'} \Delta x\right) \Delta y \Delta z. \qquad (B.3)$$

From this equation we find that

$$\rho' = \gamma' \rho. \qquad (B.4)$$

The chunk of charge (Δq) is moving with speed v' in the $-x$ direction in S', and we have learned in electrodynamics that this corresponds to a current density

$$\vec{J}' = \rho'\vec{v}' = \rho'(-v'\hat{x}) = \gamma'\rho(-v'\hat{x}), \qquad (B.5)$$

389

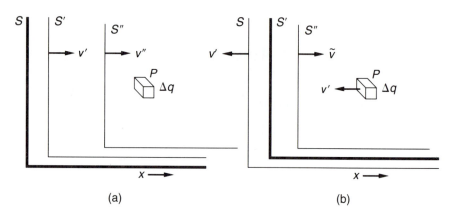

FIGURE B.1
(a) *View in frame S with Δq at rest, S' moving at speed v' and S" moving at speed v"* in the $+x$ direction. (b) *View in frame S' with S and Δq moving at speed v' in the $-x$ direction and S" moving with speed $\tilde{v}$ in the $+x$ direction.*

where we have used Eq. (B.4) in the last step. In S we had $J = 0$ and $\rho \neq 0$, so that the second expression in Eq. (B.5) shows the mixing of charge density into current density under Lorentz transformations. If we follow the same reasoning in frame S'' we obtain equations identical to Eqs. (B.4) and (B.5) with the prime replaced by a double-prime. Summarizing these results we have

$$\rho' = \gamma'\rho \quad \rho'' = \gamma''\rho \quad \vec{J}' = \gamma'\rho(-v'\hat{x}) \text{ and } \vec{J}'' = \gamma''\rho(-v''\hat{x}), \text{ when } \vec{J} = 0.$$

$$(B.6)$$

In order to find the transformation from a frame in which neither charge nor current density are zero, we will express ρ'' in terms of ρ' and J_x', i.e., we will find the transformation from S' to S''. From the first equation in Eq. (B.6), we may express ρ in terms of ρ' and, inserting that expression into the second equation, we find

$$\rho'' = \frac{\gamma''}{\gamma'} \rho'.$$

$$(B.7)$$

We leave it to the reader to verify that

$$\frac{\gamma''}{\gamma'} = \tilde{\gamma}\left(1 + \frac{\tilde{v}v'}{c^2}\right),$$

$$(B.8)$$

where $\tilde{v}\hat{x}$ is the velocity of S'' with respect to S', using the velocity addition law:

$$\tilde{v} = \frac{v'' - v'}{1 - \frac{v'v''}{c^2}},$$

$$(B.9)$$

which yields

$$v'' = \frac{\tilde{v} + v'}{1 + \dfrac{\tilde{v}v'}{c^2}}. \tag{B.10}$$

Substituting Eq. (B.8) into Eq. (B.7), we obtain

$$\rho'' = \tilde{\gamma}\left(\rho' + \frac{\tilde{v}v'}{c^2}\rho'\right) = \tilde{\gamma}\left(\rho' - \frac{\tilde{v}}{c^2}J_x'\right), \tag{B.11}$$

where we have used Eq. (B.5) in the last step. Likewise, from the equations for current density, we find

$$J_x'' = -\gamma''\rho v'' = -\gamma'' \frac{\rho'}{\gamma'}v'' = -\tilde{\gamma}\left(1 + \frac{\tilde{v}v'}{c^2}\right)v''\rho'$$

$$= -\tilde{\gamma}(\tilde{v} + v')\rho' = \tilde{\gamma}(-\tilde{v}\rho' + J_x'), \tag{B.12}$$

where we have used Eqs. (B.7), (B.8), and (B.10). Equations (B.11) and (B.12) show that the charge and current densities have been mixed in an identical manner to the mixing of time and space under Lorentz transformations for this restricted case, where the relative motion of the two frames (S' and S'') is in the same direction as the current in S'. These results may be generalized by choosing S'' to be moving in an arbitrary direction, so that $\vec{J}'$ is no longer in the direction of the relative velocity between S' and S''. It requires considerably more algebra (and the equation for $\tilde{v}_\perp$) to show that the components of the current perpendicular to the direction of relative motion do not change, and Eqs. (B.11) and (B.12), with x replaced by $\parallel$, are unchanged. Thus, one finds

$$c\rho'' = \tilde{\gamma}(c\rho' - \tilde{\beta}J_\parallel'),$$

$$J_\parallel'' = \tilde{\gamma}(J_\parallel' - \tilde{\beta}\rho'), \text{ and}$$

$$J_\perp'' = J_\perp'. \tag{B.13}$$

Consequently, **$c\rho$ and $\vec{J}$ mix exactly like ct and $\vec{r}$ under Lorentz transformations, so that they form a four-vector.**

B.2   VERIFICATION THAT A^μ IS A FOUR-VECTOR

In electrodynamics the scalar potential (V) and the vector potential ($\vec{A}$) are not uniquely defined for given $\vec{E}$ and $\vec{B}$ fields. We will verify here that, in a suitable Lorentz covariant formulation of electrodynamics, $A^\mu = (V; \vec{A})$ is a four-vector.

In order to show the four-vector nature of the potentials, we impose the Lorentz condition:

$$\partial_0 V + \vec{\nabla} \cdot \vec{A} = 0. \tag{B.14}$$

With this constraint, the scalar and vector potentials obey the wave equation, which we write in a partially covariant form as

$$\partial^\mu \partial_\mu V = (\text{const})\rho \quad \text{and} \quad \partial^\mu \partial_\mu \vec{A} = (\text{const})\vec{J}. \tag{B.15}$$

Taking the Fourier transforms (in space *and* time) of Eqs. (B.15),[1] we find that $V(k^\mu)$ and $\vec{A}(k^\mu)$, the Fourier transforms of the potentials, obey

$$V(k^\mu) = \frac{1}{k^2}(\text{const})\rho(k^\mu) \quad \text{and} \quad \vec{A}(k^\mu) = \frac{1}{k^2}(\text{const})\vec{J}(k^\mu), \tag{B.16}$$

where $\rho(k^\mu)$ and $\vec{J}(k^\mu)$ are the Fourier transforms of the components of the four-vector $J^\mu = (\rho; \vec{J})$ and, consequently, form a four-vector as well.[2] Therefore, Eqs. (B.16) show that the Fourier transforms of the potentials form a four-vector. This implies that *their* Fourier transforms form a four-vector as well; hence, V and $\vec{A}$ form a four-vector $A^\mu = (V; \vec{A})$.

Alternatively, the right-hand sides of Eqs. (B.15), which are the sources of the potentials, mix like a four-vector under Lorentz transformations; consequently, the left-hand sides of that equation must mix in the same way in order to maintain the equalities. In other words, we can say that the potentials are determined by the components of a four-vector source $J^\mu = (\rho; \vec{J})$, and consequently form a four-vector, viz., $A^\mu = (V; \vec{A})$.

Furthermore, utilizing this four-vector notation, we can rewrite the Lorentz condition, Eq. (B.14), in a manifestly covariant form as well:

$$\partial_\mu A^\mu = 0. \tag{B.17}$$

Since this manifestly Lorentz-covariant equation restricts the possible gauge transformations of the potentials, the potentials are said to be in **the Lorentz gauge**, which is a manifestly covariant gauge. This entire formulation of quantum electrodynamics is manifestly Lorentz covariant.

---

1. The operator ∂_μ becomes k_μ, etc., after a Fourier transformation.
2. The Fourier transforms of the components of a four-vector form a four-vector also since, under a Lorentz transformation, they become the same linear combinations of the original transforms as the corresponding components become of the original components.

Appendix C

Field Equations and
Noether's Theorem

C.1 CONNECTION BETWEEN LAGRANGE DENSITIES
AND FIELD EQUATIONS

We have developed our understanding of the quantum field theories of particle physics using Lagrange densities $\mathcal{L}$, which are functions of fields. We did not explore the field equations, since the fields that appear are actually particle operators in a Fock space of states (as discussed in Chapter 4) and their equations of motion were not needed for our considerations. (We did discuss the Klein-Gordon equation for boson fields in our discussion of range and mass.) Here we shall explore the connection between $\mathcal{L}$ and the field equations.

A study of mechanics shows that the theory of the motion of objects may be formulated in terms of the action (S):

$$S = \int L \, dt, \tag{C.1}$$

where L is the Lagrangian. This can be generalized to discuss fields, with the replacement of the Lagrangian by the Lagrange density. For simplicity, let us consider the case of a single scalar field $\phi(\vec{r}, t)$. The Lagrangian (L) is replaced by

$$L = \int \mathcal{L}(\phi, \partial_\mu \phi) \, d^3x, \tag{C.2}$$

where $\mathcal{L}$ is not an explicit function of space or time. The **principle of least action** tells us that the field equations may be obtained by demanding that the action be a minimum under variations of ϕ: $\phi(\vec{r}, t) \rightarrow \phi(\vec{r}, t) + \delta\phi(\vec{r}, t)$, with $\delta\phi(\vec{r}, t)$ going to zero at the (infinite) boundaries of space-time. Therefore, we demand that

$$\delta S = \delta \int \mathcal{L}(\phi, \partial_\mu \phi) \, d^4x = \int \delta\mathcal{L}(\phi, \partial_\mu \phi) d^4x = 0. \tag{C.3}$$

The variation is very much like the derivative of $\mathcal{L}$, treated as a *function* of ϕ:

$$\delta\mathcal{L}(\phi,\partial_\mu\phi) = \mathcal{L}(\phi + \delta\phi, \partial_\mu(\phi + \delta\phi)) - \mathcal{L}(\phi,\partial_\mu\phi) = \frac{\partial\mathcal{L}}{\partial\phi}\,\delta\phi + \frac{\partial\mathcal{L}}{\partial(\partial_\mu\phi)}\,\delta(\partial_\mu\phi).$$

(C.4)

Keep in mind that we are varying $\phi(\vec{r})$ infinitesimally, but not necessarily by the same amount at each point in space, so that $\delta\phi$ is a function of $\vec{r}$. It is apparent that

$$\delta(\partial_\mu\phi) = \partial_\mu(\phi + \delta\phi) - \partial_\mu\phi = \partial_\mu\delta\phi,$$

(C.5)

which may be substituted into Eq. (C.4), yielding for Eq. (C.3)

$$\delta S = \int\left(\frac{\partial\mathcal{L}}{\partial\phi}\,\delta\phi + \frac{\partial\mathcal{L}}{\partial(\partial_\mu\phi)}\,\partial_\mu\delta\phi\right)d^4x = 0.$$

(C.6)

Since $\delta\phi$ goes to zero on the boundaries, we may use integration by parts for the second term to obtain

$$\int\left(\frac{\partial\mathcal{L}}{\partial\phi} - \partial_\mu\frac{\partial\mathcal{L}}{\partial(\partial_\mu\phi)}\right)\delta\phi\,d^4x = 0.$$

(C.7)

Since $\delta\phi$ is an arbitrary function (except on the boundaries at infinity), this integral will be 0 for all possible choices of $\delta\phi$ if and only if

$$\frac{\partial\mathcal{L}}{\partial\phi} - \partial_\mu\frac{\partial\mathcal{L}}{\partial(\partial_\mu\phi)} = 0,$$

(C.8)

which are called **the Euler-Lagrange field equations**.

For a real scalar field with a quartic self-interaction, the Lagrange density may be written:

$$\mathcal{L} = \tfrac{1}{2}\partial^\mu\phi\partial_\mu\phi - \tfrac{1}{2}M^2\phi^2 - \tfrac{1}{2}\lambda\phi^4,^{[1]}$$

(C.9)

where M is the mass and λ is the self-coupling constant. The Euler-Lagrange field equations for this Lagrange density are

$$(-M^2\phi - 2\lambda\phi^3) - \partial_\mu(\partial^\mu\phi) = 0,^{[2]}$$

which may be written:

$$\partial^\mu\partial_\mu\phi + M^2\phi + 2\lambda\phi^3 = 0.$$

(C.10)

When the self-coupling is 0 ($\lambda \to 0$), Eq. (C.10) is the Klein-Gordon equation. We also see that coupling terms in the Lagrangian produce nonlinear terms in the field equations.

---

1. For a vector field (B^ν) we have $\mathcal{L} \sim \partial^\mu B^\nu\partial_\mu B_\nu - M^2 B^\nu B_\nu - \lambda(B^\nu B_\nu)^2$, so that $\mathcal{L}$ is Lorentz invariant.

2. We have used $\partial^\nu\phi = g^{\nu\mu}\partial_\mu\phi$ (where $g^{\nu\mu}$ is the inverse matrix of $g_{\nu\mu}$); the first term in $\mathcal{L}$ is quadratic in $\partial_\mu\phi$.

C.2   CONNECTION BETWEEN CONTINUOUS INVARIANCES AND CONSERVED CHARGES — NOETHER'S THEOREM

Suppose that there is a set of transformations of ϕ, with a continuous parameter α, which does not change the Lagrangian density. Using the relationships:

$$d\phi = \frac{\partial \phi}{\partial \alpha}\, d\alpha, \; d(\partial_\mu \phi) = \frac{\partial(\partial_\mu \phi)}{\partial \alpha}\, d\alpha = \partial_\mu \left(\frac{\partial \phi}{\partial \alpha} \right) d\alpha,$$

we find

$$d\mathcal{L} = \frac{\partial \mathcal{L}}{\partial \phi} \frac{\partial \phi}{\partial \alpha}\, d\alpha + \frac{\partial \mathcal{L}}{\partial(\partial_\mu \phi)}\, \partial_\mu \left(\frac{\partial \phi}{\partial \alpha} \right) d\alpha = 0. \tag{C.11}$$

We may cancel the constant $d\alpha$ in the second line and substitute for $\partial \mathcal{L}/\partial \phi$ in the first term, from the Euler-Lagrange equations Eq. (C.8), to find

$$\partial_\mu \left(\frac{\partial \mathcal{L}}{\partial(\partial_\mu \phi)} \right) \frac{\partial \phi}{\partial \alpha} + \frac{\partial \mathcal{L}}{\partial(\partial_\mu \phi)}\, \partial_\mu \left(\frac{\partial \phi}{\partial \alpha} \right) = \partial_\mu \left(\frac{\partial \mathcal{L}}{\partial(\partial_\mu \phi)} \frac{\partial \phi}{\partial \alpha} \right) = 0. \tag{C.12}$$

The last expression in Eq. (C.12) is of the form:

$$\partial_\mu J^\mu = 0, \text{ with } J^\mu \equiv \frac{\partial \mathcal{L}}{\partial(\partial_\mu \phi)} \frac{\partial \phi}{\partial \alpha}. \tag{C.13}$$

Thus, a continuous invariance produces a current obeying the continuity equation, which leads to a conserved charge $(\int J^0\, d^3 x)$, as we have shown in Chapter 4. The connection between continuous symmetries and conserved quantities was first found by Emmy Noether (a mathematician) in 1918 and is referred to as **Noether's theorem**.

Appendix D

A Contour Integral from Chapter 4

The integral to be done is

$$\Phi(\vec{r}) = \frac{g}{(2\pi)^3} \int \frac{e^{i\vec{k}\cdot\vec{r}}}{k^2 + m^2} \, d^3k. \tag{D.1}$$

We shall do the angular part of this integral using spherical coordinates. We take the polar direction (for the integral) in the $\hat{r}$ direction ($\vec{r}$ is fixed as we integrate over $\vec{k}$), so that $\vec{k}\cdot\vec{r}$ becomes $kr\cos\theta$. The integral may now be written:

$$\Phi(\vec{r}) = \frac{g}{(2\pi)^3} \int_0^{2\pi} \int_0^{\pi} \int_0^{\infty} \frac{e^{ikr\cos\theta}}{k^2 + m^2} \, k^2 \, dk \sin\theta \, d\theta \, d\phi. \tag{D.2}$$

Since the integrand is independent of ϕ, that integral yields a factor of (2π). Let us change the θ variable to $\mu \equiv \cos\theta$. Then we obtain $d\mu = -\sin\theta \, d\theta$ and (utilizing the minus sign to interchange the limits) the equation becomes

$$\Phi(\vec{r}) = \frac{g}{(2\pi)^2} \int_{-1}^{+1} \left[\int_0^{\infty} \frac{e^{ikr\mu}}{k^2 + m^2} \, k^2 \, dk \right] d\mu. \tag{D.3}$$

The integral over μ yields two terms:

$$\int_{-1}^{+1} e^{ikr\mu} \, d\mu = \frac{e^{ikr\mu}}{ikr} \bigg|_{-1}^{+1} = \frac{e^{ikr} - e^{-ikr}}{ikr}. \tag{D.4}$$

This leads to

$$\Phi(\vec{r}) = \frac{g}{(2\pi)^2 ir} \left[\int_0^{\infty} \frac{e^{ikr} k}{k^2 + m^2} \, dk + \int_{-\infty}^{0} \frac{e^{ik'r} k'}{k'^2 + m^2} \, dk' \right], \tag{D.5}$$

where, in the second term, we have introduced $k' \equiv -k$ and have used the minus sign (in front of e^{-ikr}) in Eq. (D.4) to interchange the limits. Notice that the second term in Eq. (D.5) has the same integrand as the first term but with lim-

397

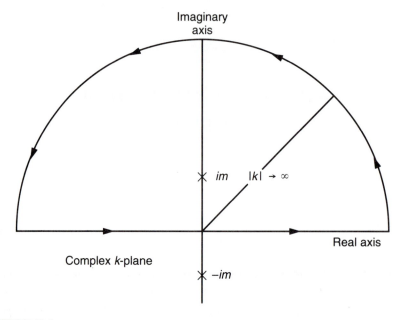

FIGURE D.1
The contour in the complex plane, whose radius $|k|$ is infinite, with $\times$'s showing the poles of the integrand.

its from $-\infty$ to 0. Consequently, we can combine these (since the integration variable's "name" is irrelevant) to obtain

$$\Phi(\vec{r}) = \frac{g}{(2\pi)^2 ir} \int_{-\infty}^{\infty} \frac{e^{ikr}k}{k^2 + m^2}\, dk = \frac{g}{(2\pi)^2 ir} \int_{-\infty}^{\infty} \frac{e^{ikr}k}{(k - im)(k + im)}\, dk,$$

(D.6)

where the denominator of the integrand has been written $(k - im)(k + im)$, to display the two poles in the complex k plane (at $\pm im$). A contour integration may be done by closing the contour in the upper half-plane as shown in Fig. D.1, since the integrand drops exponentially as the imaginary part (κ) of k goes to $+\infty$: $e^{i(k+i\kappa)r} = e^{ikr}e^{-\kappa r} \xrightarrow[\kappa \to \infty]{} 0$. Using the Cauchy integral theorem we obtain

$$\Phi(r) = \frac{g}{4\pi^2 ir} 2\pi i \operatorname*{Residue}_{k=im} \left\{ \left[\frac{e^{ikr}k}{k + im} \right] \middle/ (k - im) \right\} = \frac{g}{4\pi r} e^{-mr}. \quad \text{(D.7)}$$

Appendix E

Kinematic Relativistic
Variables

E.1 RELATIVISTIC INVARIANTS: s, t, u

Quantum theory presents us with a probability distribution of various outcomes for any given experiment; it is, therefore, especially important to combine data from many repetitions of the experiment as well as from other experiments. Very often the data for the same kinds of events, recorded in different experiments or parts of the same experiment, are in different Lorentz frames. This gives the impression that we must Lorentz transform the data before combining them. In this section we shall see how the use of **kinematic relativistic invariants**, which are quantities with the *same numerical values* **in any frame of reference**, avoids the problem of data being in different Lorentz frames. The structure of probability amplitudes as functions of these kinematic invariants is also of interest in some theoretical considerations. These invariants (s, t, and u) are often referred to as the **Mandelstam variables**.

First let us consider a two-particle system which, in some cases, may be the result of the decay of a particle (or resonance) into two particles. The most convenient Lorentz frame to work in is that in which the total three-momentum of the two particles is 0, i.e., they are moving in opposite directions with momenta of the same size, as shown in Fig. E.1. This is usually referred to as the center-of-momentum (CM) frame; in the case of an initial particle which decayed into the two particles, it would be the rest frame of that particle. We may choose the line of motion of the particles as our z axis. The only kinematic variable is, then, the size of their individual momenta (p^{CM}). Since the total energy is directly related to that momentum by

$$E_{TOTAL}^{\text{CM}} = E_1^{\text{CM}} + E_2^{\text{CM}} = \sqrt{p^{\text{CM}2} + m_1^2} + \sqrt{p^{\text{CM}2} + m_2^2}, \qquad (E.1)$$

we may use instead, as our kinematic variable, the total CM energy. (We have used the superscript CM to designate quantities in the CM frame.) Although the total energy is the fourth component of the total four-momentum, we can

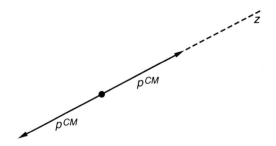

FIGURE E.1
Momentum diagram for two particles in their center-of-momentum (CM) frame. They
have momenta of the same size (p^{CM}), and here their line of motion is taken as the z
direction.

construct a *Lorentz-invariant* kinematic quantity, which corresponds to the
energy in the CM frame, but can be calculated in any frame (e.g., the lab frame):

$$s \equiv (p_{1\mu} + p_{2\mu})^2 = (E_1 + E_2)^2 - (\vec{p}_1 + \vec{p}_2)^2, \qquad \text{(E.2)}$$

where we have used the compressed notation $(p_\mu)^2$ for $p^\mu p_\mu$. Evaluation of s
in the CM frame is simple because

$$\vec{p}_1^{\,CM} + \vec{p}_2^{\,CM} = 0, \text{ so that}$$

$$s = (E_{TOTAL}^{CM})^2, \qquad \text{(E.3)}$$

which **is the same number obtained by calculating s in any frame** (i.e., by using
Eq. (E.2) in that frame). Utilizing s as our kinematic variable instead of energy,
we can then easily combine the results of many experiments without doing
Lorentz transformations. If the two particles are the products of the decay of
a particle of mass M, then conservation of energy would require that

$$s = (E_{TOTAL}^{CM})^2 = M^2. \qquad \text{(E.4)}$$

When two particles result from the decay of a resonance of mass M_R, we
would expect the reaction rate to have a peak at $s = M_R$ when plotted against
s. Furthermore, the width of the peak is directly related to the lifetime of the
resonance, as shown in Chapter 13.

A more formal way to see that s is the only independent kinematic invariant is to seek the independent invariants which can be constructed from the four-
momenta of the two particles:

$$p_{1\mu} = (E_1; \vec{p}_1) \text{ of particle 1 and}$$

$$p_{2\mu} = (E_2; \vec{p}_2) \text{ of particle 2.}$$

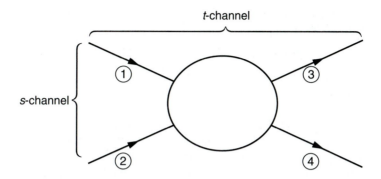

FIGURE E.2
Symbolic diagram for the reaction $1 + 2 \to 3 + 4$, shown as the "s channel," indicating the related cross-channel reaction $1 + \bar{3} \to \bar{2} + 4$ as the "t channel."

We construct kinematic invariants by taking dot products of any two linear combinations of $p_{1\mu}$ and $p_{2\mu}$. Let us consider first the dot product of each particle's four-momentum with itself:

$$(p_{1\mu})^2 = (E_1)^2 - (\vec{p}_1)^2 = (m_1)^2 \text{ and } (p_{2\mu})^2 = (E_2)^2 - (\vec{p}_2)^2 = (m_2)^2,$$

$$\text{(E.5)}$$

which are constants and, thus, not kinematic variables at all. Next consider $p_1^\mu p_{2\mu}$. This is equivalent to s, as can be shown by multiplying out the first expression for s in Eq. (E.2) to obtain

$$p_1^\mu p_{2\mu} = \tfrac{1}{2}(s - (m_1)^2 - (m_2)^2); \qquad \text{(E.6)}$$

furthermore, the dot product of any two linear combinations of $p_{1\mu}$ and $p_{2\mu}$ produces a linear combination of $(p_{1\mu})^2 = (m_1)^2$, $(p_{2\mu})^2 = (m_2)^2$, and $p_1^\mu p_{2\mu}$ which is determined by s (as we have just seen in Eq. E.6). Therefore, we cannot construct any kinematic invariants which are independent of s.

Now let us consider a two-particle scattering and its generalization to reactions with two incoming and two outgoing particles: $1 + 2 \to 3 + 4$, as shown in Fig. E.2. In general, the incoming particles will have a non-zero total three-momentum in the frame of the experiment; however, we can define the kinematic invariant s for these two particles, which represents their total energy in the CM frame, as we did in Eq. (E.2). Furthermore, conservation of energy and momentum may be expressed in four-vector notation as

$$p_{1\mu} + p_{2\mu} = p_{3\mu} + p_{4\mu}, \qquad \text{(E.7)}$$

so that we may write

$$s \equiv (p_{1\mu} + p_{2\mu})^2 = (p_{3\mu} + p_{4\mu})^2 = (E_{TOTAL}^{CM})^2. \qquad \text{(E.8)}$$

Let us consider next an invariant which combines the momenta of particles 1 and 3 (or particles 2 and 4):

$$t \equiv (p_{3\mu} - p_{1\mu})^2 = (p_{2\mu} - p_{4\mu})^2. \tag{E.9}$$

For the cases where particles 1 and 3 are of the same type (e.g., baryons, leptons, mesons, etc.) and especially if they are of the same kind (e.g., both electrons), we may think of the process as a scattering in which $p_{1\mu}$ is the four-momentum of the particle before the scattering and $p_{3\mu}$ is its four-momentum after the scattering. For simplicity, let us evaluate t in the case where particles 1 and 3 have the same mass (m) and particles 2 and 4 have the same mass (m'), which would be true for a scattering experiment. We leave it to the reader to verify[1] that conservation of energy and momentum implies that the magnitudes of the three-momenta are all equal: $p_1^{CM} = p_2^{CM} = p_3^{CM} = p_4^{CM}$. Because particles 1 and 3 have the same mass and the same size three-momenta in the CM system, their individual energies $\left(E = \sqrt{p^2 + m^2}\right)$ are the same as well, so that we may write

$$t = 0 - (\vec{p}_3^{CM} - \vec{p}_1^{CM})^2 = -\left((\vec{p}_3^{CM})^2 - 2\vec{p}_3^{CM} \cdot \vec{p}_1^{CM} + (\vec{p}_1^{CM})^2\right)$$
$$= -2(p^{CM})^2(1 - \cos\theta^{CM}), \tag{E.10}$$

where p^{CM} is the magnitude of the particles' individual three-momenta and θ^{CM} is the angle between $\vec{p}_3^{CM}$ and $\vec{p}_1^{CM}$, i.e., the angle of scattering, as shown in Fig. E.3. Thus, t is **the kinematic invariant which characterizes the angle of scattering**. Even for the cases where the indicated masses are not the same, t involves an angular variable.

In Section 13.2.2 we have discussed differential cross sections. The variables used were the angular variables θ and ϕ, which are not Lorentz invariant. As mentioned in that section, the differential cross section is independent of ϕ for unpolarized experiments, so we may integrate over ϕ to remove it from consideration. We would then obtain

$$(d\sigma)_{all\phi} = 2\pi \left(\frac{d\sigma}{d\Omega}\right)^{CM} d(\cos\theta^{CM}) = \frac{\pi}{(p^{CM})^2}\left(\frac{d\sigma}{d\Omega}\right)^{CM} dt, \tag{E.11}$$

using Eq. (E.10) to change variables. The factors multiplying dt (on the right-hand side) are then identifiable as $d\sigma/dt$, so that

$$\frac{d\sigma}{dt} = \frac{\pi}{(p^{CM})^2}\left(\frac{d\sigma}{d\Omega}\right)^{CM} \tag{E.12}$$

is the connection between the differential cross section using the invariant t as the angular variable and the usual differential cross section. Reaction rates,

---

1. As shown in Eq. (E.15), the final energy is a *monotonic* function of p'^{CM}, and for these mass relationships the equation is satisfied for $p'^{CM} = p^{CM}$. Consequently, any other value of p'^{CM} could not satisfy the equation.

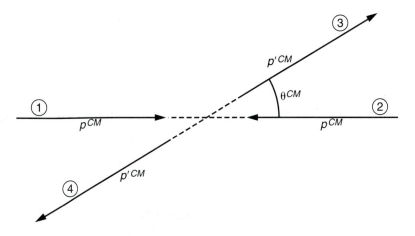

FIGURE E.3
Momentum diagram for the reaction $1 + 2 \rightarrow 3 + 4$ in the CM frame, with the scattering angle denoted as θ^{CM}. When $m_1 = m_3$ and $m_2 = m_4$, in which case all the momenta are the same size ($p'^{\mathrm{CM}} = p^{\mathrm{CM}}$), this is often thought of as a scattering (particle 3 is the scattered particle 1, likewise 4 with respect to 2).

expressed as functions of t, can be easily combined because of the Lorentz-invariant nature of t. It is usually $d\sigma/dt$ that is reported and plotted in research papers.

We have not yet completed our enumeration of the possible dot products of the particles' four-momenta (the 1,4 and 2,3 combinations are missing). Similar to the forms used so far, we define

$$u \equiv (p_{1\mu} - p_{4\mu})^2 = (p_{3\mu} - p_{2\mu})^2. \tag{E.13}$$

We have now created all the invariants that arise from the dot products of the individual four-momenta of the particles. However, s, t, and u are not all independent, and we leave it to the reader to verify that

$$s + t + u = m_1^2 + m_2^2 + m_3^2 + m_4^2. \tag{E.14}$$

It is usual to use s and t as the independent variables.

A simple analysis of the scattering process, in fact, *convinces* us that there are only **two kinematic variables needed to describe the dynamics** of the scattering. By "dynamics" we mean the **interactions**, which result from the laws of nature. Although the description of a state depends on coordinate system and reference frame, we have found that the laws of nature do not. In fact, the results of an observation in one coordinate system and in one reference frame can be transformed to yield the results in any other coordinate system and reference frame. So, without loss of generality, let us work in the CM frame and take our z axis along the direction of particle 1. Conservation of momentum then determines that particle 2 is moving in the $-z$ direction with the same size momentum (p^{CM}). Thus we have identified **our first kinematic variable** $-\boldsymbol{p}^{CM}$. Likewise, the two outgoing particles must have equal-size, opposite-direction momenta (p'^{CM}), but their direction (θ^{CM}) is not determined by conservation laws or by p^{CM}; therefore, $\boldsymbol{\theta}^{CM}$ **is our second independent kinematic variable.** You might think that p'^{CM}, the magnitude of the outgoing momenta, is another such variable, but it is determined by p^{CM} from conservation of energy:

$$\sqrt{p^{CM\,2} + m_1^2} + \sqrt{p^{CM\,2} + m_2^2} = \sqrt{p'^{CM\,2} + m_3^2} + \sqrt{p'^{CM\,2} + m_4^2}. \qquad (E.15)$$

Thus, we expect only two independent kinematic invariants, and we have found them $-s$ **(which characterizes** $\boldsymbol{p}^{CM}$**) and** t **(which characterizes** $\boldsymbol{\theta}^{CM}$**).**

It is illuminating to observe that if particles 1 and $\bar{3}$ (where the bar refers to the corresponding antiparticle) were the incoming particles and $\bar{2}$ and 4 were the outgoing ones, then Eq. (E.7) would still express conservation of energy and momentum after the appropriate replacements ($p_{2\mu} \to -p_{2\mu}$ and $p_{3\mu} \to -p_{3\mu}$) are made. The result of these replacements is that t becomes the total CM energy and s becomes the angular variable. The very structure of quantum field theories implies crossing symmetry (as discussed in Section 4.4.2), so that the amplitude for $1 + 2 \to 3 + 4$ and the amplitude for $1 + \bar{3} \to \bar{2} + 4$, as well as the amplitudes for the other process involving only these four particles, may all be represented by the same functions of s and t. The same diagrams may be used for all these processes, with the nomenclature **s-channel** for the process first under consideration (with the incoming particles designated 1 and 2 and the outgoing ones 3 and 4), **t-channel** for the related process **where** t **is the CM energy** (i.e., $1 + \bar{3} \to \bar{2} + 4$, as indicated in Fig. E.2), and **u-channel** for the related process ($1 + \bar{4} \to \bar{2} + 3$) **where** u **is the CM energy.**

E.2   RAPIDITY AND PSEUDORAPIDITY

In high-energy experiments involving large numbers of particles, the components of the velocities of the particles along the beam direction are often of interest. However, the relativistic velocity addition law is complicated by the terms in the denominator, so that transforming these speeds from one frame to another is not so simple:

$$\beta'_L = \frac{\beta_L - \beta_v}{1 - \beta_L \beta_v}, \qquad (E.16)$$

where S' is moving with speed β_v in the beam direction and the subscript L stands for "Longitudinal" (along the beam direction). This equation is reminiscent of the expression for $\tanh(a - b)$ in terms of $\tanh(a)$ and $\tanh(b)$, so we define a new quantity, **rapidity ($\boldsymbol{y}$)**, such that

$$\beta_L = \tanh(y) \text{ or } y = \tanh^{-1} \beta_L. \qquad (E.17)$$

Then, substituting Eq. (E.17) into Eq. (E.16), along with

$$\beta_v = \tanh(y_v) \text{ and } \beta_L' = \tanh(y'),$$

we find that

$$y' = y - y_v, \tag{E.18}$$

so that **rapidities merely add and subtract**. As β goes from 0 to 1, the rapidity runs from 0 to ∞. We leave it to the reader to verify that the rapidity (y) of a particle may also be expressed as

$$y = \frac{1}{2} \ln \frac{E + p_L}{E - p_L}, \tag{E.19}$$

where E is the energy of the particle and p_L is its momentum along the beam direction.

   For massless particles $E = p$ and $p_L = p \cos \theta$, so that the rapidity depends only on the direction of motion (θ) of the particle. The resulting expression is often used for rapidly moving *massive* particles as well, since θ is more easily determined experimentally, in which cases it is called the **pseudorapidity (η)**. From Eq. (E.19), we find

$$\eta = \frac{1}{2} \ln \frac{1 + \cos \theta}{1 - \cos \theta} = -\ln\left(\tan \frac{\theta}{2}\right). \tag{E.20}$$

The pseudorapidity is a good approximation to the rapidity when the particle is moving near the speed of light (so that its mass is negligible).

Appendix F

Partial Wave Analysis for Spin $\frac{1}{2}$ Particles and the Δ Resonance

F.1 SCATTERING OF A SPIN $\frac{1}{2}$ PARTICLE FROM A POTENTIAL

In Chapter 13, we discussed the partial wave expansion of the scattering amplitude and the associated phase shifts for the scattering of spinless particles by a potential. For the potential scattering of particles with spin $\frac{1}{2}$, there are two parts to the scattering amplitude, $f(\theta)$ for unflipped spin and $g(\theta, \phi)$ for spin-flip scattering. The role of f_ℓ in Eq. (13.29) is now played by two-indexed quantities $f_{\ell,j}$ with two-indexed phase shifts $\delta_{\ell,j}$. A detailed treatment yields two separate absolute squares, one from the unflipped and one from the flipped amplitude:

$$
\frac{d\sigma^{\mathrm{CM}}}{d\Omega} = \left| \sum_\ell B_\ell [(\ell + 1) f_{\ell,\ell+\frac{1}{2}} + \ell f_{\ell,\ell-\frac{1}{2}}] Y_\ell^0(\theta,\phi) \right|^2
$$

$$
+ \left| \sum_\ell B_\ell \sqrt{\ell(\ell + 1)} [f_{\ell,\ell+\frac{1}{2}} - f_{\ell,\ell-\frac{1}{2}}] Y_\ell^{2m_s}(\theta,\phi) \right|^2, \tag{F.1}
$$

$$
\text{where } f_{\ell,j} = \frac{1}{2i}(e^{2i\delta_{\ell,j}} - 1) \text{ and } B_\ell = \frac{1}{k}\sqrt{\frac{4\pi}{2\ell + 1}}.
$$

(The coefficients in the square brackets result from the Clebsch-Gordan coefficients for the coupling of ℓ to $\frac{1}{2}$ to produce the corresponding j.) **This formula may also be used to obtain the angular distribution of the scattering of a spin $\frac{1}{2}$ particle from a spin 0 particle.**

Keeping a finite number of terms (up to an ℓ determined semiclassically) yields a closed form as a function of θ and ϕ, which can then be fitted by the experimentally observed differential cross section to determine the (unknown) phase shifts. For example, using only $\ell = 0$ and $\ell = 1$, which is expected to be sufficient at low energies, this yields a differential cross section of the form

$$\frac{d\sigma}{d\Omega} = A + B \cos\theta + C \cos^2\theta,$$

where A, B, and C depend on the phase shifts. When higher partial waves are included, the differential cross section will also depend on ϕ.

F.2 π-N SCATTERING AND THE SPIN OF THE Δ RESONANCE

From the size of the total cross section at the Δ resonance, we surmise that this resonance has $j = \frac{3}{2}$. We know that either an $\ell = 1$ or an $\ell = 2$ orbital angular momentum can combine with the spin $\frac{1}{2}$ of the proton to produce a $j = \frac{3}{2}$ resonance; however, since the parity of the state depends on ℓ, an experimental determination of the particle's parity will single out one value. It has been found from experiment that the Δ has even parity (+1). The pion is a pseudoscalar (odd parity), therefore, the parity of the Δ is $(-1)^{\ell-1} = +1$, which tells us that $\ell = 1$ (P-wave).

Suppose the peak at the resonance energy (1232 MeV) is due to this $\ell = 1$, $j = \frac{3}{2}$ resonance. We can use Eq. (F.1) to predict the distribution of the angle of the momenta of the emerging pions *with respect to the incoming pion beam direction*. Keeping only the terms for $f_{1,\frac{3}{2}}$ (setting all the other $f_{\ell,j}$ equal to 0), Eq. (F.1) yields

$$\frac{d\sigma^{CM}}{d\Omega} = \left| \frac{1}{k} \sqrt{\frac{4\pi}{3}} \left[2f_{1,\frac{3}{2}} \right] Y_1^0 \right|^2 + \left| \frac{1}{k} \sqrt{\frac{4\pi}{3}} \left[\sqrt{2} f_{1,\frac{3}{2}} \right] Y_1^{\pm 1} \right|^2 \sim 2|Y_1^0|^2 + |Y_1^{\pm 1}|^2$$

$$\sim 1 + 3\cos^2\theta, \tag{F.2}$$

$$\text{for } m_s = \pm\frac{1}{2}.$$

This result may also be seen directly: The incoming pion beam (with $\ell = 1$, as we have discussed above) has $m_\ell = 0$ (along its direction of motion); therefore, the $j = \frac{3}{2}$ (Δ) state can have only $m = \pm\frac{1}{2}$. By conservation of angular momentum, the final state must also be a $|\frac{3}{2}, \pm\frac{1}{2}\rangle$ state. The products of the proton $|\frac{1}{2}, \pm\frac{1}{2}\rangle$ and pion $|1, m_\ell\rangle$ final states, which form the $|\frac{3}{2}, +\frac{1}{2}\rangle$ state, are obtained by using the Clebsch-Gordan coefficients:

$$|\tfrac{3}{2}, +\tfrac{1}{2}\rangle = C_{\frac{1}{2} + \frac{1}{2}\, 0}^{\frac{3}{2}\,\frac{1}{2}\,1} |\tfrac{1}{2}, +\tfrac{1}{2}\rangle |1, 0\rangle + C_{\frac{1}{2} - \frac{1}{2}\, 1}^{\frac{3}{2}\,\frac{1}{2}\,1} |\tfrac{1}{2}, -\tfrac{1}{2}\rangle |1, 1\rangle$$

$$= \sqrt{\tfrac{2}{3}} \uparrow Y_1^0 + \sqrt{\tfrac{1}{3}} \downarrow Y_1^1. \tag{F.3}$$

The cross section is proportional to the absolute square of this expansion, which yields the same angular distribution as obtained in Eq. (F.2).[1] The same result is obtained for the angular distribution of the $|\frac{3}{2}, -\frac{1}{2}\rangle$ state.

---

1. When checking this, remember the orthogonality of the spin up ($\uparrow$) and spin down ($\downarrow$) proton states.

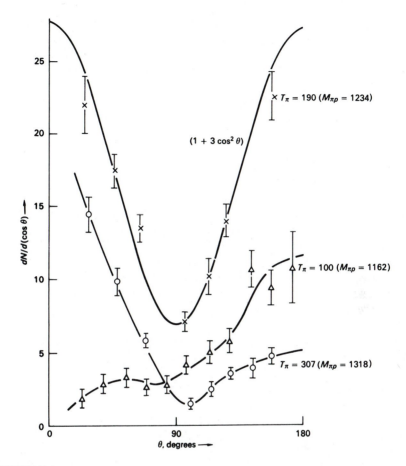

FIGURE F.1

Angular distribution of the emerging π^+'s, relative to the incident π^+ beam direction, for $\pi^+ p$ elastic scattering. Notice that the distribution is consistent with $(1 + 3\cos^2\theta)$ near 1232 MeV, as predicted for a spin $\frac{3}{2}$ resonance, and is radically different at neighboring energies.

In Fig. F.1 we see the experimental determination of the angular distribution of the emerging π^+'s for $\pi^+ p$ (pure $I = \frac{3}{2}$) elastic scattering, at energies running from below to above the Δ resonance. The dramatic change of the form of the data, as we go through the resonance, to that expected for $j = \frac{3}{2}$, is further confirmation of the spin $\frac{3}{2}$ assignment of the Δ.

Appendix G

G-Parity in the
Strong Interactions

In Chapter 12 we saw that **charge conjugation (C) is a symmetry of the strong and the electromagnetic interactions**. This discrete operation reverses the signs of the (additive) quantum numbers: charge, baryon number, lepton number, I_3 ("upness" and "downness"), strangeness, charm, topness, bottomness, etc. Systems which have 0 for all these quantum numbers will be eigenstates of C, but this is a limited set of *neutral* states. For the strong interactions, a discrete operation related to C called *G*-parity has been constructed which provides simplified selection rules for *charged* systems as well. They are *states which have 0 for all the additive quantum numbers, except possibly I_3.*

Since the u and d quarks have essentially the same mass, the group of isotopic spin rotations may be considered a symmetry of the strong interactions. Let us consider states which have 0 for all the additive quantum numbers except *Mesons only* I_3. These are charged states (whose electric charge (Q) is equal to I_3). Only states which have $I_3 = Q = 0$ are eigenstates of C, since C reverses I_3 (and the charge). This property of C implies that

$$CI_3 = -I_3 C.$$

Thus, C does not commute with $\vec{I}$. Consequently, although C transformations and I rotations are symmetries of the strong interaction, representations of I_3 and C cannot be diagonalized simultaneously (e.g., using an I-multiplet for the basis states). In particular, we cannot easily see C symmetry in experiments involving **charged particles** (eigenstates of I_3), since these particles **are not eigenstates of C**.

A discrete symmetry related to C can be constructed which *commutes* with $\vec{I}$, i.e., not only with I_3 but with $I^{\pm}$ as well; consequently, the entire isomultiplet can be an eigenmultiplet (with a single eigenvalue[1]) of the corresponding

---

1. by the same argument as was used in Chapter 3 to show the (energy) degeneracy of a J-multiplet when $[H, \vec{J}] = 0$.

symmetry operator. That symmetry is *G*-parity (*G*), and the operator may[2] be defined as

$$G = Ce^{iI_2\pi},\tag{G.1}$$

which rotates the system in *I*-space, 180° around the I_2 axis, followed by charge conjugation. **G-parity is a symmetry of the strong interaction**, since that interaction is invariant to each of the two successive operations. However, unlike *C*, **G-parity is *not* a symmetry of the electromagnetic interaction** because isorotations are not invariances of QED.

Let us consider *completely neutral* states, in which all additive quantum numbers including I_3 are 0, first. It is left to the reader to check that a π rotation about the I_2 axis corresponds to the isospace angular changes:

$$\theta, \phi \to (\pi - \theta), (\pi - \phi).^3$$

The θ has changed exactly as it would for inversion (the ϕ has not); in analogy with our space discussion of parity, we see that $Y_I^0 \to (-1)^I Y_I^0$ (since Y_I^0 is independent of ϕ). Thus, **a state with $I_3 = 0$ is an eigenstate of the isotopic rotation $e^{iI_2\pi}$ with eigenvalue $(-1)^I$**. A completely neutral state (or particle) is also an eigenstate of the charge conjugation operator, so that

$$G|I,0\rangle = c(-1)^I|I,0\rangle,\tag{G.2}$$

where *c* is the *C*-value of that state. Thus the uncharged states (or particles) are eigenstates of *G* with eigenvalue $c(-1)^I$.

The rotation by π about the I_2 direction changes a state with non-zero I_3 into one with I_3 reversed, but (since all the other additive quantum numbers are 0) *C* changes it back to the original state (up to an undefined phase). Consequently, the charged states of the *I*-multiplet are also eigenstates of *G*. The phases arising when *C* is applied to charged states are undefined and can be chosen so that the entire *I*-multiplet has the same eigenvalue for *G*:

$$G|I,I_3\rangle = c(-1)^I|I,I_3\rangle,\tag{G.3}$$

where *c* is the *C*-value of the $|I,0\rangle$ neutral state; in effect, we have

$$[G,\vec{I}] = 0.\tag{G.4}$$

In particular, consider the lightest mesons: π, η, ρ, ω, η', f_0, a_0, and ϕ, whose decay modes are listed in Table G.1.

Let us consider the pion ($I = 1$) first. Since the π^0 decays to 2γ (and *C* is conserved in electromagnetic interactions), the value of *c* for the pion is +1.[4] As shown in Chapter 6, we can take $|\pi^\pm\rangle \sim (|\pi_1\rangle \pm i|\pi_2\rangle)$ and $|\pi^0\rangle = |\pi_3\rangle$. The isorotation (in *G*) by 180° about the 2-direction yields $|\pi_1\rangle \to -|\pi_1\rangle$,

---

2. Some authors use an I_1 rotation instead of an I_2 rotation.

3. Looking in the 1-2 isoplane, we can see that a π rotation about I_2 changes $\phi \to \pi - \phi$. The rotation also reverses I_3; consequently, $\theta \to (\pi - \theta)$.

4. Each photon has $c = -1$.

TABLE G.1
Decay Modes and *G*-Parities of Nonstrange Mesons

Meson	Mass Width (or lifetime)	G-Parity	I	Final State	Branching Ratio
$\pi^{\pm}$ $J^P = 0^-$	$m = 140$ MeV $\tau = 2.6 \times 10^{-8}$ s	-1	1	$\mu^+ \nu$	99.99%
a suppressed weak decay (see Fig. 12.5 and the accompanying footnote 13)					
π^0 $J^P = 0^-$	$m = 135$ MeV $\tau = 8.4 \times 10^{-17}$ s (equivalent to $\Gamma \sim 8$ eV)	-1	1	2γ	98.8%
an electromagnetic decay					
η $J^P = 0^-$	$m = 547$ MeV $\Gamma = 1.2$ KeV	$+1$	0	2γ $3\pi^0$ $\pi^+ \pi^- \pi^0$	39% 32% 24%
$\rho(770)$ $J^P = 1^-$	$m = 768$ MeV $\Gamma = 151$ MeV	$+1$	1	2π	$\approx 100\%$
$\omega(783)$ $J^P = 1^-$	$m = 782$ MeV $\Gamma = 8$ MeV	-1	0	$\pi^+ \pi^- \pi^0$ $\pi^0 \gamma$ $\pi^+ \pi^-$	89% 8.5% 2%
$\eta'(958)$ $J^P = 0^-$	$m = 958$ MeV $\Gamma = 0.2$ MeV	$+1$	0	$\pi^+ \pi^- \eta$ $\rho^0 \gamma$ $\pi^0 \pi^0 \eta$ 2γ	44% 30% 21% 2%
$f_0(975)$ $J^P = 0^+$	$m = 974$ MeV $\Gamma = 47$ MeV	$+1$	0	2π $K\bar{K}$ 2γ	78% 22% 1×10^{-5}
$a_0(980)$ $J^P = 0^+$	$m = 983$ MeV $\Gamma = 57$ MeV	-1	1	$\eta \pi$ $K\bar{K}$ 2γ	seen seen seen
$\phi(1020)$ $J^P = 1^-$	$m = 1019$ MeV $\Gamma = 4.4$ MeV	-1	0	$K\bar{K}$ $\rho \pi$ $\pi^+ \pi^- \pi^0$	84% 13% 2%

$|\pi_2\rangle \to +|\pi_2\rangle$ and $|\pi_3\rangle \to -|\pi_3\rangle$. For C to change $|\pi^{\pm}\rangle \to |\pi^{\mp}\rangle$ we must have $C|\pi_1\rangle = +|\pi_1\rangle$ and $C|\pi_2\rangle = -|\pi_2\rangle$. Thus, the net effect of these two operations which make up G is

$$G|\pi\rangle = -1|\pi\rangle; \qquad (G.5)$$

consequently, the entire π-multiplet has negative G-parity and **a state with n pions has G-parity $G = (-1)^n$.**

The heavier mesons (shown in Table G.1) η, ρ, ω, η', f, a, and ϕ are also eigenstates of G. The G-parities of the isosinglet mesons (η, ω, η', f, and ϕ) are identical to their C-eigenvalues (c), since they are unaffected by isospin rotations. (This also follows from Eq. G.2.) The isosinglets η, η', and f have (electromagnetic) decays to 2γ, which means that $c = +1$ for these states; hence, their G-parities are $+1$ as well. The a_0 is an $I = 1$ state, which decays into 2γ, so that $c = +1$ for this state; consequently, from Eq. (G.2), its G-parity is -1. Those mesons with widths in the MeV range (ρ, ω, f, a, ϕ) decay strongly. The ρ decays strongly almost exclusively into two pions, so that its G-parity is $+1$. The G-parity of the isoscalars ω and ϕ (both with $G = -1$), which do not decay into photons, are determined from their strong-interaction decays into $\rho\pi$ or 3π systems. The $\eta\pi$ strong-decay mode of the a_0 verifies the conservation of G-parity in strong decays.

An interesting example of the use of G-parity is in an analysis of the decay of the η ($G = +1$) meson. The η has a width of 1.2 KeV, which corresponds to a very long lifetime; consequently, we infer that it does not decay strongly. It has a large branching ratio for decay to 2γ (39%), hence, it decays electromagnetically. It decays to 3π states with comparable branching ratios (32% and 24%), which leads us to suspect that these are indeed not strong decays, but electromagnetic decays (involving virtual photons) as well. This conclusion is verified by the fact that the 3-π final states have $G = -1$, so that G-parity is violated. We may ask why the η does not decay strongly into *two pions*, a decay which would conserve G-parity and has more phase space. The answer is that such a decay could not conserve both angular momentum and (ordinary space) parity. Let us examine this hypothetical decay in the rest frame of the η. Since the η is a pseudoscalar ($J^P = 0^-$) and so is the pion, the (spin 0) pions must be in a relative s state to conserve angular momentum; however, the parity of that state would be $(-1)^\ell = (-1)^0 = +1$ rather than -1.

G-parity may also be used to discuss $N\overline{N}$ states (e.g., initial states, such as $N\overline{N}$ collisions which produce mesons). For an $N\overline{N}$ state, we can use an argument similar to that presented for positronium-like states to show (see Eq. 12.14) that $c = (-1)^{\ell+s}$, so that $\boldsymbol{G = (-1)^{\ell+s+I}}$ **for $N\overline{N}$ systems.**

Appendix H

Particle Detectors

by Joey Huston
Michigan State University

The aim of this appendix is to provide a brief description of both the general types of particle detectors in use in high-energy physics today and also of the processes that occur within them. By no means is this to be considered an exhaustive review. Several excellent reviews of this type are listed in the references.[1] "Rules of thumb" (accurate to 10% or so) are presented, both to motivate a physical intuition in the reader and also to later serve as a convenient reference. An example of a large experiment, in which many types of detectors are utilized, is discussed at the end of this appendix.

The purpose of detectors in particle physics is simply to detect the particles that result from high-energy collisions. The ultimate goal is the reconstruction of the particle four-vectors, in order to understand the dynamics of the collisions. For this detection to take place, there must be some sort of interaction between the particle and the detector. This interaction can be electromagnetic, strong, or even weak in character.

H.1 INTERACTIONS

H.1.1 Electromagnetic Interactions

Charged Particles   Charged particles can interact electromagnetically with atoms in their path. The result can be the ionization or excitation of the atoms. If atomic electrons are released, then the application of an electric field can allow the electrons to be collected, yielding a signal. De-excitation of the atoms, as takes place in scintillation or Čerenkov radiation, produces photons, which can

---

1. T. Ferbel, ed., *Experimental Techniques in High Energy Physics* (Addison-Wesley, 1987). This is an excellent compendium of many important review papers on high-energy instrumentation. W. R. Leo, *Techniques for Nuclear and Particle Physics Experiments* (Springer-Verlag, 1987). This is an excellent textbook on instrumentation for particle physics which is a "must buy" for those interested in mastering the subject.

also be detected. The mean rate of electromagnetic energy loss for a charged particle is given by the Bethe-Bloch formula:[2]

$$\frac{dE}{dx} = 4\pi N_o r_e^2 m_e c^2 z^2 \; \frac{Z}{A} \frac{1}{\beta^2} \left[\ln\left(\frac{2m_e c^2 \gamma^2 \beta^2}{I} \right) - \beta^2 - \frac{\delta}{2} \right]. \qquad (H.1)$$

Here, N_o is Avogadro's number, Z and A are the atomic number and atomic weight, respectively, of the atoms in the medium, m_e is the mass of the atomic electron, z is the charge of the incident charged particle, r_e is the classical radius of the electron, and $\beta = v/c$. $(\gamma = 1/\sqrt{1 - \beta^2})$. The product $4\pi N_o r_e^2 m_e c^2$ equals 0.307 MeV cm^2 g^{-1}. X is the path length in the medium, measured in gm/cm^2. (It is often convenient to give the thickness of a material in mass transverse units, or units of grams per square centimeter (gm/cm^2). This can be converted to a length by dividing by the density of the medium (given in gm/cm^3).) I is an effective ionization potential, given approximately by $I = 16Z^{0.9}$ (eV).[3] The factor of δ in the square brackets takes into account the density effect. The greater the value of dE/dx for the material, the more ionization takes place.

At nonrelativistic energies the dE/dx formula is dominated by the $1/\beta^2$ factor and decreases with increasing velocity. The decrease in dE/dx with velocity is due to the decrease of the effective interaction time and thus to the total energy transfer. As β approaches 1, the factor inside the square brackets grows, resulting in a broad minimum for Eq. (H.1) at a value of β of approximately 0.96 (or a particle energy equal to approximately 3.5 times the rest mass). Particles with this velocity are termed "minimum ionizing." (Since dE/dx rises slowly with energy after this minimum, in practice, particles with velocities greater than this value are also termed minimum ionizing.) Since $Z/A \cong 0.5$ for most elements, the minimum value of dE/dx varies little with the medium (1–1.5 MeV-cm^2/gm). As β increases further, the $1/\beta^2$ term becomes essentially constant and dE/dx slowly rises (as 2 ln γ with the density effect eventually limiting the slope to ln γ) because of the term in the square brackets. Physically, the logarithmic increase is due to the increase in the transverse electric field as γ $(= 1/\sqrt{(1 - \beta^2)})$ increases. The larger transverse electric field leads to ionization of atoms at larger distances from the charged particle's path.

For energies below minimum ionizing, each particle exhibits a dE/dx curve which in most cases is distinct from other particles. (For particles of different mass, the same momentum implies a different value of β.) This can be used to advantage in some circumstances for particle identification. See, for example, the later discussion on the TPC (Time Projection Chamber).

The energy spectrum of the electrons created in the primary ionizations falls approximately as $1/E^2$. Some fraction of the ionized electrons have enough

---

2. See, for example, D. Perkins, *Introduction to High Energy Physics* (Addison-Wesley, 1987), or the "*Review of Particle Properties*," Physical Review D45, Part 2 (June 1992).

3. Ibid.

energy to produce additional ionizations. Thus, the total number of electrons produced can be 3–4 times the initial number of electrons ionized. Interactions between charged particles and matter can also lead to the excitations of atoms without ionization of the electrons. The subsequent de-excitation of the electron can result in the emission of photons, which can be detected. This process is known as scintillation. Detectors using scintillating materials are among the most common in high-energy physics.

Čerenkov radiation is produced whenever a particle traverses through a medium faster than the speed of light in that medium. That is,

$$v_{particle} > c/n, \tag{H.2}$$

where n is the index of refraction (frequency dependent) in that medium. A particle with a velocity satisfying the above criterion creates an electromagnetic shock wave much as a jet travelling faster than the speed of sound creates a sonic shock wave. The shock wave is conical in nature with a wake angle of $\sin\theta = 1/\beta n$ (see also the discussion in Chapter 14). A continuous spectrum of frequencies is radiated. The rate of energy loss through Čerenkov radiation is a small fraction of the total given by the Bethe-Bloch formula (1–1.5 MeV cm^2/gm) and is included in that calculation. This radiation can be very useful for determination of the particle type in detectors designed to detect this radiation (Čerenkov detector).[4]

Few detectors are composed solely of one element. The energy loss for a compound material is given by

$$\frac{1}{\rho}\frac{dE}{dx} = \sum_i \frac{w_i}{\rho_i}\frac{dE}{dx}\bigg|_i, \tag{H.3}$$

where w_i is the fraction by weight of element i and ρ_i is the density of element i.

In their passage through matter, electrons and positrons can interact with the Coulomb field of the nucleus with the resultant emission of photons. This radiation is known by its German name, *bremsstrahlung*, or "braking radiation." Bremsstrahlung photons have a $1/E$ energy spectrum. Other charged particles can emit bremsstrahlung photons, but since the emission probability is proportional to the inverse square of the particle's mass, bremsstrahlung radiation is an important effect mostly for electrons and positrons. In general, then, the total energy loss of electrons and positrons in matter is composed of two parts:

$$\frac{dE}{dx}\bigg|_{total} = \frac{dE}{dx}\bigg|_{ionization} + \frac{dE}{dx}\bigg|_{radiation}. \tag{H.4}$$

At low energies ($\ll 1$ MeV) radiation losses are small compared to collision or ionization losses, while for higher energies ($>$ few MeV) they can be comparable or dominate. (For energies larger than the minimum ionizing energy, the ionization loss is essentially constant while the radiation loss is proportional to

---

4. T. Ferbel, op. cit.; W. R. Leo, op. cit.

E.) The energy at which the ionization loss equals the radiation loss is known as the critical energy $\varepsilon_{critical}$ and can be approximated as[5]

$$\varepsilon_{critical} \approx \frac{550}{Z} \text{ (MeV)}. \tag{H.5}$$

Thus, lead has a critical energy of about 7 MeV while carbon has a critical energy of about 100 MeV. The difference is due to the Z dependence of the bremsstrahlung cross section ($\sigma \propto Z^2$).

The rate of energy loss, due to bremsstrahlung, of an electron or positron in traversing a thickness x of a material is proportional to the energy of the electron (or positron).

$$\left.\frac{dE}{dx}\right|_{radiation} = \frac{E}{X_o}. \tag{H.6}$$

The exponential solution to Eq. (H.6) leads to the interpretation of the radiation length X_o as the amount of material that reduces the mean energy of a high-energy electron by a factor e^1.

An approximate formula for the radiation length of a given material is[6]

$$X_o \approx 180 \frac{A}{Z^2} \text{ (gm/cm}^2\text{)}. \tag{H.7}$$

Thus, for lead this formula gives $X_o = 5.6$ gm/cm^2. In reality, for lead $X_o = 6.37$ gm/cm^2 (remember these rules are $\pm 10\%$), or using the density of lead (11.35 gm/cm^3), this works out to a thickness of 0.56 cm.

For a compound mixture, the equivalent radiation length (in mass transverse units) is given by

$$\frac{1}{X_o} = \sum_i \frac{w_i}{X_o^i}, \tag{H.8}$$

where the w_i are the fractions by weight of the various materials in the mixture.

Because of the $1/m$ term in the formula for dE/dx (Eq. H.1), most of the energy loss of a charged particle traversing matter is through collisions with atomic electrons rather than with the nuclei. Because of the larger mass of the nuclei, however, collisions with nuclei can greatly contribute to the transverse scattering of a charged particle. The differential cross section for the scattering of a charged particle at an angle θ into a solid angle $d\Omega$ is given by the Rutherford formula:

$$\frac{d\sigma}{d\Omega} = \frac{1}{4}\left(\frac{Zze^2}{pv}\right)^2 \frac{1}{\sin^4 \theta/2}. \tag{H.9}$$

---

5. U. Amaldi, *Physica Scripta*, 23:409 (1981). This paper is also contained in the book by Ferbel. It contains many of the rules of thumb that are given in the text. Somewhat more accurate, but sometimes more complicated, parameterizations are given in the "*Review of Particle Properties*" cited in footnote 2.

6. Ibid.

Here, p is the momentum of the incident particle, v its velocity, and z its charge. Z is the charge on the nucleus.

For small angles with respect to the incident particle direction, the total scattering is the result of many independent scatters. The resulting angular distribution is roughly Gaussian in character with an rms deviation (for a charged particle traversing one radiation length of material):

$$\theta_{rms}[radians] = \frac{21Z}{pv}. \qquad (H.10)$$

In the above equation the product pv is given in units of MeV. Because of the randomness of multiple scattering, the rms scattering angle varies with the number of radiation lengths as the square root.

This multiple scattering can limit the ability of a detector to measure the trajectory of a charged particle.

Photons   The interaction of photons with matter occurs principally through three processes:

1. photoelectric effect $\sigma \propto 1/E^3$
2. Compton effect $\sigma \propto 1/E$
3. pair production (conversion of σ is independent of E
   a photon to an e^+e^- pair)   (above 10 MeV or so)

Thus, at high energies the pair-production process dominates.

The average distance that high-energy photons traverse before converting into an e^+e^- pair is equal to $9/7\ X_o$. This distance is known as the conversion length.

Electromagnetic Showers   Except at the lowest energies, the absorption of electrons, positrons, or photons is a multistep process in which particle multiplication may occur. Consider, for example, a high-energy photon entering a block of lead from the left, as shown in Fig. H.1. Approximately 1 radiation length into the block the photon converts into an e^+e^- pair, each of which can then radiate bremsstrahlung photons which may themselves convert into e^+e^- pairs, and so on. This process of multiplication continues until the energy of the secondary photons falls below the limit for the pair production. The energy of the shower is then absorbed by the lead by the processes described earlier (ionization, Compton effect, photoelectric effect, etc.).

It is possible and convenient to describe the development of a shower by terms that are relatively independent of the specific material in which a shower is taking place. The characteristic length of an electromagnetic shower can be described in terms of the radiation length X_o, and the characteristic transverse size by the Moliere radius R_M. The Moliere radius is defined in terms of the ratio of the radiation length and the critical energy ε_c. Physically, the Moliere radius corresponds to the lateral spread of an electron beam of energy ε_c after

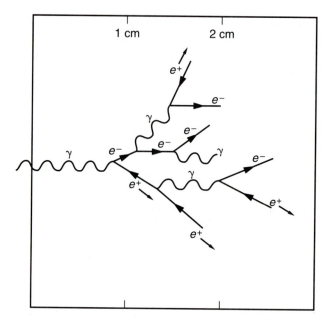

FIGURE H.1
An illustration of the interaction of a high-energy photon in a block of lead. The transverse size of the shower is greatly exaggerated for clarity, and the shower is "cut off" at an early stage with only a few characteristic interactions shown. At shower maximum, the multiplicity of photons, electrons, and positrons may be greater than 1000 for a 50-GeV incident electron.

traveling through a thickness X_o. An approximate form for the Moliere radius is[7]

$$R_M \approx \frac{21 \text{ Mev}}{\varepsilon_c} \ X_o \approx 7 \frac{A}{Z} \ (\text{gm/cm}^2). \tag{H.11}$$

Again, using lead as an example, the above formula gives $R_M = 17.7$ gm/cm^2 (about 1.5 cm).

Defined in this material-independent way, the depth for containment of 98% of a shower's energy (for the energy range 10–1000 GeV) is approximately given by

$$L(98\%) \approx 2.5 t_{max}$$

$$t_{max} \approx \ln\left[\frac{E}{\varepsilon_c} - a\right] \text{ (in radiation length } (X_o) \text{ units).} \tag{H.12}$$

$$(a = 1.1 \text{ for electrons and } 0.3 \text{ for photons})$$

_____

7. Ibid.

For electrons at 50 GeV, $L(98\%) = 19X_o$. Note that the length needed for containment increases only logarithmically with energy. The approximate depth for the shower maximum is given by t_{max}. The shower maximum is the position at which the multiplicity of photons, electrons, and positrons is at its greatest. For a 50-GeV electron showering in lead, this maximum multiplicity is greater than a thousand. The multiplicity increases roughly linearly with the energy of the particle (and inversely with the critical energy of the medium).

The radius for lateral containment of 95% of an electromagnetic shower's energy is energy independent and given by

$$R(95\%) \approx 2R_M \approx 14\frac{A}{Z} \text{ (gm/cm}^2\text{)}. \qquad (H.13)$$

Of course, the core of an electromagnetic shower is far narrower than the Moliere radius, especially in the early part of the shower's development. Before shower maximum, typically 90% of a shower's energy is contained within a transverse distance of $0.5X_o$ of the shower center.

The lateral spread of an electromagnetic shower is not due to bremsstrahlung or pair production (typical production angles are close to $1/\gamma$), but rather to two other effects:

1. multiple scattering of low-energy shower electrons and
2. transverse spread of low-energy shower photons (below the pair-production threshold).

Since the physics of electromagnetic showers is well understood, detailed simulation using Monte Carlo techniques is possible.[8] An electromagnetic shower created by a 50-GeV electron in a lead block (simulated by an EM shower Monte Carlo) is shown in Fig. H.2. Note, as expected, that most of the shower's energy is contained within a transverse distance of $2R_M$ (about 3 cm).

H.1.2   Strong Interactions

The absorption of high-energy hadrons also takes place through a showering process. Hadronic showers proceed in a fashion similar to electromagnetic showers, although the particle-production mechanism tends to be much more complicated, since the strong interaction is involved.

Consider a high-energy pion entering a block of lead from the left, as shown in Fig. H.3. At some point, approximately 1 nuclear interaction length inside the block, the pion interacts with a lead nucleus producing a number of particles in the final state $(\pi^+, \pi^-, \pi^0, p, n \ldots)$. (The probability of a hadron **not** interacting after a thickness t is given by $e^{-t/\lambda}$, where λ is the nuclear interaction length.) The π^0's decay immediately to two photons, which then produce electromagnetic showers of their own. (Thus, some fraction of a hadronic

---

8. W. R. Nelson, H. Hirayama and D. W. O. Rogers, "The EGS4 Code System, Stanford," SLAC Report-165 (1985).

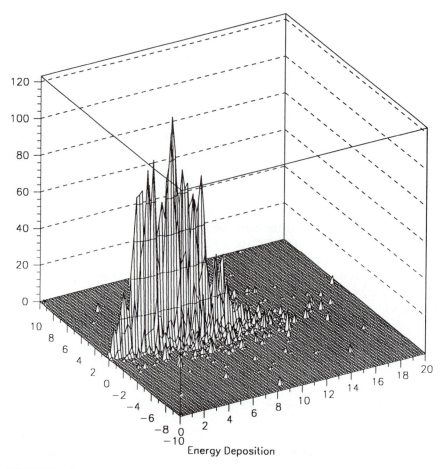

FIGURE H.2
A 50-GeV electron (entering from left) showering in a block of lead. Both the horizontal and vertical scales are in cm.

shower is electromagnetic in character. This fraction tends to increase roughly logarithmically with energy, at least over a limited range.) The charged pions may then interact with other nuclei, producing more particles, until the energy of all particles is low enough that further energy loss takes place solely through ionization. In general, the interaction with each nucleus results in the breakup of the nucleus. The low-energy charged fragments from the breakup quickly lose their energy in the surrounding medium (dE/dx is very large). The neutrons from the breakup can travel some distance before interacting. Some fraction of the energy in each interaction is used in overcoming the binding energy of the nucleus. The number of interactions that take place, and thus, the number of particles that are produced, is significantly less than in electromagnetic showers. For example, for a 50-GeV pion interacting in lead, the **total** number of sec-

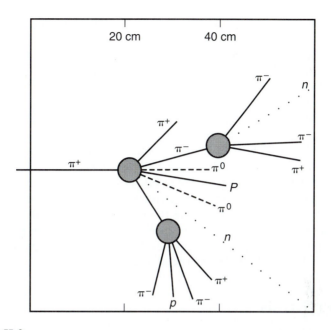

FIGURE H.3
An illustration of the interaction of a high-energy pion in a block of lead. Again, the shower is cut off at an early stage, angles are exaggerated, and only a few characteristic interactions are shown. The black circles indicate the interaction with the nuclei and the subsequent nuclear breakup. Dashed lines indicate π^0's (the resulting electromagnetic showers are not shown), and neutrons are indicated by dotted lines.

ondary pions produced is on the order of 50. Compare this to the **maximum multiplicity at shower maximum** of 1000 seen for electromagnetic showers from 50-GeV electrons in lead.

The dimensions of a hadronic shower, both lateral and longitudinal, scale roughly with the nuclear interaction length λ. The nuclear interaction length (for protons) is approximately given by $\lambda \approx 35\,A^{1/3}$ (in gm/cm^2).[9] (Since the cross section for pions is smaller than for protons, the nuclear interaction length for pions tends to be somewhat larger.) Thus, for lead the nuclear interaction length is approximately 194 gm/cm^2 (or using the density of lead, 17 cm). A length of approximately $8\lambda_{int}$ (or 136 cm) is necessary for the absorption of 98% of the energy of a 50-GeV proton. Compare this to the 10-cm length necessary for the absorption of 98% of a 50-GeV electron's energy. The depth necessary for a given containment increases logarithmically with the energy of a particle. The depth of shower maximum for hadronic showers (in absorption lengths) is given by $t_{max} = 0.2 \ln E$ (GeV).

---

9. U. Amaldi, op. cit.

The transverse size of a hadronic shower is determined primarily by the mean transverse momentum in hadronic interactions. Thus, hadronic showers tend to be much broader than electromagnetic showers, although the core is still relatively narrow. The transverse radius for 95% containment is given approximately by $R(95\%) \approx 1\lambda$ (or 17 cm for lead),[10] and is roughly energy independent. Several Monte Carlo programs for the generation of hadronic showers are available.[11] Because of the increased complexity of hadron showers compared to electromagnetic showers, the results in general are not as reliable as from electromagnetic shower Monte Carlos. However, they can still be very useful for calorimeter design and calibration. A hadronic shower created by a 100-GeV pion incident on a lead block is shown in Fig. H.4.[12] Note the differences in lateral and longitudinal size from an electromagnetic shower. (There is still significant energy deposition beyond the 10-cm transverse scale of this plot, but the dimensions were made the same as the electromagnetic shower plot for comparison purposes.) Note also the "spikiness" of a typical hadronic shower. We can see the individual interactions taking place.

H.2   DETECTORS

H.2.1   Tracking Detectors

Detectors utilizing gases for the purpose of measuring the ionization products of charged particles have proven to be extremely useful in high-energy physics. One advantage of gaseous detectors is the high mobility of the electrons and positive ions. An example of a simple gas ionization detector is shown in Fig. H.5.

A potential difference V_0 is maintained between the anode wire and the cathode. A radial electric field given by

$$E = \frac{1}{r} \frac{V_0}{\ln\left(\dfrac{r_{cyl}}{r_{wire}}\right)} \tag{H.14}$$

is thus set up. Here, r_{cyl} is the inside radius of the cylinder and r_{wire} is the radius of the anode wire. Electron-ion pairs will be created by the ionization of the atoms in the gas by the passage of charged particles through the cylinder. The electrons will be accelerated by the electric field towards the anode, the positive ions towards the cathode. The current thus induced can be measured. The number of electron-ion pairs created will be proportional to the energy deposited in the cylinder by the charged particle.

---

10. Ibid.
11. T. A. Gabriel, ORNL Report/TM-11185
12. This figure is courtesy of J. Brau and K. Furuno.

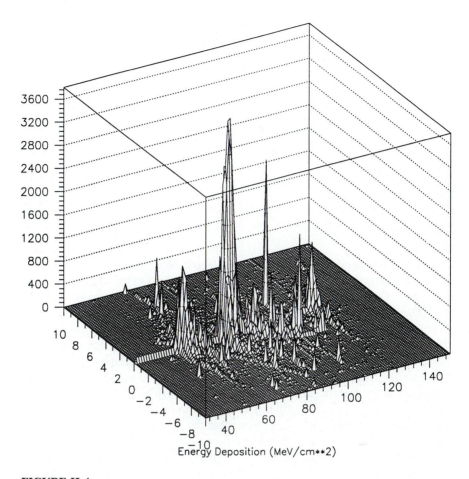

FIGURE H.4
A 100-GeV pion (entering from left) showering in a block of lead. Both the horizontal and vertical scales are in cm.

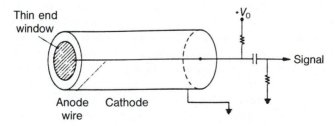

FIGURE H.5
The basic design of an ionization chamber (from Leo, footnote 1).

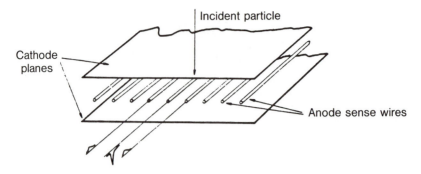

FIGURE H.6
The basic design of a multiwire proportional chamber. Each of the wires acts as an independent ionization chamber (from Leo, footnote 1).

Depending on the size of the applied electric field, the detector has several modes of operation. At low voltages, the detector works in the ionization chamber mode. The only charge collected is that for the original electron-ion pairs. At higher electric fields, the electrons produced by ionization are accelerated to high enough energies that they themselves can create further electron-ion pairs through ionization. The charge created can be much greater than the original charge (by a factor as large as 10^6), but the amount of charge is still proportional to the initial charge. This mode is known as the proportional mode. Note that most of the multiplication takes place in the regions of high electric field, i.e., near the anode wire. At still higher electric fields the proportionality is lost, and avalanches of electrons occur throughout the detector. This region is known as the Geiger-Mueller mode. Noble gases, typically argon, are normally used in ionization detectors because of their relatively low ionization potentials. The addition of other "quenching" gases is necessary, however, to prevent continuous discharges from occurring.

The amount of charge collected by a detector operating in ionization chamber mode is so low as to make measurements in a particle physics experiment very difficult. Detectors operating in the proportional mode, however, do produce practical amounts of charge. In the multiwire proportional chamber (MWPC), an array of closely spaced anode wires functions as a set of independent ionization chambers. (See Fig. H.6.) An amplifier attached to each wire allows a determination of a charged particle's position to within a wire's spacing. A large experiment might have thousands or tens of thousands of anode wires arranged in many different planes. This allows a precise determination of a particle's trajectory. If the particle is moving in a magnetic field, then a measurement of the trajectory allows a determination of the particle's momentum from the relation $r = p/qB$ (where r is the radius of curvature of the path, p the momentum of the particle, q its charge, and B the magnetic field).

A more precise measurement of a particle's position is possible with a drift chamber. In a drift chamber the anode wires are farther apart (typically 1 cm

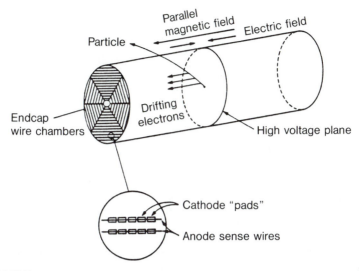

FIGURE H.7
The basic design of a Time Projection Chamber (TPC) (from Leo, footnote 1).

or greater) and additional field shaping wires provide a relatively constant electric field between the anode wires. The constant electric field allows for a constant drift velocity (as for electrons in a conductor). The measurement of the time of arrival of the ionization charge can then provide a position determination much smaller than the anode spacing (100 microns or better is possible).

The most sophisticated type of gaseous ionization detector is the time projection chamber (TPC). The TPC is essentially a three-dimensional tracking chamber that in addition to measuring the *xyz* coordinates of a charged particle also provides a measure of the level of ionization, *dE/dx*, of that particle. The TPC makes use of some of the ideas both from the MWPC and from the drift chamber. A schematic of a TPC is shown in Fig. H.7. Essentially it consists of a large cylinder with a thin cathode in the center held at a large negative high voltage (10 kV or more). At the ends of the cylinder are arrays of anode wires, and behind the anode wires are cathode pads. Electrons produced from ionization inside the cylinder drift along the constant electric field to the anode wires. One position coordinate can be obtained from the signal on the anode wires, and a second from that induced on the cathode pads. The third coordinate, that along the axis of the TPC, is obtained from the drift time. The TPC is sensitive to electrons produced along the entire path length inside the TPC, allowing a precise determination of the particle's trajectory. A magnetic field parallel to the electric field allows a momentum determination to be made and suppresses the lateral diffusion of the electrons during the drift to the anode. The measurement of the amount of charge at each point along the track indicates the value of *dE/dx* for the particle, allowing in some cases an identification of the particle type. Figure H.8 shows the values of *dE/dx* measured for

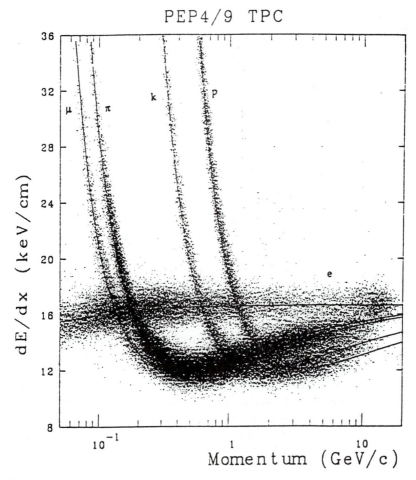

FIGURE H.8
A scatterplot of the *dE/dx* measurements for a sample of tracks measured by the PEP TPC. Note the features mentioned in the text: the rapid rise in *dE/dx* at low values of β and the slow relativistic rise of *dE/dx* as β increases past the minimum ionization point. Note also that because of their different masses, each particle follows a distinct curve at low momentum.

different particle types in a TPC as a function of the particle momentum.[13] For low β, the curves are distinct, but the *dE/dx* values become similar as β approaches 1 (as discussed earlier).

Solid-state tracking detectors, e.g., silicon strip detectors, are capable of extremely good position resolution (potentially on the order of 5–10 microns). The silicon detectors function essentially as reverse-biased diodes. Any charge

---

13. G. Lynch, Fermilab Proc. of Particle ID at High Luminosity Colliders, 485 (1989).

produced by ionization drifts away in the strong internal electric field, producing a detectable current. Because of the limitations of space, these detectors will not be discussed in detail, and the reader is referred to those sources in footnote 1.

H.2.2   Calorimeters

A calorimeter measures the energy of a particle (photon, electron, or hadron) by sampling the energy deposited by the shower in the calorimeter. The sampling is a statistical process with the number of samples (n) proportional to the energy (E). Thus, the energy resolution is proportional to $\sqrt{n}$, or $\sigma(E) \propto \sqrt{E}$. The fractional energy resolution improves as the square root of the energy resolution ($\sigma(E)/E \propto 1/\sqrt{E}$), which is one of the advantages for the use of calorimetry in the measurement of high-energy processes.

Electromagnetic Calorimeters   Since the radiation length for a medium decreases with increasing Z, most electromagnetic calorimeters utilize high Z materials, such as lead or uranium, for the sake of compactness. A calorimeter composed entirely of lead or uranium, however, would provide no way of obtaining a measurable quantity, such as charge or light, proportional to the energy deposited in the calorimeter. Most electromagnetic calorimeters utilize high Z "absorber" materials, interspersed with an "active" medium. An example of such a calorimeter structure is shown in Fig. H.9. Some fraction of the shower's energy is dissipated in the absorber, and some fraction in the active plastic scintillator.[14] The energy lost in the scintillator that appears as scintillation light can be coupled to a light detector, such as a phototube, outside the calorimeter. The energy deposited in the active medium, however, is still proportional to the total energy, and a measurement of the former yields a measurement of the latter. An approximate general form for the fractional energy resolution of a sampling calorimeter is shown below: [15]

$$\frac{\sigma(E)}{E} = \frac{c_{sampling}}{\sqrt{E}} \oplus K$$

$$c_{sampling} = 0.04\sqrt{\Delta E} \text{ (MeV)}. \tag{H.15}$$

ΔE is the energy loss per unit sampling cell for minimum ionizing particles; since most of the thickness (in gm/cm^2) of a sampling cell consists of the absorber material, ΔE depends primarily on the thickness of the absorber. The factor K will be discussed shortly.

---

14. Plastic scintillators have aromatic hydrocarbon compounds consisting of condensed benzene ring structures. Scintillation light in these compounds arises from transitions made by free valence electrons in these molecules. For more details, see Leo in footnote 1.

15. W. Fabjan and R. Wigmans, "Energy Measurement of Elementary Particles," Reports on Progress in *Physics* 52,#12,1519 (1989).

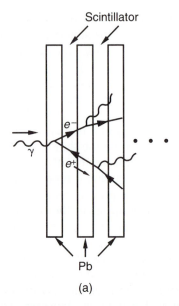

(a)

(b)

FIGURE H.9
The basic construction of an electromagnetic calorimeter consisting of plates of a heavy
absorber material interspersed with plates of the active medium, scintillator. On the bottom (b) is shown a central calorimeter module from CDF. The electromagnetic calorimeter is at the bottom of the figure, and the hadronic calorimeter (a) is at the top.

Note that the energy resolution is proportional to the square root of the sampling thickness. Most sampling calorimeters have sampling resolutions in the range from $8\%/\sqrt{E}$ to $20\%/\sqrt{E}$. Deviations from $\sqrt{E}$ scaling may be caused by noise in the electronic readout, mechanical non-uniformities, and incomplete shower containment. The dominant effects are usually parameterized by the addition, in quadrature, of a constant term K to the resolution. (If two sources of errors are uncorrelated, then we add the squares of the two errors and then take the square root to obtain the total error, rather than adding the two errors linearly.) With proper care in the design and construction of the calorimeter, this constant term can be kept to the 1–2% level.

Another example of a sampling calorimeter, the ionization calorimeter, is shown in Fig. H.10. Ionization electrons, produced by the passage of the shower through the liquid argon, are collected on the anode through the application of a strong electric field (typically 10 kV/cm). Argon is chosen for its good charge-transport properties (and its relative low cost). One major disadvantage is the necessity of operating at cryogenic temperatures. Currently, some work is being conducted to search for suitable room-temperature liquids that could replace liquid argon, such as TMP (tetramethyl pentane) and TMS (tetramethyl silane). However, calorimeters using these liquids are sensitive to impurities at the level of a few parts per billion.

Determination of the position of an electromagnetic shower can be achieved by segmenting the anode into strips or pads. (Similarly, the scintillator, in the scintillation calorimeter, can be subdivided.) A reasonable size for the segmentation is given by the Moliere radius in the calorimeter. For large calorimeters, however, this would result in too many electronic channels to be affordable, so a larger segmentation is necessary.

Some electromagnetic calorimeters utilize a relatively high Z material as both absorber and detection medium. An example is a crystal of the scintillator BaF_2. The sampling resolution for BaF_2 is on the order of $2\%/\sqrt{E}$. Another example is leaded glass, where the energy of the particle is determined by detecting the Čerenkov light radiated into the glass, and a resolution from $5–10\%/\sqrt{E}$ is possible.

Hadron Calorimeters   For a given calorimeter, hadron showers are detected with a poorer energy resolution than that for electromagnetic showers. This results mainly from the fluctuations in the amount of energy lost to nuclear binding energy effects. The fractional energy resolution for hadron calorimeters can be parameterized as[16]

$$\frac{\sigma}{E} = \sqrt{\frac{c_{int}^2 + c_{sampling}^2}{E}} \oplus K. \qquad (H.16)$$

$c_{sampling}$ has the same definition as before. c_{int} derives from the binding energy loss fluctuations and is on the order of 0.11 (0.19) for lead-scintillator (uranium-

---

16. Ibid.

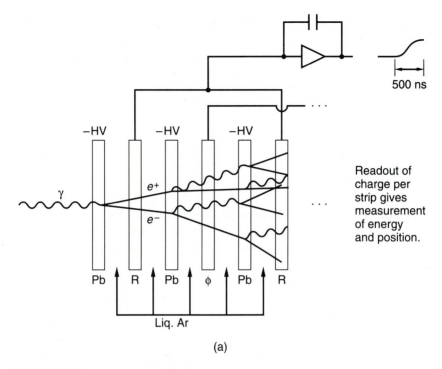

(a)

FIGURE H.10

(a) The basic construction of an electromagnetic calorimeter consisting of plates of a heavy absorber material (lead) interspersed with gaps filled with the active medium, liquid argon. A potential difference is applied between the absorber plates and the anodes, with the ionized electrons from the shower being collected on the anodes. (b) This diagram shows an exploded view of a liquid argon calorimeter from Fermilab experiment E706. There are 66 sampling cells consisting of a layer of lead, an argon gap, an anode board, and another argon gap, for a total of 30 radiation lengths. Note that the anode boards are cut into strips that measure, in alternate layers, the r position and phi position of a shower.

scintillator) calorimeters. The constant term K has contributions from the sources listed earlier for electromagnetic calorimeters and, in addition, has a contribution from the (possibly) unequal response of the calorimeter to electron and hadron showers (the "e/π" response). Knowledge of hadron calorimeters has progressed far enough that this e/π ratio can be tuned close to 1 (with the subsequent reduction of the constant term) by the judicious choice of the thicknesses of the absorber and active media in the calorimeter. Sensitivity to the neutron component of the hadronic shower turns out to be very important in this tuning.[17] With care, the constant term can be reduced to the level of a few percent.

17. J. E. Brau, *Nuclear Instruments and Methods A312*, 489 (1992) and references therein. R. Wigmans, *Nuclear Instruments and Methods A265*, 273 (1988).

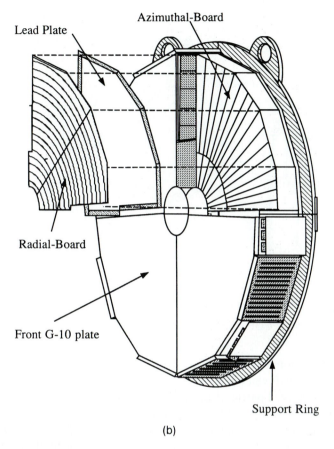

Lead Plate

Azimuthal-Board

Radial-Board

Front G-10 plate

Support Ring

(b)

FIGURE H.10 *Continued*

The numerator in the first term in Eq. (H.16) can vary from 30% to 100%, depending on the absorber material and thickness. Typical absorber materials are lead, uranium, and iron.

H.3   CDF

Most current high-energy physics experiments consist of a complex system of the types of detectors that have been described previously. As an example, consider CDF. The Collider Detector Facility, an experiment at Fermilab, was built to study interactions at the highest currently available energy. CDF is a 5000-ton magnetic detector built to study $\sqrt{s} = 1.8$ TeV (900 GeV on 900 GeV) $p\bar{p}$ collisions.[18] Events are analyzed using information from charged particle track-

18. See the compilation of articles from *Nuclear Instruments and Methods in Physics Research-A*, on the collider detector at Fermilab. North Holland Publishing (1988).

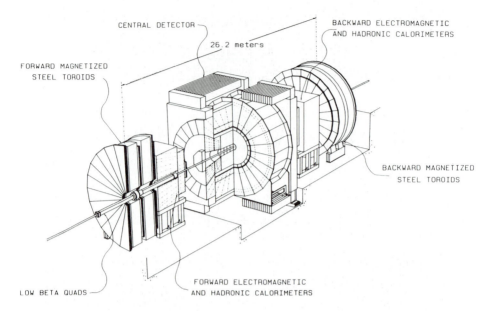

FIGURE H.11
A perspective view of CDF (from footnote 11).

ing, magnetic momentum analysis, and fine-grained calorimetry. Tracking devices, TPC's, drift chambers, and proportional wire chambers provide measurements of charged particle trajectories. The curvatures of the trajectories in the solenoidal magnetic field ($B = 1.5$ T, aligned along the collision axis) allow determination of both the sign of the charge and the momentum. The energies and positions of photons, electrons, and hadrons are measured in electromagnetic and hadronic calorimeters. A schematic of the CDF detector is shown in Fig. H.11 and a picture in Fig. H.12 (as of 1989; there have been upgrades to the detector for the 1992 run).

CDF was designed to measure the energy, momentum, and, where possible, the type of particle produced in $p\bar{p}$ collisions over as large a solid angle as possible. The detector has an approximate cylindrical symmetry. The segmentation of the detector is roughly uniform in pseudorapidity (η) and ϕ, since many physics variables, jet sizes for example, depend on η and ϕ.

Consider first the measurements from the tracking systems. The VTPC (Vertex Time Projection Chamber) consists of eight modules mounted back to back along the beam direction with a total length of 2.8 m. ($p\bar{p}$ collisions can occur within an interaction "diamond" of roughly 70-cm length centered on the nominal intersection point.) The primary purpose of the VTPC is to determine the vertex position of the interaction. The spatial coordinates obtained with the VTPC for charged tracks can also be used for momentum determination, but the small lever arm and moderate position resolution (200–500 μm) do not allow a very precise momentum determination.

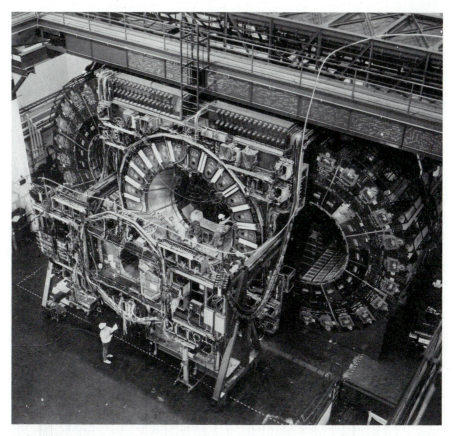

FIGURE H.12
The CDF Detector in its maintenance hall (Fermilab Visual Media Services).

The CTC (Central Tracking Chamber) is a 1.3-m radius, 3.2-m long cylindrical drift chamber with 6156 wires in 84 different radial layers. This allows for 84 separate measurements of a particle's position (spread over a "lever arm" (radial distance) of 1.3 m), each with a resolution of approximately 180 μm. The resulting uncertainty in the track curvature is given by $\sigma_c = 0.5 \times 10^{-3}$ m^{-1}, where $C = 1/($radius of curvature of track$)$.[19] The transverse momentum (p_T) is related to the curvature by the relation $p_T = 0.3\,B/C$ with B in Tesla, p_T in GeV/c, and C in m^{-1}. Thus, the uncertainty in p_T is

$$\frac{\sigma_{p_T}}{p_T^2} = \frac{\sigma_C}{0.3B} = 0.001$$

19. Ibid.

or

$$\frac{\sigma_{p_T}}{p_T} = 0.001 p_T. \tag{H.17}$$

Note how the resolution degrades with increasing momentum, in contrast to the behavior of calorimetry. The fractional momentum resolution at 50 GeV/c is 5%.

Muon trajectories (and momenta) are measured in the CTC and, in addition, by using four layers of proportional chambers outside the hadron calorimeters. Muon identification occurs with the observation that the muon deposits an energy consistent with that from a minimum ionizing particle in both the electromagnetic and the hadronic calorimeters. (The total of 4.9 interaction lengths in the calorimeters also serve to filter out hadrons, so that those particles that reach these outer chambers are mostly muons.) The muon trajectory measured in the proportional chambers outside the calorimeters is also required to be consistent with that measured in the CTC, within the uncertainties caused by multiple Coulomb scattering of the muon in the calorimeters. (In the central region, muons have to pass through $70X_o$ of material before they reach the muon chambers.)

In the central region, the electromagnetic calorimeters consist of 0.32-cm lead plates interspersed with 5-mm thick plastic scintillator with a total depth of 18 radiation lengths. (See Fig. H.9 for a picture of one "wedge" of the central calorimeter containing both an electromagnetic and a hadronic section.) From our earlier formula, Eq. (H.12), this is sufficient to contain approximately 98% of the energy of a 50-GeV electron. The hadron calorimeter consists of 2.5-cm iron plates interspersed with 1-cm plastic scintillator. The total number of interaction lengths for the electromagnetic and hadronic sections together is 4.9. While somewhat shallow compared to our earlier criterion of 8λ for 98% containment, this is still sufficient to contain 90–95% of the energy of a 50-GeV electron. The transverse segmentation for both calorimeters is accomplished in towers in pseudorapidity and ϕ (0.1 units of η by 15° in ϕ). The Moliere radius for the electromagnetic calorimeter (roughly the "size" of the shower) is 2 cm, much smaller than the segmentation ~25 cm × 50 cm at $\theta = 90°$. A finer segmentation is impractical to carry out, due both to the large cost and difficulty of implementation. However, a proportional chamber is placed after 6 radiation lengths (approximately the shower maximum position; remember, it changes only logarithmically with energy) to sample the shower at that depth and allow for better position determination. An energy resolution of $(\sigma/E)^2 = (13.5\%/\sqrt{E})^2 + (1.7\%)^2$ is obtained for the electromagnetic calorimeter. Since the energy loss per unit cell is equal to ~5 MeV in this case, we would naively expect the sampling term to be 9–10%/$\sqrt{E}$. Part of this degradation probably stems from fluctuations in the energy loss in the superconducting coil in front of the electromagnetic calorimeter. An energy resolution of $(\sigma/E)^2 = (70\%/\sqrt{E})^2 + (2–3\%)^2$ is obtained for the hadron calorimeter. Part of the constant term is due to sources mentioned earlier for the electromagnetic calorimeter. Another contribution originates because the e/π ratio for the calorimeters differs from 1.

The "endwall" hadron calorimeters consist of 5.1-cm iron plates interleaved with plastic scintillator. In the plug and forward calorimeters, the active medium consists of proportional chambers with cathode pad readout instead of plastic scintillator. The use of cathode pads allows for finer segmentation than is possible with scintillator, a desirable feature in a region where towers of constant pseudorapidity grow progressively smaller as η gets larger. Using proportional chambers for calorimeter sampling has disadvantages, though, especially because of the small sampling fraction (the fraction of the shower's energy that actually gets deposited in the active medium, in this case the gas).

In total, there are over 100,000 detector channels in CDF. A typical event contains 100 kilobytes of information.

Some examples of events in CDF are shown below. In Fig. H.13 is shown a Z^0 decaying into an e^+e^- pair. In Fig. H.13(a) the two tracks can be seen. Note how "stiff" they are due to their relatively high momentum. The energy deposited in the calorimeters is shown in Fig. H.13(b), which is a "lego" (so named because of its similarity in appearance to the popular toy) plot. In a lego plot the calorimeters are divided into bins of η and ϕ corresponding to the tower segmentation. The height of each bin indicates the corresponding transverse energy deposited in that bin. Note that the electron and positron showers are well contained within a bin or tower, i.e., an electron shower is much smaller than the tower size.

A lego plot of an event with three jets in the final state is shown in Fig. H.14. Note that each jet in general has an energy deposition in many calorimeter cells, both electromagnetic and hadronic. (The characteristic "size" of a jet is given by $\Delta R_{jet} = 0.7$, where $\Delta R = \sqrt{((\Delta\eta)^2 + (\Delta\phi)^2)}$ is the radius of the jet in η-ϕ space. This is much larger than the calorimeter segmentation of $\Delta\eta \sim 0.1$, $\Delta\phi \sim 0.2$ and is also much greater (for central rapidities or rapidities close to 0) than the characteristic sizes of individual electromagnetic and hadronic showers.)

A W particle, decaying into an electron and a neutrino, is shown in Fig. H.15. Again note the stiff electron track and accompanying shower confined to one electromagnetic tower. There is no other significant activity in the calorimeters. The neutrino, since it interacts only by the weak interaction, passes through the calorimeter without energy loss. Since the neutrino roughly balances the transverse momentum of the electron, this means that there is a missing transverse momentum or energy, typically on the order of half of the mass of the W in the event. If the calorimeters have sufficiently good energy resolution and if they are hermetic, i.e., have few cracks or holes in their coverage, this missing transverse energy can be "measured." For CDF the calorimetric resolution in missing transverse energy can be parameterized as

$$\sigma_{\not{E}_T} \approx 0.7\sqrt{\sum E_T}. \tag{H.18}$$

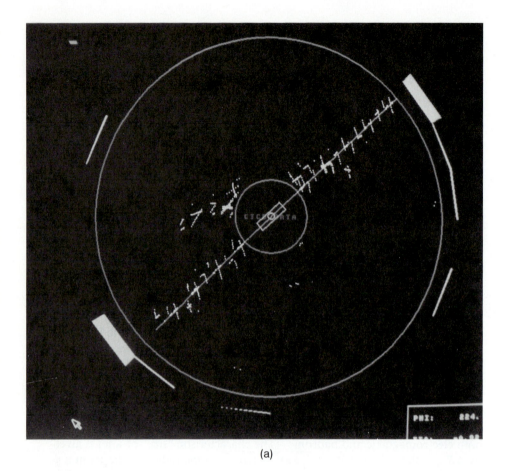

(a)

FIGURE H.13

A display of an event containing a Z^0 decaying into an e^+e^- pair. In Fig. H.13(a), the e^+ and e^- tracks are seen in the CTC. In Fig. H.13(b), a lego plot of the event is displayed, indicating the high energy electromagnetic showers from the e^+ and the e^- (Fermilab Visual Media Services).

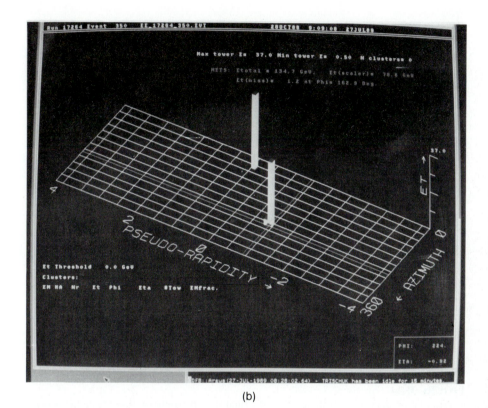

(b)

FIGURE H.13 *Continued*

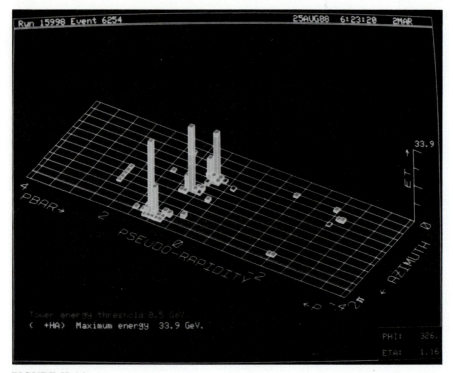

FIGURE H.14
A lego plot of an event with three jets in the final state (Fermilab Visual Media Services).
(The original photograph has the electromagnetic and hadronic energy in each tower
color-coded.)

For W decays with an electron and neutrino, $E_T \sim 50$ GeV, and so $\sigma_{\not{E}_T} \sim$
5 GeV. Since the missing transverse energy is on the order of 40 GeV, the presence of the neutrino is easily detected.

Last, Fig. H.16 shows two views (a calorimeter lego plot and a CTC view)
of an event that is a candidate for the production and decay of a top-antitop
quark pair. The top quark has decayed semileptonically into a high p_T electron
plus other particles, and the $\bar{t}$ quark has decayed semileptonically into a muon
plus other particles. (Note that the muon has left only minimum energy in the
calorimeters. Also, there is no significant jet activity in the calorimeters.) Due
to the combinatorics (the difficulty in assigning each particle to either the top
quark or the antitop quark or the beam fragmentation) and the presence of sev-

FIGURE H.15 *(see facing page)*
An event display of a W decaying into an electron and a neutrino. In Fig. H.15(a), the
electron track is shown in the CTC, along with the tower in the electromagnetic calorimeter in which the electron shower develops. The arrow indicates the direction of the
measured missing transverse energy in the event, i.e., the direction of the neutrino. In
Fig. H.15(b), a lego plot of the event is displayed (Fermilab Visual Media Services).

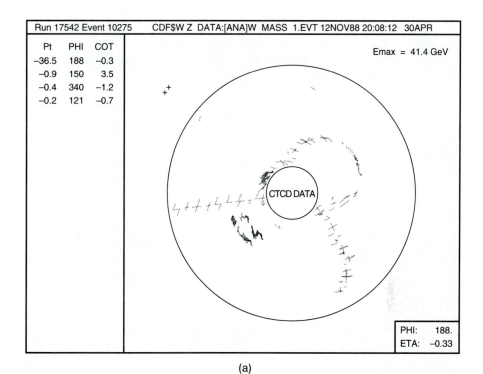

(a)

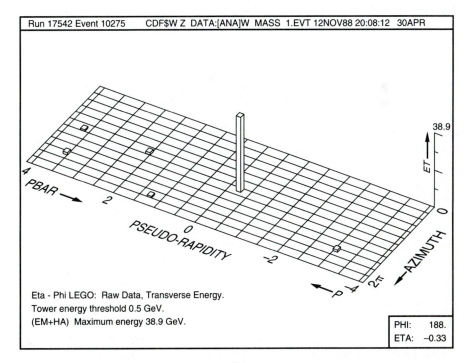

(b)

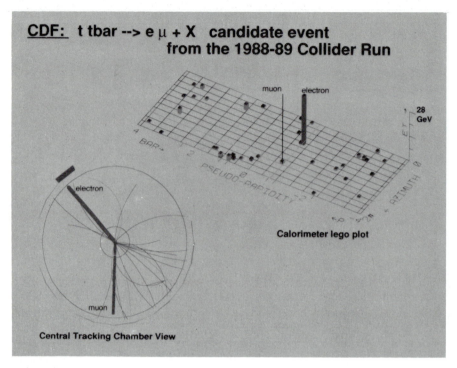

FIGURE H.16
A display of a candidate $t\bar{t}$ event (Fermilab Visual Media Services).

eral neutrinos in the final state, reconstruction of the masses of the top and antitop quarks is impossible. This signature (a high p_T electron and a high p_T muon produced roughly back to back) is consistent with the decay of a top and antitop quark, each with a mass of approximately 140 GeV/c^2. However, this event could also be due to general background processes, so confirmation of the top quark discovery has to await the production of more events of this type.

As mentioned earlier, there have been several upgrades to the CDF detector: an improvement of the muon measurement with the addition of more wire chambers (and more absorber in some regions) and the addition of silicon strip detectors around the interaction point which will allow the measurement of secondary decay vertices as, for example, from B mesons. The plug and forward calorimeters will be replaced in the 1995–96 run by calorimeters involving a new technology (developed in the last few years) involving plastic scintillator tiles and wave-shifting fiber readout.

Acknowledgments

There are many people who have helped in writing this appendix. First of all, I would like to thank Fred Lobkowicz for a careful reading of the first draft.

I would like to thank Don Groom for all his suggestions, the totality of which could create another appendix. I would like to thank Rob Hipple for help in generating the picture of the electromagnetic shower and Jim Brau and Koichiro Furuno for providing the picture of the hadronic shower.

Mathematical Tables

TABLE I
Spherical Harmonics

$$Y_l^m(\theta,\phi) = \sqrt{\frac{(2l+1)(l-m)!}{4\pi(l+m)!}}\, P_l^m(\cos\theta)e^{im\phi}$$

$$\int (Y_{l'}^{m'}(\theta,\phi))^* Y_l^m(\theta,\phi)\, d\Omega = \delta_{ll'}\delta_{mm'}$$

$$P_l^m(\cos\theta) = (-1)^m \sin^m\theta\left[\left(\frac{d}{d(\cos\theta)}\right)^m P_l(\cos\theta)\right] \quad (m \le l)$$

$$P_l(\cos\theta) = \frac{1}{2^l l!}\left[\left(\frac{d}{d(\cos\theta)}\right)^l(-\sin^2\theta)^l\right]$$

$$\int_{-1}^{+1} P_n(x)P_m(x)\, dx = \frac{2}{2n+1}\delta_{nm}$$

$$Y_l^{-m}(\theta,\phi) = (-1)^m [Y_l^m(\theta,\phi)]^* \qquad P_l^{-m}(\cos\theta) = (-1)^m \frac{(l-m)!}{(l+m)!}P_l^m(\cos\theta)$$

$l = 0 \qquad Y_0^0 = \dfrac{1}{\sqrt{4\pi}} \qquad l = 1 \qquad Y_1^0 = \sqrt{\dfrac{3}{4\pi}}\cos\theta \qquad Y_1^1 = -\sqrt{\dfrac{3}{8\pi}}\sin\theta e^{i\phi}$

$l = 2 \qquad Y_2^0 = \sqrt{\dfrac{5}{16\pi}}(3\cos^2\theta - 1) \qquad Y_2^1 = -\sqrt{\dfrac{15}{8\pi}}\sin\theta\cos\theta e^{i\phi}:$

$\qquad\qquad Y_2^2 = \sqrt{\dfrac{15}{32\pi}}\sin^2\theta e^{2i\phi}$

$l = 3 \qquad Y_3^0 = \sqrt{\dfrac{7}{16\pi}}(5\cos^3\theta - 3\cos\theta) \qquad Y_3^1 = -\sqrt{\dfrac{21}{64\pi}}\sin\theta(5\cos^2\theta - 1)e^{i\phi}$

$\qquad\qquad Y_3^2 = \sqrt{\dfrac{105}{32\pi}}\sin^2\theta\cos\theta e^{2i\phi} \qquad Y_3^3 = -\sqrt{\dfrac{35}{64\pi}}\sin^3\theta e^{3i\phi}$

TABLE II
Clebsch-Gordan Coefficients*

$\frac{1}{2} \times \frac{1}{2}$ $C^{J\ \frac{1}{2}\ \frac{1}{2}}_{M\,m_1\,m_2}$

	J:	1	1	0	1
	M:	+1	0	0	−1
m_1	m_2				
$+\frac{1}{2}$	$+\frac{1}{2}$	1			
$+\frac{1}{2}$	$-\frac{1}{2}$		$\sqrt{\frac{1}{2}}$	$\sqrt{\frac{1}{2}}$	
$-\frac{1}{2}$	$+\frac{1}{2}$		$\sqrt{\frac{1}{2}}$	$-\sqrt{\frac{1}{2}}$	
$-\frac{1}{2}$	$-\frac{1}{2}$				1

$1 \times \frac{1}{2}$ $C^{J\ 1\ \frac{1}{2}}_{M\,m_1\,m_2}$

	J:	$\frac{3}{2}$	$\frac{3}{2}$	$\frac{1}{2}$	$\frac{3}{2}$	$\frac{1}{2}$	$\frac{3}{2}$
	M:	$+\frac{3}{2}$	$+\frac{1}{2}$	$+\frac{1}{2}$	$-\frac{1}{2}$	$-\frac{1}{2}$	$-\frac{3}{2}$
m_1	m_2						
+1	$+\frac{1}{2}$	1					
+1	$-\frac{1}{2}$		$\sqrt{\frac{1}{3}}$	$\sqrt{\frac{2}{3}}$			
0	$+\frac{1}{2}$		$\sqrt{\frac{2}{3}}$	$-\sqrt{\frac{1}{3}}$			
0	$-\frac{1}{2}$				$\sqrt{\frac{2}{3}}$	$\sqrt{\frac{1}{3}}$	
−1	$+\frac{1}{2}$				$\sqrt{\frac{1}{3}}$	$-\sqrt{\frac{2}{3}}$	
−1	$-\frac{1}{2}$						1

1×1 $C^{J\ 1\ 1}_{M\,m_1\,m_2}$

	J:	2	2	1	2	1	0	2	1	2
	M:	+2	+1	+1	0	0	0	−1	−1	−2
m_1	m_2									
+1	+1	1								
+1	0		$\sqrt{\frac{1}{2}}$	$\sqrt{\frac{1}{2}}$						
0	+1		$\sqrt{\frac{1}{2}}$	$-\sqrt{\frac{1}{2}}$						
+1	−1				$\sqrt{\frac{1}{6}}$	$\sqrt{\frac{1}{2}}$	$\sqrt{\frac{1}{3}}$			
0	0				$\sqrt{\frac{2}{3}}$	0	$-\sqrt{\frac{1}{3}}$			
−1	+1				$\sqrt{\frac{1}{6}}$	$-\sqrt{\frac{1}{2}}$	$\sqrt{\frac{1}{3}}$			
0	−1							$\sqrt{\frac{1}{2}}$	$\sqrt{\frac{1}{2}}$	
−1	0							$\sqrt{\frac{1}{2}}$	$-\sqrt{\frac{1}{2}}$	
−1	−1									1

*The phase convention is that of Condon and Shortley, *The Theory of Atomic Spectra*, Cambridge University Press, 1951.

TABLE II *Continued*

$\frac{3}{2} \times 1$		$C^{J\ \frac{3}{2}\ 1}_{M\,m_1\,m_2}$											
J:		$\frac{5}{2}$	$\frac{5}{2}$	$\frac{3}{2}$	$\frac{5}{2}$	$\frac{3}{2}$	$\frac{1}{2}$	$\frac{5}{2}$	$\frac{3}{2}$	$\frac{1}{2}$	$\frac{5}{2}$	$\frac{3}{2}$	$\frac{5}{2}$
M:		$+\frac{5}{2}$	$+\frac{3}{2}$	$+\frac{3}{2}$	$+\frac{1}{2}$	$+\frac{1}{2}$	$+\frac{1}{2}$	$-\frac{1}{2}$	$-\frac{1}{2}$	$-\frac{1}{2}$	$-\frac{3}{2}$	$-\frac{3}{2}$	$-\frac{5}{2}$
m_1	m_2												
$+\frac{3}{2}$	$+1$	1											
$+\frac{3}{2}$	0		$\sqrt{\frac{2}{5}}$	$\sqrt{\frac{3}{5}}$									
$+\frac{1}{2}$	$+1$		$\sqrt{\frac{3}{5}}$	$-\sqrt{\frac{2}{5}}$									
$+\frac{3}{2}$	-1				$\sqrt{\frac{1}{10}}$	$\sqrt{\frac{2}{5}}$	$\sqrt{\frac{1}{2}}$						
$+\frac{1}{2}$	0				$\sqrt{\frac{3}{5}}$	$\sqrt{\frac{1}{15}}$	$-\sqrt{\frac{1}{3}}$						
$-\frac{1}{2}$	$+1$				$\sqrt{\frac{3}{10}}$	$-\sqrt{\frac{8}{15}}$	$\sqrt{\frac{1}{6}}$						
$+\frac{1}{2}$	-1							$\sqrt{\frac{3}{10}}$	$\sqrt{\frac{8}{15}}$	$\sqrt{\frac{1}{6}}$			
$-\frac{1}{2}$	0							$\sqrt{\frac{3}{5}}$	$-\sqrt{\frac{1}{15}}$	$-\sqrt{\frac{1}{3}}$			
$-\frac{3}{2}$	$+1$							$\sqrt{\frac{1}{10}}$	$-\sqrt{\frac{2}{5}}$	$\sqrt{\frac{1}{2}}$			
$-\frac{1}{2}$	-1										$\sqrt{\frac{3}{5}}$	$\sqrt{\frac{2}{5}}$	
$-\frac{3}{2}$	$+0$										$\sqrt{\frac{2}{5}}$	$-\sqrt{\frac{3}{5}}$	
$-\frac{3}{2}$	-1												1

$\frac{3}{2} \times \frac{1}{2}$		$C^{J\ \frac{3}{2}\ \frac{1}{2}}_{M\,m_1\,m_2}$							
J:		2	2	1	2	1	2	1	2
M:		$+2$	$+1$	$+1$	0	0	-1	-1	-2
m_1	m_2								
$+\frac{3}{2}$	$+\frac{1}{2}$	1							
$+\frac{3}{2}$	$-\frac{1}{2}$		$\sqrt{\frac{1}{4}}$	$\sqrt{\frac{3}{4}}$					
$+\frac{1}{2}$	$+\frac{1}{2}$		$\sqrt{\frac{3}{4}}$	$-\sqrt{\frac{1}{4}}$					
$+\frac{1}{2}$	$-\frac{1}{2}$				$\sqrt{\frac{1}{2}}$	$\sqrt{\frac{1}{2}}$			
$-\frac{1}{2}$	$+\frac{1}{2}$				$\sqrt{\frac{1}{2}}$	$-\sqrt{\frac{1}{2}}$			
$-\frac{1}{2}$	$-\frac{1}{2}$						$\sqrt{\frac{3}{4}}$	$\sqrt{\frac{1}{4}}$	
$-\frac{3}{2}$	$+\frac{1}{2}$						$\sqrt{\frac{1}{4}}$	$-\sqrt{\frac{3}{4}}$	
$-\frac{3}{2}$	$-\frac{1}{2}$								1

TABLE II *Continued*

$2 \times \frac{1}{2}$						$C^{J\ 2\ \frac{1}{2}}_{M m_1 m_2}$					
	J:	$\frac{5}{2}$	$\frac{5}{2}$	$\frac{3}{2}$	$\frac{5}{2}$	$\frac{3}{2}$	$\frac{5}{2}$	$\frac{3}{2}$	$\frac{5}{2}$	$\frac{3}{2}$	$\frac{5}{2}$
	M:	$+\frac{5}{2}$	$+\frac{3}{2}$	$+\frac{3}{2}$	$+\frac{1}{2}$	$+\frac{1}{2}$	$-\frac{1}{2}$	$-\frac{1}{2}$	$-\frac{3}{2}$	$-\frac{3}{2}$	$-\frac{5}{2}$
m_1	m_2										
$+2$	$\frac{1}{2}$	1									
$+2$	$-\frac{1}{2}$		$\sqrt{\frac{1}{5}}$	$\sqrt{\frac{4}{5}}$							
$+1$	$+\frac{1}{2}$		$\sqrt{\frac{4}{5}}$	$-\sqrt{\frac{1}{5}}$							
$+1$	$-\frac{1}{2}$				$\sqrt{\frac{2}{5}}$	$\sqrt{\frac{3}{5}}$					
0	$+\frac{1}{2}$				$\sqrt{\frac{3}{5}}$	$-\sqrt{\frac{2}{5}}$					
0	$-\frac{1}{2}$						$\sqrt{\frac{3}{5}}$	$\sqrt{\frac{2}{5}}$			
-1	$+\frac{1}{2}$						$\sqrt{\frac{2}{5}}$	$-\sqrt{\frac{3}{5}}$			
-1	$-\frac{1}{2}$								$\sqrt{\frac{4}{5}}$	$\sqrt{\frac{1}{5}}$	
-2	$+\frac{1}{2}$								$\sqrt{\frac{1}{5}}$	$-\sqrt{\frac{4}{5}}$	
-2	$-\frac{1}{2}$										1

Bibliography

SCIENTIFIC AMERICAN & PHYSICS TODAY ARTICLES (through 1992)*

Readings from Scientific American Particles and Fields with an introduction by W. J. Kaufman, III, W. H. Freeman (1980).

Abbott, L., *"The Mystery of the Cosmological Constant,"* Scientific American, May 1982, p. 106.

Adair, R. K., *"A Flaw in a Universal Mirror,"* Scientific American, February 1988, p. 50.

Barrow, J. D., and J. Silk, *"The Structure of the Early Universe,"* Scientific American, April 1980, p. 118.

Bloom, E. D., and G. J. Feldman, *"Quarkonium,"* Scientific American, May 1982, p. 66.

Boyd, R. N., and I. Tanihata, *"Physics with Radioactive Nuclear Beams,"* Physics Today, June 1992, p. 44.

Breuker, H., H. Drevermann, C. Grab, A. A. Rademakers, and H. Stone, *"Tracking and Imaging Elementary Particles,"* Scientific American, August 1991, p. 58.

Carrigan, R. A., Jr., and W. P. Trower, *"Superheavy Magnetic Monopoles,"* Scientific American, April 1982, p. 106.

Chew, G. F., M. Gell-Mann, and A. H. Rosenfeld, *"Strongly Interacting Particles,"* Scientific American, February 1964, p. 74.

Cline, D. B., *"Beyond Truth and Beauty: A Fourth Family of Particles,"* Scientific American, August 1988, p. 60.

Cline, D., and C. Rubbia, *"Antiproton-Proton Colliders and Intermediate Bosons,"* Physics Today, August 1980, p. 44.

Cronin, J. W., and M. S. Greenwood, *"CP Symmetry Violation,"* Physics Today, July 1982, p. 38.

---

*Mostly compiled by Grace Yong.

Dawson, J. M., *"Plasma Particle Accelerators,"* Scientific American, March 1989, p. 54.

DeWitt, B. S., *"Quantum Gravity,"* Scientific American, December 1983, p. 112.

Dimopoulos, S., S. A. Raby, and F. Wilczek, *"Unification of Couplings,"* Physics Today, October 1991, p. 25.

Drell, S. D., *"Electron-Positron Annihilation and the New Particles,"* Scientific American, June 1985, p. 50.

Freedman, D. Z., and P. van Nieuwenhuizen, *"Supergravity and the Unification of the Laws of Physics,"* Scientific American, February 1978, p. 126.

Freedman, D. Z., and P. van Nieuwenhuizen, *"The Hidden Dimensions of Spacetime,"* Scientific American, March 1985, p. 74.

Fulcher, L. P., J. Rafelski, and A. Klein, *"The Decay of the Vacuum,"* Scientific American, December 1979, p. 150.

Gabrielse, G., *"Extremely Cold Antiprotons,"* Scientific American, December 1992, p. 78.

Georgi, H., and S. L. Glashow, *"Unified Theory of Elementary-Particle Forces,"* Physics Today, September 1980, p. 30.

Georgi, H., *"A Unified Theory of Elementary Particles and Forces,"* Scientific American, April 1981, p. 48.

Georgi, H., *"Flavor SU(3) Symmetries in Particle Physics,"* Physics Today, April 1988, p. 29.

Glashow, S. L., *"Quarks with Color and Flavor,"* Scientific American, October 1975, p. 38.

Goldman, T., R. J. Hughes, and M. M. Nieto, *"Gravity and Antimatter,"* Scientific American, March 1988, p. 48.

Golub, R., W. Mampe, J. M. Pendlebury, and P. Ageron, *"Ultracold Neutrons,"* Scientific American, June 1979, p. 34.

Green, M. B., *"Superstrings,"* Scientific American, September 1986, p. 48.

Greenberger, D. M., and A. W. Overhauser, *"The Role of Gravity in Quantum Theory,"* Scientific American, May 1980, p. 66.

Greiner, W., and H. Stöcker, *"Hot Nuclear Matter,"* Scientific American, January 1985, p. 76.

Gross, D. J., *"Asymptotic Freedom,"* Physics Today, January 1987, p. 39.

Gutbrod, H., and H. Stöcker, *"The Nuclear Equation of State,"* Scientific American, November 1991, p. 58.

Haber, H. E., and Gordon L. Kane, *"Is Nature Supersymmetric?"* Scientific American, June 1986, p. 52.

Harari, H., *"The Structure of Quarks and Leptons,"* Scientific American, April 1983, p. 56.

Ishikawa, K., *"Glueballs,"* Scientific American, November 1982, p. 142.

Jackson, J. D., M. Tigner, and S. Wojcicki, *"The Superconducting Supercollider,"* Scientific American, March 1986, p. 66.

Jacob, M., and P. Landshoff, *"The Inner Structure of the Proton,"* Scientific American, March 1980, p. 66.

Johnson, K. A., *"The Bag Model of Quark Confinement,"* Scientific American, July 1979, p. 100.

Koshiba, M.-T., *"Observational Neutrino Astrophysics,"* Physics Today, December 1987, p. 38.

Krauss, L. M., *"Dark Matter in the Universe,"* Scientific American, December 1986, p. 58.

Krisch, A. D., *"The Spin of the Proton,"* Scientific American, May 1979, p. 68.

Langacker, P., and A. K. Mann, *"The Unification of Electromagnetism with the Weak Force,"* Physics Today, December 1989, p. 22.

Lederman, L. M., *"The Upsilon Particle,"* Scientific American, October 1978, p. 72.

Lederman, L. M., *"The Tevatron,"* Scientific American, March 1991, p. 48.

Linde, A., *"Particle Physics and Inflationary Cosmology,"* Physics Today, September 1987, p. 61.

LoSecco, J. M., F. Reines, and D. Sinclair, *"The Search for Proton Decay,"* Scientific American, June 1985, p. 54.

McKee, C. F., and W. H. Press, *"Theoretical Astrophysics,"* Physics Today, April 1991, p. 69.

Moe, M. K., and S. P. Rosen, *"Double-Beta Decay,"* Scientific American, November 1989, p. 48.

Nagle, D. E., M. B. Johnson, and D. F. Measday, *"Pion Physics at the Meson Factories,"* Physics Today, April 1987, p. 56.

Nambu, Y., *"The Confinement of Quarks,"* Scientific American, November 1986, p. 48.

Perl, M. L., and W. J. Kirk, *"Heavy Leptons,"* Scientific American, March 1978, p. 50.

Perl, M. L., *"Popular and Unpopular Ideas in Particle Physics,"* Physics Today, December 1986, p. 24.

Rebbi, C., *"The Lattice Theory of Quark Confinement,"* Scientific American, February 1983, p. 54.

Rebbi, C., *"Solitons,"* Scientific American, February 1979, p. 92.

Rees, J. R., *"The Stanford Linear Collider,"* Scientific American, October 1989, p. 58.

Rubin, V. C., *"Dark Matter in Spiral Galaxies,"* Scientific American, June 1983, p. 96.

Sadoulet, B., and J. W. Cronin, *"Particle Astrophysics,"* Physics Today, April 1991, p. 53.

Schramm, D. N., and G. Steigman, *"Particle Accelerators Test Cosmological Theory,"* Scientific American, June 1988, p. 66.

Schwarz, J. H., *"Superstrings,"* Physics Today, November 1987, p. 33.

Schwitters, R. F., *"Fundamental Particles with Charm,"* Scientific American, October 1977, p. 56.

Sessler, A. M., *"New Particle Acceleration Techniques,"* Physics Today, January 1988, p. 26.

Spergel, D. N., and N. G. Turok, *"Textures and Cosmic Structure,"* Scientific American, March 1992, p. 52.

Stodolsky, L., *"Neutrino and Dark-Matter Detection at Low Temperature,"* Physics Today, August 1991, p. 24.

't Hooft, G., *"Gauge Theories of the Forces between Elementary Particles,"* Scientific American, June 1980, p. 104.

Veltman, M. J. G., *"The Higgs Boson,"* Scientific American, November 1986, p. 76.

Weinberg, S., *"Unified Theories of Elementary-Particle Interaction,"* Scientific American, July 1974, p. 50.

Weinberg, S., *"The Decay of the Proton,"* Scientific American, June 1981, p. 64.

Wilczek, F., *"The Cosmic Asymmetry between Matter and Antimatter,"* Scientific American, December 1980, p. 82.

Wilson, R. R., *"The Next Generation of Particle Accelerators,"* Scientific American, January 1980, p. 42.

Wolfenstein, L., and E. W. Beier, *"Neutrino Oscillations and Solar Neutrinos,"* Physics Today, July 1989, p. 28.

BOOKS AT A SIMILAR LEVEL

Cahn, R. N., and G. Goldhaber, *The Experimental Foundations of Particle Physics*, Cambridge University Press, 1989.

Crease, R. P., and C. C. Mann, *The Second Creation*, Macmillan Publishing Company, 1986.

Edmonds, A. R., *Angular Momentum in Quantum Mechanics*, Princeton University Press, 1957.

Frauenfelder, H., and E. M. Henley, *Subatomic Physics, Second Edition*, Prentice Hall, 1991.

Gottfried, K., and V. F. Weisskopf, *Concepts of Particle Physics, Volumes I and II*, Oxford University Press, 1986.

Griffiths, D., *Introduction to Elementary Particles*, Harper & Row Publishers, Inc., 1987.

Halzen, F., and A. Martin, *Quarks and Leptons*, John Wiley & Sons, 1984.

Kane, G., *Modern Particle Physics*, Addison-Wesley, 1987.

Lederman, L. N., and D. N. Schramm, *From Quarks to the Cosmos*, Scientific American, 1989.

Leo, W. R., *Techniques for Nuclear and Particle Physics Experiments*, Springer-Verlag, 1987.

Lipkin, H. J., *Lie Groups for Pedestrians*, North Holland Publishing Co., 1966.

Neeman, Y., and Y. Kirsh, *The Particle Hunters*, Cambridge University Press, 1986.

Pais, A., *Inward Bound*, Oxford University Press, 1986.

Perkins, D. H., *Introduction to High Energy Physics, Third Edition*, Addison-Wesley, 1987.

Pickering, A., *Constructing Quarks*, Edinburgh University Press, 1984.
Tinkham, M., *Group Theory and Quantum Mechanics*, McGraw-Hill, 1964.
Wigner, E. P., *Group Theory*, Academic Press, 1959.

MISCELLANEOUS

The simplified form of the nucleon state for Problem 7 in Chapter 14 was found
by:
Franklin, J., *Phys. Rev.* 172, 1807 (1968).
Rolnick, W. B., *Phys. Rev. D* 25, 2439 (1982).
Rolnick, W. B., *Phys. Rev. D* 26, 1804 (1982).

Credits

Figure 4.3: Copyright © 1984 JF Cartier.

Figure 6.1: From "Elementary Particles" by Murray Gell-Mann and E. P. Rosenbaum, *Scientific American*, July 1957. Reprinted with permission.

Figures 6.4, 15.7, 15.12, 15.15, 15.17, 15.19, 15.20, F.1: From Donald H. Perkins, *Introduction to High Energy Physics*, © 1987 Addison-Wesley Publishing Co., Inc., Reading, MA. Reprinted with permission.

Figure 13.11: From Aachen-Berlin-CERN, *Physics Letters* **12**, 356 (1964), in Gasiorowicz, *Elementary Particle Physics*, p. 314. Copyright © 1966 John Wiley & Sons, Inc. Reprinted with permission of John Wiley & Sons, Inc.

Figure 13.12: From "Production of pion resonances in $\pi^+ + p$ interactions," in Alff et al., *Phys. Rev. Letters* **9**, 322 (1962). Reprinted with permission of the American Institute of Physics.

Figure 14.2: Courtesy of Argonne National Laboratory.

Figures 14.3, 14.11, 14.12, 15.1, H.9, H.10, H.11, H.12, H.13a&b, H.14, H.15a&b, H.16: Courtesy of Fermilab Visual Media Services.

Figure 14.4: Courtesy of Lawrence Berkeley Laboratory.

Figures 14.8 and 14.10: Courtesy of the Stanford Linear Accelerator Center and the U.S. Department of Energy.

Figure 14.13: Courtesy of CERN.

Figure 15.4: Courtesy of Professor David Schramm, University of Chicago.

Figure 15.9: From Halzen/Martin, *Quarks and Leptons*, p. 229. Copyright © 1984 John Wiley & Sons, Inc. Reprinted with permission of John Wiley & Sons, Inc.

Index

A

A^μ 66, 130, 175, 391–392
a_0 meson 413–414
$|a\rangle_{in}$ and $|a\rangle_{out}$ 13
α (fine structure constant) *ifc*†, 71, 148ff
α_1 (for $U_Y(1)$) 176, 352–353
α_2 (for $SU_L(2)$) 176, 352–353
α_3 or α_s (for $SU_c(3)$) *ifc*†, 148ff, 152ff, 352–353
α_5 (for $SU(5)$) 353, 357
α_g 363
α_s $((34 \text{ GeV})^2)$ 153
Abelian 82, 86
Absorption of γ rays 269
Accelerators 277ff
  cyclic 277ff
  linear 281, 282ff
Aces 110
Action 63, 393–394
  Euclidean 337ff
Active transformations 24
Additive quantum numbers 207
Adjoint representation 93, 94, 141
AGS 284, 309
Amplitude
  probability 14
  scattering or decay 13, 245
Angular momentum 16, 23, 32ff, 387–388
  addition of 42ff
  commutation relations 32, 387
  matrices for $\ell = 1$ 39, 387
  orbital 30, 33
  raising and lowering or ladder operators 35, 388
  spin 23, 32, 39ff
  total 23, 24
Annihilation 60, 65, 66, 68, 69
Anomalies 202
Anticommutation relations 55, 65
Anticommutator 55
Antimatter. *See* Antiparticle
Antineutrino 163ff, 169
Antiparticle 7, 59, 60, 68, 111–112
Antiproton 286
Antiproton accumulator 286
Antishielding 148ff
Antisymmetric 114ff, 116
Antitriplet ($\bar{3}$) 93, 110–112
Antiunitary 230, 235
Associated production 103
Associativity 81
Asymptotic freedom 153
Asymptotic series 72
Atomic weight and baryon number 5
Axial vector 62, 209
Axion 337, 343, 379

B

B meson 123
  and CP violation 227–228
Background 276

---

† *ifc* stands for *inside front cover* and *PG* for the *Particle Glossary* located on the endsheet facing the *ifc*.

---

† *ifc* stands for *inside front cover* and *PG* for the *Particle Glossary* located on the endsheet facing the *ifc*.

---

† *ifc* stands for *inside front cover* and *PG* for the *Particle Glossary* located on the endsheet facing the *ifc*.

† *ifc* stands for *inside front cover* and *PG* for the *Particle Glossary* located on the endsheet facing the *ifc*.

† ifc stands for inside front cover and PG for the Particle Glossary located on the endsheet
facing the ifc.

† *ifc* stands for *inside front cover* and *PG* for the *Particle Glossary* located on the endsheet facing the *ifc*.

---

† *ifc* stands for *inside front cover* and *PG* for the *Particle Glossary* located on the endsheet facing the *ifc*.

---

† *ifc* stands for *inside front cover* and *PG* for the *Particle Glossary* located on the endsheet facing the *ifc*.

---

† *ifc* stands for *inside front cover* and *PG* for the *Particle Glossary* located on the endsheet facing the *ifc*.

---

† *ifc* stands for *inside front cover* and *PG* for the *Particle Glossary* located on the endsheet facing the *ifc*.

---

† *ifc* stands for *inside front cover* and *PG* for the *Particle Glossary* located on the endsheet facing the *ifc*.